T0134505

Lecture Notes in Computer Science 14190

Founding Editors

Gerhard Goos
Juris Hartmanis

The series Lecture Notes in Computer Science (LNCS), including its subseries Lecture Notes in Artificial Intelligence (LNAI) and Lecture Notes in Bioinformatics (LNBI), has established itself as a medium for the publication of new developments in computer science and information technology research, teaching, and education.

LNCS enjoys close cooperation with the computer science R & D community, the series counts many renowned academics among its volume editors and paper authors, and collaborates with prestigious societies. Its mission is to serve this international community by providing an invaluable service, mainly focused on the publication of conference and workshop proceedings and postproceedings. LNCS commenced publication in 1973.

Gernot A. Fink · Rajiv Jain · Koichi Kise ·
Richard Zanibbi
Editors

Document Analysis and Recognition – ICDAR 2023

17th International Conference
San José, CA, USA, August 21–26, 2023
Proceedings, Part IV

 Springer

Editors
Gernot A. Fink
TU Dortmund University
Dortmund, Germany

Rajiv Jain
Adobe
College Park, MN, USA

Koichi Kise
Osaka Metropolitan University
Osaka, Japan

Richard Zanibbi
Rochester Institute of Technology
Rochester, NY, USA

ISSN 0302-9743   ISSN 1611-3349 (electronic)
Lecture Notes in Computer Science
ISBN 978-3-031-41684-2   ISBN 978-3-031-41685-9 (eBook)
https://doi.org/10.1007/978-3-031-41685-9

This Springer imprint is published by the registered company Springer Nature Switzerland AG
The registered company address is: Gewerbestrasse 11, 6330 Cham, Switzerland

Foreword

We are delighted to welcome you to the proceedings of ICDAR 2023, the 17th IAPR International Conference on Document Analysis and Recognition, which was held in San Jose, in the heart of Silicon Valley in the United States. With the worst of the pandemic behind us, we hoped that ICDAR 2023 would be a fully in-person event. However, challenges such as difficulties in obtaining visas also necessitated the partial use of hybrid technologies for ICDAR 2023. The oral papers being presented remotely were synchronous to ensure that conference attendees interacted live with the presenters and the limited hybridization still resulted in an enjoyable conference with fruitful interactions.

ICDAR 2023 was the 17th edition of a longstanding conference series sponsored by the International Association of Pattern Recognition (IAPR). It is the premier international event for scientists and practitioners in document analysis and recognition. This field continues to play an important role in transitioning to digital documents. The IAPR-TC 10/11 technical committees endorse the conference. The very first ICDAR was held in St Malo, France in 1991, followed by Tsukuba, Japan (1993), Montreal, Canada (1995), Ulm, Germany (1997), Bangalore, India (1999), Seattle, USA (2001), Edinburgh, UK (2003), Seoul, South Korea (2005), Curitiba, Brazil (2007), Barcelona, Spain (2009), Beijing, China (2011), Washington, DC, USA (2013), Nancy, France (2015), Kyoto, Japan (2017), Sydney, Australia (2019) and Lausanne, Switzerland (2021).

Keeping with its tradition from past years, ICDAR 2023 featured a three-day main conference, including several competitions to challenge the field and a post-conference slate of workshops, tutorials, and a doctoral consortium. The conference was held at the San Jose Marriott on August 21–23, 2023, and the post-conference tracks at the Adobe World Headquarters in San Jose on August 24–26, 2023.

We thank our executive co-chairs, Venu Govindaraju and Tong Sun, for their support and valuable advice in organizing the conference. We are particularly grateful to Tong for her efforts in facilitating the organization of the post-conference in Adobe Headquarters and for Adobe's generous sponsorship.

The highlights of the conference include keynote talks by the recipient of the IAPR/ICDAR Outstanding Achievements Award, and distinguished speakers Marti Hearst, UC Berkeley School of Information; Vlad Morariu, Adobe Research; and Seiichi Uchida, Kyushu University, Japan.

A total of 316 papers were submitted to the main conference (plus 33 papers to the ICDAR-IJDAR journal track), with 53 papers accepted for oral presentation (plus 13 IJDAR track papers) and 101 for poster presentation. We would like to express our deepest gratitude to our Program Committee Chairs, featuring three distinguished researchers from academia, Gernot A. Fink, Koichi Kise, and Richard Zanibbi, and one from industry, Rajiv Jain, who did a phenomenal job in overseeing a comprehensive reviewing process and who worked tirelessly to put together a very thoughtful and interesting technical program for the main conference. We are also very grateful to the members of the Program Committee for their high-quality peer reviews. Thank you to our competition chairs, Kenny Davila, Chris Tensmeyer, and Dimosthenis Karatzas, for overseeing the competitions.

The post-conference featured 8 excellent workshops, four value-filled tutorials, and the doctoral consortium. We would like to thank Mickael Coustaty and Alicia Fornes, the workshop chairs, Elisa Barney-Smith and Laurence Likforman-Sulem, the tutorial chairs, and Jean-Christophe Burie and Andreas Fischer, the doctoral consortium chairs, for their efforts in putting together a wonderful post-conference program.

We would like to thank and acknowledge the hard work put in by our Publication Chairs, Anurag Bhardwaj and Utkarsh Porwal, who worked diligently to compile the camera-ready versions of all the papers and organize the conference proceedings with Springer. Many thanks are also due to our sponsorship, awards, industry, and publicity chairs for their support of the conference.

The organization of this conference was only possible with the tireless behind-the-scenes contributions of our webmaster and tech wizard, Edward Sobczak, and our secretariat, ably managed by Carol Doermann. We convey our heartfelt appreciation for their efforts.

Finally, we would like to thank for their support our many financial sponsors and the conference attendees and authors, for helping make this conference a success. We sincerely hope those who attended had an enjoyable conference, a wonderful stay in San Jose, and fruitful academic exchanges with colleagues.

August 2023

David Doermann
Srirangaraj (Ranga) Setlur

Preface

Welcome to the proceedings of the 17th International Conference on Document Analysis and Recognition (ICDAR) 2023. ICDAR is the premier international event for scientists and practitioners involved in document analysis and recognition.

This year, we received 316 conference paper submissions with authors from 42 different countries. In order to create a high-quality scientific program for the conference, we recruited 211 regular and 38 senior program committee (PC) members. Regular PC members provided a total of 913 reviews for the submitted papers (an average of 2.89 per paper). Senior PC members who oversaw the review phase for typically 8 submissions took care of consolidating reviews and suggested paper decisions in their meta-reviews. Based on the information provided in both the reviews and the prepared meta-reviews we PC Chairs then selected 154 submissions (48.7%) for inclusion into the scientific program of ICDAR 2023. From the accepted papers, 53 were selected for oral presentation, and 101 for poster presentation.

In addition to the papers submitted directly to ICDAR 2023, we continued the tradition of teaming up with the International Journal of Document Analysis and Recognition (IJDAR) and organized a special journal track. The journal track submissions underwent the same rigorous review process as regular IJDAR submissions. The ICDAR PC Chairs served as Guest Editors and oversaw the review process. From the 33 manuscripts submitted to the journal track, 13 were accepted and were published in a Special Issue of IJDAR entitled "Advanced Topics of Document Analysis and Recognition." In addition, all papers accepted in the journal track were included as oral presentations in the conference program.

A very prominent topic represented in both the submissions from the journal track as well as in the direct submissions to ICDAR 2023 was handwriting recognition. Therefore, we organized a Special Track on Frontiers in Handwriting Recognition. This also served to keep alive the tradition of the International Conference on Frontiers in Handwriting Recognition (ICFHR) that the TC-11 community decided to no longer organize as an independent conference during ICFHR 2022 held in Hyderabad, India. The handwriting track included oral sessions covering handwriting recognition for historical documents, synthesis of handwritten documents, as well as a subsection of one of the poster sessions. Additional presentation tracks at ICDAR 2023 featured Graphics Recognition, Natural Language Processing for Documents (D-NLP), Applications (including for medical, legal, and business documents), additional Document Analysis and Recognition topics (DAR), and a session highlighting featured competitions that were run for ICDAR 2023 (Competitions). Two poster presentation sessions were held at ICDAR 2023.

As ICDAR 2023 was held with in-person attendance, all papers were presented by their authors during the conference. Exceptions were only made for authors who could not attend the conference for unavoidable reasons. Such oral presentations were then provided by synchronous video presentations. Posters of authors that could not attend were presented by recorded teaser videos, in addition to the physical posters.

Three keynote talks were given by Marti Hearst (UC Berkeley), Vlad Morariu (Adobe Research), and Seichi Uchida (Kyushu University). We thank them for the valuable insights and inspiration that their talks provided for participants.

Finally, we would like to thank everyone who contributed to the preparation of the scientific program of ICDAR 2023, namely the authors of the scientific papers submitted to the journal track and directly to the conference, reviewers for journal-track papers, and both our regular and senior PC members. We also thank Ed Sobczak for helping with the conference web pages, and the ICDAR 2023 Publications Chairs Anurag Bharadwaj and Utkarsh Porwal, who oversaw the creation of this proceedings.

August 2023

Gernot A. Fink
Rajiv Jain
Koichi Kise
Richard Zanibbi

Organization

General Chairs

David Doermann   University at Buffalo, The State University of New York, USA

Srirangaraj Setlur   University at Buffalo, The State University of New York, USA

Executive Co-chairs

Venu Govindaraju   University at Buffalo, The State University of New York, USA

Tong Sun   Adobe Research, USA

PC Chairs

Gernot A. Fink   Technische Universität Dortmund, Germany (Europe)

Rajiv Jain   Adobe Research, USA (Industry)

Koichi Kise   Osaka Metropolitan University, Japan (Asia)

Richard Zanibbi   Rochester Institute of Technology, USA (Americas)

Workshop Chairs

Mickael Coustaty   La Rochelle University, France

Alicia Fornes   Universitat Autònoma de Barcelona, Spain

Tutorial Chairs

Elisa Barney-Smith   Luleå University of Technology, Sweden

Laurence Likforman-Sulem   Télécom ParisTech, France

Competitions Chairs

Kenny Davila   Universidad Tecnológica Centroamericana,
                                    UNITEC, Honduras
Dimosthenis Karatzas   Universitat Autònoma de Barcelona, Spain
Chris Tensmeyer   Adobe Research, USA

Doctoral Consortium Chairs

Andreas Fischer   University of Applied Sciences and Arts Western
                                    Switzerland
Veronica Romero   University of Valencia, Spain

Publications Chairs

Anurag Bharadwaj   Northeastern University, USA
Utkarsh Porwal   Walmart, USA

Posters/Demo Chair

Palaiahnakote Shivakumara   University of Malaya, Malaysia

Awards Chair

Santanu Chaudhury   IIT Jodhpur, India

Sponsorship Chairs

Wael Abd-Almageed   Information Sciences Institute USC, USA
Cheng-Lin Liu   Chinese Academy of Sciences, China
Masaki Nakagawa   Tokyo University of Agriculture and Technology,
                                    Japan

Industry Chairs

Andreas Dengel   DFKI, Germany
Véronique Eglin   Institut National des Sciences Appliquées (INSA)
                                    de Lyon, France
Nandakishore Kambhatla   Adobe Research, India

Publicity Chairs

Sukalpa Chanda   Østfold University College, Norway
Simone Marinai   University of Florence, Italy
Safwan Wshah   University of Vermont, USA

Technical Chair

Edward Sobczak   University at Buffalo, The State University of
                             New York, USA

Conference Secretariat

University at Buffalo, The State University of New York, USA

Program Committee

Senior Program Committee Members

Srirangaraj Setlur   Apostolos Antonacopoulos
Richard Zanibbi   Lianwen Jin
Koichi Kise   Nicholas Howe
Gernot Fink   Marc-Peter Schambach
David Doermann   Marcal Rossinyol
Rajiv Jain   Wataru Ohyama
Rolf Ingold   Nicole Vincent
Andreas Fischer   Faisal Shafait
Marcus Liwicki   Simone Marinai
Seiichi Uchida   Bertrand Couasnon
Daniel Lopresti   Masaki Nakagawa
Josep Llados   Anurag Bhardwaj
Elisa Barney Smith   Dimosthenis Karatzas
Umapada Pal   Masakazu Iwamura
Alicia Fornes   Tong Sun
Jean-Marc Ogier   Laurence Likforman-Sulem
C. V. Jawahar   Michael Blumenstein
Xiang Bai   Cheng-Lin Liu
Liangrui Peng   Luiz Oliveira
Jean-Christophe Burie   Robert Sabourin
Andreas Dengel   R. Manmatha
Robert Sablatnig   Angelo Marcelli
Basilis Gatos   Utkarsh Porwal

Program Committee Members

Harold Mouchere
Foteini Simistira Liwicki
Vernonique Eglin
Aurelie Lemaitre
Qiu-Feng Wang
Jorge Calvo-Zaragoza
Yuchen Zheng
Guangwei Zhang
Xu-Cheng Yin
Kengo Terasawa
Yasuhisa Fujii
Yu Zhou
Irina Rabaev
Anna Zhu
Soo-Hyung Kim
Liangcai Gao
Anders Hast
Minghui Liao
Guoqiang Zhong
Carlos Mello
Thierry Paquet
Mingkun Yang
Laurent Heutte
Antoine Doucet
Jean Hennebert
Cristina Carmona-Duarte
Fei Yin
Yue Lu
Maroua Mehri
Ryohei Tanaka
Adel M. M. Alimi
Heng Zhang
Gurpreet Lehal
Ergina Kavallieratou
Petra Gomez-Kramer
Anh Le Duc
Frederic Rayar
Muhammad Imran Malik
Vincent Christlein
Khurram Khurshid
Bart Lamiroy
Ernest Valveny
Antonio Parziale

Jean-Yves Ramel
Haikal El Abed
Alireza Alaei
Xiaoqing Lu
Sheng He
Abdel Belaid
Joan Puigcerver
Zhouhui Lian
Francesco Fontanella
Daniel Stoekl Ben Ezra
Byron Bezerra
Szilard Vajda
Irfan Ahmad
Imran Siddiqi
Nina S. T. Hirata
Momina Moetesum
Vassilis Katsouros
Fadoua Drira
Ekta Vats
Ruben Tolosana
Steven Simske
Christophe Rigaud
Claudio De Stefano
Henry A. Rowley
Pramod Kompalli
Siyang Qin
Alejandro Toselli
Slim Kanoun
Rafael Lins
Shinichiro Omachi
Kenny Davila
Qiang Huo
Da-Han Wang
Hung Tuan Nguyen
Ujjwal Bhattacharya
Jin Chen
Cuong Tuan Nguyen
Ruben Vera-Rodriguez
Yousri Kessentini
Salvatore Tabbone
Suresh Sundaram
Tonghua Su
Sukalpa Chanda

Mickael Coustaty
Donato Impedovo
Alceu Britto
Bidyut B. Chaudhuri
Swapan Kr. Parui
Eduardo Vellasques
Sounak Dey
Sheraz Ahmed
Julian Fierrez
Ioannis Pratikakis
Mehdi Hamdani
Florence Cloppet
Amina Serir
Mauricio Villegas
Joan Andreu Sanchez
Eric Anquetil
Majid Ziaratban
Baihua Xiao
Christopher Kermorvant
K. C. Santosh
Tomo Miyazaki
Florian Kleber
Carlos David Martinez Hinarejos
Muhammad Muzzamil Luqman
Badarinath T.
Christopher Tensmeyer
Musab Al-Ghadi
Ehtesham Hassan
Journet Nicholas
Romain Giot
Jonathan Fabrizio
Sriganesh Madhvanath
Volkmar Frinken
Akio Fujiyoshi
Srikar Appalaraju
Oriol Ramos-Terrades
Christian Viard-Gaudin
Chawki Djeddi
Nibal Nayef
Nam Ik Cho
Nicolas Sidere
Mohamed Cheriet
Mark Clement
Shivakumara Palaiahnakote
Shangxuan Tian

Ravi Kiran Sarvadevabhatla
Gaurav Harit
Iuliia Tkachenko
Christian Clausner
Vernonica Romero
Mathias Seuret
Vincent Poulain D'Andecy
Joseph Chazalon
Kaspar Riesen
Lambert Schomaker
Mounim El Yacoubi
Berrin Yanikoglu
Lluis Gomez
Brian Kenji Iwana
Ehsanollah Kabir
Najoua Essoukri Ben Amara
Volker Sorge
Clemens Neudecker
Praveen Krishnan
Abhisek Dey
Xiao Tu
Mohammad Tanvir Parvez
Sukhdeep Singh
Munish Kumar
Qi Zeng
Puneet Mathur
Clement Chatelain
Jihad El-Sana
Ayush Kumar Shah
Peter Staar
Stephen Rawls
David Etter
Ying Sheng
Jiuxiang Gu
Thomas Breuel
Antonio Jimeno
Karim Kalti
Enrique Vidal
Kazem Taghva
Evangelos Milios
Kaizhu Huang
Pierre Heroux
Guoxin Wang
Sandeep Tata
Youssouf Chherawala

Reeve Ingle
Aashi Jain
Carlos M. Travieso-Gonzales
Lesly Miculicich
Curtis Wigington
Andrea Gemelli
Martin Schall
Yanming Zhang
Dezhi Peng
Chongyu Liu
Huy Quang Ung
Marco Peer
Nam Tuan Ly
Jobin K. V.
Rina Buoy
Xiao-Hui Li
Maham Jahangir
Muhammad Naseer Bajwa

Oliver Tueselmann
Yang Xue
Kai Brandenbusch
Ajoy Mondal
Daichi Haraguchi
Junaid Younas
Ruddy Theodose
Rohit Saluja
Beat Wolf
Jean-Luc Bloechle
Anna Scius-Bertrand
Claudiu Musat
Linda Studer
Andrii Maksai
Oussama Zayene
Lars Voegtlin
Michael Jungo

Program Committee Subreviewers

Li Mingfeng
Houcemeddine Filali
Kai Hu
Yejing Xie
Tushar Karayil
Xu Chen
Benjamin Deguerre
Andrey Guzhov
Estanislau Lima
Hossein Naftchi
Giorgos Sfikas
Chandranath Adak
Yakn Li
Solenn Tual
Kai Labusch
Ahmed Cheikh Rouhou
Lingxiao Fei
Yunxue Shao
Yi Sun
Stephane Bres
Mohamed Mhiri
Zhengmi Tang
Fuxiang Yang
Saifullah Saifullah

Paolo Giglio
Wang Jiawei
Maksym Taranukhin
Menghan Wang
Nancy Girdhar
Xudong Xie
Ray Ding
Mélodie Boillet
Nabeel Khalid
Yan Shu
Moises Diaz
Biyi Fang
Adolfo Santoro
Glen Pouliquen
Ahmed Hamdi
Florian Kordon
Yan Zhang
Gerasimos Matidis
Khadiravana Belagavi
Xingbiao Zhao
Xiaotong Ji
Yan Zheng
M. Balakrishnan
Florian Kowarsch

Mohamed Ali Souibgui
Xuewen Wang
Djedjiga Belhadj
Omar Krichen
Agostino Accardo
Erika Griechisch
Vincenzo Gattulli
Thibault Lelore
Zacarias Curi
Xiaomeng Yang
Mariano Maisonnave
Xiaobo Jin
Corina Masanti
Panagiotis Kaddas
Karl Löwenmark
Jiahao Lv
Narayanan C. Krishnan
Simon Corbillé
Benjamin Fankhauser
Tiziana D'Alessandro
Francisco J. Castellanos
Souhail Bakkali
Caio Dias
Giuseppe De Gregorio
Hugo Romat
Alessandra Scotto di Freca
Christophe Gisler
Nicole Dalia Cilia
Aurélie Joseph
Gangyan Zeng
Elmokhtar Mohamed Moussa
Zhong Zhuoyao
Oluwatosin Adewumi
Sima Rezaei
Anuj Rai
Aristides Milios
Shreeganesh Ramanan
Wenbo Hu

Arthur Flor de Sousa Neto
Rayson Laroca
Sourour Ammar
Gianfranco Semeraro
Andre Hochuli
Saddok Kebairi
Shoma Iwai
Cleber Zanchettin
Ansgar Bernardi
Vivek Venugopal
Abderrhamne Rahiche
Wenwen Yu
Abhishek Baghel
Mathias Fuchs
Yael Iseli
Xiaowei Zhou
Yuan Panli
Minghui Xia
Zening Lin
Konstantinos Palaiologos
Loann Giovannangeli
Yuanyuan Ren
Shubhang Desai
Yann Soullard
Ling Fu
Juan Antonio Ramirez-Orta
Chixiang Ma
Truong Thanh-Nghia
Nathalie Girard
Kalyan Ram Ayyalasomayajula
Talles Viana
Francesco Castro
Anthony Gillioz
Huawen Shen
Sanket Biswas
Haisong Ding
Solène Tarride

Contents – Part IV

Posters: Handwriting

A Shallow Graph Neural Network with Innovative Node Updating for Online Handwritten Stroke Classification

Yan-Rong Wang[1], Da-Han Wang[1(✉)], Xiao-Long Yun[2], Yan-Ming Zhang[2], Fei Yin[2], and Shunzhi Zhu[1]

[1] Fujian Key Laboratory of Pattern Recognition and Image Understanding, School of Computer and Information Engineering, Xiamen University of Technology, Xiamen 361024, China
wangdh@xmut.edu.cn

[2] National Laboratory of Pattern Recognition (NLPR), Institute of Automation of Chinese Academy of Sciences, Beijing 100086, China

Abstract. Stroke classification is important to the layout analysis of online handwritten documents. Due to the diversity of writing styles and the complexity of layout structure, stroke classification is challenging. Graph neural networks (GNNs) is one of the most effective frameworks for stroke classification. However, GNNs has the problem of node over-compression caused by the deep structure of GNNs, which will lead to loss of node information and hence may deteriorate the performance of stroke classification. In this paper, we propose a shallow graph neural network model that is capable of retaining long-term receptive field by constructing a more reasonable graph through edge classification before the node classification step. Moreover, a novel node learning method is used to integrate edge features into nodes, where edge features not only participate in the calculation of node attention weight as in previous GNN based methods, but also participate in the final node integration. Experiments on the IAMonDo dataset show that our proposed method achieves an accuracy of 97.71% that is superior to existing state-of-the-art methods, demonstrating the effectiveness of the proposed method.

Keywords: Stroke classification · Online handwritten documents · Graph Neural Network · Node updating · Edge classification

1 Introduction

With the extensive application of handwriting devices, electronic handwritten document is becoming increasingly widespread. The digitization of electronic handwritten document is conducive to its preservation and retrieval. Electronic handwritten document is composed of a series of strokes, containing coordinates, time, pen pressure and other information. Stroke classification aims to classify the strokes into the text/non-text classes or multiple classes including text, graphics, tables, list, formula et al. It inherently is the task of layout analysis that is very important to the performance of online handwritten document digitization systems. However, due to the diversity of writing style and the complexity of document structure, stroke classification is challenging and has entered its bottleneck in improving the performance recently.

G. A. Fink et al. (Eds.): ICDAR 2023, LNCS 14190, pp. 3–19, 2023.
https://doi.org/10.1007/978-3-031-41685-9_1

Many works have been proposed to push forward the study of stroke classification. In the past, the temporal characteristics of strokes is considered as the main basis for stroke classification. Researchers use recurrent neural network (RNN) and its variants to model the stroke classification tasks [1–5, 20, 21]. According to the Markov property of the stroke sequence, some works use probabilistic graphical models (PGMs) to classify strokes [6–9, 22–28].

The development of graph neural networks (GNNs) brings more choices to solve the stroke classification task [10–12, 29]. However, previous study [33] has shown that, with the deepening of graph neural network, the information incorporated by nodes increases exponentially and the node information is over-compressed, that is easy to cause information loss. How to solve this problem with high quality and more efficiently requires more research efforts. We find that shallow graph neural network structures can avoid the above problems. The key to reducing the number of network layers is a more effective node information fusion mechanism and a reasonable graph structure that enables the network to expand the receptive field without the superposition of layers.

In this paper, we propose a shallow graph neural network model with the strategy of obtaining a reasonable graph construction by edge classification before the node classification step. And then, the multi-scale fusion mechanism is adopted to integrate the subgraph features, global features and initial node features for node classification. There are two kinds of network layers in our framework: edge learning layer and node learning layer that are used to update edges and nodes of the graph, respectively. It is observed that, in previous works, edge features only participate in the calculation of node attention coefficients but are ignored in the final node updates. Inspired by [30] where edge features also participate in the final node integration, we propose a novel node learning method to better utilize edge features. We combine an edge with one of its corresponding nodes to form a composite structure named node-edge (NE) structure, and use this structure to calculate the attention mechanism and update nodes.

The main contributions in this paper are summarized as follows:

1) We propose a shallow graph neural network framework to handle stroke classification tasks. The strategy is to use a reasonable graph construction to expand the receptive field of nodes, so as to avoid the node compression problem caused by network layer superposition.
2) We propose a novel node learning method with the NE structure that contains an edge and one of the edge's nodes. The NE structure can be used for both node attention score calculation and node update.
3) Our proposed framework achieves state-of-the-art accuracy on the IAMonDo dataset [13] and the effectiveness of each module is proved by ablation experiments.

The rest of this paper is organized as follows: Sect. 2 introduces some related work. Section 3 specifies our framework and method. Section 4 shows experimental data and results analysis. Section 5 concludes our work.

2 Related Work

2.1 Stroke Classification

A stroke contains a series of coordinates, time, pen pressure and other information. Three properties are embodied in strokes: temporal continuity, spatial continuity and markov property. We will introduce the past work in three categories: recurrent neural network (RNN) and its variants, probabilistic graphical models (PGMs), and graph neural networks (GNNs).

Due to its temporal continuity, recurrent neural network (RNN) and its variants are proposed for stroke classification. Degtyarenko et al. [14] proposed a new structure, hierarchical recurrent neural network (HRNN). They divided the document into three levels of processing: points, strokes, and objects. The model was built by layer-by-layer integration from low-level features to high-level features. Grygoriev et al. [15] proposed a novel fragment pooling technique for feature integration between hierarchical levels based on HRNN. They achieved the state-of-the-art accuracy in multi-class stroke classification task on the IAMonDo dataset [13].

Considering the Markov property of strokes, probabilistic graphical models (PGMs) was proposed for the stroke classification task. Bishop et al. [6] applied multi-layer perceptron (MLP) and hidden markov model (HMM) to stroke classification. Zhou et al. [7] and Delaye et al. [8] proposed the combination of support vector machine (SVM) and Markov Random Field (MRF) or Conditional Random Field (CRF). Li et al. [9] replaced SVM with neural network for potential energy learning in the combination of SVM and CRF.

Yun et al. [10] applied graph attention network to stroke classification of flowchart by enhancing the utilization of edge features in the model. To improve the influence of edges in graph structure, Ye et al. [11] proposed an edge graph attention network. By updating edge features, nodes could incorporate more information between node pairs. Yun et al. [12] proposed the Instance GNN for stroke classification, which is composed of node classification, edge classification and node representation. The Instance GNN improves the performance of stroke classification and clustering by post-processing.

Although these GNNs based works have achieved great progresses in stroke classification, the over-compression problem has not been fully considered and still remains an open problem deserving further study. Moreover, it is also observed that, in the node learning/updating procedure of GNNs, edge features only participate in the calculation of node attention coefficients but are ignored in the final node integration. The edge features are not fully utilized and need to be better integrated into node features.

2.2 Graph Natural Networks

As a kind of non-European structure data, graphs cannot be learned using traditional convolution neural network until the rise of graph neural networks (GNNs).

Defferrand et al. [16] map the graph in the spatial domain to the spectral domain for convolution operation by Laplace transform. Hamilton et al. [17] proposed graphsage network (GraphSage) with graph sampling and message aggregation.

To reflect the difference of neighbor's influence on nodes, Velickovic et al. [18] introduce self-attention mechanism into graph neural network. Graph attention network (GAT) calculates the different weights between the target node and its neighbors, then aggregating information. As an important component of a graph, edges are critical to graph learning as well. Gong et al. [19] proposed a new structure by stacking layers with node feature update and edge feature update alternately. Wang et al. [30] proposed the edge-integrated attention mechanism, where edge features not only participate in the calculation of node attention weights, but also participate in the final node integration. This work is most relevant to our node learning method. Our work is the first to adopt this strategy to the stroke classification task, and innovatively use this strategy to construct a better graph to overcome the over-compression problem for stroke classification.

3   Proposed Method

3.1   Problem Formulation

Given a set of documents $\{d^{(n)}, i = 1, \ldots, N\}$ as the training set and each document consists of a group of strokes with different numbers $\{s_t^{(n)}, t = 1, \ldots, T^{(n)}\}$, where N is the number of documents and $T^{(n)}$ is the number of strokes of n_{th} document. There is a set of labels equal to the number of strokes and correspond with strokes one by one $\{l_t^{(n)} \in L, t = 1, \ldots, T^{(n)}\}$, where L is the number of strokes' classes. Our task is to design a structure to predict strokes in the test set after learning from the training set.

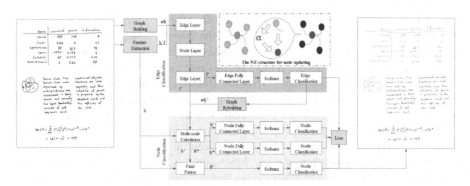

Fig. 1. An illustration of the training and prediction process for one input document.

3.2   Framework

Figure 1 illustrates the framework of our model. In this work, we construct the electronic handwritten document as an initial undirected graph composed of manual node features, edge features and adjacency matrix, and then send it to the edge classification branch to obtain a more reasonable graph for node fusion and classification. We will introduce the framework in three parts: constructing undirected graph, manual extraction of stroke features and our model.

Graph Construction. In this work, an electronic handwritten document is constructed as an undirected graph $G(V, E)$. Each stroke is considered as a graph node $i \in V$, and the relationship between a pair of strokes is considered as graph edge $(i, j) \in E$. Following the work [12, 29], we use the spatial context and temporal context of strokes to construct the adjacency matrix of the graph. Specifically, four types of contexts are considered: spatial radius, temporal radius, spatial nearest and temporal nearest. If a pair of strokes have adjacency, whether spatial adjacency or temporal adjacency, there is an edge between these two strokes. In addition, each node has a self-connected edge.

Spatial Radius. If the spatial distance between two strokes is less than the threshold k_s, the pair of these two strokes is considered to have a spatial adjacency relationship.

Temporal Radius. If the temporal distance between two strokes is less than the threshold k_t, the pair of these two strokes is considered to have a temporal adjacency relationship.

Spatial Nearest. If the spatial distance between stroke j and stroke i is the top k among the spatial distance between all other strokes and stroke i, this pair of stroke i and stroke j is considered to have a spatial adjacency relationship. We define the threshold top k as k_{tops}.

Temporal Nearest. If the temporal distance between stroke j and stroke i is the top k among the temporal distance between all other strokes and stroke i, this pair of stroke i and stroke j is considered to have a temporal adjacency relationship. We define the threshold top k as k_{topt}.

Feature Extraction. Each node is described by a set of stroke features, including geometric features and contextual features. Each edge is described by a set of features extracting from stroke pairs. Different from the previous work [11], we expanded 30 node features referencing [32]. Table 1 lists the additional node features differing from [11]. Edge feature design and feature normalization are completely based on [11].

$$
\begin{aligned}
x' &= sign(x)\sqrt{|x|}, \\
x'' &= \frac{x'-\mu}{\sigma}.
\end{aligned}
\tag{1}
$$

NE Graph Neural Network. We propose a shallow graph neural network model called NEGNN with the strategy of obtaining a reasonable graph construction first. The proposed method mainly contains two branches: the edge classification and the node classification branch. The edge classification branch transforms the original graph structure into a more reasonable graph structure. It consists of two edge learning layers and one node learning layer. In the node classification branch, the multi-scale fusion mechanism is adopted to integrate the subgraph features, global features and initial node features. It is composed of a multi-scale calculation layer and a final node fusion layer. These two layers are variants of node learning layer. In the stroke classification task, the previous node learning method only used edge features for node attention calculation, ignoring the edge features in the final node update. Inspired by [30], an innovative node learning method is proposed with the node-edge (NE) structure that is composed of an edge and one of the edge's nodes. It can be used for not only node attention calculation but also the final node update. The architecture of the proposed NEGNN is shown in Fig. 1.

Table 1. Extended node features extracted from stroke i.

#	Description
1	Number of sampling points
2	Ratio of duration to stroke length
3	The ratio of the length of the first and last lines to the stroke length
4	Standard deviation of sampling point from the line between the first and last point
5	Slope of the connection between the first and last points
6	Ratio of stroke length to number of sample points
7	The ratio of the first and last lines to the number of sampling points
8	Rotation angle of minimum circumscribed rectangle
9	Width of the minimum circumscribed rectangle
10	Height of the minimum circumscribed rectangle
11	Diagonal length of minimum circumscribed rectangle
12	Width of bounding box
13	Height of bounding box
14	Diagonal length of bounding box
15	Bounding box diagonal angle
16	Ratio of stroke length to bounding box diagonal angle
17	Minimum circumscribed radius
18	Variance of the distance between the sampling point and the center of the minimum circumscribed circle
19	Variance of distance between sampling point and centroid
20	Area of bounding box
21	Area of minimum circumscribed rectangle
22	Area of minimum circumscribed circle
23	Ratio of convex hull area to bounding box area
24	Ratio of convex hull area to minimum circumscribed circle area
25	Ratio of convex hull area to minimum circumscribed rectangle area
26	The ratio of the stroke projection on the x-axis to the line between the first and last points
27	Maximum angle composed of sampling points
28	Minimum angle composed of sampling points
29	Average angle of sample point composition
30	Distance between the center of the minimum circumscribed circle and the center of mass

The original graph structure are described by three inputs: node features $H = \{h_i, i \in V\}, h_i \in \mathbb{R}^P$, edge features $F = \{f_{ij}, (i, j) \in E\}, f_{ij} \in \mathbb{R}^Q$, and adjacency matrix

$A = \{a_{ij}, (i, j) \in E\}$, $a_{ij} \in (0, 1)$, where P and Q are the numbers of node features and edge features.

The edge classification branch transforms the original graph structure into a more reasonable graph structure. We first send the initial graph structure to the branch, and then obtain the updated edge features and classify the edges. The result of edge classification is the new adjacency matrix. Different from [11], the first layer of our network is not the node learning layer, but the edge learning layer. There are two reasons why we prioritize edge learning. Firstly, we extend the node feature without extending the edge feature so that edge learning first would mine more context information from the extended node features. Secondly, we expect to extract the edge features and update the adjacency matrix before the node classification, after which we do not update the edge features any more. Hence, more edge update layers would be more desired in the purpose of using as few layers as possible to make the network shallow. In the edge-node-edge structure, edge features can be updated twice within three layers, while the node-edge-node-edge structure requires 4 layers. The output, a more reasonable graph, is composed of the original node features, the updated edge features and the updated adjacency matrix.

In the node classification branch, we fuse node information and classify nodes using the reasonable graph construction. In this branch, two variants of node learning layer are applied. The first node learning layer called multi-scale calculation is used to calculate the features of the subgraph and the global features. The subgraph features are calculated by the normal attention coefficient. The global features are calculated by removing the low attention coefficient and recalculated the weight of remaining scores for the whole graph. These two features will also be classified separately. Finally, we use another node learning layer called final fusion for multi-scale features fusion, and then carry out the final node classification. The motivation for this design stems from the diversity of tables, e.g., some tables have lines and some do not. For the former kind of tables, the subgraph features like the border of tables will quickly determine that the subgraph is a table. For the latter kind, the global features, such as the global arrangement of the text, can help determine whether the subgraph is a table.

A novel node learning method is adopted in both edge classification and node classification modules. Although [11, 12, 29] used edge features in the node update, they only used it in the calculation of attention weight. Inspired by [30], we propose the NE structure to both calculate the node attention coefficients and update nodes. For example, NE_{ij} is calculated by edge features f_{ij} and node features h_j. As for the edge learning method, we refer to [11]. The edge learning method, node learning method and other details are defined in the following.

Edge Learning Method. Edge features are learned from two parts, one from the learning between node features and another from the learning of input edge features. It is defined as

$$
\begin{aligned}
r_{ij} &= \sigma\left(W_h\left[h_i \oplus h_j \oplus \left|h_i - h_j\right|\right]\right), \\
t_{ij} &= \sigma\left(W_f f_{ij}\right), \\
f_{ij'} &= \sigma\left(W_r\left[r_{ij} \oplus t_{ij}\right]\right),
\end{aligned}
\tag{2}
$$

where $\oplus$ is the concatenation operator, $|\cdot|$ is elementwise absolute function, $\sigma(\cdot)$ is leakyReLU activation function, $W_h \in \mathbb{R}^{3P \times P}$, $W_f \in \mathbb{R}^{Q \times P}$, $W_r \in \mathbb{R}^{2P \times Q'}$ are learnable parameters, P is the number of input node features, Q and Q' is the number of input and output edge features respectively. In the last edge layer, we will update the adjacency matrix according to the edge classification results. It is defined as

$$a_{ij} = argmax\left(f_{ij}^*\right), \tag{3}$$

where f_{ij}^* is edge features after edge fully connected layer and softmax function.

Node Learning Method with NE Structure. The NE structure is calculated at the edge f_{ij} and node h_j, and then applied to the calculation of attention coefficient and the final node update. The node learning method of node update layer is defined as

$$
\begin{aligned}
NE_{ij} &= \sigma\left(W_r\left(\sigma\left(W_h h_j\right) \oplus \sigma\left(W_f f_{ij} + b\right)\right)\right), \\
s_{ij} &= \sigma\left(v^T NE_{ij}\right), \\
\alpha_{ij} &= \frac{e^{s_{ij}}}{\sum_{k \in N_i} e^{s_{ik}}}, \\
h_{i'} &= \sigma\left(\sum_{j \in N_i} \alpha_{ij} NE_{ij}\right), \\
h_{i'} &= \oplus_{k=1}^{K} \sigma\left(\sum_{j \in N_i} \alpha_{ij}^k NE_{ij}^k\right),
\end{aligned}
\tag{4}
$$

where $W_h \in \mathbb{R}^{P \times P'}$, $W_f \in \mathbb{R}^{Q \times P'}$, $W_r \in \mathbb{R}^{2P' \times P'}$, $b \in \mathbb{R}^{P'}$, $v \in \mathbb{R}^{P'}$ are learnable parameters and P' is the number of output node features, N_i is the neighborhood of node i and K is the number of attention heads. Note that, the parameters here are independent of the edge learning layer. In multi-scale calculation layer, the node learning method for calculation of subgraph features is as described above while the global one is defined as

$$
\begin{aligned}
s_{ij}^* &= a^{\alpha_{ij}}, \\
\alpha_{ij}^* &= \frac{e^{s_{ij}^*}}{\sum e^{s_{ij}^*}}, \\
h_i'^* &= \sigma\left(\sum \alpha_{ij}^* NE_{ij}\right), \\
h_i'^* &= \oplus_{k=1}^{K} \sigma\left(\sum \alpha_{ij}^{*k} NE_{ij}^k\right),
\end{aligned}
\tag{5}
$$

where a is a hyperparameter controlling the concentration of attention. In the final fusion layer, the NE structure is defined as

$$
\begin{aligned}
h_{cat} &= h_i \oplus h_{i'} \oplus h_i'^*, \\
NE_{ij} &= \sigma\left(W_r\left(\sigma\left(W_{cat} h_{catj}\right) \oplus \sigma\left(W_f f_{ij} + b\right)\right)\right),
\end{aligned}
\tag{6}
$$

where $W_{cat} \in \mathbb{R}^{(P+2P_{mnl}) \times P_{cls}}$, $W_f \in \mathbb{R}^{Q' \times P_{cls}}$, $W_r \in \mathbb{R}^{2P_{cls} \times P_{cls}}$, $b \in \mathbb{R}^{P_{cls}}$, $v \in \mathbb{R}^{P_{cls}}$ are learnable parameters, P_{mnl} is the number of node features from multi-scale calculation layer, Q' is the number of edge features from the last edge learning layer and P_{cls} is the number of node classes. The remaining learning method is consistent with node update layer.

Others. The edge fully connected layer and node fully connected layer are simple linear transformations. For the input x,

$$y = Wx + b, \tag{7}$$

where $W \in \mathbb{R}^{C \times C}$, $b \in \mathbb{R}^C$ are learnable parameters, C and C' are input dimension and output dimension of x.

3.3 Network Training

We use the standard cross entropy as the loss function to optimize the training parameters. The node classification loss and the edge classification loss are considered.

For node classification, given the output probability $p \in \mathbb{R}^{T^{(n)} \times P_{cls}}$ and the ground truth label $y \in \{0, 1, \ldots, L\}^{T^{(n)}}$,

$$\text{Loss}_{\text{node}}(W) = -\frac{1}{T^{(n)}} \sum_{t=1}^{T^{(n)}} \sum_{l=1}^{L} \delta(y_t = l) \log p_t(l; W). \tag{8}$$

For edge classification, if two strokes come from the same layout, for example, both from the same table, the edge of these two strokes is considered to be positive edge, otherwise negative. Given the output probability $p \in \mathbb{R}^{T_i \times T_i \times 2}$ and the ground truth label $e \in \{0, 1\}^{T_i \times T_i}$,

$$\text{Loss}_{\text{edge}}(W) = -\frac{1}{T^{(n)} * T^{(n)}} \sum_{i=1}^{T^{(n)}} \sum_{j=1}^{T^{(n)}} \sum_{l=1}^{2} \delta(e_{ij} = l) \log p_{ij}(l; W). \tag{9}$$

The total loss consists of four parts. Edge classification loss, two kinds of node classification loss from multi-scale calculation layer and final node classification loss from final fusion layer.

$$Loss(W) = \text{Loss}_{\text{edge}}(W) + \text{Loss}_{\text{sub}}(W) + \text{Loss}_{\text{global}}^{*}(W) + \text{Loss}_{\text{node}}(W). \tag{10}$$

We implement the NEGNN training algorithm with pytorch [31]. More training settings is shown in Table 2.

4 Experiments

To evaluate the performance of our NEGNN model, we conduct stroke multi-classification experiments on the IAMonDo dataset [13]. We compare the experimental results with previous works, and verify the effect of the main modules in our model by ablation experiments. In order to make a reasonable comparison with the previous works, we repeat each group of experiments for 10 times, and take its mean and standard deviation for comparison among different models.

Table 2. Hyperparameters for all experiments.

Hyperparameter	Value
Initial learning rate	0.0008
Dropout rate	0.1
Batch size (number of nodes)	64
Max epoch	100
Number of attention heads (K) [node layer, multi-scale calculation]	[1, 1]
Number of node output features [node layer, multi-scale calculation]	[32, 32]
Number of edge output features	[64, 64]
Base number for multi-scale attention (a)	100
Spatial Radius threshold (k_s)	30
Temporal Radius threshold (k_t)	5
Spatial nearest threshold (k_{tops})	-
Temporal nearest threshold (k_{topt})	-

4.1 Dataset

IAMonDo dataset is a publicly available collection of freely handwritten English online documents written by more than 200 writers, including 1000 documents and 329849 strokes. Its layout includes text blocks, formulas, tables, lists, diagram et al. The publisher divided the dataset into five collections, each with about 200 documents. According to the official guidelines, we adopt set0 and set1 as the training set, while set2 and set3 are used as verification set and test set respectively. Table 3 lists details of each set.

Table 3. Statistics of the IAMOnDo dataset number of strokes per category.

Category	Training	Validation	Test
Graphic	37496	15481	17488
Text	79812	39796	40469
Table	13044	6562	6883
List	6337	3474	3115
Math	6659	3412	2972

4.2 Evaluation Metrics

We use accuracy to evaluate our model and report the accuracy for each class. Overall accuracy and the accuracy for each class are defined as

$$
accuracy = \frac{\sum_{n=1}^{N} \sum_{t=1}^{T^{(n)}} \delta\left(\hat{y}_t^{(n)} = y_t^{(n)}\right)}{\sum_{n=1}^{N} T^{(n)}},
$$
$$
accuracy(c) = \frac{\sum_{n=1}^{N} \sum_{t=1}^{T^{(n)}} \delta\left(y_t^{(n)} = c\right) \delta\left(\hat{y}_t^{(n)} = y_t^{(n)}\right)}{\sum_{n=1}^{N} T^{(n)} \delta\left(y_t^{(n)} = c\right)}.
\tag{11}
$$

4.3 Comparison with Previous Methods

Table 4. Performance of different methods for multi-class stroke classification.

Method	Graphic	Text	Table	List	Math	Overall
A. Delaye [24]	95.85	97.25	77.64	74.73	84.28	93.46
EGAT [11]	97.11 ± 0.38	98.35 ± 0.24	89.70 ± 1.86	76.15 ± 1.78	88.43 ± 1.53	95.81 ± 0.29
HRNN [14]	97.68	98.71	97.37	**88.99**	93.03	97.25 ± 0.25
HCRNN [15]	97.52	**99.12**	**97.91**	87.61	**94.85**	97.58 ± 0.21
NEGNN (ours)	**98.07 ± 0.52**	98.97 ± 0.09	95.82 ± 1.72	88.31 ± 3.04	92.62 ± 1.68	**97.71 ± 0.20**

Table 4 illustrates that our model has achieved the state-of-the-art accuracy for the task of multi-class stroke classification. Since the work of HRNN and HCRNN only reported the accuracy of each class under the highest accuracy, we cannot obtain its average level. Compared with EGAT with 10 layers and 8 attention heads, our model only uses 3 basic fully connected layers and 5 main learning layers with 1 attention head to achieve this overall accuracy increasing on 1.9% up to 97.71%. In addition, the performance of NEGNN in accuracy of each class is better than EGAT thoroughly. Compared with the HRNN and HCRNN, the proposed NEGNN also achieves superior performance in overall.

We perform experiments on a computer with Intel(R) Xeon(R) Silver 4208 CPU(2.10GHz) and Tesla V100-PCIE-32GB.

Table 5 presents the number of parameters, training time, test speed and layers. In our work, the number of layers refers to layers used for node fusion after graph structural adjustment, that is, the number of layers for node classification branches.

4.4 Ablation Study

We conducted ablation experiments from four aspects to determine the contribution of each part: extended node features, update of adjacency matrix before node classification branch, multi-scale fusion mechanism and the proposal of NE structure. Ablation

Table 5. The number of parameters, training time, test speed and layers.

Method	Parameters	Training time	Test speed(s/doc)	Number of layers
EGAT [11]	1.63M	1680s	0.0033	10
NEGNN (ours)	0.034M	80s/epoch	0.0089	2

experimental result is shown in Table 6, Table 7, Table 8 and Table 9. The parameters of all experiments are shown in Table 2.

Before analyzing the experiments results, a brief description of Table 6 and Table 7 is necessary. "Adj" in the table means update of the adjacency matrix. "Sub Loss" in the table denotes that the subgraph features participate in loss calculation. "Sub Fusion" in the table denotes that the subgraph features participate in calculation of final fusion layer. "Global Loss" and "Global Fusion" in the table is similar to the above two phrase while the protagonist becomes the global one. "E. F" represents whether to use extended node features. "NE", "EGAT [+]", and "EGAT [cat]" denote that the node update method adopts the NE structure, EGAT, or the structure like EGAT by replacing $h_i + h_j$ with $h_i \oplus h_j$.

Table 6. Methods and its associated modules. The NE structure is adopted for all these settings.

#	Adj	Sub Loss	Sub Fusion	Global Loss	Global Fusion	E. F	Accuracy
1	✓	✓	✓	✓	✓	✓	**97.71 ± 0.20**
2	✓	✓	✓	✓	✓		97.35 ± 0.12
3		✓	✓	✓	✓	✓	97.4 ± 0.26
4	✓			✓	✓	✓	97.6 ± 0.18
5	✓	✓	✓			✓	96.67 ± 0.22
6	✓		✓		✓	✓	97.28 ± 0.17

Table 7. Different node learning methods.

#	NE	EGAT [+]	EGAT [cat]	Accuracy
1	✓			**97.71 ± 0.20**
2		✓		90.64 ± 1.15
3			✓	90.79 ± 0.6

Extended Node Features. As shown in Table 6, comparing group 1 and group 2, we can see that it brings 0.36% growth by using extended node features. Without the use of extended features, our model also improves the accuracy of EGAT by 1.54%.

Update of Adjacency Matrix. As shown in Table 6, Compared with the group 1 and group 3, it can be seen that using adjacency matrix update method is slightly better than not. Two groups of ablation experiments under extreme initial graph structure have been added. We set up two groups of hyperparameters separately for its ablation experiment. In the case of $k_{tops} = 2$ and $k_{topt} = 2$, the stroke in the initial construction graph can be only connected with one neighbor, because the nearest one is itself. In the same way, we also set another extreme set of super parameters $k_{tops} = maxT$ and $k_{topt} = maxT$, where $maxT$ is the maximum number of strokes. Table 8 proves that updating the adjacency matrix is better than not. In order to prove that the new adjacency matrix is a more reasonable graph structure, we made statistics on the edge accuracy of the initial graph and the edge accuracy of the new adjacency matrix. As shown in Table 9, the edge accuracy has increased from 68.52% to 94.59%, indicating that the updated graph structure is more reasonable.

Table 8. Performance of adjacency matrix updating under different hyperparameters.

#	Adj	k_{topk}	k_{topt}	Accuracy
1	✓	2	2	**96.91 ± 0.38**
2		2	2	96.02 ± 0.36
3	✓	MaxT	MaxT	**96.5 ± 0.39**
4		MaxT	MaxT	96.35 ± 0.43

Table 9. Statistics on the edge accuracy of the initial graph structure and the updated graph structure.

Graph Structure	Edge Accuracy
Initial	68.52
Updated	94.59

Multi-scale Fusion Mechanism. As shown in Table 6, compared with group 1 and group 4, the difference in accuracy is only 0.11%. It indicates that subgraph features have a small contribution to the final node classification, but it did work. Compared with group 1 and group 5, accuracy increased by 1.04% with global features. Compared with group 1 and group 6, the multi-scale features participating in loss calculation is beneficial to increase the accuracy of the model by 0.43% than not.

NE Structure. As shown in Table 7, the use of NE structure is nearly 7% higher than the use of two forms of EGAT structure. Obviously, the proposal of NE structure is the key for the whole model to reach state-of-the-art accuracy.

4.5 Error Analysis

We visualized the experimental results with the highest accuracy. Figure 2 shows some examples of error. Through observation, we have got two inspirations, which may be helpful for future work. Firstly, we believe that semantic information would be considered in future work. For example, according to different writing habits, the margin between the graphics and the text beside the graphics would be large or small. Whether the text belonging to a graphic or another layout can be distinguished by semantics. Secondly, pay more attention to strokes at the junction of two different layouts. On the one hand, this part of the stroke itself is easy to integrate the information of other modules due to the wrong edge, resulting in classification errors. On the other hand, its classification errors will also affect other strokes in its field. The improvement of edge classification accuracy may help to classify such strokes.

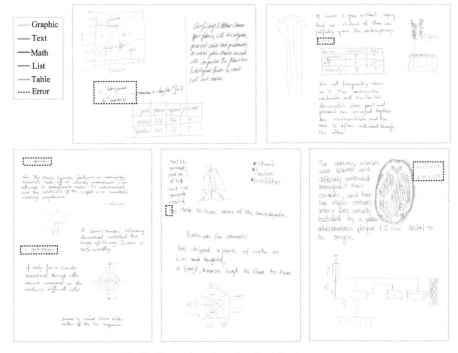

Fig. 2. Examples of stroke classification errors.

5 Conclusion

In this paper, we propose a shallow graph neural network model with the strategy of obtaining reasonable graph construction first by edge classification. The receptive field of nodes is perceived through the reasonable graph structure, rather than the superposition of network layers. Therefore, the problem of nodes being over-compressed can be solved.

For node classification, multi-scale node information fusion mechanism is adopted so that subgraph features, global features and original features can be all integrated into final node features. Specially, we have innovated the node learning method in order to integrate edge features into nodes more sufficiently. In our proposed node learning method, edge features not only participate in the calculation of node attention weight, but also participate in the final node integration. Experiments on the IAMonDo dataset show that our proposed method achieves an accuracy of 97.71%. That is superior to existing state-of-the-art methods, demonstrating the effectiveness of the proposed method. In the ablation experiments, we proved the effectiveness of extended feature, adjacency matrix update, multi-scale fusion mechanism and the use of NE structure.

In the feature, several potential directions can be explored. Through error analysis, we propose that the addition of semantics would help identify the relevance of the two modules. For example, the text near the drawing belongs to a graphic description or a text block. Another direction is how to improve the accuracy of edge classification. A high accuracy of edge classification can improve the graph construction, which plays an important role in the learning of graph neural network.

Acknowledgement. This work is supported by National Natural Science Foundation of China (No. 61773325, 62276258), Industry University Cooperation Project of Fujian Science and Technology Department (No. 2021H6035), and the Science and Technology Planning Project of Fujian Province (No. 2020Y9064), and Fu-Xia-Quan National Independent Innovation Demonstration Project (No. 2022FX4).

References

1. Otte, S., Krechel, D., Liwicki, M., et al.: Local feature based online mode detection with recurrent neural networks. In: 2012 International Conference on Frontiers in Handwriting Recognition, pp. 533–537. IEEE (2012)
2. Indermühle, E., Frinken, V., Bunke, H.: Mode detection in online handwritten documents using BLSTM neural networks. In: 2012 International Conference on Frontiers in Handwriting Recognition, pp. 302–307. IEEE (2012). Author, F., Author, S., Author, T.: Book title. 2nd edn. Publisher, Location (1999)
3. Khomenko, V., Volkoviy, A., Degtyarenko, I., et al.: Handwriting text/non-text classification on mobile device. In: The Fourth International Conference on Artificial Intelligence and Pattern Recognition (AIPR), pp. 42–49 (2017)
4. Polotskyi, S., Deriuga, I., Ignatova, T., Melnyk, V., Azarov, H.: Improving online handwriting text/non-text classification accuracy under condition of stroke context absence. In: Rojas, I., Joya, G., Catala, A. (eds.) IWANN 2019. LNCS, vol. 11506, pp. 210–221. Springer, Cham (2019). https://doi.org/10.1007/978-3-030-20521-8_18
5. Van Phan, T., Nakagawa, M.: Text/non-text classification in online handwritten documents with recurrent neural networks. In: 2014 14th International Conference on Frontiers in Handwriting Recognition, pp. 23–28. IEEE (2014)
6. Bishop, C.M., Svensen, M., Hinton, G.E.: Distinguishing text from graphics in on-line handwritten ink. In: Ninth International Workshop on Frontiers in Handwriting Recognition, pp. 142–147. IEEE (2004)
7. Zhou, X.D., Liu, C.L.: Text/non-text ink stroke classification in Japanese handwriting based on markov random fields. In: Ninth International Conference on Document Analysis and Recognition (ICDAR 2007), vol. 1, pp. 377–381. IEEE (2007)

8. Delaye, A., Liu, C.-L.: Text/non-text classification in online handwritten documents with conditional random fields. In: Liu, C.-L., Zhang, C., Wang, L. (eds.) Pattern Recognition, pp. 514–521. Springer Berlin Heidelberg, Berlin, Heidelberg (2012). https://doi.org/10.1007/978-3-642-33506-8_63

9. Ye, J.Y., Zhang, Y.M., Liu, C.L.: Joint training of conditional random fields and neural networks for stroke classification in online handwritten documents. In: 2016 23rd International Conference on Pattern Recognition (ICPR), pp. 3264–3269. IEEE (2016)

10. Yun, X.-L., Zhang, Y.-M., Ye, J.-Y., Liu, C.-L.: Online handwritten diagram recognition with graph attention networks. In: Zhao, Y., Barnes, N., Chen, B., Westermann, R., Kong, X., Lin, C. (eds.) Image and Graphics: 10th International Conference, ICIG 2019, Beijing, China, August 23–25, 2019, Proceedings, Part I, pp. 232–244. Springer International Publishing, Cham (2019). https://doi.org/10.1007/978-3-030-34120-6_19

11. Ye, J.Y., Zhang, Y.M., Yang, Q., et al.: Contextual stroke classification in online handwritten documents with graph attention networks. In: 2019 International Conference on Document Analysis and Recognition (ICDAR), 993–998. IEEE (2019)

12. Yun, X.L., Zhang, Y.M., Yin, F., et al.: Instance GNN: a learning framework for joint symbol segmentation and recognition in online handwritten diagrams. IEEE Trans. Multimedia **24**, 2580–2594 (2021)

13. Indermühle, E., Liwicki, M., Bunke, H.: IAMonDo-database: an online handwritten document database with non-uniform contents. In: Proceedings of the 9th IAPR International Workshop on Document Analysis Systems, pp. 97–104 (2010)

14. Degtyarenko, I., Deriuga, I., Grygoriev, A., et al.: Hierarchical recurrent neural network for handwritten strokes classification. In: ICASSP 2021–2021 IEEE International Conference on Acoustics, Speech and Signal Processing (ICASSP), pp. 2865–2869. IEEE (2021)

15. Grygoriev, A., Degtyarenko, I., Deriuga, I., Polotskyi, S., Melnyk, V., Zakharchuk, D., Radyvonenko, O.: HCRNN: a novel architecture for fast online handwritten stroke classification. In: Lladós, J., Lopresti, D., Uchida, S. (eds.) Document Analysis and Recognition – ICDAR 2021: 16th International Conference, Lausanne, Switzerland, September 5–10, 2021, Proceedings, Part II, pp. 193–208. Springer International Publishing, Cham (2021). https://doi.org/10.1007/978-3-030-86331-9_13

16. Defferrard, M., Bresson, X., Vandergheynst, P.: Convolutional neural networks on graphs with fast localized spectral filtering. In: Advances in Neural Information Processing Systems, pp. 3844–3852 (2016)

17. Hamilton, W., Ying, Z., Leskovec, J.: Inductive representation learning on large graphs. In: Advances in Neural Information Processing Systems, vol. 30 (2017)

18. Veličković, P., Cucurull, G., Casanova, A., et al.: Graph attention networks. arXiv preprint arXiv:1710.10903 (2017)

19. Gong, L., Cheng, Q.: Exploiting edge features for graph neural networks. In: Proceedings of the IEEE/CVF Conference on Computer Vision and Pattern Recognition, pp. 9211–9219 (2019)

20. Polotskyi, S., Radyvonenko, O., Degtyarenko, I., et al.: Spatio-temporal clustering for grouping in online handwriting document layout analysis with GRU-RNN. In: 2020 17th International Conference on Frontiers in Handwriting Recognition (ICFHR), pp. 276–281. IEEE (2020)

21. Van Phan, T., Nakagawa, M.: Combination of global and local contexts for text/non-text classification in heterogeneous online handwritten documents. Pattern Recogn. **51**, 112–124 (2016)

22. Li, X.H., Yin, F., Liu, C.L.: Printed/handwritten texts and graphics separation in complex documents using conditional random fields. In: 2018 13th IAPR International Workshop on Document Analysis Systems (DAS), pp. 145–150. IEEE (2018)

23. Wang, C., Mouchere, H., Viard-Gaudin, C., et al.: Combined segmentation and recognition of online handwritten diagrams with high order markov random field. In: 2016 15th International Conference on Frontiers in Handwriting Recognition (ICFHR), pp. 252–257. IEEE (2016)

24. Delaye, A., Liu, C.L.: Multi-class segmentation of free-form online documents with tree conditional random fields. Int. J. Doc. Anal. Recogn. (IJDAR) 17(4), 313–329 (2014)

25. Inatani, S., Van Phan, T., Nakagawa, M.: Comparison of MRF and CRF for Text/Non-text classification in Japanese Ink Documents. In: 2014 14th International Conference on Frontiers in Handwriting Recognition, pp, 684–689. IEEE (2014)

26. Delaye, A., Liu, C.L.: Graphics extraction from heterogeneous online documents with hierarchical random fields. In: 2013 12th International Conference on Document Analysis and Recognition, pp, 1007–1011. IEEE (2013)

27. Delaye, A., Liu, C.L.: Contextual text/non-text stroke classification in online handwritten notes with conditional random fields. Pattern Recogn. 47(3), 959–968 (2014)

28. Delaye, A., Liu, C.L.: Context modeling for text/non-text separation in free-form online handwritten documents. Doc. Recogn. Retrieval XX SPIE 8658, 98–109 (2013)

29. Ye, J.Y., Zhang, Y.M., Yang, Q., et al.: Joint stroke classification and text line grouping in online handwritten documents with edge pooling attention networks. Pattern Recogn. 114, 107859 (2021)

30. Wang, Z., Chen, J., Chen, H.: EGAT: edge-featured graph attention network. In: Farkaš, I., Masulli, P., Otte, S., Wermter, S. (eds.) Artificial Neural Networks and Machine Learning – ICANN 2021: 30th International Conference on Artificial Neural Networks, Bratislava, Slovakia, September 14–17, 2021, Proceedings, Part I, pp. 253–264. Springer International Publishing, Cham (2021). https://doi.org/10.1007/978-3-030-86362-3_21

31. Paszke, A., Gross, S., Massa, F., et al.: Pytorch: an imperative style, high-performance deep learning library. In: Advances in Neural Information Processing Systems, vol. 32 (2019)

32. Indermühle, E.: Analysis of digital link in electronic documents. Verlag nicht ermittelbar (2012)

33. Alon, U., Yahav, E.: On the bottleneck of graph neural networksand its practical implications. arXiv preprint arXiv:2006.05205 (2020)

Improving Handwritten OCR with Training Samples Generated by Glyph Conditional Denoising Diffusion Probabilistic Model

Haisong Ding[1(✉)], Bozhi Luan[2], Dongnan Gui[2], Kai Chen[1(✉)], and Qiang Huo[1]

[1] Microsoft Research Asia, Beijing, China
dinghs11@mail.ustc.edu.cn, chenkai.cn@hotmail.com, qianghuo@microsoft.com
[2] University of Science and Technology of China, Hefei, China
{lbz0075,gdn2001}@mail.ustc.edu.cn

Abstract. Constructing a highly accurate handwritten OCR system requires large amounts of representative training data, which is both time-consuming and expensive to collect. To mitigate the issue, we propose a denoising diffusion probabilistic model (DDPM) to generate training samples. This model conditions on a printed glyph image and creates mappings between printed characters and handwritten images, thus enabling the generation of photo-realistic handwritten samples with diverse styles and unseen text contents. However, the text contents in synthetic images are not always consistent with the glyph conditional images, leading to unreliable labels of synthetic samples. To address this issue, we further propose a progressive data filtering strategy to add those samples with a high confidence of correctness to the training set. Experimental results on IAM benchmark task show that OCR model trained with augmented DDPM-synthesized training samples can achieve about 45% relative word error rate reduction compared with the one trained on real data only.

Keywords: handwritten OCR · handwritten image generation · denoising diffusion probabilistic model

1 Introduction

In recent years, researchers in handwritten optical character recognition (OCR) area are continuously making progress by leveraging advanced model architectures (e.g., [7, 9, 10, 28, 33, 48, 49]). However, it is still a challenging problem, due to the cursive nature of handwritten strokes and diverse writing styles. To achieve excellent recognition accuracy for a handwritten OCR system, large amounts of labeled handwritten images are required. The handwritten image dataset should

---

B. Luan and D. Gui—This work was done when Bozhi Luan and Dongnan Gui worked as interns in MMI Group, Microsoft Research Asia, Beijing, China.

G. A. Fink et al. (Eds.): ICDAR 2023, LNCS 14190, pp. 20–37, 2023.
https://doi.org/10.1007/978-3-031-41685-9_2

be representative enough to cover diverse writing styles and text contents. Obviously, collecting and labeling such a dataset are both time-consuming and expensive, and existing training data are limited in terms of style and content coverage. For example, as observed in [31], in the popular IAM dataset [34], a limited number of samples are collected for each writer, and some words are only written by a few writers. To handle this content-style data representation issue, one potential solution is to train a handwritten image generator to synthesize training samples for handwritten OCR. For any given text and a writer style, the generator should be able to synthesize photo-realistic handwritten images that can match the content of the input text and style of the writer.

In the past several years, generative adversarial network (GAN) [13,27,36] based handwritten image generation methods have achieved promising results. Most of GAN-based handwritten image generation approaches adopt a text-to-image framework [1,3,5,11,12,23,25,47,52]. Given an input text and a writer embedding, it is able to generate a photo-realistic handwritten image that matches the content of the input text and style of the writer. However, using text as input is not sufficiently flexible to embed various contents such as adjacent character interval and character arrangements [31]. By rendering text to a printed glyph image, SLOGAN [31] proposed to use an image-to-image framework for handwritten image generation. It is able to generate flexible contents by rearranging characters on the input glyph image. It is noted that these GAN-based approaches all rely on guidance from an external handwritten recognizer trained on real data, which implies that the ability of GANs is limited to directly learn the mapping from texts or printed glyph images to handwritten images without external recognition model guidance.

Recently, denoising diffusion probabilistic models (DDPMs) [18,44] achieve superior performances compared with other generation techniques on image generation tasks, including text-to-image generation [6,42,43,45] and image-to-image generation [29,46]. For handwritten image generation task, [30] investigated a text-to-image DDPM for online handwritten generation and achieved promising results. [16] proposed a writer dependent glyph conditional DDPM (GC-DDPM) for offline handwritten Chinese character generation. GC-DDPM conditions on a printed glyph image and creates mappings between printed Chinese characters and handwritten images. Training from samples of a small Chinese character set, the GC-DDPM is capable of generating photo-realistic handwritten samples of unseen Chinese character categories. In [16], the DDPM is trained on a large-scale handwritten Chinese character database, where the number of training samples for each writer is relatively sufficient. In this paper, we investigate GC-DDPM on the offline English handwritten image generation task. We conduct experiments on the popular IAM dataset [34] with limited training samples and content-style representation coverage. We find that even with limited training data, the GC-DDPM is still able to generate photo-realistic handwritten images.

Since no explicit recognition model guidance is adopted in GC-DDPM, during sampling, the model can generate noisy samples where the synthesized images do not match the text contents in glyph conditional images. Directly adding these samples to the training set for OCR can degrade the recognition performance. To address this problem, inspired by the self-training framework in automatic speech recognition [21,39] and OCR [50,51], we propose a progressive data filtering strategy to add samples with a high confidence of correctness to the training set. Experimental results on IAM benchmark task show that the performance of the OCR model can be significantly improved when trained with augmented DDPM-synthesized samples.

The remainder of this paper is organized as follows. In Sect. 2, we briefly review related works. In Sect. 3, we adopt the GC-DDPM approach in [16] for offline English handwritten image generation task and introduce the progressive data filtering strategy. Experimental results are presented in Sect. 4. Finally, we conclude this study in Sect. 5.

2   Related Works

2.1   GAN-Based Handwritten Image Generation Approaches

Handwritten image generation aims to synthesize offline handwritten images given input texts. Most of the previous approaches directly use a text-to-image framework based on GANs. For example, [1] proposed a GAN-based handwritten word image generator with additional guidance from an external handwritten recognizer trained on real data. [25] further proposed a handwritten generation model to synthesize handwritten word images that can match the given conditional writing styles. ScrabbleGAN [11] and HiGAN [12] used fully-convolutional generators, which can generate images of words and text lines with arbitrary lengths by making the image width proportional to the length of input text. In [5], the image widths are also automatically learned based on the input text and style. The text-to-image handwritten image generation framework is further improved with advanced generation model architectures such as self-attention and deformable convolution layers [3,23,47].

Since only text inputs are leveraged, the generation model needs to learn the mapping from text embedding to handwritten strokes, which is quite difficult. Besides pure text, JokerGAN [52] proposed to leverage an additional text line clue about the existence of "below the baseline" and "above the mean line" characters to improve the generation model. By rendering text to a printed glyph image using a standard typeset font, the resulting glyph image obviously contains more information than text. Using the glyph image as input, SLOGAN [31] proposed to use an image-to-image framework for handwritten image generation. It is able to generate flexible contents by changing the positions of characters and adjusting space interval between adjacent characters in glyph images. Besides using text or glyph image as input, [15] proposed to synthesize handwritten images from online handwritten samples based on StyleGAN [27].

Many of the above-mentioned approaches leveraged synthesized handwritten images as training data to boost the performances of handwritten OCR systems, and achieved substantial improvements (e.g., [15,23,31]). In this paper, we investigate DDPM to synthesize handwritten images to augment the training data for handwritten OCR.

2.2 Diffusion Model

DDPMs [18,44] have been extremely popular in image generation tasks. DDPM defines a Markov chain of T diffusion steps. In a forward diffusion process, it slowly corrupts data by adding random noises, then a reverse diffusion process is learned to recreate data from Gaussian noise. It is shown in [6] that DDPMs can outperform GANs on image synthesis. In [38], DDPMs for text-to-image synthesis are explored. The model is able to generate photo-realistic images that match the content of conditional text with the help of a classifier-free guidance [19]. Furthermore, in [42,43], DDPMs have demonstrated powerful capabilities to generate high-quality images given input texts. DDPMs are also successfully applied to other tasks such as image-to-image generation (e.g., [46]). [30] investigated DDPM for online handwritten generation and achieved promising results. [16] proposed a GC-DDPM for offline handwritten Chinese character generation. The GC-DDPM conditions on a printed glyph image and learns the mappings between printed Chinese character images and handwritten ones. It is able to generate photo-realistic handwritten images of unseen Chinese character categories. In this paper, we adopt GC-DDPM to generate offline English handwritten images.

3   Our Approach

3.1   GC-DDPM for Handwritten Image Generation

Given an input text and a writer ID (denoted as $\mathbf{w}$), we adopt a writer dependent GC-DDPM [16] to generate photo-realistic handwritten images that match the content of the text and style of the writer. For each input text, we directly render it to a printed glyph image using a standard glyph font. It is more suitable to use a glyph image as input because it contains much more information about the shapes of individual characters than pure text. We denote the glyph image as $\mathbf{g}$. As shown in Fig. 1a, let $\mathbf{x}_0$ denote a data sampled from a real distribution, i.e., $\mathbf{x}_0 \sim q(\mathbf{x})$ with a corresponding writer ID $\mathbf{w}$ and glyph image $\mathbf{g}$. In the forward diffusion process, small amounts of Gaussian noise are added to $\mathbf{x}$ in T steps, producing a sequence $\{\mathbf{x}_t\}_{t=1}^{T}$ calculated as follows:

$$q(\mathbf{x}_t|\mathbf{x}_{t-1}) = \mathcal{N}(\mathbf{x}_t; \sqrt{1-\beta_t}\mathbf{x}_{t-1}, \beta_t\mathbf{I}), \quad \mathbf{x}_t = \sqrt{\alpha_t}\mathbf{x}_{t-1} + \sqrt{1-\alpha_t}\epsilon_t, \quad (1)$$

where $\epsilon_t \sim \mathcal{N}(\mathbf{0}, \mathbf{I})$, $\beta_t \in (0,1)$ and $\alpha_t = 1 - \beta_t$. It is easy to calculate that

$$q(\mathbf{x}_t|\mathbf{x}_0) = \mathcal{N}(\mathbf{x}_t; \sqrt{\bar{\alpha}_t}\mathbf{x}_0, (1-\bar{\alpha}_t)\mathbf{I}), \quad \mathbf{x}_t = \sqrt{\bar{\alpha}_t}\mathbf{x}_0 + \sqrt{1-\bar{\alpha}_t}\epsilon, \quad (2)$$

where $\epsilon \sim \mathcal{N}(\mathbf{0}, \mathbf{I})$, $\bar{\alpha}_t = \prod_{i=1}^{t} \alpha_i$. When $T \to \infty$, $\bar{\alpha}_T \to 0$, and $\mathbf{x}_T \in \mathcal{N}(\mathbf{0}, \mathbf{I})$. A nice property of the forward diffusion process is that the reverse conditional probability is Gaussian when conditioned on $\mathbf{x}_0$:

$$q(\mathbf{x}_{t-1}|\mathbf{x}_t, \mathbf{x}_0) = \mathcal{N}(\mathbf{x}_{t-1}; \tilde{\boldsymbol{\mu}}(\mathbf{x}_t, \mathbf{x}_0), \tilde{\beta}_t\mathbf{I}), \quad (3)$$

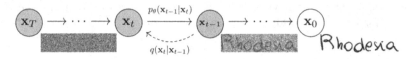

(a) The Markov chain of forward diffusion process $q(\mathbf{x}_t|\mathbf{x}_{t-1})$ and the learned reverse diffusion process $p_\theta(\mathbf{x}_{t-1}|\mathbf{x}_t)$ of DDPM.

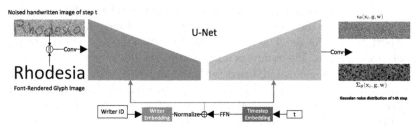

(b) The architecture of the U-Net to estimate $\boldsymbol{\epsilon}_\theta(\mathbf{x}_t, \mathbf{g}, \mathbf{w})$ and $\boldsymbol{\Sigma}_\theta(\mathbf{x}_t, \mathbf{g}, \mathbf{w})$. Figures are adapted from [16].

Fig. 1. Illustration of GC-DDPM framework for handwritten image generation.

where

$$\tilde{\boldsymbol{\mu}}(\mathbf{x}_t, \mathbf{x}_0) = \frac{1}{\sqrt{\alpha_t}}(\mathbf{x}_t - \frac{1-\alpha_t}{\sqrt{1-\bar{\alpha}_t}}\boldsymbol{\epsilon}_t), \quad \tilde{\beta}_t = \frac{1-\bar{\alpha}_{t-1}}{1-\bar{\alpha}_t} \cdot \beta_t . \tag{4}$$

Moreover, the reverse process as shown in Fig. 1a will also be a Gaussian when β_t is sufficiently small. Therefore, we can learn a model p_θ to approximate the reverse process conditioned on $\mathbf{g}$ and $\mathbf{w}$:

$$p_\theta(\mathbf{x}_{t-1}|\mathbf{x}_t, \mathbf{g}, \mathbf{w}) = \mathcal{N}(\mathbf{x}_{t-1}; \boldsymbol{\mu}_\theta(\mathbf{x}_t, \mathbf{g}, \mathbf{w}), \boldsymbol{\Sigma}_\theta(\mathbf{x}_t, \mathbf{g}, \mathbf{w})) . \tag{5}$$

Following [16,37], $\boldsymbol{\mu}_\theta(\mathbf{x}_t, \mathbf{g}, \mathbf{w})$ and $\boldsymbol{\Sigma}_\theta(\mathbf{x}_t, \mathbf{g}, \mathbf{w})$ are re-parameterized as

$$\boldsymbol{\mu}_\theta(\mathbf{x}_t, \mathbf{g}, \mathbf{w}) = \frac{1}{\sqrt{\alpha_t}}\left(\mathbf{x}_t - \frac{1-\alpha_t}{\sqrt{1-\bar{\alpha}_t}}\boldsymbol{\epsilon}_\theta(\mathbf{x}_t, \mathbf{g}, \mathbf{w})\right) \tag{6}$$

$$\boldsymbol{\Sigma}_\theta(\mathbf{x}_t, \mathbf{g}, \mathbf{w}) = \exp\left(\boldsymbol{\nu}_\theta(\mathbf{x}_t, \mathbf{g}, \mathbf{w})\log\beta_t + (1-\boldsymbol{\nu}_\theta(\mathbf{x}_t, \mathbf{g}, \mathbf{w}))\log\tilde{\beta}_t\right) .$$

A neural network is trained to estimate $\boldsymbol{\epsilon}_\theta(\mathbf{x}_t, \mathbf{g}, \mathbf{w})$ and $\boldsymbol{\Sigma}_\theta(\mathbf{x}_t, \mathbf{g}, \mathbf{w})$. We use the same hybrid objective function as in [37]. After the reverse process is learned, conditioned on $\mathbf{g}$ and $\mathbf{w}$, we are able to draw samples $\mathbf{x}_0$ according to Eqn. (5), starting with a Gaussian noise $\mathbf{x}_T \sim \mathcal{N}(\mathbf{0}, \mathbf{I})$.

We adopt the same U-Net architecture in GC-DDPM [6,16] in our task. As shown in Fig. 1b, $\mathbf{x}_t$ and $\mathbf{g}$ are normalized to a fixed size. Then they are concatenated together and used as input to the U-Net. Time step t is embedded with sinusoidal embedding, and then processed with a 2-layer feed-forward network (FFN). Writer information $\mathbf{w}$ is embedded with a learnable embedding, followed by L2-normalization: $\mathbf{z} = \mathbf{w}/\|\mathbf{w}\|_2$. Finally, they are added together and fed to layers in U-Net using a feature-wise linear modulation (FiLM) operator [40].

In DDPM, classifier-free guidance [19] is an effective approach to improve generation quality. Following [16], a content guidance scale γ and a style content scale η are used. During sampling, $\epsilon_\theta(\mathbf{x}_t, \mathbf{g}, \mathbf{w})$ is directly replaced with

$$\tilde{\epsilon}_\theta(\mathbf{x}_t, \mathbf{g}, \mathbf{w}) = \epsilon_\theta(\mathbf{x}_t, \mathbf{g}, \mathbf{w}) + \gamma\epsilon_\theta(\mathbf{x}_t, \mathbf{g}, \emptyset) \tag{7}$$
$$+ \eta\epsilon_\theta(\mathbf{x}_t, \emptyset, \mathbf{w}) - (\gamma + \eta)\epsilon_\theta(\mathbf{x}_t\emptyset, \emptyset) .$$

Here $\epsilon_\theta(\mathbf{x}_t, \mathbf{g}, \emptyset)$, $\epsilon_\theta(\mathbf{x}_t, \emptyset, \mathbf{w})$ and $\epsilon_\theta(\mathbf{x}_t, \emptyset, \emptyset)$ are trained together with $\epsilon_\theta(\mathbf{x}_t, \mathbf{g}, \mathbf{w})$ using the same U-Net, where $\mathbf{w}$, $\mathbf{g}$ or both are replaced with a special token $\emptyset$.

Works in [16] synthesize offline Chinese character images which are of fixed height and width. While in English handwritten image generation task, the widths of the generated images should be inferred from the text and writer style. To achieve this goal, we prepare handwritten images of words and short phrases to train GC-DDPM, where the maximum aspect ratio of images is set as 8. First, images are resized to a height of 64 while keeping aspect ratio. Then, images are padded to a width of 512 with black pixels on both left and right margins. The glyph images are also processed to the size of 64×512 using the same procedure. In experiments, we find that during sampling, GC-DDPM will learn the width of black margins based on input text and writer style. It will generate handwritten images with clear black margins robustly. Therefore, to get the final handwritten sample, we use a simple image processing method to remove the padded black margins.

3.2 Progressive Data Filtering Strategy

By conditioning on glyph images and writer IDs, we expect GC-DDPM to learn the mapping from glyph images to handwritten ones and generate high-quality training data to improve handwritten OCR systems. However, we notice that the generated images are not always consistent with the glyph conditional images. Adding these data to the training set would degrade the performance and robustness of the handwritten OCR system. To alleviate this problem, we propose a progressive data filtering strategy to remove these noisy samples.

In text-to-image generation tasks, a dot product score between text and image embeddings is used as a metric to select generated samples to improve generation quality [42]. In the self-training framework, for each unlabeled sample with a pseudo label, a confidence score is calculated. Only data with high confidence scores are added to the training set [21, 39, 50, 51]. Inspired by these works, we design a metric to estimate the "confidence of correctness" for each generated handwritten image. Then samples with a high confidence of correctness are added to the training set progressively.

Let $\mathbf{R} = \{(\boldsymbol{x}_i, \boldsymbol{y}_i)\}$ denote a real dataset with handwritten image $\boldsymbol{x}_i$ and ground truth label $\boldsymbol{y}_i$. Let $\mathbf{S} = \{(\tilde{\boldsymbol{x}}_j, \tilde{\boldsymbol{y}}_j)\}$ denote DDPM-generated synthetic dataset with handwritten image $\tilde{\boldsymbol{x}}_j$ and corresponding conditional text $\tilde{\boldsymbol{y}}_j$. First, we train an initial OCR model $\boldsymbol{M}$ using $\mathbf{R}$ only. Then the trained model is used

---

Algorithm 1: Progressive data filtering strategy

Input: Real data $\mathbf{R} = \{(\boldsymbol{x}_i, \boldsymbol{y}_i)\}$, synthetic data $\mathbf{S} = \{(\tilde{\boldsymbol{x}}_j, \tilde{\boldsymbol{y}}_j)\}$, initial selected synthetic data $\mathbf{S}' = \{\}$, number of progressive data filtering rounds N, data filtering threshold τ

1 Train model M using $\mathbf{R}$;
2 **for** $n \leftarrow 1$ *to* N **do**
3      $\mathbf{S}' = \{(\tilde{\boldsymbol{x}}_j, \tilde{\boldsymbol{y}}_j) \in \mathbf{S} \mid c(\tilde{\boldsymbol{x}}_j, \tilde{\boldsymbol{y}}_j; M) \geq \tau\}$;
4   Train model M using $\mathbf{R} \cup \mathbf{S}'$ starting from random weight initialization;
5 **return** $M, \mathbf{S}'$;

---

to calculate the confidence of correctness score for each $\tilde{\boldsymbol{x}}_j$ in $\mathbf{S}$ as follows:

$$c(\tilde{\boldsymbol{x}}_j, \tilde{\boldsymbol{y}}_j; M) = \frac{\mathcal{L}(\tilde{\boldsymbol{x}}_j, \hat{\boldsymbol{y}}_j; M)}{\mathcal{L}(\tilde{\boldsymbol{x}}_j, \tilde{\boldsymbol{y}}_j; M)} , \tag{8}$$

where $\mathcal{L}(\boldsymbol{x}, \boldsymbol{y}; M) = -\log p(\boldsymbol{y}|\boldsymbol{x}; M)$ is the negative log posterior probability calculated using recognizer M, and $\hat{\boldsymbol{y}}_j$ is the decoding result of $\tilde{\boldsymbol{x}}_j$ using M. Obviously, if $\tilde{\boldsymbol{y}}_j = \hat{\boldsymbol{y}}_j$, the score equals to 1, meaning that the recognizer's prediction is consistent with the conditional text. Then the confidence of $\tilde{\boldsymbol{x}}$ matching $\tilde{\boldsymbol{y}}$ is high. If $\mathcal{L}(\tilde{\boldsymbol{x}}_j, \hat{\boldsymbol{y}}_j; M) \ll \mathcal{L}(\tilde{\boldsymbol{x}}_j, \tilde{\boldsymbol{y}}_j; M)$, the score is close to 0, and the confidence of $\tilde{\boldsymbol{x}}$ matching $\tilde{\boldsymbol{y}}$ is low. In practice, a threshold $\tau \in (0, 1]$ is used, and $(\tilde{\boldsymbol{x}}_j, \tilde{\boldsymbol{y}}_j)$ with $c(\tilde{\boldsymbol{x}}_j, \tilde{\boldsymbol{y}}_j; M) \geq \tau$ is included in a selected set $\mathbf{S}'$. Then, a new OCR model can be trained with $\mathbf{R} \cup \mathbf{S}'$. After that, the scores can be re-calculated using the new model. This process can be repeated for multiple rounds until the performance of the OCR model does not improve further. The whole progressive data filtering strategy is summarized in Algorithm 1.

4 Experiments

4.1 Experimental Setup

We conduct experiments on the IAM dataset [34]. It contains 13,353 isolated text line images and 115,320 word images written by 657 different writers. We use the RWTH Aachen partition as in [22] in experiments. Following [6], diffusion step number T is set as 1,000 with a linear noise schedule. During training, $\mathbf{w}$ and $\mathbf{g}$ are randomly set to $\emptyset$ with probability 10%, independently. When $\mathbf{g} = \emptyset$, a blank glyph image will be used; when $\mathbf{w} = \emptyset$, a special embedding will be used. During sampling, we use DDIM [20] sampling method with 50 steps to save sampling time. As for the handwritten OCR system, the same CTC-based [14] model in [31] is used without leveraging external language models, and $\hat{\boldsymbol{y}}_j$ in Eqn. (8) is decoded with the best path decoding algorithm. The total number of parameters of the OCR model is 14M. We mainly conduct experiments on the IAM word benchmark, except in Sect. 4.6 where we conduct experiments on IAM text line benchmark. A word error rate (WER) of 19.47% and a character

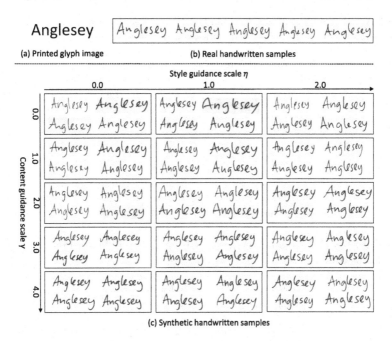

Fig. 2. Real samples of words "Anglesey" written by writer 333 and synthetic hand-written images generated with different guidance scales. Glyph conditional image is shown in (a).

error rate (CER) of 7.27% is achieved when trained on real IAM word data only, which is similar to the baseline result (19.12% WER and 7.39% CER) presented in [31].

4.2 Effect of Classifier-Free Guidance Scales in GC-DDPM

Works in [16,19] show that the classifier guidance scale is able to control the trade-off between the quality and diversity of generated samples. Figure 2 (b) shows real samples of "Anglesey" written by writer 333. Clearly, the style and position of individual characters vary each time the same writer writes them. Figure 2 (c) visualizes synthetic samples generated with different guidance scales. With higher content guidance scales, the generated samples become less diverse, which is consistent with the observation in [16]. For example, in real samples, the "le" in "Anglesey" is either separately written, or consecutively written with a single stroke. Synthetic samples with lower content guidance scales successfully capture both writing variants. Whereas samples with higher content guidance scales only capture the consecutively written one. We also observe that the variance in generated image widths becomes smaller when sampled with higher content guidance scales. As for the style guidance scales, since the writer ID is already a distinctive guidance, the sampling qualities with different scales are similar.

To evaluate the behavior of guidance scales in generating training data for handwritten OCR, we try $\gamma, \eta \in \{0.0, 0.5, 1.0\}$ and synthesize the whole IAM

Table 1. WER and CER of handwritten OCR models on IAM word testing set trained with synthetic dataset generated with different guidance scales.

Style scale η	Content scale γ	WER (%)	CER (%)
0.0	0.0	**20.17**	**7.50**
	0.5	21.06	7.94
	1.0	21.93	8.23
0.5	0.0	20.35	7.52
	0.5	21.14	7.91
	1.0	21.67	8.14
1.0	0.0	20.40	7.59
	0.5	20.41	7.56
	1.0	21.25	7.91
IAM training set		19.47	7.27

Table 2. WER and CER of handwritten OCR models on IAM word testing set trained with different training sets.

Training set	WER(%)	CER(%)
IAM training set	19.47	7.27
+ Synth-IAM-Words	11.57	3.88
+ Synth-EN-Words	14.78	5.14
+ Synth-EN-Words-WI	14.83	5.18

word training set using the exact word corpus and writer IDs. The number of synthetic images equals to the number of images in the IAM training set. Then, we use synthetic data only to train an OCR model and evaluate its recognition performance on the real IAM word testing set. As shown in Table 1, with the same η, WER increases as γ becomes higher. This shows that the diversity of generated images is important when synthesizing training data for OCR. The best recognition performance is achieved with both guidance scales set as 0. The best WER is only absolute 0.7% worse than that trained on real dataset, which demonstrates the high quality of DDPM-generated handwritten images. Based on these observations, we set $\gamma = 0.0$, $\eta = 0.0$ in the following experiments.

In Fig. 3, we show generated samples conditioned by different writer IDs with input words that are seen in IAM training set and out-of-vocabulary words. It is clear that GC-DDPM is able to synthesize these words while mimicking the writing styles (e.g., cursive, slant, stroke pattern) of the conditional writers. It is noted that although "Z" only appears four times in the training set, the GC-DDPM still can generate high quality handwritten images.

4.3 Augment Training Set with Synthetic Images for OCR

Next, we use synthetic handwritten images to boost handwritten OCR performance. To evaluate the quality of generated samples of seen/unseen words, three sets of synthetic handwritten images are generated in our experiments:

nominating	*opposed*	*delegations*	*shopgirls*
nominating	opposed	delegations	shopgirls
Labour	*institution*	*Relations*	*will*
Labour	institution	Relations	will
temperatures	*temperatures*	*temperatures*	*temperatures*	*temperatures*
violently	*violently*	*violently*	*violently*	*violently*
comprehensive	*comprehensive*	*comprehensive*	*comprehensive*	*comprehensive*
famille	*famille*	*famille*	*famille*	*famille*
certes	*certes*	*certes*	*certes*	*certes*
Zimbabwe	*Zimbabwe*	*Zimbabwe*	*Zimbabwe*	*Zimbabwe*

Fig. 3. Synthetic handwritten images conditioned by different writer IDs with different words. Top: real samples with corresponding words from IAM writers 001, 002, 023 and 027, respectively. Middle: synthetic samples of words that are seen in IAM training set. Bottom: synthetic samples of out-of-vocabulary words.

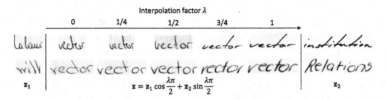

Fig. 4. Synthetic handwritten images of word "vector" generated using writer style interpolations between $\mathbf{z}_1$ and $\mathbf{z}_2$.

- Synth-IAM-Words: Since the IAM training set is insufficient in terms of content and style coverage, we use GC-DDPM to generate handwritten images for each writer. For each of the 442 writers observed in the training set, we synthesize a handwritten image for each and every word in the entire IAM word training corpus. As a result, the Synth-IAM-Words dataset is 442 times the size of the original training set in terms of the number of samples.
- Synth-EN-Words: Samples in Synth-IAM-Words only contain words that have been observed in training set. To investigate the quality of synthesized handwritten samples of unseen words, following [11,31], an external "English words"[1] corpus is used. It contains 466,550 unique words, 98.9% of which are not observed in the IAM training set. To generate a diverse dataset, for each word, we synthesize 8 samples conditioning on 8 randomly selected writer IDs. As a result, Synth-EN-Words dataset contains about 3.7M samples.
- Synth-EN-Words-WI: The above two datasets are generated with trained writer IDs. The GC-DDPM is also able to generate unseen styles using writer style interpolations [16]. To achieve this, given two normalized writer

---

embedding z_1 and z_2, a new embedding z can be obtained with spherical interpolation [42]: $z = z_1 \cos \frac{\lambda\pi}{2} + z_2 \sin \frac{\lambda\pi}{2}$ with interpolation factor $\lambda \in [0, 1]$. Figure 4 shows handwritten samples generated using writer style interpolations. It is clear that as λ increases from 0 to 1, the style of generated samples gradually shifts from z_1 to z_2. To evaluate the quality of synthesized handwritten samples with interpolated styles, for each word in "English words" corpus, we also synthesize 8 samples conditioning on 8 randomly calculated writer interpolations. We use $\lambda = 1/2$. We name this dataset "Synth-EN-Words-WI". It also contains 3.7M samples.

Table 2 lists the performances of handwritten OCR models on IAM word testing set trained with different training sets. It is clear that augmenting the training set with synthetic handwritten images can significantly boost the recognition performances. Specifically, a 40.6% WER reduction (WERR) and a 46.6% CER reduction (CERR) are achieved with Synth-IAM-Words dataset. It shows that the generated Synth-IAM-Words can successfully alleviate the insufficient content and style coverage problem in training set and achieves significant OCR performance improvements. Augmenting the training set with Synth-EN-Words and Synth-EN-Words-WI achieves similar recognition accuracy improvements (about 24% WERR and 29% CERR), which suggests that the synthesized qualities of Synth-EN-Words and Synth-EN-Words-WI are similar.

In our experiments, we construct synthetic datasets using two corpora, the IAM corpus and an external "English words" corpus. Words in IAM corpus are seen in the training of DDPM, whereas most of words in "English words" corpus are unseen. We find that the quality of synthesized data of words in "English words" is worse than that of words in IAM corpus. This shows that the synthesized data quality of unseen words is worse than that of seen words.

To further show the quality difference between Synth-IAM-Words and Synth-EN-Words/Synth-EN-Words-WI, we treat their conditional texts as ground truths and evaluate the WER using the OCR model trained on IAM training set. A 22% WER is observed on Synth-IAM-Words, while the WERs on Synth-EN-Words and Synth-EN-Words-WI are 73% and 72%, which are significantly higher. There are two potential reasons for this observation: (a) the generalization ability of the OCR model trained with IAM training set is limited when recognizing unseen words, and (b) the quality of DDPM-synthesized handwritten images of unseen words is worse than seen words. As a comparison, we evaluate the OCR model on a subset of IAM word testing set containing unseen words and achieve a 40.8% WER. Therefore, both reasons contribute to the high WER. Based on these analyses, we conclude that the text contents in synthetic images are not always consistent with the glyph conditional images, leading to unreliable labels of synthetic data. Adding noisy data to the training set could degrade the performance of the OCR model. Next, we will leverage the progressive data filtering strategy to remove synthetic data with unreliable labels.

Table 3. WER and CER of handwritten OCR models on IAM word testing set trained on augmented synthetic datasets with progressive data filtering strategy.

Synthetic dataset	N	$\tau = 1.0$		$\tau = 0.7$		Use all samples	
		WER(%)	CER(%)	WER(%)	CER(%)	WER(%)	CER(%)
Synth-IAM-Words	1	12.70	4.60	12.40	4.45	11.57	**3.88**
	2	11.84	4.23	11.73	4.16		
	3	11.77	4.22	**11.54**	4.14		
Synth-EN-Words	1	15.33	5.38	14.53	5.09	14.78	5.14
	2	14.38	5.01	14.13	4.86		
	3	14.12	4.88	**13.85**	**4.83**		
Synth-EN-Words-WI	1	15.18	5.30	14.39	4.96	14.83	5.18
	2	14.43	4.99	14.20	4.96		
	3	14.15	4.92	**14.06**	**4.85**		

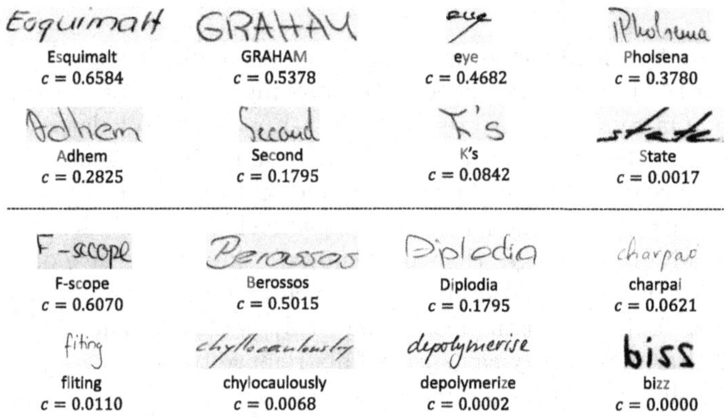

Fig. 5. Samples of synthetic handwritten images in (top) Synth-IAM-Words and (bottom) Synth-EN-Words sets that are removed using progressive data filtering. Wrongly generated characters are highlighted in red. Confidence of correctness scores are listed below. (Color figure online)

4.4 Effect of Progressive Data Filtering Strategy

To evaluate the effect of the progressive data filtering strategy, we set the number of data filtering rounds $N = 3$, and try $\tau = \{1.0, 0.7\}$. We conduct experiments on three synthetic dataset and results are listed in Table 3. We also list the baseline results when all synthetic samples are added to the training set. The performances of OCR models improve with more progressive data filtering rounds. We also try to use an additional 4th round, but no further performance improvements are observed. Compared with $\tau = 1.0$, better performances are achieved using $\tau = 0.7$. This implies that $\tau = 0.7$ achieves a better tradeoff between numbers of high-quality and noisy samples. Compared with using all generated samples, progressive data filtering achieves slightly better WER and

Fig. 6. Visual comparisons with Alonso *et al.* [1], ScrabbleGAN [11] and SLOGAN [31]. The words generated from top to bottoms are: olibrius, inventif, bonjour, ionique, malade, golf, ski. The writer IDs of our generated samples are 135, 111, 011, 023, 001, 002, 027, respectively.

worse CER on Synth-IAM-Words. After 3 rounds, about 90% and 91% of samples in Synth-IAM-Words are added to the training set with $\tau = 1.0$ and $\tau = 0.7$, respectively. On Synth-EN-Words and Synth-EN-Words-WI datasets, progressive data filtering achieves much better results than using all generated samples. With progressive data filtering, about 55% and 59% of the data are added to the training set with $\tau = 1.0$ and $\tau = 0.7$, respectively. Figure 5 shows samples of synthetic handwritten images that are removed using progressive data filtering. The generated images can contain errors, such as missing or repeating certain characters, or failing to distinguish some easily confused characters. The proposed data filtering strategy can successfully remove these error samples. These results show that DDPM-generated samples of unseen words are much noisier than samples of seen words, and the progressive data filtering strategy is helpful to remove noisy samples and achieve better OCR performance.

4.5 Comparison with Previous Methods

Figure 6 shows visual comparisons with previous methods. Our GC-DDPM approach can generate photo-realistic handwritten images with fewer artifacts. In ScrabbleGAN [11] and SLOGAN [31], the width of the generated image is determined by the length of input text or the width of glyph conditional image. Our approach is able to generate images with variable widths according to the text content and writer style. GC-DDPM can successfully mimic the unique style of the conditional writer styles. For example, writer 111 usually writes "t" similarly with "T", and the generated stroke of "t" in "inventif" is also similar to "T".

Besides visual comparison, we also compare the synthetic data quality using FID metric. Following GANwriting [25] and SLOGAN [31], we generate 400 unique out-of-vocabulary (OOV) words and calculate an averaged FID [17] score

Table 4. Comparison with previous methods on IAM word testing set. No lexicons and language models are applied.

Method	Synthetic data	WER(%)	CER(%)
Kang *et al.* [22]	No	16.39	6.43
Learn to Augment [32] + AFDM [4]	No	13.35	5.13
SLOGAN (Baseline) [31]	No	19.12	7.39
Dutta *et al.* [8]	Font-based	12.61	4.88
Kang *et al.* [26]	Font-based	17.26	6.75
SLOGAN [31]	GAN-based	14.97	5.95
SLOGAN [31] + Learn to Augment [32]	GAN-based	12.90	4.94
Ours (Real)	No	19.47	7.27
Ours (Real + filtered synthetic data)	DDPM-based	**10.72**	**3.75**
Ours (Filtered synthetic data only)	DDPM-based	11.55	4.07

Table 5. Comparison of FID on out-of-vocabulary word images.

Method	GANwriting [25]	SLOGAN [31]	GC-DDPM (ours)
FID	125.87	97.81	**86.93**

of each handwriting style[2]. As shown in Table 5, we achieve an FID score of 86.93 which is better than that of both GANwriting and SLOGAN.

Since the goal of our approach is to generate handwritten images to augment the training set for handwritten OCR, we compare our approach with other synthetic data augmented OCR systems on the IAM word benchmark. To push the OCR performance to limit, we add all samples in Synth-IAM-Words, filtered Synth-EN-Words and Synth-EN-Words using $\tau = 0.7$ to augment the IAM training data. As shown in Table 4, we achieve a 10.72% WER and a 3.75% CER, which are much better than previous methods using font rendered or GAN-based synthetic images. Compared with using real IAM training data alone, we achieve a relative 45% WERR and a relative 48% CERR. We also conduct an experiment of using only filtered synthetic data, and achieve a 11.55% WER and a 4.07% CER, which are significantly better than using IAM real training data only.

4.6   Experiments on IAM Text Line Dataset

Finally, we conduct experiments on the IAM text line benchmark. Although the GC-DDPM is trained on images with maximum aspect ratio of 8, it can generate text lines as shown in Fig. 7. We use the same CTC-based OCR model as in IAM word experiments, and achieve a 7.05% CER. For synthetic data,

---

[2] According to the authors of GANwriting, the exact list of 400 OOV words is no longer available. Therefore, we follow their advice to build our own OOV word list.

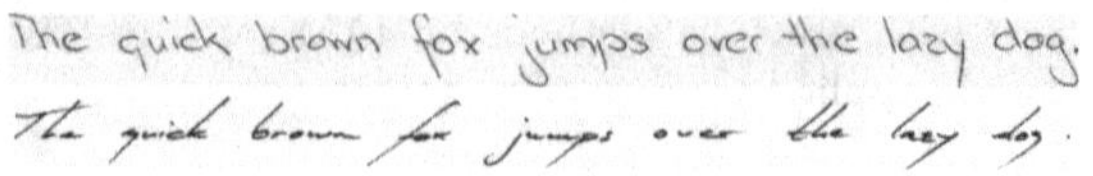

Fig. 7. Samples of synthetic handwritten text line images of a sentence: "The quick brown fox jumps over the lazy dog." .

Table 6. Comparison with previous methods on IAM text line testing set. No lexicons and language models are applied.

Method	Synthetic data	WER(%)	CER(%)
Puigcerver [41]	No	18.40	5.80
Michael et al. [35]	No	–	5.24
Wick et al. [48]	No	–	5.67
Dutta et al. [8]	Font-based	17.82	5.70
Barrere et al. [2]	Font-based	16.31	4.76
Kang et al. [24]	Font-based	15.45	4.67
Wick et al. (CTC) [49]	Font-based	16.85	4.99
Wick et al. [49]	Font-based	12.20	3.96
TrOCR$_{\text{SMALL}}$ [28]	Font-based	–	4.22
Ours (Real)	No	22.11	7.05
Ours (Real + filtered synthetic data)	DDPM-based	13.08	4.13

besides using the filtered synthetic word dataset, we also synthesize handwritten text line samples using the IAM training line corpus, and filter these samples using progressive data filtering. As shown in Table 6, we finally achieve a 4.13% CER, which is slightly better than TrOCR$_{\text{SMALL}}$. It should be noted that works in [28,49] use advanced sequence-to-sequence framework for OCR. [49] achieves a 4.99% CER with CTC-based model. TrOCR$_{\text{SMALL}}$ also leverages pre-trained encoder and decoder with 62M total parameters. We only use a simple CTC-based OCR model with 14M parameters, without using any image pre-processing technique and pre-trained models.

5   Conclusion

In this paper, we investigate GC-DDPM to generate handwritten images to augment training data for handwritten OCR. The proposed GC-DDPM is able to generate photo-realistic handwritten samples with diverse styles and text contents. However, we find that the text contents in synthetic images are not always consistent with the glyph conditional images, especially in images with out-of-vocabulary words. Therefore, we further propose a progressive data filtering method to remove samples with noisy labels. Experiments on both IAM word and text line benchmarks show that the performance of the OCR model trained with augmented DDPM-synthesized samples can perform much better than the one trained on real data only.

References

1. Alonso, E., Moysset, B., Messina, R.O.: Adversarial generation of handwritten text images conditioned on sequences. In: Proceedings of ICDAR, pp. 481–486 (2019)
2. Barrere, K., Soullard, Y., Lemaitre, A., Coüasnon, B.: A light Transformer-based architecture for handwritten text recognition. In: Proceedings of DAS, pp. 275–290 (2022)
3. Bhunia, A.K., Khan, S.H., Cholakkal, H., Anwer, R.M., Khan, F.S., Shah, M.: Handwriting transformers. In: Proceedings of ICCV, pp. 1066–1074 (2021)
4. Bhunia, A.K., Das, A., Bhunia, A.K., Kishore, P.S.R., Roy, P.P.: Handwriting recognition in low-resource scripts using adversarial learning. In: Proceedings of CVPR, pp. 4767–4776 (2020)
5. Davis, B.L., Morse, B.S., Price, B.L., Tensmeyer, C., Wigington, C., Jain, R.: Text and style conditioned GAN for the generation of offline-handwriting lines. In: Proceedings of BMVC (2020)
6. Dhariwal, P., Nichol, A.Q.: Diffusion models beat GANs on image synthesis. In: Proceedings of NeurIPS, vol. 34, pp. 8780–8794 (2021)
7. Diaz, D.H., Qin, S., Ingle, R.R., Fujii, Y., Bissacco, A.: Rethinking text line recognition models. CoRR abs/2104.07787 (2021)
8. Dutta, K., Krishnan, P., Mathew, M., Jawahar, C.V.: Improving CNN-RNN hybrid networks for handwriting recognition. In: Proceedings of ICFHR, pp. 80–85 (2018)
9. d'Arce, R., Norton, T., Hannuna, S., Cristianini, N.: Self-attention networks for non-recurrent handwritten text recognition. In: Proceedings of ICFHR, pp. 389–403 (2022)
10. Etter, D., Rawls, S., Carpenter, C., Sell, G.: A synthetic recipe for OCR. In: Proceedings of ICDAR, pp. 864–869 (2019)
11. Fogel, S., Averbuch-Elor, H., Cohen, S., Shai Mazor, R.L.: ScrabbleGAN: semi-supervised varying length handwritten text generation. In: Proceedings of CVPR, pp. 4323–4332 (2020)
12. Gan, J., Wang, W.: HiGAN: handwriting imitation conditioned on arbitrary-length texts and disentangled styles. In: Proceedings of AAAI, pp. 7484–7492 (2021)
13. Goodfellow, I.J., et al.: Generative adversarial nets. In: Proceedings of NIPS, pp. 2672–2680 (2014)
14. Graves, A., Fernández, S., Gomez, F.J., Schmidhuber, J.: Connectionist temporal classification: labelling unsegmented sequence data with recurrent neural networks. In: Proceedings of ICML, pp. 369–376 (2006)
15. Guan, M., Ding, H., Chen, K., Huo, Q.: Improving handwritten OCR with augmented text line images synthesized from online handwriting samples by style-conditioned GAN. In: Proceedings of ICFHR, pp. 151–156 (2020)
16. Gui, D., Chen, K., Ding, H., Huo, Q.: Zero-shot generation of training data with denoising diffusion probabilistic model for handwritten Chinese character recognition. In: Proceedings of ICDAR (2023)
17. Heusel, M., Ramsauer, H., Unterthiner, T., Nessler, B., Hochreiter, S.: GANs trained by a two time-scale update rule converge to a local nash equilibrium. In: Proceedings of NIPS, vol. 30, pp. 6626–6637 (2017)

18. Ho, J., Jain, A., Abbeel, P.: Denoising diffusion probabilistic models. In: Proceedings of NeurIPS, vol. 33, pp. 6840–6851 (2020)
19. Ho, J., Salimans, T.: Classifier-free diffusion guidance. In: Proceedings of NeurIPS, Workshop on Deep Generative Models and Downstream Applications (2021)
20. Song, J., Chenlin Meng, S.E.: Denoising diffusion implicit models. In: Proceedings of ICLR (2021)
21. Kahn, J., Lee, A., Hannun, A.Y.: Self-training for end-to-end speech recognition. In: Proceedings of ICASSP, pp. 7084–7088 (2020)
22. Kang, L., Riba, P., Rusiñol, M., Fornés, A., Villegas, M.: Distilling content from style for handwritten word recognition. In: Proceedings of ICFHR, pp. 139–144 (2020)
23. Kang, L., Riba, P., Rusiñol, M., Fornés, A., Villegas, M.: Content and style aware generation of text-line images for handwriting recognition. IEEE Trans. Pattern Anal. Mach. Intell. **44**, 8846–8860 (2022)
24. Kang, L., Riba, P., Rusiñol, M., Fornés, A., Villegas, M.: Pay attention to what you read: non-recurrent handwritten text-line recognition. Pattern Recogn. **129**, 108799 (2022)
25. Kang, L., Riba, P., Wang, Y., Rusiñol, M., Fornés, A., Villegas, M.: GANwriting: content-conditioned generation of styled handwritten word images. In: Proceedings of ECCV, vol. 23, pp. 273–289 (2020)
26. Kang, L., Rusiñol, M., Fornés, A., Riba, P., Villegas, M.: Unsupervised adaptation for synthetic-to-real handwritten word recognition. In: Proceedings of WACV, pp. 3491–3500 (2020)
27. Karras, T., Laine, S., Aila, T.: A style-based generator architecture for generative adversarial networks. In: Proceedings of CVPR, pp. 4401–4410 (2019)
28. Li, M., e al.: TrOCR: transformer-based optical character recognition with pre-trained models. CoRR abs/2109.10282 (2022)
29. Lugmayr, A., Danelljan, M., Romero, A., Yu, F., Timofte, R., Gool, L.V.: Repaint: inpainting using denoising diffusion probabilistic models. In: Proceedings of CVPR, pp. 11451–11461 (2022)
30. Luhman, T., Luhman, E.: Diffusion models for handwriting generation. CoRR abs/2011.06704 (2020)
31. Luo, C., Zhu, Y., Jin, L., Li, Z., Peng, D.: SLOGAN: handwriting style synthesis for arbitrary-length and out-of-vocabulary text. IEEE Trans. Neural Netw. Learn. Syst., 1–13 (2022)
32. Luo, C., Zhu, Y., Jin, L., Wang, Y.: Learn to augment: joint data augmentation and network optimization for text recognition. In: Proceedings of CVPR, pp. 13743–13752 (2020)
33. Ly, N.T., Nguyen, H.T., Nakagawa, M.: 2D self-attention convolutional recurrent network for offline handwritten text recognition. In: Proceedings of ICDAR, pp. 191–204 (2021)
34. Marti, U., Bunke, H.: The IAM-database: an English sentence database for offline handwriting recognition. Int. J. Doc. Anal. Recogn., pp. 39–46 (2002)
35. Michael, J., Labahn, R., Grüning, T., Zöllner, J.: Evaluating sequence-to-sequence models for handwritten text recognition. In: Proceedings of ICDAR, pp. 1286–1293 (2019)
36. Mirza, M., Osindero, S.: Conditional generative adversarial nets. Comput. Sci., pp. 2672–2680 (2014)
37. Nichol, A.Q., Dhariwal, P.: Improved denoising diffusion probabilistic models. In: Proceedings of ICML, pp. 8162–8171 (2021)

38. Nichol, A.Q., et al.: Glide: towards photorealistic image generation and editing with text-guided diffusion models. In: Proceedings of ICML, pp. 16784–16804 (2022)
39. Park, D.S., et al.: Improved noisy student training for automatic speech recognition. In: Proceedings of Interspeech, pp. 2817–2821 (2020)
40. Perez, E., Strub, F., de Vries, H., Dumoulin, V., Courville, A.C.: FiLM: visual reasoning with a general conditioning layer. In: Proceedings of AAAI, pp. 3942–3951 (2018)
41. Puigcerver, J.: Are multidimensional recurrent layers really necessary for handwritten text recognition? In: Proceedings of ICDAR, pp. 67–72 (2017)
42. Ramesh, A., Dhariwal, P., Nichol, A., Chu, C., Chen, M.: Hierarchical text-conditional image generation with CLIP latents. CoRR abs/2204.06125 (2022)
43. Saharia, C., et al.: Photorealistic text-to-image diffusion models with deep language understanding. CoRR abs/2205.11487 (2022)
44. Sohl-Dickstein, J., Weiss, E.A., Maheswaranathan, N., Ganguli, S.: Deep unsupervised learning using nonequilibrium thermodynamics. In: Proceedings of ICML, pp. 2256–2265 (2015)
45. Song, Y., Sohl-Dickstein, J., Kingma, D.P., Kumar, A., Ermon, S., Poole, B.: Score-based generative modeling through stochastic differential equations. In: Proceedings of ICLR (2021)
46. Wang, T., et al.: Pretraining is all you need for image-to-image translation. CoRR abs/2205.12952 (2022)
47. Wang, Y., Wang, H., Sun, S., Wei, H.: An approach based on Transformer and deformable convolution for realistic handwriting samples generation. In: Proceedings of ICPR, pp. 1457–1463 (2022)
48. Wick, C., Zöllner, J., Grüning, T.: Transformer for handwritten text recognition using bidirectional post-decoding. In: Proceedings of ICDAR, pp. 112–126 (2021)
49. Wick, C., Zöllner, J., Grüning, T.: Rescoring sequence-to-sequence models for text line recognition with CTC-prefixes. In: Proceedings of DAS, pp. 260–274 (2022)
50. Wolf, F., Fink, G.A.: Combining self-training and minimal annotations for handwritten word recognition. In: Proceedings of ICFHR, pp. 300–315 (2022)
51. Wolf, F., Fink, G.A.: Self-training of handwritten word recognition for synthetic-to-real adaptation. In: Proceedings of ICPR, pp. 3885–3892 (2022)
52. Zdenek, J., Nakayama, H.: JokerGAN: memory-efficient model for handwritten text generation with text line awareness. In: Proceedings of ACM Multimedia, pp. 5655–5663 (2021)

Improved Learning for Online Handwritten Chinese Text Recognition with Convolutional Prototype Network

Yi Chen[1,2], Heng Zhang[1], and Cheng-Lin Liu[1,2(✉)]

[1] State Key Laboratory of Multimodal Artificial Intelligence Systems (MAIS), Institution of Automation, Chinese Academy of Sciences, Beijing 100190, China
{yi.chen,liucl}@nlpr.ia.ac.cn, heng.zhang@ia.ac.cn
[2] School of Artificial Intelligence, University of Chinese Academy of Sciences, Beijing 100049, China

Abstract. Segmentation-based handwritten text recognition has the advantage of character interpretability but needs a character classifier with high classification accuracy and non-character rejection capability. The classifier can be trained on both character samples and string samples but real string samples are usually insufficient. In this paper, we proposed a learning method for segmentation-based online handwritten Chinese text recognition with a convolutional prototype network as the underlying classifier. The prototype classifier is inherently resistant to non-characters, and so, can be trained with character and string samples without the need of data augmentation. The learning has two stages: pre-training on character samples with a modified loss function for improving non-character resistance, and weakly supervised learning on both character and string samples for improving recognition performance. Experimental results on the CASIA-OLHWDB and ICDAR2013-Online datasets show that the proposed method can achieve promising recognition performance without training data augmentation.

Keywords: online handwritten text recognition · text segmentation · convolutional prototype network · weakly supervised learning

1 Introduction

As people are using more and more mobile digitizing devices such as iPad, e-books, interactive whiteboards, and smartphones, online handwritten documents are generated constantly. To facilitate the storage and retrieval of online handwriting, research in online handwriting recognition has recently been gaining more attention. Despite the already conducted intensive study in the past decades on online handwritten character and text recognition, it is still necessary to give further in-depth studies on character recognition accuracy and interpretability. In the context of handwriting recognition, we mean interpretability

G. A. Fink et al. (Eds.): ICDAR 2023, LNCS 14190, pp. 38–53, 2023.
https://doi.org/10.1007/978-3-031-41685-9_3

as the ability of producing character segmentation and confidence in recognition, to be consistent with human reading. In real application scenarios, there are many challenges to robust Chinese online handwritten text recognition [1], which poses the challenge of large character sets and as well challenges common with general online handwriting recognition. Writing styles from different groups of people vary greatly for the writer's age, educational background, and personal habits. When writing online texts, even the same person can output a strong variability in writing styles because of the different input speed, sloppiness, postures (sitting, lying, or walking), and devices (mobile phones, personal computers, and iPad). Different from offline handwriting without time information, there are often some delayed or post-correction strokes during writing. The large character set in Chinese handwriting makes the model design more difficult than the model for alphabetic languages because the coverage of a large number of character classes in the model entails more complicated training with more training samples carrying a large number of pattern variations. Even if there is enough training data in deep learning, confident recognition of handwriting is still a problem that has not been completely solved.

Many methods have been proposed for online and offline handwritten text line recognition. The methods can be categorized into two groups: explicit segmentation based and implicit segmentation based. In explicit segmentation-based methods [3–6], the input data is usually first over-segmented into a sequence of primary components each of which is a character or part of a character. Consecutive components are combined into candidate character patterns and given candidate recognition classes to construct the segmentation-recognition candidate lattice. Each path on the lattice corresponds to a segmentation recognition candidate and the optimal path is obtained by dynamic programming as the final text recognition result. For path scoring, candidate character classification and some contexts e.g. geometric and linguistic models are integrated to improve text line recognition. In implicit segmentation-based methods, hidden Markov model (HMM) [7–9] is mostly used for character recognition. Adopting the sliding window approach in HMM, the text line is first equidistantly split into frames to be concatenated into characters for recognition. But the relationship between contextual features is difficult to be modeled. Some end-to-end learned deep neural networks have been applied to explicit segmentation based [10] and implicit segmentation based [11] methods, but there are still some insufficiencies: hard to achieve both high accuracy and good interpretability, requiring a large number of training samples or training data augmentation/synthesis.

Aiming to overcome the need for large training data and improve the interpretation of online handwritten Chinese text recognition, we propose a recognition method based on the segmentation-recognition framework with an improved learning strategy. In the recognition model, we use the recently proposed convolutional prototype network (CPN) [13] as the base character classifier to replace the popular CNN because prototype learning can improve the intra-class compactness in feature representation. For enhancing non-character rejection, we propose a character-level learning algorithm to pre-train the CPN model and the prototype loss (PL) is similar to the maximum likelihood regularization in [14]. In weakly-supervised learning on string samples, the pre-trained CPN character model is fine-tuned with the negative log-likelihood (NLL) loss [4] on the candidate lattice. Besides, to better exploit character samples for CPN training and recognition, we propose a new character sample normalization method based on geometric information in text lines. We conducted experiments on two online handwriting datasets CASIA-OLHWDB and ICDAR2013-Online. The experimental results demonstrate the effectiveness of the proposed method compared with the baselines, and the benefits of our CPN-based character recognition i.e. the CPN character model can output confident results to character recognition and non-character rejection without non-character samples given in pre-training and synthetic text lines in weakly supervised learning.

The rest of this paper is organized as follows: Sect. 2 reviews some related work. Section 3 describes our proposed method with character normalization and CPN character model learning. Section 4 presents experimental results on public datasets, and Sect. 5 draws concluding remarks

2 Related Work

2.1 Online Chinese Handwriting Recognition

Early online Chinese handwriting recognition methods were usually based on the segmentation-recognition framework [3,4], mainly relying on character classification model [12]. In recent ten years, the convolutional neural network has shown superior performance on image classification [15] and has also been applied to online Chinese handwritten character recognition [16], where online handwritten characters are represented as offline images e.g. normalization-cooperated direction-decomposed feature maps [17]. Liu et al. [18] proposed a stroke sequence-dependent deep convolutional neural network (SSDCNN) by incorporating the natural sequence information of strokes and eight-directional features in a natural way. Ren et al. [19] regarded online handwriting as time series and presented a novel recurrent neural network (RNN) with two virtual unidirectional RNN for online handwritten Chinese character recognition. In the end-to-end learning framework, Xie et al. [20] proved a character model based on integrated CNN and RNN to be effective for online handwritten Chinese text recognition. Chen et al. [21] proposed a compact CNN-RNN with a small footprint and low computation cost trained by a connectionist temporal classification (CTC) criterion with a multi-step training strategy. For online Chinese signature segmentation and character recognition, Qin et al. [22] used a progressive multitask learning network (PMLNet) consisting of a dual channel stroke feature extraction block (DSF-Block), a stacked transformer encoder block (STE-Block) and a progressive multitask learning block (PML-Block).

2.2 Prototype-Based Handwriting Recognition

Research in prototype learning for large-set handwritten character recognition seems to have peaked in the early 1990s [23]. The first prototype-based handwritten Chinese character recognition method is available in [25] with a modified LVQ3 algorithm to optimize the reference vector. Prototype learning algorithms were reviewed in [2] for the nearest-neighbor (NN) classifier and their performance were evaluated for handwritten character recognition. Different prototype selection methods were studied for recognizing online handwritten characters in [26]. Enriching each prototype as a binary discriminant function with a threshold, one-vs-all training [27] was used for prototype classifiers in both multiclass and binary classification. A two-stage prototype generation technique [28] was introduced for handwriting digit recognition. Crossmodal prototype learning was first proposed by Ao et al. [29] for zero-shot online handwritten character recognition, where handwritten characters can be recognized by learning from a few handwritten or even printed samples. To overcome the lack of robustness for CNN, the CPN model was first proposed for open-set pattern recognition with convolutional prototype learning (CPL) [30] and improved the intra-class compactness of the feature representation. Then Gao et al. [31] applied the CPN model for end-to-end Chinese handwritten text recognition combining the prototype loss and CTC (Connectionist Temporal Classification) loss in one-vs-all prototype learning. Motivated by the work in [31], Yu et al. [32] proposed an efficient Chinese handwritten text line recognition method based on prototype learning with feature-level sliding windows for classification.

In this paper, we first use the CPN model as our character classier in the segmentation-recognition framework for handwriting recognition. Compared with other deep learning-based handwriting recognition, convolutional prototype learning can get an effective and robust feature representation for character recognition and rejection. Different from CPN-based end-to-end text recognition [31], our method can give comparable character accuracy and simultaneously output character segmentation results without data augmentation.

3   Methodology

3.1   Overview

As shown in Fig. 1, our online Chinese handwritten text recognition method follows the segmentation-recognition framework [4,6] and uses a convolutional prototype network (CPN) for character classification. The original data is first over-segmented [33] into component sequences and successive components are combined into candidate character patterns. All the candidate patterns are recognized by the character classifier to construct the candidate lattice, and the optimal path on the lattice is dynamically searched with the integration of language model and character classification for path scoring. The optimal path with character sequence and segmentation boundaries is the final recognition result. Before character model training and pattern recognition, we use statistics of the relative character position and size counted from training text lines to normalize character samples. For segmented character classification/recognition, current methods of feature extraction from online data and from offline images yield comparable accuracy [10,12], while offline images offer an advantage of stroke

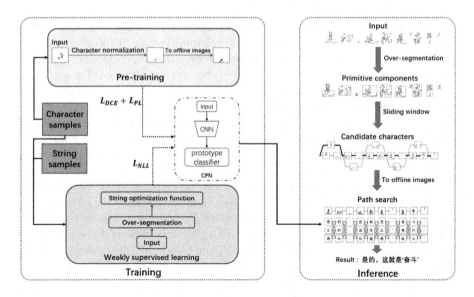

Fig. 1. An illustration of our overall framework.

order invariance. So the normalized online character patterns are converted to offline images for feature representation and classification. The CPN model is first pre-trained on isolated character samples from C character classes and then fine-tuned on weakly labeled text lines.

3.2 Character Normalization

In Chinese handwriting, as shown in Fig. 2(a) and (b), it is very difficult to recognize some appearances of similar characters e.g. punctuations """' and "," which can not be distinguished without character positions and shape sizes in text lines. Inspired by this analysis, we propose a character normalization method before feature representation and classification. As shown in Fig. 2, the normalized characters are more easily distinguishable for recognition.

For character normalization, we define two random variables i.e. z_d and z_h confirming to Gaussian distribution for each character class c ($c \in C$):

$$\begin{cases} z_{d_c} \sim \mathcal{N}_{d_c}(\mu_{d_c}, \sigma_{d_c}^2), & (1) \\ z_{h_c} \sim \mathcal{N}_{h_c}(\mu_{h_c}, \sigma_{h_c}^2), & (2) \end{cases}$$

where z_{d_c} is the normalized distance distribution between median lines of character samples and their located text lines, z_{h_c} is the normalized height distribution of character samples, and values of μ_{d_c}, σ_{d_c}, μ_{h_c}, σ_{h_c} are estimated on training

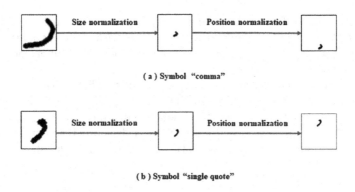

(a) Symbol "comma"

(b) Symbol "single quote"

Fig. 2. Illustration of character normalization.

text lines. For each character sample to be normalized, random sampling is performed on z_{d_c} and z_{s_c} to estimate the normalized distance d_c and normalized character height h_c. Using d_c, h_c, and the width/height ratio of the character sample, character normalization is performed successfully as shown in Fig. 2.

Both in training and inference, the character image is normalized considering its relative size and position in the text line. In training, the size and position information of each character class are estimated from the training dataset of text line images, while in inference, the size and position information of the candidate character are calculated based on the test text line image.

3.3 Character Classification Model

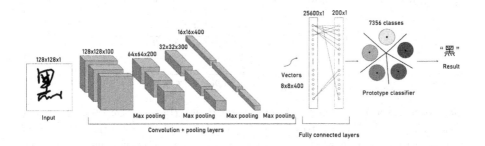

Fig. 3. Character classifier based on prototype learning.

The character classification model i.e. CPN is illustrated in Fig. 3, where the character image with size 128×128 is encoded by CNN for feature representation and prototype matching is used for character classification. Our CNN contains a stack of 12 convolutional layers with 3×3 kernel. The number of channels in each convolutional layer is enlarged gradually from 50 to 400 across all the convolutional layers. A maxpooling layer is used after every 3 convolutional layers to reduce the size of feature maps and enlarge the receptive fields. At last, the obtained feature maps are embedded to a vector with dimensionality 400 and then fed into a fully-connected layer to get a compact 200 dimensional feature vector as the final representation.

The feature of each character sample x output by CNN with parameter θ is denoted as $f_\theta(x)$. In our work, for simplicity, we use only one prototype for each character class c denoted as p_c ($c \in C$). Euclidean distance $d_c(x) = ||f_\theta(x) - p_c||_2^2$ is used to describe the dissimilarity between each character sample and class prototype, then our character classification function can be denoted as:

$$\arg \min_{c \in C} d_c(x), \tag{3}$$

where the feature of x is compared with all prototypes and classified by the nearest-neighbor criterion.

For model learning, a confidence measure is desired to approximate the class posterior probability $p(c|x)$ proportional to an exponential function [34]:

$$p(c|x) \propto exp(-d_c(x)). \tag{4}$$

In the segmentation-recognition framework, there are many non-character patterns and some samples out of the C character classes i.e. outlier class denoted as c^o. We set a learnable dissimilarity score $d_{c^o}(x) = T$ for outlier class, and the outlier class posterior probability $p(c^o|x)$ can be computed as:

$$p(c^o|x) \propto exp(-d_{c^o}(x)) = exp(-T). \tag{5}$$

Then we can get a confidence measure based on the normalization of Eq. (4) and Eq. (5) equivalent to the Softmax form of $(C + c^o)$ classes [35]:

$$\begin{aligned} p(c|x) &= \frac{exp(-d_c(x) + T)}{1 + \sum_{c \in C} exp(-d_c(x) + T)} \\ &= \frac{exp(-d_c(x))}{\sum_{c \in (C+c^o)} exp(-d_c(x))}, \end{aligned} \tag{6}$$

where parameter T can be set to a constant or jointly optimized together with θ and $\{p_c\}_{c \in C}$. If T is set to a constant, it is equivalent to setting the radius of a sphere in the feature space. Then a character sample is judged as a non-character if the distance to the nearest prototype is greater than the radius, and vice versa. Considering that the feature extraction layer and classification layer of CPN are learned together, T can be set to be learnable and optimized simultaneously with the entire network.

3.4   Character Model Training

Character Model Pre-Training. Based on our proposed confidence measure in Eq. (6), the character model is pre-trained on isolated samples with character classification loss i.e. distance based cross-entropy loss (DCE):

$$L_{DCE} = -log p(c|x) = -log \frac{exp(-d_c(x) + T)}{1 + \sum_{c \in C} exp(-d_c(x) + T)}, \tag{7}$$

which encourages character sample x nearest to the prototype from the genuine class.

To prevent the model from over-fitting and improve the character model learning, we also use prototype loss (PL) according to [13] as regularization:

$$L_{PRE} = L_{DCE} + \lambda \cdot L_{PL}, \tag{8}$$

where $L_{PL} = d_c(x) = ||f_\theta(x) - p_c||_2^2$, and λ is a hyper-parameter weighting the PL loss. Then intra-class compact and inter-class separable feature representations can be learned with L_{PRE} loss.

Weakly Supervised Learning. After over-segmentation and candidate pattern recognition, the training text line sample X with ground truth G is converted to the candidate lattice, which contains the segmentation path S and recognition string Y. On the candidate lattice, the energy function $E(Y, S, X)$ [4] is defined as:

$$E(Y, S, X) = - \sum_{i=0}^{|S|-1} F(y_{s_i}, s_i, X), \tag{9}$$

where s_i is the i-th candidate character pattern in S with length $|S|$, and y_{s_i} is the character recognition result of s_i. In Eq. (9), the feature function $F(y_{s_i}, s_i, X)$ is defined by integrating character classification and the 5-gram language model:

$$F(y_{s_i}, s_i, X) = log p(y_{s_i}|s_i) + \omega \cdot log P_{LM}(y_{s_i}|y_{s_{i-1}} \cdots y_{s_{i-4}}), \tag{10}$$

where ω is the weight of LM score, and $p(y_{s_i}|s_i)$ is given by CPN character model as in Eq. (6). Then the posterior probability $P(Y, S|X)$ can be computed by normalizaiton of the exponential function of $E(Y, S, X)$:

$$P(Y, S|X) = \frac{exp(-E(Y, S, X))}{\sum_{Y_0, S_0} exp(-E(Y_0, S_0, X))}. \tag{11}$$

For character model parameter estimation, we use the negative log-likelihood (NLL) loss [4] to maximize the posterior probability $P(Y = G|X)$ for weakly supervised learning.

$$L_{WSL} = -log P(Y = G|X), \tag{12}$$

where $P(Y = G|X)$ is computed according to Eq. (11):

$$P(Y = G|X) = \frac{\sum_S exp(-E(Y = G, S, X))}{\sum_{Y_0, S_0} exp(-E(Y_0, S_0, X))}. \tag{13}$$

There is an exponential number of segmentation-recognition paths on the lattice, and so the computation of Eq. (13) is performed by the forward-backward algorithm [4,6]. The loss function L_{WSL} is optimized by stochastic gradient descent (SGD) with parameters updated iteratively on training samples.

4 Experiments

4.1 Datasets

We evaluated the performance of our approach on a large database of online Chinese handwriting i.e. CASIA-OLHWDB [36] and the ICDAR2013-Online Chinese handwriting recognition competition dataset (ICDAR2013-Online) [37]. We conduct a series of ablation studies to explore the effects of our models and also compare our method with state-of-the-art. **CASIA-OLHWDB**, containing both isolated characters and string samples, is divided into a training set of 816 writers and a test set of 294 writers. There are $3,129,496$ isolated character samples of $7,356$ classes and $41,710$ handwritten text lines on CASIA-OLHWDB. The string test set contains $10,510$ text lines from $1,020$ text pages, including $269,674$ characters of $2,631$ classes. **ICDAR2013-Online** competition dataset contains $3,432$ online handwritten Chinese text lines from 60 writers.

4.2 Implementation Details

We implement experiments based on the framework of Pytorch with 4 NVIDIA RTX 24G GPUs. The SGD optimizer is applied to train our model with a learning rate initialized to 1×10^{-2} in pre-training and 1×10^{-6} in weakly supervised learning respectively. The learning rate will be exponentially decayed throughout the whole training. The hyper-parameters λ and ω are set to 1×10^{-5} and 0.4. Parameter T is initialized to 10 and optimized with other parameters. All prototypes are initialized as zero vectors. The pre-training stops at 80 epochs and the weakly supervised training stops at 30 epochs.

4.3 Ablation Experiment

In this part, we design several ablation experiments to prove the effectiveness of our proposed method without language model, which will be used in comparison with state-of-the-art methods.

Effectiveness of Character Normalization. In this experiment, we use CNN and CPN respectively as the character classifier with/without character normalization. For simplicity, character models are only trained on isolated samples and tested on CASIA-OLHWDB and ICDAR2013-Online datasets. Text line recognition is evaluated by Correct Rate (CR) and Accurate Rate (AR) [4].

Visual features of normalized characters are combined with some geometric information i.e. character position and size in text lines, so text line recognition results are improved with character normalization for both CNN and CPN models as shown in Table 1. The following experiments are all performed with character normalization.

Table 1. Effectiveness of character normalization for CNN and CPN models (CharNorm).

Methods	CASIA-OLHWDB		ICDAR2013-Online	
	AR	CR	AR	CR
CNN	80.86	83.33	77.92	81.07
CNN+CharNorm	84.28	86.22	81.67	83.57
CPN	83.17	86.42	80.25	83.79
CPN+CharNorm	**87.26**	**89.19**	**84.34**	**86.66**

Effectiveness of Prototype Learning. We use the same isolated character data as above to train CNN and CPN respectively for text line recognition. To further prove the advantages of CPN, we added non-character training, which means that non-character samples are added to train character classifiers in pre-training with isolated character samples together. These non-character samples are constructed from parts of character samples and a combination of multiple character samples.

One CNN model is trained with a mixture of character and non-character (hundred thousand) samples, and the other CNN is trained without non-character samples. Two CPN models are also trained respectively with and without non-character samples. Experimental results are shown in Table 2. CPN combines the advantages of both discriminant and generative models by learning prototypes and convolutional network parameters together, and so performs better than CNN.

Table 2. Effectiveness of prototype learning with/without non-character training (NonCharTr).

Methods	CASIA-OLHWDB		ICDAR2013-Online	
	AR	CR	AR	CR
CNN	84.28	86.22	81.67	83.57
CPN	87.29	89.19	84.34	86.66
CNN+NonCharTr	86.82	88.17	84.23	86.19
CPN+NonCharTr	**87.74**	**89.59**	**84.68**	**87.09**

Effectiveness of Weakly Supervised Learning. In this experiment, we first pre-train CPN and CNN using character normalization on character samples respectively with and without non-character samples. Then weakly supervised learning is performed by fine-tuning pre-trained models on string samples together with character samples, which are regarded as text line samples for

Table 3. Effectiveness of weakly supervised learning (WSL) for models with different configurations.

Methods	CASIA-OLHWDB		ICDAR2013-Online	
	AR	CR	AR	CR
CNN	84.28	86.22	81.67	83.57
CNN+WSL	85.16	87.34	83.09	85.66
CNN+NonCharTr+WSL	89.17	91.62	87.14	89.23
CPN	87.29	89.19	84.34	86.66
CPN+WSL	89.66	91.71	87.62	89.49
CPN+NonCharTr+WSL	**89.97**	**91.98**	**87.71**	**89.97**

robust learning. The experimental results are shown in Table 3. The objective of our weakly supervised learning is to maximize the posterior probability of ground truth given the text line sample, so the recognition performance of all character classification models is improved by weakly supervised learning. After weakly supervised learning, the CPN model can still get the best recognition results benefiting by joint parameter optimization of the convolutional network, prototypes, and dissimilarity score T for the outlier class. Benefiting from learnable dissimilarity score T for outlier class, CPN even trained without non-character samples can perform better than CNN trained with additional non-character samples and get recognition accuracies close to CPN trained with non-character samples.

Table 4. Comparison with existing methods on the CASIA-OLHWDB competition set, results marked by "*" denote using implicit language model.

Methods	Without LM		With LM	
	AR	CR	AR	CR
Shi et al. [38]	87.67	89.58	–	–
Wang et al. [3]	–	–	92.97	93.76
Zhou et al. [39]	–	–	94.69	95.32
Xie et al. [20]	91.38	92.29	95.50	96.09
Xie et al. [20]	**93.31***	**94.47***	**97.23***	**97.50***
Ours	89.66	91.71	94.77	95.47

4.4 Comparison with the State-of-the-Art Methods

Comparison results with existing methods are shown in Tables 4 and 5 (LM: n-gram language model), where we use a 5-gram statistical language model [5] for context fusion.

On the CASIA-OLHWDB dataset, our method achieves recognition performance of AR 94.77% and CR 95.47% as shown in Table 4. Xie et al. achieved a state-of-the-art performance with a multi-spatial-context fully convolutional recurrent network and implicit language model. To be fair, we also list their performance without using the implicit language model, and our method can perform comparably with state-of-arts. It is worth mentioning that under the framework of semi-Markov conditional random fields, Zhou et al. [33] used additional geometry models and achieved a similar performance to ours.

Table 5. Comparison with existing methods on the ICDAR2013-Online competition set, results marked by "*" denote using the implicit language model, and the "†" denotes using the powerful Transformer-based language model.

Methods	Without LM		With LM	
	AR	CR	AR	CR
Shi et al. [38]	83.60	85.14	–	–
Xie et al. [20]	86.85	87.82	91.81	92.67
Xie et al. [20]	88.88*	90.17*	96.50	97.15
Peng et al. [40]	91.24	91.81	–	–
Peng et al. [40]	**95.05†**	**95.46†**	97.36†	97.63†
Peng et al. [10]	94.46	94.67	**97.89†**	**98.06†**
Ours	87.62	89.49	95.37	95.55

We also compare the performance of different methods on the ICDAR2013-Online dataset in Table 5, and our recognition accuracy results are AR95.37%, CR95.55%. The method proposed by Peng et al. [10] got a state-of-the-art result regardless of whether the language model was used or not, which is due to the fact that they synthesized string samples to make their model learn better and they used a powerful Transformer-based language model. In addition, they also designed a context regularization based on RNN layers. Our proposed method is still potential and competitive because we can still achieve promising performance without any data augmentation techniques when training models with prototype learning.

4.5   Further Visualization and Analysis

Some recognition results are shown in Fig. 4, which present the superiority of our CPN character model learning. As shown in Fig. 4(1, 2), our CPN-based method can give more correct recognition results. But there are still some recognition errors uncorrected. In Fig. 4(2), strokes from neighboring characters are combined into one component in over-segmentation and therefore wrongly recognized. In Fig. 4(3), the error in CPN-based recognition can be corrected with a more powerful language model. The above analysis of recognition errors indicates our further work on text segmentation and language modeling.

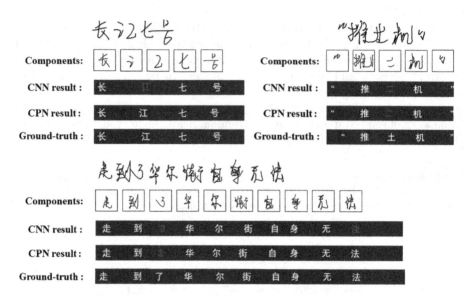

Fig. 4. Recognition visualization for experimental analysis.

5 Conclusions

In this paper, we propose a new segmentation-recognition method for online Chinese handwritten text recognition using a convolutional prototype network (CPN) for character classification. Different from previous character model learning for candidate lattice generation from over-segmentation components, our CPN model can be pre-trained only on closed-set character samples and still get superior recognition accuracy compared with the classifier trained using additional non-character samples. During weakly supervised prototype learning on the text lines, only limited training data is needed and the resulting character classifier can also perform well compared with end-to-end recognition models, which must be learned on big data. In the future, the recognition model can benefit from training with augmented data and combining with more powerful language model and geometric context model.

Acknowledgement. This work has been supported by the National Key Research and Development Program Grant 2020AAA0109700, the National Natural Science Foundation of China (NSFC) grant 61936003.

References

1. Liu, C.-L., Jäger, S., Nakagawa, M.: Online recognition of Chinese characters: the state-of-the-art. IEEE Trans. Pattern Anal. Mach. Intell. **26**(2), 198–213 (2004)
2. Liu, C.-L., Nakagawa, M.: Evaluation of prototype learning algorithms for nearest-neighbor classifier in application to handwritten character recognition. Pattern Recognit. **34**(3), 601–615 (2001)

3. Wang, D.-H., Liu, C.-L., Zhou, X.-D.: An approach for real-time recognition of online Chinese handwritten sentences. Pattern Recognit. **45**(10), 3661–3675 (2012)

4. Zhou, X.-D., Wang, D.-H., Tian, F., Liu, C.-L., Nakagawa, M.: Handwritten Chinese/Japanese text recognition using semi-Markov conditional random fields. IEEE Trans. Pattern Anal. Mach. Intell. **35**(10), 2413–2426 (2013)

5. Wu, Y.-C., Fei, Y., Liu, C.-L.: Improving handwritten Chinese text recognition using neural network language models and convolutional neural network shape models. Pattern Recognit. **65**, 251–264 (2017)

6. Wang, Z.-X., Wang, Q.-F., Yin, F., Liu, C.-L.: Weakly supervised learning for over-segmentation based handwritten Chinese text recognition. In: ICFHR 2020, pp. 157–162 (2020)

7. Su, T.-H., Zhang, T., Guan, D.-J., Huang, H.-J.: Off-line recognition of realistic Chinese handwriting using segmentation-free strategy. Pattern Recognit. **42**(1), 167–182 (2009)

8. Jiang, Z.-W., Ding, X.-Q., Liu, C., Wang, Y.-W.: A novel short merged off-line handwritten Chinese character string segmentation algorithm using hidden Markov model. In: ICDAR 2011, pp. 668–672 (2011)

9. Jayech, K., Mahjoub, M.A., Amara, N.E.B.: Synchronous multi-stream hidden Markov model for offline Arabic handwriting recognition without explicit segmentation. Neurocomputing **214**, 958–971 (2016)

10. Peng, D.-Z., et al.: Recognition of handwritten Chinese text by segmentation: a segment-annotation-free approach. IEEE Trans. Multimed. (2022)

11. Wang, Z.-R., Jun, D., Wang, J.-M.: Writer-aware CNN for parsimonious HMM-based offline handwritten Chinese text recognition. Pattern Recognit. **100**, 107102 (2020)

12. Liu, C.-L., Yin, F., Wang, D.-H., Wang, Q.-F.: Online and offline handwritten Chinese character recognition: benchmarking on new databases. Pattern Recognit. **46**(1), 155–162 (2013)

13. Yang, H.-M., Zhang, X.-Y., Yin, F., Yang, Q., Liu, C.-L.: Convolutional prototype network for open set recognition. IEEE Trans. Pattern Anal. Mach. Intell. **44**(5), 2358–2370 (2022)

14. Liu, C.-L., Sako, H., Fujisawa, H.: Effects of classifier structures and training regimes on integrated segmentation and recognition of handwritten numeral strings. IEEE Trans. Pattern Anal. Mach. Intell. **26**(11), 1395–1407 (2004)

15. Ciresan, D.C., Meier, U., Schmidhuber, J.: Multi-column deep neural networks for image classification. In: CVPR 2012, pp. 3642–3649

16. Zhang, X.-Y., Bengio, Y., Liu, C.-L.: Online and offline handwritten Chinese character recognition: a comprehensive study and new benchmark. Pattern Recognit. **61**, 348–360 (2017)

17. Liu, C.-L.: Normalization-cooperated gradient feature extraction for handwritten character recognition. IEEE Trans. Pattern Anal. Mach. Intell. **29**(8), 1465–1469 (2007)

18. Liu, X., Hu, B.-T., Chen, Q.-C., Xiang-Ping, W., You, J.-H.: Stroke sequence-dependent deep convolutional neural network for online handwritten chinese character recognition. IEEE Trans. Neural Networks Learn. Syst. **31**(11), 4637–4648 (2020)

19. Ren, H.-Q., Wang, W.-Q., Xi-Wen, Q., Cai, Y.-Q.: A new hybrid-parameter recurrent neural network for online handwritten Chinese character recognition. Pattern Recognit. Lett. **128**, 400–406 (2019)

20. Xie, Z.-C., Sun, Z.-H., Jin, L.-W., Ni, H., Lyons, T.J.: Learning spatial-semantic context with fully convolutional recurrent network for online handwritten Chinese text recognition. IEEE Trans. Pattern Anal. Mach. Intell. **40**(8), 1903–1917 (2018)
21. Chen, K., et al.: A compact CNN-DBLSTM based character model for online handwritten Chinese text recognition. In: ICDAR 2017, pp. 1068–1073
22. Qin, X.-H., Zhang, H.-Y., Ke, X., Shen, Z.-H., Qi, S.-M., Liu, K.: Progressive multitask learning network for online Chinese signature segmentation and recognition. In: ICFHR 2022, pp. 153–167
23. Lee, S.-W., Song, H.-H.: Optimal design of reference models for large-set handwritten character recognition. Pattern Recognit. **27**(9), 1267–1274 (1994)
24. Niu, S.-C., et al.: Towards stable test-time adaptation in dynamic wild world. ICLR (2023)
25. Liu, C.-L., Kim, I.-J., Kim, J.H.: High accuracy handwritten Chinese character recognition by improved feature matching method. In: ICDAR 1997, pp. 1033–1037
26. Raghavendra, B.S., Narayanan, C.K., Sita, G., Ramakrishnan, A.G., Sriganesh, M.: Prototype learning methods for online handwriting recognition. In: ICDAR 2005, pp. 287–291
27. Liu, C.-L.: One-vs-all training of prototype classifier for pattern classification and retrieval. In: ICPR 2010, pp. 3328–3331
28. Impedovo, S., Mangini, F.M., Barbuzzi, D.: A novel prototype generation technique for handwriting digit recognition. Pattern Recognit. **47**(3), 1002–1010 (2014)
29. Ao, X., Zhang, X.-Y., Liu, C.-L.: Cross-modal prototype learning for zero-shot handwritten character recognition. Pattern Recognit. **131**, 108859 (2022)
30. Yang, H.-M., Zhang, X.-Y., Yin, F., Liu, C.-L.: Robust classification with convolutional prototype learning. In: CVPR 2018, pp. 3474–3482
31. Gao, L.-K., Zhang, H., Liu, C-L.: Handwritten text recognition with convolutional prototype network and most aligned frame based CTC training. ICDAR (1), pp. 205–220 (2021)
32. Yu, M.-M., Zhang, H., Yin, F., Liu, C.-L.: An efficient prototype-based model for handwritten text recognition with multi-loss fusion. In: ICFHR 2022, pp. 404–418
33. Heng Zhang, Cheng-Lin Liu: A Lattice-Based Method for Keyword Spotting in Online Chinese Handwriting. ICDAR 2011: 1064–1068
34. Liu, C.-L., Nakagawa, M.: Precise Candidate Selection for Large Character Set Recognition by Confidence Evaluation. IEEE Trans. Pattern Anal. Mach. Intell. **22**(6), 636–642 (2000)
35. Jeffrey A. Barnett: Computational Methods for A Mathematical Theory of Evidence. Classic Works of the Dempster-Shafer Theory of Belief Functions 2008: 197–216
36. Liu, C.-L., Yin, F., Wang, D.-H., Wang, Q.-F.: CASIA online and offline Chinese handwriting databases. In: ICDAR 2011, pp. 37–41 (2011)
37. Yin, F., Wang, Q.-F., Zhang, X.-Y., Liu, C.-L.: ICDAR 2013 Chinese handwriting recognition competition. In: ICDAR 2013, pp. 1464–1470
38. Shi, B.-G., Bai, X., Yao, C.: An end-to-end trainable neural network for image-based sequence recognition and its application to scene text recognition. IEEE Trans. Pattern Anal. Mach. Intell. **39**(11), 2298–2304 (2017)

39. Zhou, X.-D., Zhang, Y.-M., Tian, F., Wang, H.-A., Liu, C.-L.: Minimum-risk training for semi-Markov conditional random fields with application to handwritten Chinese/Japanese text recognition. Pattern Recognit. **47**(5), 1904–1916 (2014)
40. Peng, D.-Z., Jin, L.-W., Wu, Y.-Q., Wang, Z.-P., Cai, M.-X.: A fast and accurate fully convolutional network for end-to-end handwritten Chinese text segmentation and recognition. In: ICDAR 2019, pp. 25–30

Vision Conformer: Incorporating Convolutions into Vision Transformer Layers

Brian Kenji Iwana[1]([⊠])[ID] and Akihiro Kusuda[2]

[1] Kyushu University, Fukuoka, Japan
iwana@ait.kyushu-u.ac.jp
[2] Nara Institute of Science and Technology, Nara, Japan

Abstract. Transformers are popular neural network models that use layers of self-attention and fully-connected nodes with embedded tokens. Vision Transformers (ViT) adapt transformers for image recognition tasks. In order to do this, the images are split into patches and used as tokens. One issue with ViT is the lack of inductive bias toward image structures. Because ViT was adapted for image data from language modeling, the network does not explicitly handle issues such as local translations, pixel information, and information loss in the structures and features shared by multiple patches. Conversely, Convolutional Neural Networks (CNN) incorporate this information. Thus, in this paper, we propose the use of convolutional layers within ViT. Specifically, we propose a model called a Vision Conformer (ViC) which replaces the Multi-Layer Perceptron (MLP) in a ViT layer with a CNN. In addition, to use the CNN, we proposed to reconstruct the image data after the self-attention in a reverse embedding layer. Through the evaluation, we demonstrate that the proposed convolutions help improve the classification ability of ViT.

Keywords: Transformer · Vision Transformer · Convolutional Neural Network · Character Recognition

1 Introduction

Recently, there has been a sudden rise in the popularity of Transformers [32] in pattern recognition. They were originally proposed and popularized in the Natural Language Processing (NLP) domain [11, 26, 32]. Fundamentally, Transformers are feed-forward neural networks that utilize self-attention, inner-layer residual connections, and a specialized token-based encoding. Typically, shown in Fig. 1a, Transformers are designed with multiple Transformer blocks, each containing a Multi-Head Self-Attention (MHSA) layer and a fully-connected layer, and are

This research was partially supported by MEXT-Japan (Grant No. JP23K16949).
A. Kusuda—Equal contribution.

G. A. Fink et al. (Eds.): ICDAR 2023, LNCS 14190, pp. 54–69, 2023.
https://doi.org/10.1007/978-3-031-41685-9_4

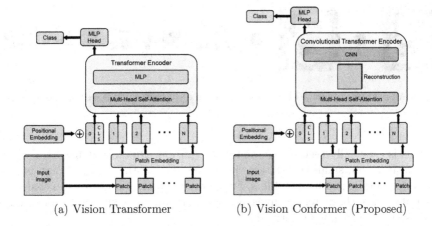

(a) Vision Transformer   (b) Vision Conformer (Proposed)

Fig. 1. Comparison of the overall structure of ViT and the proposed method.

trained with an encoder and a decoder network. The self-attention layer uses an attention mechanism to relate the weighted positions of an input sequence with the other positions of the input. This self-attention helps the fully-connected nodes to put more emphasis on the pairwise relationships of the input.

The popularity of Transformers has also spread to image recognition tasks. Part of the trend in image-based Transformers usage is due to the success of Transformers in image classification and object detection. For example, in seminal works, Dosovitskiy et al. [12] proposed the Vision Transformer (ViT) which breaks images into patches to be used as tokens, and Chen et al. [3] and Parmar et al. [25] use a sequence Transformer to do row-wise generation of images. More recently, Transformer-based models hold the state-of-the-art results on many computer vision benchmarks, including ImageNet [9,39], COCO [36,38], CIFAR10/100 [12,27,31], and more.

Introduced by ViT, a common mechanism in image-based Transformers is to break the input image into fixed-sized patches and embed the patches that are linearly transformed into 1D vectors. This is done to adapt the images to be used with traditionally sequential word piece token representations. The subsequent patch-based token representations are used by the self-attention and fully-connected layers of the Transformer.

One problem with ViT is that it is possible for the information to be lost in the fixed-sized patch-based representation. Since the patches are arbitrarily divided, there is no consideration for the objects, features, or structures that could exist across or between multiple patches. This is because the self-attention mechanism and fully connected layers only consider the patches as a whole. It should be noted, that one solution to this problem is the use of hierarchical transformers that use different-sized patches [24]. In addition, while self-attention does create global relationships, the relationships are only pairwise relationships and can only rely on the positional embedding for global structure.

Fig. 2. An example of how ViT turns an image of a character into a sequence of patches.

This is especially true for character recognition. Due to patches being fixed size and location, small translations in the location of the character can have a large effect on the patch embedding. Furthermore, as shown in Fig. 2, in character recognition, the character occupies multiple patches, and consideration for cross-patch structures is not performed.

Conversely, Convolutional Neural Networks (CNN) [21] have been shown to overcome these issues. The use of convolutional layers and max pooling allows for some translation invariance [21]. Unlike fixed-size patch-based Transformers, the embedding for the fully-connected layers in a CNN can represent large overlapping receptive fields.

CNNs have shown to be a powerful tool in computer vision and pattern recognition [29]. Thus, there is a desire to combine the advantages of CNNs and ViT. Consequentially, many convolutional variations of ViT have been proposed. Many Transformers, such as Convolutional vision Transformer (CvT) [33], Compact Convolutional Transformer (CCT) [15], and Convolution and Self-Attention Network (CoAtNet) [9] realize the advantages of feature extraction using convolutional layers. Therefore, they combine CNNs with ViT by using convolutional layers to improve the embeddings for Transformers.

In order to take advantage of the feature extraction, slight translation invariance, and structure-preserving abilities of CNNs, we propose to incorporate a convolutional layer inside the Transformer block. As shown in Fig. 3, the spatial outputs of the self-attention layer are reconstructed into an image and then provided to convolutional layers. After the convolutional layers, a fully-connected layer is used, much like a typical Transformer. Another interpretation of this would be replacing the Multi-Layer Perceptron (MLP) of ViT with a CNN.

The contributions of this paper are as follows:

- We propose a new convolutional vision Transformer called a Vision Conformer (ViC). The proposed ViC replaces the MLP of ViT with a CNN.
- In order to adapt the Transformer block to use a convolution, we introduce a reverse embedding layer to reconstruct the patches from the vector embeddings and a reconstruction module to recompose the patches into a matrix for the convolutional layers.
- The proposed method is evaluated on the character recognition task. We show results on three common character recognition datasets, MNIST, EMNIST, and KMNIST. We compare the proposed method to five state-of-the-art convolutional transformers and ViT.

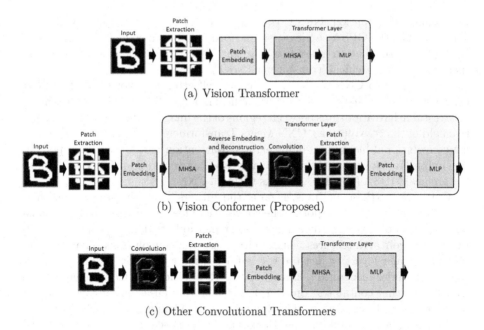

(a) Vision Transformer

(b) Vision Conformer (Proposed)

(c) Other Convolutional Transformers

Fig. 3. Comparison of the steps taken in ViT, ViC, and other convolutional transformers like CCT and CvT.

The code is publicly available at https://github.com/uchidalab/vision-conformer.

2 Related Work

The most common way to use convolutions with transformer or self-attention-based models is to use convolutional layers to extract features for the embeddings. For example, Coccomini et al. [7] used EfficientNet [30] to extract features for the patches of ViT and Cross-Attention Multi-Scale Vision Transformer (CrossViT) [2]. CvT [33] uses convolutions to generate the token embeddings. Similarly, Convolution-enhanced image Transformers (CeiT) [37] use depth-wise convolutions to locally enhance the tokens of a ViT. Hassani et al. proposed the CCT [15] that replaces the standard ViT patch embedding layer with a convolutional layer. CoAtNet [9] uses a hybrid of a CNN and Transformer by including convolutional layers for the lower layers and Transformer blocks for the upper layers. In another work, Chu et al. [4] propose a modified positional embedding for ViT called Conditional Positional Embeddings (CPE). In their proposed Conditional Positional encoding Vision Transformer (CPVT), the CPEs are constructed using convolutions. Pooling-based Vision Transformer (PiT) [18] uses convolutions with stride 2 to downsample the spatial embeddings of ViT.

In addition, there are other ways convolutions are used in ViT-based models. For example, Convolutional Vision Transformer (ConViT) [10] proposes Gated

Positional Self-Attention (GPSA) that can be initialized as a convolutional layer. Through this, the GPSA can learn similar properties as a convolution. In Zhang et al. [40], a convolution is used to aggregate hierarchical transformer blocks in their Nested Hierarchical Transformer (NesT).

There are also CNN models that utilize self-attention. For example, CNNs using self-attention have been used for medical imaging [22,34]. The self-attention helps establish relationships between regions of the image that is beyond the receptive field of the convolutions. CNN Meets Transformer (CMT) [14] uses alternating convolutional blocks and lightweight self-attention blocks with local perception units. Transformer in Convolutional Neural Networks (TransCNN) [23] utilize hierarchical MHSA (H-MHSA) blocks within a CNN.

Compared to these methods, as far as the authors know, the proposed method is the only method to incorporate the convolutional layers directly into the transformer layers of a transformer model. Most methods that incorporate convolutions into their architectures usually either use the convolutions separate from the transformer layers or just use the self-attention as part of a CNN.

In time series recognition, namely speech recognition, the Conformer [13] is a transformer that contains a convolution between the self-attention layer and the fully-connected layer. Due to the similarity of the Conformer and the proposed method, we adopt a similar name. However, the proposed method was developed without inspiration or relation to a Conformer. And unlike the Conformer, the proposed method is based on ViT and requires extra consideration for the patch embeddings and image-based convolutions.

3 Vision Transformers (ViT)

Transformers are neural network models that were originally designed for NLP [32]. They are constructed from Transformer blocks consisting of a self-attention layer followed by a fully-connected layer with residual connections between layers. In the traditional NLP Transformers, text is modeled using sequences of word part tokens. These tokens are embedded into vectors and fed to the Transformer.

ViT [12] is a Transformer that was adapted for image recognition. The novel idea of ViT is that they proposed to use small image patches from the input images as tokens instead of the traditional word part tokens. Another difference is the traditional Transformer is trained in an encoder-decoder structure. Comparatively, ViT only uses an encoder during training. In this section, we will provide background and describe the important features of ViT.

3.1 Image Tokenization

In order to adapt images to be used as sequences of token embeddings, ViT breaks the image into fixed-sized patches, as shown in Fig. 1. The patches are serialized into a sequence. Each element of the sequence is flattened and embedded into a vector using a trainable linear projection. The result is a sequence

$\mathbf{X} = \mathbf{x}_1, \ldots, \mathbf{x}_t, \ldots, \mathbf{x}_T$, where $\mathbf{x}_t$ is a vector embedding of each patch and T is the number of patches. Because the patches are serialized, structures that span multiple patches are arbitrarily split based on the patch size.

Furthermore, a 1D positional embedding is added to the patch embedding. The positional embedding indicates the position of the token in the sequence. The purpose of the positional embedding is to retain the positional information of the patches.

3.2   Classification Tokenization

In addition to the patch embedding a special token is used to indicate the embedding to use for the classifier. This special classification token is prepended as $\mathbf{x}_{\mathrm{CLS}}$ to $\mathbf{X}$ as the first element of the sequence, i.e. $\mathbf{X} = \mathbf{x}_{\mathrm{CLS}}, \mathbf{x}_1, \ldots, \mathbf{x}_t, \ldots, \mathbf{x}_T$. In each layer of ViT, the first element of the sequence of embeddings is the classification token and the subsequent elements are the patch tokens. For classification, an MLP head is used as a classifier by attaching a fully-connected layer to the output embedding of the topmost classification token.

3.3   Multi-Head Self-Attention

Self-attention is a special case of attention where elements of the input sequence are weighted and multiplied with themselves. Specifically, the input sequences are copied into a query $\mathbf{Q}$, key $\mathbf{K}$, and value $\mathbf{V}$ and weighted separately. The elements of query $\mathbf{Q}$ and key $\mathbf{K}$ are multiplied and become the attention mechanism for value $\mathbf{V}$. In Transformers, Scaled Dot-Product Attention is used for the self-attention mechanism. Namely, the Scaled Dot-Product Attention is defined as:

$$\mathrm{Attention}(\mathbf{Q}, \mathbf{K}, \mathbf{V}) = \mathrm{softmax}\left(\frac{\mathbf{Q}\mathbf{K}^T}{\sqrt{d}}\right)\mathbf{V}, \tag{1}$$

where $1/\sqrt{d}$ is a scaling factor by the number dimensions d of the input sequences. The idea of self-attention is that the important pairwise relationships between tokens should be emphasized in the representation for the fully-connected layer.

Multi-head attention is attention that uses more than one parallel attention blocks. In Transformers, for each $\mathbf{Q}$, $\mathbf{K}$, and $\mathbf{V}$, multiple self-attention layers are used and the results are concatenated. This is done to jointly attend different combinations of pairwise matches simultaneously.

3.4   Multi-Layer Perceptron (MLP)

After the multi-head self-attention layer, ViT and other Transformers use a fully-connected MLP layer. The input to the MLP layer is the sequence output by self-attention, including the vector embedding related to the classification token. Also, between each layer Layer Normalization [1] and a residual connection [16] is used.

Algorithm 1. Reverse Embedding, Reconstruction, and CNN

Require: $\mathbf{Z}^{(in)} = \mathbf{z}_{CLS}, \mathbf{z}_1, \ldots, \mathbf{z}_t, \ldots, \mathbf{z}_T$
Ensure: $\mathbf{Z}'(out) = \mathbf{z}'_{CLS}, \mathbf{z}'_1, \ldots, \mathbf{z}'_t, \ldots, \mathbf{z}'_T$
1: $\mathbf{z}_{CLS}, \mathbf{Z}_{patch} \leftarrow \mathbf{Z}^{(in)}$ ▷ Separate the patch embeddings
2: $\mathbf{Z}_{patch} \leftarrow \mathrm{Linear}(\mathbf{Z}_{patch})$ ▷ Reverse embedding
3: $\mathbf{P} \leftarrow \mathrm{Reshape}(\mathbf{Z}_{patch})$
4: $\mathbf{I} \leftarrow \mathrm{SpatialConcatenate}(\mathbf{P})$ ▷ Reconstruction
5: **for** $c \leftarrow 1, C$ **do**   ▷ CNN
6:     $\mathbf{I}' \leftarrow \mathrm{Convolution}(\mathbf{I})$
7: **end for**
8: $\mathbf{P}' \leftarrow \mathrm{PatchExtraction}(\mathbf{I}')$
9: $\mathbf{Z}'_{patch} \leftarrow \mathrm{Flatten}(\mathbf{P}')$ ▷ Patch embedding
10: $\mathbf{Z}'_{patch} \leftarrow \mathrm{Linear}(\mathbf{Z}'_{patch})$
11: $\mathbf{Z}' \leftarrow \mathrm{Concatenate}(\mathbf{z}_{CLS}, \mathbf{Z}'_{patch}) + \mathbf{Z}^{(in)}$
12: $\mathbf{Z}'(out) = \mathrm{MLP}(\mathbf{Z}')$

---

4 Vision Conformer (ViC)

The proposed Vision Conformer (ViC) is modeled on ViT [12]. As shown in Fig. 1, we adopt a similar structure, including the same patch tokenization and embedding and self-attention. The difference between ViT and ViC is that we propose to replace the MLP of ViT with a CNN.

Figure 4 details the proposed ViC Encoder block. Similar to ViT, the embedded patches are input and Layer Normalization [1] is applied. Next, MHSA is used. The residual connection is then added to the output of MHSA and Layer Normalization is applied again. After Layer Normalization, the hidden vector $\mathbf{Z}^{(in)} = \mathbf{z}_{CLS}, \mathbf{z}_1, \ldots, \mathbf{z}_t, \ldots, \mathbf{z}_T$ is split into the classification embedding $\mathbf{z}_{CLS}$ and the patch embeddings $\mathbf{z}_1, \ldots, \mathbf{z}_t, \ldots, \mathbf{z}_T$. This is done because the classification embedding $\mathbf{z}_{CLS}$ is not part of the image representation.

From there, we introduce a Reverse Embedding step, a Reconstruction module, a CNN, and a Patch Embedding step that are applied to the patch embeddings $\mathbf{Z}_{patch}$. Finally, the hidden embeddings are fed to the fully-connected MLP.

The input of this encoder block is the same patch embeddings with positional encodings as ViT. Furthermore, the block can be stacked L number of times. To use the proposed method for classification, an MLP head is used on the final $\mathbf{z}_{CLS}$ embedding.

4.1 Image Reconstruction

In order to use a 2D CNN within a Transformer block, the output of MHSA needs to be a matrix. Thus, we reconstruct the image structure from the patch embeddings. To do this, we introduce two techniques, *Reverse Embedding* and *Reconstruction*. The process of Reverse Embedding and Reconstruction is the reverse operation of patch embedding used on the input of ViC. Specifically, the Reverse Embedding returns the latent vectors back to the original shape of

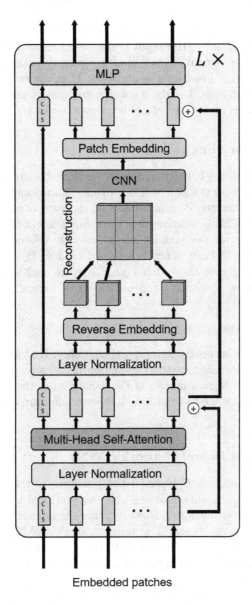

Fig. 4. Details of a ViC Encoder block within the proposed ViC. The ViC Encoder is similar to a standard Transformer Encoder, except that after the Multi-Head Self-Attention layer, the patch embeddings are reconstructed into an image and a CNN is used instead of the standard MLP. To reconstruct the image, a Reverse Embedding layer is used. Note, special tokens such as the classification token are passed directly to the fully connected layer.

the patches and the Reconstruction module recombines the patches into a single matrix.

Algorithm 1 shows the steps required to reconstruct the image from the token embeddings. In Algorithm 1, the input is the token embeddings $\mathbf{Z}^{(in)}$ from the output of MHSA. The output $\mathbf{Z}'(out)$ is either passed to the next ViC layer or is provided to the MLP head for classification. $\mathbf{P}$ is the reconstructed patches from the patch embeddings, $\mathbf{I}$ is the reconstructed image, and $\mathbf{P}'$ and $\mathbf{I}'$ are the corresponding patches and image after the convolutions, respectively.

4.2   Reverse Embedding

First, Reverse Embedding is used to convert the latent vectors from the MHSA into the original shape of the patches. In Reverse Embedding, a trainable linear layer is used to restore the dimensionality of the reconstruction, as shown in Algorithm 1 Line 2. This is used because the dimensions of the embedding d are defined by a hyperparameter and are not necessarily related to the patch size. After the embedding of each patch is mapped to a $P_h P_w$-dimensional vector, where P_h is the height of the patch and P_w is the width, the embedding is reshaped to patch $\mathbf{p}_t \in \mathbf{P}$ with the dimensions of the original patch size (P_h, P_w).

4.3   Reconstruction

Next, a feature map $\mathbf{I}$ is constructed from the patches. Each of the reconstructed patches in $\mathbf{P}$ is concatenated spatially in the position corresponding to the original patch location to form a matrix of the same size as the input image. With feature map $\mathbf{I}$, it is possible to treat it as if it were a feature map in a traditional CNN.

4.4   Convolutional Neural Network (CNN)

A CNN is used inside the ViC Encoder block instead of the typical MLP. The CNN used in the experiments contains one convolutional layer ($C = 1$, where C is the number of convolutions). However, there is no restriction on the CNN used in the ViC Encoder block. The purpose of the CNN is to use information across patches instead of discrete flattened patches like the MLP in ViT. Furthermore, the convolutional layers are able to help extract features from the image representation.

4.5   Patch Embedding

To continue using the ViC Encoder block as a Transformer block, the output of the CNN needs to be restored to a sequence of token embeddings. Thus, we perform the same patch embedding procedure as the input of ViT and the proposed ViC. Namely, another trainable linear projection is used to create the patch embeddings $\mathbf{Z}'_{\text{patch}}$.

Finally, the classification token z_{CLS} is prepended back to form the full $\mathbf{Z}'$. $\mathbf{Z}'$ can now be used with the MLP like a standard Transformer.

5   Experimental Results

5.1   Architecture Settings

For the experiments, the proposed method uses three of the ViC blocks ($L = 3$) shown in Fig. 4. Each block uses MHSA with four heads. In addition, each block has one convolutional layer with 3×3 convolutions at stride 1 and 32 filters each. As suggested by Dosovitskiy et al. [12], Gaussian Error Linear Units (GeLU) [17] is used as the activation function for all of the trainable layers, including the convolutional layers. For the patch encodings, we use 4×4 pixel patches and a patch encoding latent space with 256 dimensions.

The proposed method and all of the comparative evaluations use the same training scheme. The networks are trained with batch size 256 for 500 epochs using an Adam optimizer [19] with an initial learning rate of 0.001 and a weight decay of 0.0005. The size of the input, number of input channels, and number of classes is determined by each dataset. In addition, we use the pre-defined training and test sets that were determined by the dataset authors.

5.2   Comparative Evaluations

In order to evaluate the proposed method, we compare it to other ViT models that incorporate convolutions in some aspect. For a fair comparison, the shared hyperparameters of each of the comparison methods were set to match the proposed method. Namely, three Transformer blocks with four heads are used. In addition, all of the comparative methods use the same 4×4 pixel patches and a 256 dimensional linear projection embedding. They are all trained for the same 500 epochs with Adam optimizer and an initial learning rate of 0.001 and a weight decay of 0.0005. All of the networks are trained without pre-training or data augmentation. The following comparative evaluations were performed:

- **Vision Transformer (ViT)**. This is the baseline used to demonstrate the usefulness of the proposed method. The ViT evaluation uses all of the same hyperparameters, except for the convolutional layers and reverse embedding layers.
- **Compact Convolutional Transformer (CCT)** [15]. CCT uses two convolutional layers instead of the traditional embedding layer in ViT. The implementation uses convolutions with kernel size 3 at stride 1 like the proposed method. In addition, CCT uses Sequence Pooling for the MLP head used for classification.
- **Convolution and Self-Attention Network (CoAtNet)** [9]. CoAtNet combines a CNN with ViT by having the lower layers be CNN blocks while the higher layers be Transformer blocks. We use the C-T-T-T version of CoAt-Net for comparison as it consists of one convolutional block, one depth-wise convolutional block [28], and three Transformer blocks.

- **Convolutional vision Transformer (CvT)** [33]. CvT uses convolutions in two parts of the model. First, convolutions are used in the token embeddings. Second, convolutions are used for the projections for the self-attention layer. Again, for the experiments, the hyperparameters used were set to match the proposed method.
- **Nested Hierarchical Transformer (NesT)** [40]. NesT incorporates a convolution into the aggregation function of hierarchical Transformer blocks. NesT-T is used for the evaluation which includes three hierarchical layers of 8, 4, and 1 Transformer blocks each.
- **Pooling-based Vision Transformer (PiT)** [18]. PiT uses strided convolutions to downsample the patch token embeddings. To match the proposed method, one Transformer block is used between each downsampling, for a total of three Transformer blocks.

Table 1. Average Test Accuracy (%) of Five Trainings

Model	MNIST	EMNIST	KMNIST
ViC (Proposed)	**99.03**	87.74	**95.86**
ViT	98.34	86.81	92.92
CoAtNet	98.82	86.85	94.56
CCT	98.68	87.75	94.47
CvT	98.72	**88.02**	94.86
NesT	98.80	85.02	94.68
PiT	98.77	87.39	94.91

5.3   Results on MNIST

Dataset. The Modified National Institute of Standards and Technology (MNIST) database [21] is a standard benchmark dataset. It is made of 28 × 28 pixel, grayscale, isolated handwritten digits. There are 10 classes, "0" to "9." MNIST has a pre-defined training set of 60,000 images and a test set of 10,000 images.

Results. The results of the experiments are shown in Table 1. The results are the mean of training each model five times. This is done to increase the reliability of the results. In the table, it can be observed that the proposed ViC was able to achieve a higher accuracy than ViT on MNIST. In addition, it did remarkably better than the comparison convolutional Transformers. ViC had a 99.03% accuracy, whereas all of the others had less than 99%. It should be noted that the accuracies are lower than some state-of-the-art methods in literature. This is because all of the comparisons were evaluated from scratch without the use of techniques such as data augmentation, pre-trained weights, parameter searches, etc.

5.4 Results on EMNIST

Dataset. Extended MNIST (EMNIST) [8] is an extension of MNIST that includes alphabet characters in addition to digits. For the experiments, we use the "balanced" subset. The subset includes 112,800 training images and 18,800 test images. There are 47 classes, 10 digits and 37 letters. The letters include uppercase and lowercase letters in distinct classes, but with certain letters merged due to ambiguity between upper and lowercase. The merged letters are "C," "I," "J," "K," "L," "M," "O," "P," "S," "U," "V," "W," "X," "Y," and "Z." EMNIST is used because it is similar to MNIST, but is a more difficult problem.

Results. In Table 1, the results for EMNIST are also shown. For EMNIST, CvT and CCT had higher accuracies than the proposed method. However, the proposed method still outperformed the standard ViT. The CvT evaluation performed better than all of the comparison methods.

Fig. 5. Two examples from each class of KMNIST

5.5 Results on KMNIST

Dataset. The final character recognition dataset used is Kuzushiji-MNIST (KMNIST) dataset [5]. KMNIST consists of 10 classes of kurzushiji, or Japanese cursive. The 10 classes represent 10 of the classical Japanese Hiragana characters. An example from each class is shown in Fig. 5. Similar to MNIST, the images are 28×28 grayscale isolated characters. There are 60,000 training images and 10,000 test images. KMNIST is also used as a baseline with a more difficult character recognition task.

Results. KMNIST had the largest discrepancy between the proposed method and ViT. ViT only had an average accuracy of 92.92%, whereas all of Transformers that include convolutions had 94% or higher. The proposed method performed the best at 95.86% accuracy. One possible explanation for the increase in accuracy, especially over the original ViT, is that there are more variations of characters within classes of KMNIST. In Fig. 5, for some characters, there are large deformations. To some extent, adding convolutions provides some translation invariance [21].

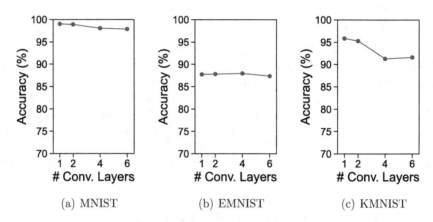

(a) MNIST   (b) EMNIST   (c) KMNIST

Fig. 6. The test accuracy comparing the depth of the CNN. These are the average of 5 trainings.

6   Ablation Study

In order to see the effects of the convolutional layers, we perform an ablation study that examines the change in performance while varying the number of convolutional layers in the ViC layers. The experimental setup is the same as described in Sect. 5. The results of the experiments are shown in Fig. 6. On MNIST and KMNIST, adding additional convolutional layers decreased the accuracy. EMNIST had a slight increase in accuracy at two and four layers, but it decreased again at six. Thus, in general, increasing the number of layers further would not provide additional benefit.

7   Application to General Object Recognition

The proposed method is not limited to only character recognition. It can be extended to any image recognition task. Thus, we demonstrate the ability to use the proposed method on general object recognition.

7.1   Datasets

In order to demonstrate the use of the proposed method on other tasks, we use four datasets. The first two datasets are the CIFAR10 and CIFAR100 datasets [20]. They are natural scene object datasets with 10 classes and 100 classes, respectively. The next dataset is the STL10 dataset [6]. The STL10 dataset is similar to CIFAR10, but contains larger images of 96 × 96 pixels. The final dataset is the FashionMNIST dataset [35]. This dataset contains images in a similar style to MNIST, except instead of digits, it is made of articles of clothing.

Table 2. General Image Average Test Accuracy (%) of Five Trainings

Model	CIFAR10	CIFAR100	FashionMNIST	STL10
ViC (Proposed)	**80.23**	53.14	90.02	**58.21**
ViT	76.60	**55.58**	87.86	57.58
CoAtNet	63.44	32.83	88.30	53.39
CCT	77.89	50.26	89.51	51.90
CvT	77.99	46.33	89.18	46.94
NesT	67.96	38.10	88.34	51.55
PiT	69.80	41.51	**90.19**	51.55

7.2 Results

The results are shown in Table 2. Similar to character recognition, the proposed method performed better overall when compared to ViT and convolutional Transformers. In every case except CIFAR100, the proposed ViC did better than ViT. However, PiT had a higher accuracy than the proposed method on FashionMNIST.

8 Conclusion

In this paper, we presented a new neural network model that combines ViT with a CNN called a Vision Conformer (ViC). Unlike CNNs, ViT has much less image-specific inductive bias [12]. This is because the original transformer was designed for discrete sequences. Therefore, in order to add consideration for image structures and local pixel relationships, we introduced convolutions into ViT.

Specifically, we proposed replacing the MLP inside the Transformer block of ViT with a CNN. In order to be able to do this, the internal representation should be an image-like feature map. Thus, we propose a process of reverse embedding and reconstruction. In the reverse embedding, we perform the opposite operation of the patch embedding process. Namely, the latent token embeddings from MHSA are embedded into a flattened patch-sized space using a trainable linear projection. Next, the embeddings are reshaped into patches and reconstructed into a feature map using spatial concatenation.

We evaluated the proposed method on character recognition. Namely, we demonstrated that the proposed method was effective on the MNIST, EMNIST, and KMNIST datasets. Through the experiments, we demonstrate that the proposed method can improve the classification ability of ViT. In addition, we showed that similar results were found on general image recognition tasks. In the future, work will be done exploring the complexity and improving upon the CNN within the proposed ViC.

References

1. Ba, J.L., Kiros, J.R., Hinton, G.E.: Layer normalization. arXiv preprint arXiv:1607.06450 (2016)
2. Chen, C.F., Fan, Q., Panda, R.: CrossViT: cross-attention multi-scale vision transformer for image classification. arXiv preprint arXiv:2103.14899 (2021)
3. Chen, M., et al.: Generative pretraining from pixels. In: ICML, pp. 1691–1703 (2020)
4. Chu, X., et al.: Conditional positional encodings for vision transformers. arXiv preprint arXiv:2102.10882 (2021)
5. Clanuwat, T., Bober-Irizar, M., Kitamoto, A., Lamb, A., Yamamoto, K., Ha, D.: Deep learning for classical Japanese literature. arXiv preprint arXiv:1812.01718 (2018)
6. Coates, A., Ng, A., Lee, H.: An analysis of single-layer networks in unsupervised feature learning. In: AISTATS, pp. 215–223 (2011)
7. Coccomini, D., Messina, N., Gennaro, C., Falchi, F.: Combining efficientNet and vision transformers for video deepfake detection. arXiv preprint arXiv:2107.02612 (2021)
8. Cohen, G., Afshar, S., Tapson, J., Van Schaik, A.: EMNIST: extending MNIST to handwritten letters. In: IJCNN, pp. 2921–2926 (2017)
9. Dai, Z., Liu, H., Le, Q.V., Tan, M.: CoatNet: marrying convolution and attention for all data sizes. arXiv preprint arXiv:2106.04803 (2021)
10. d'Ascoli, S., Touvron, H., Leavitt, M., Morcos, A., Biroli, G., Sagun, L.: ConViT: improving vision transformers with soft convolutional inductive biases. arXiv preprint arXiv:2103.10697 (2021)
11. Devlin, J., Chang, M.W., Lee, K., Toutanova, K.: BERT: pre-training of deep bidirectional transformers for language understanding. arXiv preprint arXiv:1810.04805 (2018)
12. Dosovitskiy, A., et al.: An image is worth 16x16 words: transformers for image recognition at scale. In: ICLR (2020)
13. Gulati, A., et al.: Conformer: convolution-augmented transformer for speech recognition. In: Interspeech (2020). https://doi.org/10.21437/interspeech.2020-3015
14. Guo, J., et al.: CMT: convolutional neural networks meet vision transformers. arXiv preprint arXiv:2107.06263 (2021)
15. Hassani, A., Walton, S., Shah, N., Abuduweili, A., Li, J., Shi, H.: Escaping the big data paradigm with compact transformers. arXiv preprint arXiv:2104.05704 (2021)
16. He, K., Zhang, X., Ren, S., Sun, J.: Deep residual learning for image recognition. In: CVPR (2016). https://doi.org/10.1109/cvpr.2016.90
17. Hendrycks, D., Gimpel, K.: Gaussian Error Linear Units (GELUs). arXiv preprint arXiv:1606.08415 (2016)
18. Heo, B., Yun, S., Han, D., Chun, S., Choe, J., Oh, S.J.: Rethinking spatial dimensions of vision transformers. arXiv preprint arXiv:2103.16302 (2021)
19. Kingma, D.P., Ba, J.: Adam: a method for stochastic optimization. arXiv preprint arXiv:1412.6980 (2014)
20. Krizhevsky, A.: Learning multiple layers of features from tiny images (2009)
21. Lecun, Y., Bottou, L., Bengio, Y., Haffner, P.: Gradient-based learning applied to document recognition. Proc. IEEE 86(11), 2278–2324 (1998). https://doi.org/10.1109/5.726791

22. Li, M., Hsu, W., Xie, X., Cong, J., Gao, W.: SACNN: self-attention convolutional neural network for low-dose CT denoising with self-supervised perceptual loss network. IEEE Trans. Medical Imaging **39**(7), 2289–2301 (2020). https://doi.org/10.1109/tmi.2020.2968472

23. Liu, Y., Sun, G., Qiu, Y., Zhang, L., Chhatkuli, A., Van Gool, L.: Transformer in convolutional neural networks. arXiv preprint arXiv:2106.03180 (2021)

24. Liu, Z., et al.: Swin transformer: hierarchical vision transformer using shifted windows. In: ICCV (2021). https://doi.org/10.1109/iccv48922.2021.00986

25. Parmar, N., et al.: Image transformer. In: ICML, pp. 4055–4064 (2018)

26. Radford, A., Wu, J., Child, R., Luan, D., Amodei, D., Sutskever, I., et al.: Language models are unsupervised multitask learners. OpenAI Blog **1**(8), 9 (2019)

27. Ridnik, T., Sharir, G., Ben-Cohen, A., Ben-Baruch, E., Noy, A.: ML-Decoder: scalable and versatile classification head. arXiv preprint arXiv:2111.12933 (2021)

28. Sandler, M., Howard, A., Zhu, M., Zhmoginov, A., Chen, L.C.: MobileNetV2: inverted residuals and linear bottlenecks. In: CVPR (2018). https://doi.org/10.1109/cvpr.2018.00474

29. Schmidhuber, J.: Deep learning in neural networks: an overview. Neural Netw. **61**, 85–117 (2015). https://doi.org/10.1016/j.neunet.2014.09.003

30. Tan, M., Le, Q.: EfficientNet: rethinking model scaling for convolutional neural networks. In: ICML, pp. 6105–6114 (2019)

31. Touvron, H., Cord, M., Sablayrolles, A., Synnaeve, G., Jégou, H.: Going deeper with image transformers. In: ICCV (2021)

32. Vaswani, A., et al.: Attention is all you need. In: NeurIPS, pp. 5998–6008 (2017)

33. Wu, H., et al.: CvT: introducing convolutions to vision transformers. In: ICCV (2021)

34. Wu, Y., Ma, Y., Liu, J., Du, J., Xing, L.: Self-attention convolutional neural network for improved MR image reconstruction. Inf. Sci. **490**, 317–328 (2019). https://doi.org/10.1016/j.ins.2019.03.080

35. Xiao, H., Rasul, K., Vollgraf, R.: Fashion-MNIST: a novel image dataset for benchmarking machine learning algorithms. arXiv preprint arXiv:1708.07747 (2017)

36. Xu, M., et al.: End-to-end semi-supervised object detection with soft teacher. In: ICCV (2021)

37. Yuan, K., Guo, S., Liu, Z., Zhou, A., Yu, F., Wu, W.: Incorporating convolution designs into visual transformers. arXiv preprint arXiv:2103.11816 (2021)

38. Yuan, L., et al.: Florence: a new foundation model for computer vision. arXiv preprint arXiv:2111.11432 (2021)

39. Zhai, X., Kolesnikov, A., Houlsby, N., Beyer, L.: Scaling vision transformers. arXiv preprint arXiv:2106.04560 (2021)

40. Zhang, Z., Zhang, H., Zhao, L., Chen, T., Arik, S., Pfister, T.: Nested hierarchical transformer: towards accurate, data-efficient and interpretable visual understanding. In: AAAI (2022)

Modeling Cross-layer Interaction for Chinese Calligraphy Style Classification

Zhigang Li[1], Li Liu[1(✉)], Taorong Qiu[1], Yue Lu[2], and Ching Y. Suen[3]

[1] School of Mathematics and Computer Sciences, Nanchang University,
Nanchang 330031, China
liuli_033@163.com
[2] School of Communication and Electronic Engineering, East China Normal
University, Shanghai 200241, China
[3] Centre for Pattern Recognition and Machine Intelligence, Concordia University,
Montreal H3G 1M8, Canada

Abstract. Chinese calligraphy style classification plays a significant role in Chinese calligraphy study. It is a fine-grained classification problem since the difference among different styles is extremely subtle. We propose a novel convolutional neural network equipped with the cross-layer interaction module to address the issue of Chinese calligraphy style classification in this paper. In our proposed network, a multi-scale attention mechanism is first presented, with which the input image can be characterized at multiple levels. Then we model the interaction between any two layers in the network using Hadamard product. In addition, for each input image, we generate its profile image, which is fed to the network together with the input image. In order to evaluate the effectiveness of the proposed network, we conduct extensive experiments on two datasets. The results show that modeling cross-layer interaction is beneficial for the fine-grained Chinese calligraphy style classification task. The multi-scale attention mechanism can highlight the informative part of the image at multiple scales, which can boost the classification performance. Since the profile image can give clues about the stroke compactness of the characters, it is useful in capturing the subtle difference among different styles. The proposed network achieves the accuracies of 98.62% and 95.92% on the two datasets respectively, which compares favorably with state-of-the-art methods.

Keywords: Chinese calligraphy style classification · Cross-layer interaction · Multi-scale attention · Profile image

1 Introduction

Chinese calligraphy is an important part of Chinese culture. It is renowned all over the world for its beauty and elegance. In recent years, more and more calligraphy works have been scanned and stored in digital libraries like CADAL (China Academic Digital Associative Library). Considering the large quantities of scanned calligraphy works, a concern is raised naturally: How to employ the machine learning techniques to promote the calligraphy-related research?

G. A. Fink et al. (Eds.): ICDAR 2023, LNCS 14190, pp. 70–84, 2023.
https://doi.org/10.1007/978-3-031-41685-9_5

In this paper, we mainly address the issue of Chinese calligraphy style classification, which plays a vital role in Chinese calligraphy study. As stated in [1], the problem of Chinese calligraphy style classification is often confused with Chinese calligraphy font classification. Regarding Chinese calligraphy font, it refers to a broad categorization of Chinese scripts. There are mainly five Chinese calligraphy fonts, viz. seal, clerical, cursive, semi_cursive and standard. However, Chinese calligraphy styles are formed by famous Chinese calligraphers. For instance, there are Yan style, Ou style, Liu style and Zhao style, which are formed by the four famous calligraphers: Yan Zhenqing, Ouyang Xun, Liu Gongquan, and Zhao Mengfu, respectively. The characteristic of each style is largely determined by the personality trait of the corresponding calligrapher. Compared with Chinese calligraphy font, the difference among different Chinese calligraphy styles is subtle which is hard for untrained eyes to discern as can been seen from Fig. 1. Therefore, Chinese calligraphy style classification is a challenging fine-grained image classification problem.

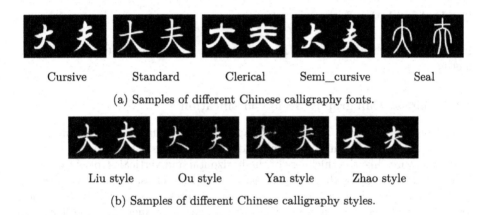

(a) Samples of different Chinese calligraphy fonts.

(b) Samples of different Chinese calligraphy styles.

Fig. 1. Comparison between Chinese calligraphy font and Chinese calligraphy style.

We propose a novel convolutional neural network equipped with the cross-layer interaction module to solve the Chinese calligraphy style classification problem in this paper. As stated above, Chinese calligraphy style classification is a fine-grained classification problem. In this way, it is of great significance to extract discriminative features from the image. A lot of methods have been proposed to deal with the issue of fine-grained image classification. For example, the informative regions are first segmented from the image in [2]. For each region, discriminative features are then extracted. Despite the promising results obtained, segmenting informative regions from the image is a non-trivial task itself. In this paper, we employ the attention mechanism to focus on the informative part of the image. Since different styles may vary at different scales, we propose a multi-scale attention mechanism, with which the input image can be characterized at multiple levels. As indicated in [3–5], it is insufficient to employ a single convolutional network layer to describe the image for fine-grained classification. So we employ multiple layers to depict the image. To be more specific, we fully exploit the interaction between different layers to characterize the image based on Hadamard product. The final image representation is generated by concatenating the interactions between different combinations of two layers. Since different styles are often different in terms of stroke compactness, we generate the profile image for each input image. The profile image is fed to the network together with the input image. In order to validate the effectiveness of the proposed network, we conduct extensive comparison and ablation experiments on two datasets. The proposed network demonstrates promising results, which outperforms state-of-the-art methods.

The remainder of the paper is organized as follows. Related work is summarized in Sect. 2. We detail the proposed method in Sect. 3. Experiments are elaborated in Sect. 4. Section 5 concludes the paper and provides suggestions for future work.

2 Related Works

In this section, we briefly review the works that are most relevant to our proposed method.

2.1 Chinese Calligraphy Style Classification

To address the issue of Chinese calligraphy style classification, early methods adopt handcrafted feature-based methods. For example, Zhang et al. [6] first extract features like the thickness of the horizontal and vertical strokes from the image, and then employ the Bayesian classifier for style classification. In [7], three types of character features, namely, position features, proportion features and projection features, are extracted from the image. Subsequently, these features are fed to the SVM classifier for classification.

Deep learning is introduced to solve the problem of Chinese calligraphy style classification in [8], where three convolutional neural networks (CNN), viz. Local Convolution Neural Network (LCNN), Global Convolution Neural Network (GCNN), Two Pathway Convolution Neural Network (TPCNN), are proposed. In [1], a novel CNN structure is presented, which is equipped with the squeeze-and-excitation (SE) [9] block and Haar wavelet modules. Since SE block only highlights the channel-wise feature maps, it is replaced with the Convolution Block Attention Module (CBAM) [10] in [11], with CBAM focusing not only channel but also the spatial dimension. Liu et al. [12] present a siamese network to deal with the issue of Chinese calligraphy style classification. The siamese network is composed of two streams sharing weights. Each stream is further extended as a classification network. The siamese network is trained to jointly minimize two types of loss: constrastive loss and cross-entropy loss, which are complementary to each other. In [13], a sword-like model called SwordNet is presented to solve the Chinese calligraphy style classification problem. The skip connections are added in the network to enhance the model's generalization ability.

2.2   Fine-Grained Image Classification

Since Chinese calligraphy style classification is a fine-grained classification problem, we present a brief overview of the fine-grained image classification methods found in the literature.

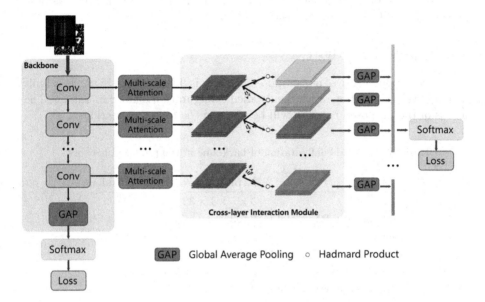

Fig. 2. Architecture of the proposed network.

As indicated in [15], extracting discriminative image features plays a vital role in fine-grained classification. A hierarchical bilinear pooling (HBP) framework is proposed in [15] to solve the fine-grained classification problem. Considering that HBP models feature interaction on the whole image, including the noisy background, it may reduce the discrimnative power of the extracted features. To address this issue, Tan et al. [16] present a model called Hierarchical Bilinear Pooling with Aggregated Slack Mask (HBPASM). It can generate a RoI-aware image feature representation, which outperforms the HBP model. In [17], a channel interaction network (CIN) is proposed. To be more specific, a self-channel interaction (SCI) module is presented for a single image to explore channel-wise correlation within the image. Besides, a contrastive channel interaction (CCI) module is proposed for an image pair to model the cross-sample channel interaction. Ruan et al. [18] propose a Spatial Attentive Comparison Network (SACN), which can deal with the fine-grained classification problem with only a few samples. Three modules are involved in the proposed network, viz. feature extraction module, selective-comparison similarity module, and classification module. A weakly supervised approach for fine-grained image classification is presented in [19]. It also consists of three components: object detection, ObjectMask, and classification. He et al. [20] exploit the layered triplet loss to solve the fine-grained image classification problem. Different from the commonly-used triplet loss that selects samples based on only a single criterion, the loss function is designed with the coarse to fine scheme in this work. In [21], a multilayer feature fusion (MFF) network with parallel convolutional block (PCB) mechanism is proposed. Compared with the original convolutional blocks, PCB has more effective residual connection ability in extracting the region of interest.

3   Proposed Method

The overall architecture of our proposed network is illustrated in Fig. 2. It is built on a lightweight backbone with the detailed configurations shown in Table 1. The details of the proposed network will be given in the following sections.

Table 1. Detailed configuration of backbone in the proposed network.

Layer	Kernel	Stride	Padding
Conv	$5 \times 5 \times 1 \times 32$	1	2
MaxPooling + BN + ReLU	3×3	1	
Conv	$5 \times 5 \times 32 \times 32$	1	2
MaxPooling + BN + ReLU	3×3	1	
Conv	$5 \times 5 \times 32 \times 64$	1	2
MaxPooling + BN + ReLU	3×3	1	
Conv	$5 \times 5 \times 64 \times 32$	1	2
AvgPooling + BN + ReLU	3×3	1	
Global Average Pooling			

3.1   Profile Image Generation

Given an image I with the height and width being H and W respectively, we generate its profile image I' as follows:

$$I' = \mathbb{H}_v \bullet \mathbb{H}_h \tag{1}$$

where $\mathbb{H}_v$ and $\mathbb{H}_h$ denote the vertical and horizontal projections of the image I respectively. $\bullet$ represents the outer product.

For clarity, a sample image is shown in Fig. 3(a). Its vertical and horizontal projections are demonstrated in Fig. 3(b) and Fig. 3(c) respectively. The final profile image is illustrated in Fig. 3(d). One can observe that the profile image is informative which can give some clues about the stroke compactness of the image. So we feed the profile image together with the original image to the network. In detail, the profile image is stacked with the original image along the channel dimension as shown in Fig. 2.

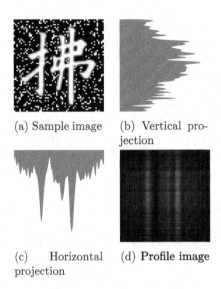

(a) Sample image   (b) Vertical projection

(c) Horizontal projection   (d) **Profile image**

Fig. 3. Generating profile image for a sample image.

3.2 Multi-scale Attention

In order to highlight the informative part of the image, we present a multi-scale attention module as shown in Fig. 4. It can process the input at multiple scales, which is shown to be useful for fine-grained feature learning.

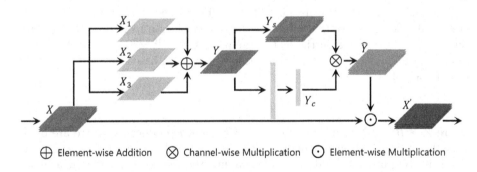

⊕ Element-wise Addition   ⊗ Channel-wise Multiplication   ⊙ Element-wise Multiplication

Fig. 4. The proposed multi-scale attention module.

More specifically, given a feature map $X \in R^{C \times W \times H}$ from a layer of the network, where C, W and H denote the channel dimension, width and height of the feature map respectively, we apply multi-scale convolutional filters with kernel sizes of 1×1, 3×3, 5×5, which generate three output feature maps $X_1 \in R^{1 \times W \times H}$, $X_2 \in R^{1 \times W \times H}$ and $X_3 \in R^{1 \times W \times H}$. Afterwards, they are aggregated as Y through element-wise addition.

To emphasize the discriminative information both in spatial and channel dimensions, we attach Y with two branches, with one branch corresponding to channel attention and the other branch spatial attention. In detail, the spatial attention is achieved by a convolutional operation with C convolutional filters with the kernel size of 1×1. The obtained spatial attention map is denoted as $Y_s \in R^{C \times W \times H}$. Regarding the channel attention, we first stretch Y as a one-dimensional feature, and then feed it to a fully-connected layer to obtain the channel attention map $Y_c \in R^{C \times 1 \times 1}$. Given the spatial attention map Y_s and the channel attention map Y_c, the final attention map $\hat{Y}$ is generated as follows:

$$\hat{Y} = \sigma(Y_s \otimes Y_c) \qquad (2)$$

where $\sigma(\cdot)$ represents the sigmoid function, and $\otimes$ denotes the channel-wise multiplication.

Hence we can apply the attention map $\hat{Y}$ to the feature map X as follows:

$$X' = X \odot \hat{Y} \qquad (3)$$

where $\odot$ is the element-wise multiplication, and X' is the enhanced feature map.

3.3 Cross-layer Interaction

Since different layers of the convolutional neural network play different roles in representing an image, it is usually insufficient to employ a single layer for image representation in the fine-grained classification tasks. Therefore, we employ multiple layers in our work. To be more specific, we make use of the cross-layer interaction in generating image representation.

Formally, since the feature maps from different layers are usually different in spatial resolution, we first exploit a 1×1 convolution to normalize them to the same resolution. Given the feature maps $X \in R^{C \times W \times H}$ and $Y \in R^{C \times W \times H}$ from two different layers, we compute their interaction based on the Hadamard product as follows:

$$Z = X \circ Y \qquad (4)$$

where $Z \in R^{C \times W \times H}$ and $\circ$ denotes the Hadamard product. As there are only a few layers in the proposed network, the interaction between any two different layers is computed. In addition, we compute the interaction between a layer and itself, which has shown to be useful according to preliminary experiments. Afterwards, we apply the global average pooling to the interactions of different combinations of layers. Then the generated features are concatenated to form the final image representation as illustrated in Fig. 2.

During training, we employ two cross-entropy losses to jointly optimize the network as demonstrated in Fig. 2. In greater details, the total loss is formally defined as follows:

$$L_{total} = L_1 + \alpha \, L_2 \qquad (5)$$

where L_1 and L_2 denote the two cross-entropy losses, respectively. α is the weight coefficient which can be employed to balance the two losses.

4   Experiments

In order to validate the effectiveness of our proposed network, we have conducted extensive experiments. The datasets employed in our study are first introduced. Then the ablation experiments are performed to demonstrate the efficacy of each module in the proposed network. Finally, we compare the proposed approach with several state-of-the-art methods.

4.1   Datasets

We test the performance of the proposed network on two datasets, and their details are listed below:

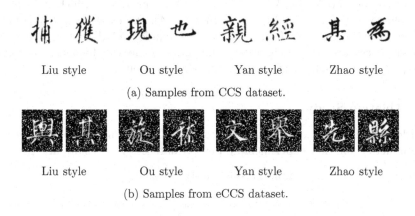

Liu style   Ou style   Yan style   Zhao style

(a) Samples from CCS dataset.

Liu style   Ou style   Yan style   Zhao style

(b) Samples from eCCS dataset.

Fig. 5. Samples from the two datasets.

(1) CCS (Chinese Calligraphy Style) dataset
    This dataset is made publicly available in Zhang et al.'s work [1]. It is composed of 3, 200 character images with four styles: Ou style, Yan style, Liu style and Zhao style. Hence there are 800 images for each style. The size of each image is 64 × 64. All the images are in BMP format. Several samples from this dataset are illustrated in Fig. 5(a).
    (2) eCCS (extended Chinese Calligraphy Style) dataset
    To the best of our knowledge, there are no other public datasets for Chinese calligraphy style classification. As the CCS dataset is small, it may be hard to test the generalization ability of the proposed method. To address this issue, we have expanded the CCS dataset and created the eCCS dataset. To be more specific, we collect more samples for each style. In this way, the number of images for each style has been extended to 1, 800. Afterwards, we apply Gaussian noise

with the mean 0 and variance 0.04 to each image in the dataset to better simulate the real-life scenario. Overall, the newly created eCCS dataset consists of 7,200 images. Several samples from this dataset are illustrated in Fig. 5(b), from which one can observe that this dataset is more challenging compared with the CCS dataset.

For the CCS dataset, we follow the train-validation-test split employed in [1]. The eCCS dataset is split into training, validation and test sets with a ratio of 8 : 1 : 1. We conducted five random splits in this study, and the average classification accuracy was used as the evaluation metric.

4.2 Implementation Details

The network is trained with the Adam optimizer. The learning rate is initially set as 0.01 and is decayed by a factor of 10 after 50 epochs. The network is trained for 60 epochs. The size of the minibatch is 64 and the momentum is set as 0.9. As shown in Eq. (5), a weight coefficient α is introduced to balance the two losses. We set $\alpha = 1.25$ in our study, which has shown to be effective in the preliminary experiments. All the experiments are implemented with the PyTorch framework and we employ NVIDIA GeForce GTX 1080 graphics card for acceleration. The source code will be made available at https://github.com/chnlzg/Calligraphy-classification.

4.3 Ablation Study

In order to verify the effectiveness of each module in our proposed network, we conducted ablation experiments on eCCS dataset. The results are given in

Table 2. Ablation study results.

Backbone	Two losses	Profile image	Multi-scale attention	Cross-layer interaction	Accuracy(%)
✓					90.46
✓	✓				93.97
✓		✓			92.53
✓			✓		93.04
✓				✓	92.85
✓	✓	✓			94.10
✓	✓		✓		94.54
✓	✓			✓	94.38
✓		✓	✓		94.18
✓		✓		✓	93.67
✓			✓	✓	94.08
✓	✓	✓	✓		94.84
✓	✓	✓		✓	94.89
✓	✓		✓	✓	94.86
✓		✓	✓	✓	94.38
✓	✓	✓	✓	✓	**95.92**

Table 2. As shown in this table, employing the backbone only results in unsatisfactory performance. The classification accuracy of the network is increased to 93.97%, 92.53%, 93.04% and 92.85% when the two loss, profile image generation, multi-scale attention, and cross-layer interaction modules are introduced individually. The results indicate that each module is helpful in learning fine-grained features. In addition, the performance of the proposed network is further boosted when more modules are considered. When all the modules are involved, an accuracy of 95.92% was achieved, confirming that all the modules are complementary to each other.

4.4   Effect of the Proposed Multi-scale Attention Model

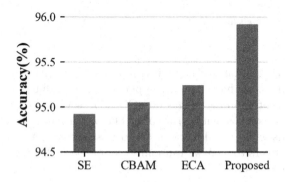

Fig. 6. Comparison of the proposed multi-scale attention model with several well-known attention models.

To show the effectiveness of the proposed multi-scale attention model, we compare it with several well-known attention models including SE, CBAM and ECA [24]. The comparison results are shown in Fig. 6. We can see that the proposed multi-scale attention model outperforms the other attention models, which suggests that highlighting the image at multiple scales is beneficial for improving the fine-grained classification performance.

4.5   Comparison with State-of-the-art Methods

We further compared our proposed network with several state-of-the-art Chinese calligraphy style classification methods. In detail, Dai et al.'s method [8], Zhang et al.'s method [1], Liu et al.'s method [12], Chen et al.'s method [14] and Zhang et al.'s method [11] are employed for comparison. Besides, we compare our proposed approach with two well-known fine-grained image classification methods, namely, Yu et al.'s method [15] and Tan et al.'s method [16]. Moreover, two popular generic image classification methods: Li et al.'s method [22] and Howard et al.'s method [23] are also employed for comparison. The comparison results on the two datasets are shown in Tables 3 and 4, respectively. In addition to the classification accuracy, we also give the number of parameters required in each method.

Table 3. Comparison of the proposed model with many state-of-the-art methods on CCS dataset.

Models	Accuracy(%)	Number of parameters
Dai et al. [8]	93.19	41,560,260
Zhang et al. [1]	97.88	286,448
Liu et al. [12]	82.63	16,328,906
Chen et al. [14]	96.75	11,178,584
Zhang et al. [11]	98.50	289,002
Yu et al. [15]	96.12	73,962,628
Tan et al. [16]	97.12	33,993,604
Li et al. [22]	89.25	24,113,692
Howard et al. [23]	82.87	1,662,492
Proposed method	**98.62**	600,892

Considering that obtaining enough training samples may be difficult under some circumstances, we have tested the performance of different methods with a limited number of training images on eCCS dataset. To be more specific, we reduce the training set size with a rate of r. The performance comparison among different methods with $r = 0.8$ and $r = 0.6$ is shown in Table 5. For instance, when $r = 0.8$, it refers to that we randomly sample 80% of the images from each class on the training set.

Table 4. Comparison of the proposed method with many state-of-the-art methods on eCCS dataset.

Models	Accuracy(%)
Dai et al. [8]	70.43
Zhang et al. [1]	90.22
Liu et al. [12]	81.38
Chen et al. [14]	89.32
Zhang et al. [11]	90.41
Yu et al. [15]	91.74
Tan et al. [16]	90.71
Li et al. [22]	74.21
Howard et al. [23]	68.59
Proposed method	**95.92**

From the tables, the following observations can be made:

(1) The generic image classification methods perform poorly on the two datasets. In contrast, the two fine-grained classification methods perform much better, which confirms that Chinese calligraphy style classification is a fine-grained classification problem. To achieve effective Chinese calligraphy style classification, fine-grained features are required.

(2) Compared with CCS dataset, the performance of all the methods deteriorate on the more challenging eCCS dataset. For example, Zhang et al's method [11] performs well on CCS dataset, yet an accuracy drop of 8.09% is observed on eCCS dataset. Our proposed method achieves promising results on both of the two datasets, which compares favorably with all the methods. Besides, the number of parameters in our proposed network is reasonable.

Table 5. Comparison of the proposed method with many state-of-the-art methods on eCCS dataset with different training set reduction rates.

	Accuracy(%)	
Models	$r=0.8$	$r=0.6$
Dai et al. [8]	72.22	72.08
Zhang et al. [1]	87.50	83.80
Liu et al. [12]	77.22	75.00
Chen et al. [14]	85.82	83.97
Zhang et al. [11]	88.10	85.11
Yu et al. [15]	90.41	88.32
Tan et al. [16]	88.72	88.26
Li et al. [22]	72.09	65.95
Howard et al. [23]	62.45	57.15
Proposed method	**94.54**	**93.23**

(3) From Table 5, we can see that the classification accuracy of all the methods is decreased when the size of the training set becomes small. Yet, our proposed approach can still maintain satisfying performance. For instance, the proposed method achieves an accuracy of 93.23% with $r = 0.6$, greatly outperforming the other methods.

4.6   Error Analysis

In order to gain a better understanding of our proposed method, we present the confusion matrix as shown in Fig. 7. We can see that Ou style suffers from the lowest classification accuracy. It is heavily confused with Zhao style. The possible reason is that Ouyang Xun's calligraphy has a great impact on that of Zhao Mengfu, resulting in the great resemblance of their calligraphy. Furthermore, the second largest confusion lies between Liu style and Ou style. It is mainly because Liu Gongquan has learned a lot from Ouyang Xun. Hence their calligraphy look so much alike which leads to misclassification. Several misclassified samples are shown in Fig. 8.

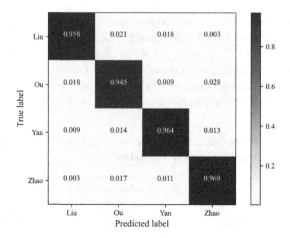

Fig. 7. Confusion matrix of our proposed network.

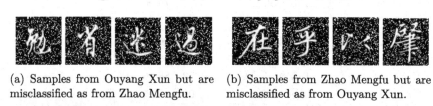

(a) Samples from Ouyang Xun but are misclassified as from Zhao Mengfu.

(b) Samples from Zhao Mengfu but are misclassified as from Ouyang Xun.

(c) Samples from Liu Gongquan but are misclassified as from Ouyang Xun.

(d) Samples from Ouyang Xun but are misclassified as from Liu Gongquan.

Fig. 8. Several misclassified samples.

5 Conclusions

In this paper, we address the issue of Chinese calligraphy style classification, which plays a vital role in Chinese calligraphy study. It is a challenging problem due to that the difference among different styles is subtle. We propose a novel convolutional neural network with the cross-layer interaction module to deal with this fine-grained classification problem. In order to characterize the image at multiple scales, we present a multi-scale attention mechanism. Instead of employing a single layer for image representation, we exploit multiple layers and model the interaction between different layers, which has shown to be useful for fine-grained image classification. Besides, we generate the profile image for each input image, and feed the profile image together with the input image to the network. The network is trained to jointly optimize two cross-entropy losses, which have shown to be complementary to each other. In order to validate the proposed approach, extensive experiments have been conducted on two datasets. The efficacy of all the modules in the proposed network has been demonstrated in the ablation experiments. The proposed approach has achieved the accuracies of 98.62% and 95.92% on the two datasets respectively, which compare favorably with state-of-the-art methods. It can achieve promising results even with a small training set. However, the datasets employed in our study only consist of four calligraphy styles. We will extend the datasets by collecting more styles in the future and will investigate the performance of our proposed method on the extended datasets.

References

1. Zhang, J., Guo, M., Fan, J.: A novel CNN structure for fine-grained classification of Chinese calligraphy styles. Int. J. Doc. Anal. Recogn. **22**(2), 177–88 (2019)
2. Peng, Y., He, X., Zhao, J.: Object-part attention model for fine-grained image classification. IEEE Trans. Image Process. **27**(3), 1487–1500 (2017)
3. He, M., Cheng, Q., Qi, G.: Weakly supervised semantic and attentive data mixing augmentation for fine-grained visual categorization. IEEE Access **10**, 35814–35823 (2022)
4. Guang, J., Liang, J.: CMSEA: compound model scaling with efficient attention for fine-grained image classification. IEEE Access **10**, 18222–18232 (2022)
5. Melnyk, P., You, Z., Li, K.: A high-performance CNN method for offline handwritten Chinese character recognition and visualization. Soft. Comput. **24**(11), 7977–7987 (2020)
6. Zhang, X., Nagy, G.: Style comparisons in calligraphy. In: Document Recognition and Retrieval XIX, pp. 177–186 (2012)
7. Zhang, Y., Liu, Y., He, J., Zhang, J.: Recognition of calligraphy style based on global feature descriptor. In: Proceedings of the International Conference on Multimedia and Expo (ICME), pp. 1–6 (2013)
8. Dai, F., Tang, C., Lv, J.: Classification of calligraphy style based on convolutional neural network. In: Proceedings of the International Conference on Neural Information Processing (ICONIP), pp. 359–370 (2018)
9. Hu, J., Shen, L., Sun, G.: Squeeze-and-excitation networks. In: Proceedings of the International Conference on Computer Vision and Pattern Recognition (CVPR), pp. 7132–7141 (2018)
10. Woo, S., Park, J., Lee, J.-Y., Kweon, I.S.: CBAM: convolutional block attention module. In: Ferrari, V., Hebert, M., Sminchisescu, C., Weiss, Y. (eds.) ECCV 2018. LNCS, vol. 11211, pp. 3–19. Springer, Cham (2018). https://doi.org/10.1007/978-3-030-01234-2_1
11. Zhang, J., Yu, W., Wang, Z., Li, J., Pan, Z.: Attention-enhanced CNN for Chinese calligraphy styles classification. In: Proceedings of the International Conference on Virtual Reality (ICVR), pp. 352–358 (2021)
12. Liu, L., et al.: Multi-loss Siamese convolutional neural network for Chinese calligraphy style classification. In: Proceedings of the International Conference on Neural Information Processing (ICONIP), pp. 425–432 (2021)
13. Li, X., Wang, J., Zhang, H., Huang, Y., Huang, H.: SwordNet: Chinese character font style recognition network. IEEE Access **10**, 8388–8398 (2022)

14. Chen, J., Mu, S., Xu, S., Ding, Y.: HENet: forcing a network to think more for font recognition. In: Proceedings of the International Conference on Advanced Information Science and System (AISS), pp. 1–5 (2021)
15. Yu, C., Zhao, X., Zheng, Q., Zhang, P., You, X.: Hierarchical bilinear pooling for fine-grained visual recognition. In: Ferrari, V., Hebert, M., Sminchisescu, C., Weiss, Y. (eds.) ECCV 2018. LNCS, vol. 11220, pp. 595–610. Springer, Cham (2018). https://doi.org/10.1007/978-3-030-01270-0_35
16. Tan, M., Wang, G., Zhou, J., Peng, Z., Zheng, M.: Fine-grained classification via hierarchical bilinear pooling with aggregated slack mask. IEEE Access 7, 117944–117953 (2019)
17. Gao, Y., Han, X., Wang, X., Huang, W., Scott, M.: Channel interaction networks for fine-grained image categorization. In: Proceedings of the AAAI Conference on Artificial Intelligence, pp. 10818–10825 (2020)
18. Ruan, X., Lin, G., Long, C., Lu, S.: Few-shot fine-grained classification with spatial attentive comparison. Knowl.-Based Syst. 218, 106840 (2021)
19. Chen, J., Hu, J., Li, S.: Learning to locate for fine-grained image recognition. Comput. Vis. Image Underst. 206, 103184 (2021)
20. He, G., Li, F., Wang, Q., Bai, Z., Xu, Y.: A hierarchical sampling based triplet network for fine-grained image classification. Pattern Recogn. 115, 107889 (2021)
21. Wang, L., He, K., Feng, X., Ma, X.: Multilayer feature fusion with parallel convolutional block for fine-grained image classification. Appl. Intell. 52(3), 2872–2883 (2022)
22. Li, X., Wang, W., Hu, X., Yang, J.: Selective kernel networks. In: Proceedings of the International Conference on Computer Vision and Pattern Recognition (CVPR), pp. 510–519 (2019)
23. Howard, A., et al.: Searching for MobileNetV3. In: Proceedings of the International Conference on Computer Vision (ICCV), pp. 1314–1324 (2019)
24. Wang, Q., Wu, B., Zhu, P., Li, P., Zuo, W., Hu, Q.: ECA-Net: efficient channel attention for deep convolutional neural networks. In: Proceedings of the International Conference on Computer Vision and Pattern Recognition (CVPR), pp. 11531–11539 (2020)

Exploring Semantic Word Representations for Recognition-Free NLP on Handwritten Document Images

Oliver Tüselmann[(✉)] and Gernot A. Fink

Department of Computer Science, TU Dortmund University, 44227 Dortmund, Germany
{oliver.tuselmann,gernot.fink}@cs.tu-dortmund.de

Abstract. A semantic analysis of documents offers a wide range of practical application scenarios. Thereby, the combination of handwriting recognizer and textual NLP models constitutes an intuitive solution. However, due to the difficulty of recognizing handwriting and the error propagation problem, optimized architectures are required. Recognition-free approaches proved to be robust, but often produce poorer results compared to recognition-based methods. In our opinion, a major reason for this is that recognition-free approaches do not use largely pre-trained semantic word embeddings, which proves to be one of the most powerful method in the textual domain. To overcome this limitation, we explore and evaluate several semantic embeddings for word image representation. We are able to show that context-based embedding methods are well suited for static word representations and that they are more predictive at word image level compared to classical static embedding methods. Furthermore, our recognition-free approach with pre-trained semantic information outperforms recognition-free as well as recognition-based approaches from the literature on several Named Entity Recognition benchmark datasets.

1 Introduction

Due to the combination of visual and textual properties, the semantic analysis of handwritten document images constitutes both an exciting and challenging field of research. Even though the focus of the Document Image Analysis community has been on visual rather than semantic tasks in the past, the community is steadily shifting towards the semantic analysis and understanding of document images [1, 20, 24, 34–36]. Thereby, classical Natural Language Processing (NLP) tasks like Named Entity Recognition (NER) [1, 38], Named Entity Linking [35] and Question Answering [24, 36] have already been investigated for handwritten document images.

An intuitive approach for realizing NLP tasks on handwritten document images is to combine the advances from the visual and textual domain, using a two-stage model [38]. Thereby, a Handwritten Text Recognizer (HTR) transfers a given document into a textual representation and the outcome is processed by a textual NLP model. Unfortunately, despite advances in machine learning,

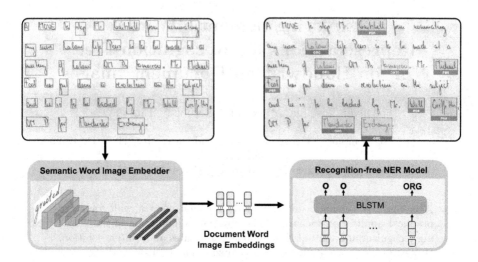

Fig. 1. An Overview of our proposed recognition-free NLP approach on word-segmented handwritten document images with NER as the downstream task.

HTR approaches are still not perfect and can cause many recognition errors [38]. Several publications show that recognition errors have a strong negative impact on the performance of NLP models, mainly caused by error propagation [14,38]. To overcome this limitation, recognition-free end-to-end architectures are favored for documents that are difficult to recognize [24].

Even though recognition-free approaches can alleviate the error propagation problem, they are outperformed by two-stage recognition-based approaches on several semantic tasks [24,38]. In our opinion, this is mainly due to the fundamental drawback of not using pre-trained semantic word embeddings, which is one of the most powerful advantages of the NLP domain [40]. To overcome this limitation, we explore and evaluate which textually pre-trained semantic embeddings from the NLP domain are best suited for representing semantic information in word images. Furthermore, we incorporates these semantic embeddings into a recognition-free NLP framework for handwritten document images (see Fig. 1) and evaluate the performance on several NER datasets.

The remainder of this paper is organized as follows. Section 2 introduces related work in the fields of semantic word embeddings and NER on handwritten document images. In Sect. 3, we present our recognition-free NLP framework and specifically focus on textually pre-trained semantic word embeddings for word image representation. We evaluate these representations and the framework for NER on handwritten document images in Sect. 4. Finally, we summarize our results in Sect. 5.

2   Related Work

This section reviews related work regarding the main concepts used in our proposed recognition-free NLP framework. We provide an overview of syntactic and semantic word embedding methods and show how they are predicted from word image level. We further present related work in the field of NER on document images.

2.1   Word Embeddings

Processing textual words using electronic devices, requires a transformation of these words into numeric representations. Current methods realize such a transformation by using word embeddings. They find their application throughout all NLP tasks and many other domains [31]. Thereby, the use of specialized embedding techniques lead to a significant performance improvement in a wide variety of areas, including NLP [31] and Document Image Analysis tasks [24,32,36]. Even though there are numerous embedding methods, we will only consider semantic and syntactic word embedding approaches in the following.

The majority of semantic word embedding approaches are based on the distributional hypothesis [15]. This hypothesis states that words occurring in similar contexts tend to have similar meanings. Approaches can be roughly divided into static [4,25] and context-based methods [2,8,27]. Static approaches generate embeddings independently of their context and thus map a word always to the same vector representation [4,25]. These methods have the fundamental drawback of ignoring the fact that a word can have various meanings in different contexts. In recent years, several context-based embeddings approaches have been published [2,8,27]. These approaches are trained on language modeling tasks and rely on recurrent neural networks [2,27] or transformer-based architectures [8]. The change from static to context sensitive embeddings led to better results in almost all tasks in the NLP domain [10]. For a detailed overview of semantic word embeddings in the textual domain, see [31].

While semantic information refers to the meaning of a word, syntactic information represents its structural properties. Even though syntactic word embeddings seem to have a minor importance in the field of textual semantic analysis tasks, they are commonly used in the Document Image Analysis domain [32,34,36]. Syntactic word embeddings (e.g. Pyramidal Histogram of Character [3]) are often used in the field of handwritten word images to allow a similarity comparison between a textual query and a word image [3,32,34].

2.2   Word Image Mapping

Currently, methods based on Convolutional Neural Networks (CNNs) are most suitable for obtaining semantic and syntactic word embeddings at the word image level [20,37,39]. A variety of approaches have been published for realizing a syntactic representation on word image level [19,32,39]. Whereas semantic embedding approaches follow a unified strategy by predicting textually pre-trained embeddings for word images [20,34,37,39]. First approaches in this area map word images into a textually pre-trained semantic space by using a two-stage CNN-based approach [37,39]. Thereby, the word images are converted into a feature representation and afterwards mapped into the semantic space. End-to-end approaches are able to outperform two-stage architectures on semantic word image mapping [20,34]. Recently, the realization of a combined syntactic and semantic word image representation has been investigated [20,34].

2.3 Named Entity Recognition

Named Entity Recognition (NER) is a sequence labeling task with a long tradition in NLP [40]. The goal of this task is to extract named entities (e.g. places, person, organizations) from an unstructured text. Traditional approaches mainly rely on handcrafted rules, dictionaries or ontologies [40]. Today, methods using neural architectures outperform traditional ones [2,8,21]. Especially, the combination of a Long Short-Term Memory (LSTM) and a Conditional Random Field (CRF) yields state-of-the-art scores on many benchmarking datasets [21]. Similar to many other NLP tasks, the use of pre-trained semantic word embeddings leads to a considerable performance gain on most benchmarks [2,8]. For a detailed overview of NER in the textual domain, see [40].

There is a wide range of applications in the field of NER on document images. In the following we focus on approaches that work directly on word image level and not on already transcribed text. Publications in this field can be grouped according to their focus on machine-printed [9,14] and handwritten document images [1,30,33,38]. A further categorization of the works can be made on the basis of segmentation-free [6,11] and segmentation-based [1,30,33] approaches. Thereby, segmentation-free approaches work on the entire document image, whereas segmentation-based approaches require a line or word segmentation. A combination of a CNN and an LSTM has proven to be particularly successful for segmentation-based NER approaches [1,30,33]. Furthermore, it has been shown that integrating additional information (e.g. part-of-speech tags) [30] or using an attention mechanism [1] can lead to further improvements in this domain. Tueselmann et al. showed recently, that a two-stage architecture consisting of an HTR and a textual NER model is able to outperform end-to-end approaches on several NER datasets [38].

3   Method

In this section, we present our recognition-free NLP framework for word segmented handwritten document images (see Fig. 1). The approach consists of a textually pre-trained semantic word embedding, a word image mapper and a recognition-free NLP model. Thereby, the word image mapper processes the word images in the order in which they occur on a pre-segmented document image and predicts a semantic word embedding for each of them. Afterwards, these embeddings are transferred to a recognition-free NLP model (e.g. NER), which fulfills the appropriate task. This framework closely follows the two-stage recognition-based approach as proposed in [38], however, we avoid an explicit recognition step and obtain the semantic word representations directly on word image level.

3.1 Semantic Word Embeddings

Semantic word embeddings play an important role in tasks related to text understanding and lead to considerable improvements in almost all areas of NLP [31]. Especially, context-based approaches achieved major performance gains [8,27]. In the field of handwritten document image analysis, however, only static word embeddings have been used so far [20,37,39]. The main reason for this is most probably that already the mapping of context-independent embeddings poses a major challenge [37]. Recently, Ethayarajh showed in [10] that contextualized semantic representations (e.g. BERT) contain powerful types of context-independent embeddings in their first layers. These representations are able to outperform traditional context-independent approaches on many static semantic benchmarks [10]. Given these new insights, we evaluate in this work whether these outcomes can be transferred to the word image domain. Furthermore, we investigate which word embedding approaches from the textual domain are best suited for obtaining a powerful semantic word image representation. In the following, we provide a short overview of word embedding methods that we consider in our evaluation.

For our recognition-free NLP framework, we evaluate static [4,17,26] as well as contextualized [2,8,27] semantic embedding approaches. A classical static method is GloVe [26] which determines its semantic representations by using coincidence statistics between a target word and its context words defined by a fixed context window. This approach has the major disadvantage of being unable to predict embeddings for words that were not part of the training. To overcome this limitation, subword-based approaches like FastText (FT) [4] and BytePair [17] have been published which split words into subwords and combine their embeddings into a single representation. The drawback of static methods is that the word order is not taken into account. Context-based methods are used to encode this type of information. The training of these models focuses on language modeling. First approaches like ELMO [27] and Flair [2] use LSTM-based architectures. A fundamental difference between these two approaches is that Flair processes the textual input purely character based while ELMO uses a mixture of character and static word embeddings. State-of-the-art methods like BERT [8] are based on transformers and subword-based representations. Furthermore, we consider combinations of semantic representations in our evaluation, as they often lead to performance improvements in the textual domain [13].

3.2 Word Image Representation

For obtaining semantic word image representations, we use the same modified ResNet architecture (Attribute-ResNet) as proposed in [34]. The Attribute-

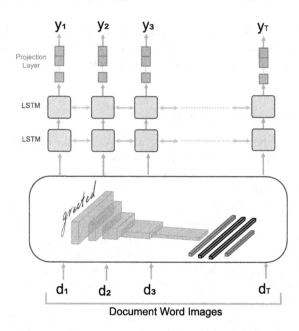

Fig. 2. Our proposed architecture for realizing a robust recognition-free NER system incorporating semantic information.

ResNet uses a ResNet34 architecture [16] for feature extraction, whereby the global average pooling layer at the end of the network is replaced with a Temporal Pyramid Pooling (TPP) layer. The output of the TPP layer is transferred into a three-layered Fully-Connected Network (FCN). This FCN has as many neurons in the last layer as there are dimensions in the word representation to be predicted (e.g. FastText = 300). Except for the final layer, the ReLU activation function is applied to the output of all layers in the network.

3.3 Named Entity Recognition

The NER approach roughly follows the architecture proposed by Toledo et al. [33]. Figure 2 provides an overview of our model. The first step of our approach is the prediction of semantic word image representations for each word image from the document $(d_1, ..., d_T)$. We further capture relations among these representations by using a two-layered Bidirectional-LSTM (BLSTM). Finally, a linear layer is applied to each hidden layer of the BLSTM in order to obtain a named entity tag for each word image $(y_1, ..., y_T)$.

4   Experiments

We evaluate the semantic quality of word embeddings for handwritten word images by using an efficient strategy from the textual NLP domain, which consists of an intrinsic and an extrinsic evaluation [29]. In this context, an intrinsic evaluation involves tasks that are simple and fast to compute and allows inference about the performance on real-world tasks. An extrinsic evaluation, on the other hand, focuses on the actual task (e.g. NER, QA) and is thus more time-consuming.

For our intrinsic and extrinsic experiments, we describe the evaluation datasets, implementation details as well as evaluation protocols. We further present and discuss the results of the two evaluations in this section.

4.1   Datasets

For our experiments, both intrinsic and extrinsic evaluation datasets are required. In order to compare with approaches from the literature, we use the IAM-DB, GNHK and sGMB datasets for our intrinsic evaluation. Similar to [38], we use the IAM-DB, sGMB, and George Washington datasets for our extrinsic evaluation. Moreover, the HW-Synth dataset is used for pre-training the word image mapper.

IAM-DB. The IAM Database [23] is a major benchmark for a variety of handwritten document image tasks. The documents contain modern English sentences written by a total of 657 different people. The database consists of 1539 scanned text pages containing a total of 13353 text lines and 115320 words. Tueselmann et al. manually annotated the dataset with named entity labels and proposed an optimized semantic split into train, validation and test data [38]. There are two versions of this dataset available with different label sets containing 6 and 18 classes.

HW-Synth. The HW-Synth (HW) dataset [18] provides a collection of synthetically rendered word images. The dataset is often used for pre-training handwritten models. The word images are generated by True Type Fonts that resemble handwriting. The vocabulary consists of the 12000 most common words from the English language. For each word, 50 training and 4 test images are generated. The font is randomly sampled from over 300 publicly available fonts.

GNHK. The GoodNotes Handwriting Kollection (GNHK) dataset [22] includes unconstrained camera-captured document images of English handwritten notes. It consists of 687 documents containing a total of 9363 text lines and 39026 words. The official partitioning divides the data into training and test sets with a ratio of 75% and 25%, respectively.

SGMB. The synthetic Groningen Meaning Bank (sGMB) dataset [6] consists of synthetically generated handwritten document pages obtained from the corpus of the Groningen Meaning Bank [5]. The dataset provides unstructured English text and splits the data into 38048 training, 5150 validation and 18183 test word images. The label set consists of the following categories: *Geographical Entity, Organization, Person, Geopolitical Entity and Time indicator.*

George Washington. The George Washington (GW) dataset [28] consists of 20 pages of correspondences between George Washington and his associates dating from 1755. The documents were written by a single person in historical English. The word images are labeled with the following categories: *Cardinal, Date, Location, Organization* and *Person.*

4.2 Implementation Details

The semantic network follows the same training and optimization strategy as described in [34]. To obtain gold standard semantic embeddings for our word images, we used the Flair framework [2]. Thereby, we used the uncased, base model of BERT and the default English models for ELMO, BytePair and GloVe. For the Flair embeddings, the pre-trained forward and backward English models are used and for FastText the Common Crawl English model [12]. Furthermore, the PHOC representation consists of layers $2, 3, 4, 5$ and an alphabet with characters $a - z$ and $0 - 9$. It is important to note that for all embeddings, we have lower-cased the transcriptions and followed the same alphabet as used for PHOC. In our experiments we realize a combined representation of semantic approaches by concatenating their embeddings.

The BLSTM model of our NER architecture uses a hidden layer size of 256 and a dropout of 0.5. For optimization we use the Cross Entropy Loss and the ADAM optimizer. The learning rate is initially set to 0.001 and divided by two whenever the training loss does not decrease in a certain range within 10 epochs. We follow the label smoothing approach proposed by [7]. There is no sentence segmentation and all word images of a document are processed simultaneously.

4.3 Evaluation Protocol

Since we evaluate the use of various textual semantic embeddings for word image representation intrinsically as well as extrinsically, several metrics and protocols are required. For this purpose, we use syntactic and semantic metrics for our intrinsic evaluation and NER task for our extrinsic evaluation.

Intrinsic Evaluation. For an intrinsic evaluation of the word image representation methods introduced in Sect. 3.1, a semantic as well as syntactic metric is required. We use the exact same metrics and protocols as described in [20,34]. Thereby, word spotting [3,20,32] is used as the syntactic and Word Analogy (WA) [25] as the semantic quality measure. Word spotting is a retrieval-based

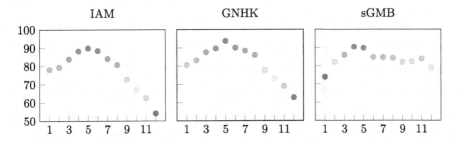

Fig. 3. Inspecting the quality of each layer (1-12) in the BERT model for use as static word embedding on the IAM, GNHK and sGMB dataset. The quality is determined using the WA score measured in accuracy [%].

task, which obtains a ranking of word images from a collection of document images based on its similarity w.r.t. a given query. There exists a variety of different query types with Query-by-Example (QbE) and Query-by-String (QbS) being the most prominent ones. In QbE applications, the query is a word image, whereas in QbS it is a textual string representation. Mean Average Precision (mAP) is the de-facto standard metric for evaluating retrieval tasks.

In the WA task, three words a, b and c are given and the goal is to infer the fourth word d that satisfies the following condition: a is to b as c is to d. We use the collection of human-defined WA examples proposed in [25]. Note, that questions which contain words that are not part of the test corpus of a dataset are excluded from the evaluation. The accuracy of correctly predicted analogies is used as the final semantic evaluation score.

Named Entity Recognition. We use the macro F1-score with the exact same protocol as described in [38]. The F1-score can be interpreted as a weighted average of precision (P) and recall (R) and is formally defined as shown in Eq. 1. In macro F1 the precision, recall and F1-scores are calculated per class and are finally averaged. It is important to note that we exclude the non-entity (O) class in our evaluation.

$$F1 = 2 * \frac{precision * recall}{precision + recall} \tag{1}$$

4.4   Intrinsic Evaluation

We evaluate the capability of various textual semantic word representations to represent semantic in word images. For this purpose, we first present and evaluate our method for extracting static embeddings from context-based approaches. Afterwards, we determine the quality of the semantic word representations introduced in Sect. 3.1 on the gold standard annotations of each dataset using WA. Finally, we evaluate the prediction of semantic word representations at word-image level both semantically and syntactically.

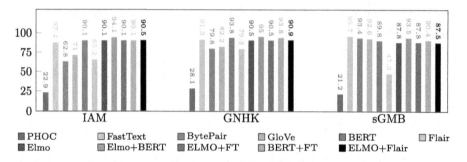

Fig. 4. WA scores for different word embedding methods from the NLP domain on the gold standard annotations of the IAM, GNHK and sGMB datasets. The results are given in accuracy [%].

Table 1. Performances on the four evaluated datasets using accuracy [%] for the WA task (semantic) and mAP for QbE and QbS word spotting (syntactic).

Method	IAM			GW			sGMB			GNHK		
	QbE	QbS	WA	QbE	QbS	WA	QbE	QbS	WA	QbE	QbS	WA
PHOC	91.9	96.2	23.9	96.7	96.8	–	95.7	94.2	20.3	81.5	81.8	28.1
FastText (FT)	86.5	72.0	80.5	95.3	79.5	–	89.7	63.5	75.6	75.2	53.2	70.2
BytePair	87.0	72.2	58.6	94.8	82.2	–	94.2	71.0	85.7	73.4	50.7	60.3
GloVe	87.1	72.2	67.7	96.2	78.6	–	95.0	71.3	85.9	76.9	53.7	66.1
BERT	89.2	74.8	85.1	96.6	81.3	–	95.4	74.6	79.6	77.6	55.3	67.4
Flair	87.4	85.8	49.2	94.7	92.4	–	95.8	82.5	35.8	77.2	67.2	38.4
ELMO	87.5	78.5	86.8	96.4	91.1	–	94.5	75.9	78.6	74.7	58.6	73.6
ELMO + BERT	88.5	78.9	**88.9**	92.7	81.5	–	94.2	76.0	77.9	77.7	61.1	**77.3**
ELMO + FT	87.3	78.6	87.4	96.2	90.5	–	94.4	75.7	77.7	78.1	62.0	76.9
BERT + FT	88.4	74.3	85.1	95.7	83.6	–	95.5	74.2	79.1	78.8	57.8	60.3
ELMO + Flair	90.2	85.0	74.8	96.3	93.7	–	94.6	77.9	54.1	78.0	65.8	55.0

For obtaining static embeddings from context-based approaches, we utilize the findings of Ethayarajh [10] and use the layer from the context-based model that provides the best static characteristics. Figure 3 visualizes the word analogy scores of each layer within the context-based BERT model. The results show that the performances of the individual layers differ considerably. Thereby, the first layers seem to be able to realize a powerful static word representation. Whereby, the fifth layer proves to be most suitable due to its performance on all three datasets. From the fifth layer onwards, the quality decreases and the last layers seem to be rather context-sensitive and thus poorly represent static information.

Figure 4 visualizes the WA scores for our considered semantic word representations on the gold standard annotations of the three datasets introduced in Sect. 4.1. The static embeddings extracted from the context-based approaches (BERT, ELMO) clearly demonstrate improved or similar scores compared to

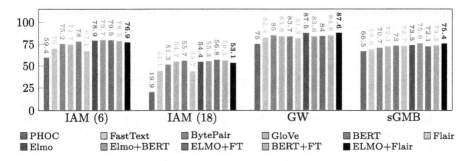

Fig. 5. NER results for predicted word embeddings at the word image level. We report the macro F1 scores [%] for the examined semantic word embeddings on the four evaluation datasets.

the static approaches (FastText, GloVe, BytePair). The combination of semantic embeddings seems to be promising, especially the combination of the ELMO and BERT embedding achieves good results on all datasets. Flair is a purely character-based embedding and leads to comparatively low scores in our evaluation.

Interestingly, the static approaches lead to high WA scores on the sGMB dataset. That is primarily due to the different examples in the WA task for each dataset, since only examples are considered in which the result of the analogy occurs as a word image in the test set. Since the sGMB dataset consists of several news texts, the analogies comprise more than 90% of pure relations between countries and cities. Those relations seem to be very well encoded in the static embeddings (FastText, BytePair, GloVe). This raises the question regarding the usefulness of intrinsic metrics for evaluating semantic quality and whether the focus should rather be on downstream tasks when evaluating semantic representations.

The results obtained for predicting the semantic embeddings at word image level generally follow the trends observed in the WA scores on the textual gold standard data. The BERT and ELMO representations improved the QbS and QbE scores and thus encode better syntactic information. Especially, the ELMO embedding appears to be much more suitable based on its mixture of character and word representation. Flair can achieve high syntactic scores, however, the performance on the semantic evaluation measure is quite low. Similar to the gold standard annotation, the combination of ELMO and BERT is able to achieve high semantic scores on almost all datasets.

4.5 Extrinsic Evaluation

We use the challenging and well-known NER task for our extrinsic evaluation. Figure 5 provides the performances of our intrinsically evaluated semantic embeddings on several NER datasets measured in macro F1 score [%]. The embeddings used so far in the literature for building recognition-free NLP

Table 2. NER performances for the evaluated datasets measured in precision (P), recall (R) and macro-F1 (F1) scores.

Method	IAM (6)			IAM (18)			GW			sGMB		
	P	R	F1	P	R	F1	P	R	F1	P	R	F1
Annotation-NER [38]	87.3	87.6	87.5	68.5	61.0	63.5	96.5	84.7	89.6	81.9	79.2	80.2
HTR-NER [38]	83.3	71.0	76.4	64.8	47.5	53.6	86.9	78.3	81.3	80.1	**72.7**	**75.8**
Rowtula et al. [30]	65.5	47.6	54.6	36.9	28.0	30.3	76.4	59.8	66.6	62.7	58.1	60.1
Toledo et al. [33]	50.2	31.4	37.4	35.4	13.4	18.0	72.5	33.5	45.3	44.3	35.3	38.8
Ours (ELMO+BERT)	**86.4**	**74.6**	**79.7**	**78.1**	**51.2**	**55.3**	**96.2**	79.5	**83.0**	**80.6**	72.0	75.6

approaches (PHOC and FastText) perform rather poorly on these datasets compared to our newly introduced semantic representations. While the Flair embedding can only achieve comparatively low values particularly on the IAM dataset, the BERT and ELMO representations achieve good performances. Especially the combination of semantic embeddings proves to be promising and leads to the highest scores on all datasets. There is a correlation between the WA scores from the intrinsic evaluation and the F1 scores achieved on the NER task, however, it is not possible to generally conclude that a higher WA score leads to improved results on the downstream task.

To compare our recognition-free NER model with approaches from the literature, we use a combination of ELMO and BERT as the semantic representation. The results are shown in Table 2. Thereby, Annotation-NER is a recognition-based approach that works on the gold standard annotations of the datasets and thus reflects the NER performances under perfect recognition. The results show that our approach obtains considerably superior scores compared to the recognition-free approaches from the literature ([30,33]). This demonstrates the importance of using pre-trained semantic information. Moreover, except on the sGMB dataset, our approach is able to outperform the purely recognition-based approach of [38] (HTR-NER). Thereby, our approach obtains a similar performance on the sGMB dataset and the recognition-based approach benefits from low recognition errors due to the synthetic nature of this dataset. In the case of the IAM dataset, it should be noted that the word image mapper was pretrained on the word spotting split of the dataset and thus a potential test set leak could exist.

4.6 Discussion

Further interesting research questions are, what is the best way to incorporate semantic information into our architecture and whether this information is beneficial. For this purpose, we examine three approaches. The first approach is the same as in the previous sections. Here, we train the word mapping network separately from the downstream task and subsequently freeze the pre-trained network while training on the downstream tasks. Thus, the parameters in the Attribute-ResNet are not adjusted during the training process. The second

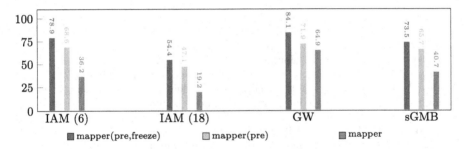

Fig. 6. Examine whether the pre-training (pre) of the image mapper (mapper) is helpful and how to integrate it most effectively into the NLP model.

approach also trains the semantic model separately, however, during training of the downstream task, the parameters of the Attribute-ResNet can be adjusted. The last approach is an end-to-end approach, which does not rely on a semantically pre-trained network and is similar to the approach of [33]. Whereas the Attribute-ResNet is used instead of the PHOCNet [32].

The results clearly show that a pre-training of the Attribute-ResNet is extremely important. Furthermore, the results show that changing the parameters of the Attribute-ResNet during the training of the downstream task is counterproductive. This is probably due to the fact that the datasets are quite small and thus quickly lead to overfitting when the large number of parameters in the ResNet are adjustable.

5   Conclusions

In this work, we present a recognition-free framework for NLP tasks on word-segmented handwritten document images. Our approach focuses on the prediction of textually pre-trained semantic embeddings for word images. For this purpose, we intrinsically evaluated both static and context-based approaches and demonstrate that the context-based approaches and especially their combination are often more suitable than the previously used static embeddings such as FastText. In our extrinsic evaluation on several Named Entity Recognition datasets, we can support the findings from the intrinsic evaluations and show that our approach can outperform both recognition-free as well as recognition-based approaches from the literature.

References

1. Adak, C., Chaudhuri, B.B., Lin, C., Blumenstein, M.: Detecting named entities in unstructured Bengali manuscript images. In: Proceedings International Conference on Document Analysis and Recognition, pp. 196–201. Sydney, Australia (2019)
2. Akbik, A., Blythe, D., Vollgraf, R.: Contextual string embeddings for sequence labeling. In: Proceedings International Conference on Computational Linguistics, pp. 1638–1649. Santa Fe, NM, USA (2018)

3. Almazán, J., Gordo, A., Fornés, A., Valveny, E.: Word spotting and recognition with embedded attributes. IEEE Trans. Pattern Anal. Mach. Intell. **36**(12), 2552–2566 (2014)
4. Bojanowski, P., Grave, E., Joulin, A., Mikolov, T.: Enriching word vectors with subword information. Trans. Assoc. Comput. Linguist. **5**, 135–146 (2017)
5. Bos, J., Basile, V., Evang, K., Venhuizen, N., Bjerva, J.: The Groningen meaning bank. In: Proceedings Joint Symposium on Semantic Processing, pp. 463–496. Trento, Italy (2013)
6. Carbonell, M., Fornés, A., Villegas, M., Lladós, J.: A neural model for text localization, transcription and named entity recognition in full pages. Pattern Recogn. Lett. **136**, 219–227 (2020)
7. Carbonell, M., Villegas, M., Fornés, A., Lladós, J.: Joint recognition of handwritten text and named entities with a neural end-to-end model. In: International Workshop on Document Analysis Systems, pp. 399–404. Vienna, Austria (2018)
8. Devlin, J., Chang, M., Lee, K., Toutanova, K.: BERT: pre-training of deep bidirectional transformers for language understanding. In: Annual Conference of the North American Chapter of the Association for Computational Linguistics, pp. 4171–4186. Minneapolis, MN, USA (2019)
9. Ehrmann, M., Romanello, M., Bircher, S., Clematide, S.: Introducing the CLEF 2020 HIPE shared task: Named entity recognition and linking on historical newspapers. In: European Conference on Information Retrieval, pp. 524–532. Lisbon, Portugal (2020)
10. Ethayarajh, K.: How contextual are contextualized word representations? Comparing the geometry of BERT, ELMo, and GPT-2 embeddings. In: Proceedings of the Conference on Empirical Methods in Natural Language Processing, pp. 55–65. Hong Kong (2019)
11. Fornés, A., et al.: ICDAR2017 competition on information extraction in historical handwritten records. In: Proceedings International Conference on Document Analysis and Recognition, pp. 1389–1394. Kyoto, Japan (2017)
12. Grave, E., Bojanowski, P., Gupta, P., Joulin, A., Mikolov, T.: Learning word vectors for 157 languages. In: Proceedings International Conference on Language Resources and Evaluation. Miyazaki, Japan (2018)
13. Gupta, P., Jaggi, M.: Obtaining better static word embeddings using contextual embedding models. In: Joint Conference of the Annual Meeting of the Association for Computational Linguistics and the International Joint Conference on Natural Language Processing, pp. 5241–5253. Bangkok, Thailand (2021)
14. Hamdi, A., Jean-Caurant, A., Sidère, N., Coustaty, M., Doucet, A.: Assessing and minimizing the impact of OCR quality on named entity recognition. In: International Conference on Theory and Practice of Digital Libraries, pp. 87–101. Lyon, France (2020)
15. Harris, Z.S.: Distributional structure. Word **10**(2–3), 146–162 (1954)
16. He, K., Zhang, X., Ren, S., Sun, J.: Deep residual learning for image recognition. In: Conference on Computer Vision and Pattern Recognition, pp. 770–778. Las Vegas, NV, USA (2016)
17. Heinzerling, B., Strube, M.: BPEmb: Tokenization-free pre-trained subword embeddings in 275 languages. In: Proceedings International Conference on Language Resources and Evaluation. Miyazaki, Japan (2018)
18. Krishnan, P., Jawahar, C.V.: Generating synthetic data for text recognition. CoRR abs/1608.04224 (2016)
19. Krishnan, P., Jawahar, C.V.: HWNet v2: an efficient word image representation for handwritten documents. Int. J. Doc. Anal. Recogn. **22**, 387–405 (2019)

20. Krishnan, P., Jawahar, C.V.: Bringing semantics into word image representation. Pattern Recogn. **108**, 107542 (2020)

21. Lample, G., Ballesteros, M., Subramanian, S., Kawakami, K., Dyer, C.: Neural architectures for named entity recognition. In: Annual Conference of the North American Chapter of the Association for Computational Linguistics, pp. 260–270. San Diego, CA, USA (2016)

22. Lee, A.W.C., Chung, J., Lee, M.: GNHK: a dataset for English handwriting in the wild. In: Proceedings International Conference on Document Analysis and Recognition, pp. 399–412. Lausanne, Switzerland (2021)

23. Marti, U., Bunke, H.: The IAM-database: an English sentence database for offline handwriting recognition. Int. J. Doc. Anal. Recogn. **5**(1), 39–46 (2002)

24. Mathew, M., Gómez, L., Karatzas, D., Jawahar, C.V.: Asking questions on handwritten document collections. Int. J. Doc. Anal. Recogn. **24**, 235–249 (2021)

25. Mikolov, T., Chen, K., Corrado, G., Dean, J.: Efficient estimation of word representations in vector space. In: International Conference on Learning Representations. Scottsdale, AZ, USA (2013)

26. Pennington, J., Socher, R., Manning, C.D.: GloVe: global vectors for word representation. In: Proceedings of the Conference on Empirical Methods in Natural Language Processing, pp. 1532–1543. Doha, Qatar (2014)

27. Peters, M.E., et al.: Deep contextualized word representations. In: Annual Conf. of the North American Chapter of the Association for Computational Linguistics, pp. 2227–2237. New Orleans, LA, USA (2018)

28. Rath, T.M., Manmatha, R.: Word spotting for historical documents. Int. J. Doc. Anal. Recogn. **9**(2–4), 139–152 (2007)

29. Resnik, P., Lin, J.: The Handbook of Computational Linguistics and Natural Language Processing, chap. 11, pp. 271–295 (2010)

30. Rowtula, V., Krishnan, P., Jawahar, C.V.: PoS tagging and named entity recognition on handwritten documents. In: International Conference on Natural Language Processing. Patiala, India (2018)

31. Sezerer, E., Tekir, S.: A survey on neural word embeddings. CoRR abs/2110.01804 (2021)

32. Sudholt, S., Fink, G.A.: PHOCNet: a deep convolutional neural network for word spotting in handwritten documents. In: Proceedings of the International Conference on Frontiers in Handwriting Recognition, pp. 277–282. Shenzhen, China (2016)

33. Toledo, J.I., Carbonell, M., Fornés, A., Lladós, J.: Information extraction from historical handwritten document images with a context-aware neural model. Pattern Recogn. **86**, 27–36 (2019)

34. Tüselmann, O., Brandenbusch, K., Chen, M., Fink, G.A.: A weighted combination of semantic and syntactic word image representations. In: Proceedings of the International Conference on Frontiers in Handwriting Recognition, pp. 285–299. Hyderabad, India (2022)

35. Tüselmann, O., Fink, G.A.: Named entity linking on handwritten document images. In: International Workshop on Document Analysis Systems, pp. 199–213. La Rochelle, France (2022)

36. Tüselmann, O., Müller, F., Wolf, F., Fink, G.A.: Recognition-free question answering on handwritten document collections. In: Proceedings of the International Conference on Frontiers in Handwriting Recognition, pp. 259–273. Hyderabad, India (2022)

37. Tüselmann, O., Wolf, F., Fink, G.A.: Identifying and tackling key challenges in semantic word spotting. In: Proceedings of the International Conference on Frontiers in Handwriting Recognition, pp. 55–60. Dortmund, Germany (2020)
38. Tüselmann, O., Wolf, F., Fink, G.A.: Are end-to-end systems really necessary for NER on handwritten document images? In: Proceedings of the International Conference on Document Analysis and Recognition, pp. 808–822. Lausanne, Switzerland (2021)
39. Wilkinson, T., Brun, A.: Semantic and verbatim word spotting using deep neural networks. In: Proceedings of the International Conference on Frontiers in Handwriting Recognition, pp. 307–312. Shenzhen, China (2016)
40. Yadav, V., Bethard, S.: A survey on recent advances in named entity recognition from deep learning models. In: Proceedings of the International Conference on Computational Linguistics, pp. 2145–2158. Santa Fe, NM, USA (2018)

OCR Language Models with Custom Vocabularies

Peter Garst[(✉)], Reeve Ingle, and Yasuhisa Fujii

Google, Mountain View, CA 94303, USA
{pgarst,reeveingle,yasuhisaf}@google.com

Abstract. Language models are useful adjuncts to optical models for producing accurate optical character recognition (OCR) results. One factor which limits the power of language models in this context is the existence of many specialized domains with language statistics very different from those implied by a general language model - think of checks, medical prescriptions, and many other specialized document classes. This paper introduces an algorithm for efficiently generating and attaching a domain specific word based language model at run time to a general language model in an OCR system. In order to best use this model the paper also introduces a modified CTC beam search decoder which effectively allows hypotheses to remain in contention based on possible future completion of vocabulary words. The result is a substantial reduction in word error rate in recognizing material from specialized domains.

Keywords: OCR · Language model · Fine tuning

1 Introduction

Optical character recognition (OCR) is a fundamental tool enabling many applications, such as visual search, document digitization, understanding and translating scene text, and support for the visually impaired [4,18,19,22,26,30].

OCR systems range from the very general, supporting arbitrary input in many of the world's writing systems [9,23,27], to the very specific, for example for bank checks [8,15] or license plates [1,11]. The aim of this paper is to describe a system which allows general OCR systems to be quickly and easily configured for specialized tasks at run time, in many cases providing the benefits of a custom system engineered for a specific application at very small cost. The lever for making this change is the language model.

It has been clear for decades that language models can improve OCR results by estimating the prior probability of OCR outputs [5,24]. Originally this was accomplished by postprocessing the output of OCR systems, in effect applying a spelling checker of some sort to the output. More recent systems often integrate language models into their decoders [9,12,20].

Independent of their application in OCR systems, there has been a long history of creating language models with some specificity. Adaptation of speech

G. A. Fink et al. (Eds.): ICDAR 2023, LNCS 14190, pp. 101–115, 2023.
https://doi.org/10.1007/978-3-031-41685-9_7

recognition systems and handwriting recognition systems to individual users has been studied for a long time, and in many cases modifications to the associated language models has been part of that [7,16]. Many of these systems create an adapted language model as a sum or interpolation of two models of the same type.

Standalone language models, used for tasks like question answering or summarization, also benefit from specialization or fine tuning [6,10]. There is, for example, a language model trained to work well on radiology reports [29].

The combination of task specific language models with successful general purpose OCR engines leads to many specialized applications, such as recognition of receipts, invoices, tax forms, medical prescriptions or notes, and many others. Some of the specialized language models used in other applications require extensive training, but in order to provide fast and simple run time configuration the models discussed here require only a vocabulary list and some frequency information for the words in the vocabulary.

The goal of this paper is to define language models and an OCR decoder architecture which efficiently solves the specialization problem for many applications. These are our contributions:

- We define simple language models for words and regular expressions which may contain domain specific vocabularies
- We provide tools to generate these models from domain text, and also allow flexible user configuration
- These models may be quickly added to existing general purpose language models at run time
- We modify the CTC decoder to support these models with a limited kind of lookahead

2   Custom Vocabulary Models

2.1   Baseline System

Test images for this work may be either line images or full page images. In the full page image case, the full recognition system includes some preliminary material which finds text lines in the image and feeds those to a line recognizer. If the test set contains line images, then the line recognizer may consume them directly. Only the line recognizer varies between the baseline and experimental systems, so we will focus on that and treat the line segmentation code as a constant part of the environment.

The baseline line recognizer is a general purpose OCR system using a CTC [14] beam search decoder [12]. The input line image is divided into frames (possibly with an overlapping sliding window), and at each frame the beam search maintains a list of hypotheses, each of which is an assignment of a character label to each preceding frame. In general multiple frames are mapped to one output character, with a special blank label to indicate a transition from a character to an identical character.

Each hypothesis has a score. At each frame, the decoder accepts the optical model score for each possible label in that frame. The decoder generates a new hypothesis for each preceding hypothesis and each possible label for the new frame. The score for the new hypothesis combines the score for the preceding hypothesis; the optical score for the proposed new label; the cost of a character unigram prior; transition costs for new characters, blank labels, and repeated characters; and, in the first frame for a new character, the cost of a character language model. The parameter values are optimized to minimize the character error rate on a development set with a black-box optimization [13].

The decoder maintains a list of the best scoring hypotheses, keeping only the best N, and also pruning those too far away from the best. N is called the beam width, and is typically 30 for the baseline recognizer.

The subject of this paper is the last component, the language model score. The baseline system includes a character based language model which estimates the probability of each possible next character in the search, given the left context of the characters already present in the hypothesis. At the frame in which the CTC search transitions to a new character, a weighted negative log probability of the new character is added to the score. In the baseline system the same pretrained language model is applied to all input.

2.2   Custom Vocabularies

We wish to specify a set of words which are likely to appear in input images, and boost the score of any hypothesis in the beam search which contains one. There are a number of properties of the vocabulary to specify:

- The algorithm supports both literal words and regular expressions.
- The literal words may be case sensitive or insensitive.
- Vocabulary entries may optionally be anchored to the start or end of word.
- Each vocabulary entry has a weight.

There are two reasons for drawing a distinction between literal words and regular expressions. The first is just implementation efficiency. The second is that the scoring algorithms, which we will discuss below, work better for fixed length vocabulary items. In the current system that is just the literal words, but this should be applied to other fixed length regular expressions as well.

As we will see the scoring algorithm includes a number of hyperparameters as well. A vocabulary with these items specified is the essential information a user must supply, along with the input data, to benefit from this algorithm. Vocabulary sizes in tests so far have ranged from a handful of items to a few tens of thousands of items.

The tools used in the tests below can generate the vocabulary from sample text, so in general this should not be a burdensome requirement. The user is free to specify some or all of the vocabulary if there are words of particular importance in an application.

Ideally one would retune the CTC parameters and language model weights after adding a custom vocabulary, but that would not be consistent with the goal of adding new vocabularies at runtime with low latency, and in practice good results appear not to require this.

2.3 Designing Appropriate Vocabularies

In some applications the vocabulary may be clear. In processing prescriptions, the medication names are the words the user is most concerned to recognize correctly. In other applications the user may have a body of data from a specific domain, but the appropriate vocabulary is not clear.

The essential element of designing a vocabulary is choosing appropriate weights for the words. If the user already has a specific vocabulary this is all that is required. Otherwise, we may process a body of text from the domain, finding the common words and calculating their weights, and use some cutoff on the weight values to choose the vocabulary. We use three factors in choosing the weight for a word.

1. The length of the word. The scoring formula adds a value proportional to the length of the word, to give a per-character change in the score, but this does not fully capture the effect of word length. Short words tend to have many more possible confusions in the text than long words, so if short and long words have the same per-character score delta there will be more short false positives than long ones. Thus the weight for longer words should be higher. In the PubMed dataset below "palmitoylation" is a common word which is not easily confusable with others - it benefits from a high weight.
2. The empirical word distribution. A body of text produces an empirical probability distribution for the words. Frequent words should get higher weights.
3. The language model distribution. We have a base language model which can estimate the probability distribution for the next character, given the left context. For each word in a body of text this leads to a language model score for the word.

We have experimented with a number of functions of these factors and settled on a simple form:

$$(c_0 + c_1 \cdot length + c_2 \cdot (frequency/lm\_score)), \tag{1}$$

where the c_n are parameters chosen to minimize the OCR error rate. That is, the weight is higher for long words, and for words which are frequent in the domain text but do not score well in the baseline language model.

We have done a black-box optimization [13] to choose these parameters in a number of data sets and chosen values which work well in a variety of cases, although there is some difference in the optimal values for different kinds of tasks. Ideally one would do a fresh training for each data set, but that may not be feasible with fast run time configuration. In the future weight formulas which account for the coverage by the vocabulary of the target text and the confusability of the vocabulary items may provide more precise choice of the parameters.

2.4   Language Model State

The language model represents regular expressions and literal words separately as finite state machines. Classes available in the OpenFST toolkit [3] represent the machines. The regular expressions in the vocabulary are separately compiled, with final state weights representing the weight of the expression in the vocabulary. These are then combined into a single state machine and optimized. Each state has two scores: one is the weight of the regular expression with that as a valid final state, if any; and the other is the best weight for which the state represents a prefix.

The literal words are handled similarly, with the vocabulary compiled into a trie, or prefix tree. Each node in the trie has two numbers attached: one is the weight of the word which ends at that node, or 0 if there is none; and the other is the weight of the best word for which the node is a prefix. These two state machines, plus the baseline character model and a number of hyperparameters, comprise the custom vocabulary model.

Each hypothesis in the beam search includes a language model state, which depends just on the textual transcription of that hypothesis, not the way it is divided into frames. The essential properties of the state are that it can generate a score, and that appending a new character leads us to a new state. The state includes independent components for the states of the base character model, the regular expressions, and the literal words.

Given a hypothesis which contains a sequence of characters "abcd" in its transcript, we must consider that a vocabulary word may start with the character a, or with the character b, and so on. Thus the decoder state for the literal words, for example, will be a vector of trie states from the trie representing the literal vocabulary. If we use Trie("abcd") to represent the state of the trie we get by traversing the string "abcd" from the start state, then the literal word portion of the decoder state will be [Trie("abcd"), Trie("bcd"), Trie("cd"), Trie("d")]. As we decode many of these strings will be invalid - that is, not a valid prefix for any of the vocabulary words - so the vector will in practice be pretty small. Thus the full decoder language model state contains the state of the character model, whatever that may be; a vector of states from the literal vocabulary trie; and a vector of states from the regular expression state machine:

$$(C, V_L, V_R) \tag{2}$$

where C is the character model state, V_L is the vector of valid trie states, and V_R is the vector of valid regular expression states.

If the model configuration anchors the vocabulary words to the word start position there will be many fewer valid states active at any one time. For the literal words we could also use the Aho-Corasick algorithm [2,17] to generate a single state machine valid for any starting position in the string.

2.5   Scoring the Language Model

Suppose we have a hypothesis in the beam search and we wish to transition to a new character c. We have the language model state for the hypothesis and

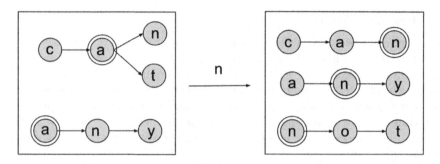

Fig. 1. State transition for the literal word part of the language model state. In this example the vocabulary contains can, cat, any and not. The first box is the literal word state after seeing "ca", and the second box after we accept "n". In the second state "can" contributes to the base score, and all the other words in contention contribute to the best score.

the new character, and we must produce the new language model state and the score, containing both the base score for the material we have seen and the best score representing possible future word completions.

The underlying character model produces a score which is part of both base and best scores, and transitions its part of the state C to a new character model state however it chooses.

For the word part of the state, each element of V_L transitions to a new trie state with the addition of the character c. In some cases the transition will be invalid and the state will drop out; in the other cases the trie generates both a base and a best score. If it is consistent with the model configuration, we may also add a new state to the vector by transitioning from the start state of the trie with the character c. All these actions together generate the word part of the next state.

The base score for this word transition is the optimal value among all the base values for valid trie transitions. It represents the most valuable word in the vocabulary completed by the new character c. Similarly, the best score is the optimal value among all the best scores for the valid transitions, reflecting the most valuable partially completed vocabulary word. These word base and best scores are part of the base and best scores for the whole model. Figure 1 shows an example.

The actions for the regular expression state list is similar. Adding this part in, at the end we have a new language model state, containing a new character model state and new word and regular expression state vectors; and we have base and best scores for all the parts together.

2.6   Dual Criterion Beam Search

We have seen that there are two scores we attach to a hypothesis during the beam search, representing what we have seen and what we hope to see. Because beam search capacity is a limited resource, neither alone is a reliable guide to which hypotheses we should keep. If we rely only on the current scores we may eliminate a vocabulary word because of some poorly formed characters in the middle, reducing the recall on the in vocabulary words in the image. If we rely only on the hopeful scores we may force out some hypotheses with mediocre current scores, and then the hopeful hypotheses may drop out anyway if the data goes in a different direction. Figure 2 visualizes this problem. Thus we keep hypotheses in the beam search if the base score is good, or if the best score is good.

The dual criterion beam search is presented in Algorithm 1.

Algorithm 1. Dual criterion beam search for one frame

Input: hypotheses from the previous frame, and candidate characters for this frame

1. Generate a new set of hypotheses using the scoring algorithm outlined above, based on the hypotheses from the previous frame and the possible characters for the new frame. Each hypothesis will have a base score and a best score.
2. Pick out the top up to N hypotheses, using the base score, subject to a constraint on the width of the beam. These will remain in the beam for the following frame. In the base system these are all the hypotheses kept by the algorithm.
3. Pick out up to M additional hypotheses, based on the best score, subject to the constraint that the best score is no worse than the worst base score hypothesis in the original set of N. The union of these two sets is the set of hypotheses presented to the next frame.

Most of the benefit of the custom vocabulary comes from the language model, but in some cases the dual criterion beam search provides an additional improvement.

2.7   Performance Considerations

This algorithm is useful in the context of a running OCR service, for which users wish to specify at run time a custom vocabulary which applies to some group of input images. As such, there are two latency figures of concern.

The additional computation required to maintain the additional beam hypotheses and to score the finite state machines associated with the custom vocabulary is negligible for common vocabulary sizes compared to the effort required to generate the optical model scores. If at some point greater efficiency becomes important the Aho-Corasick algorithm could be used to simplify and streamline the processing.

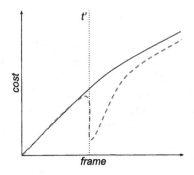
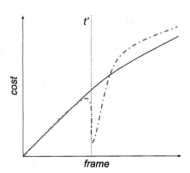
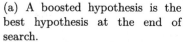

(a) A boosted hypothesis is the best hypothesis at the end of search.

(b) A boosted hypothesis is not the best hypothesis at the end of search.

Fig. 2. Conceptual diagram to show the need of dual criterion beam search. The black solid line is the best hypothesis without the cost bonus. The red and blue dashed lines are hypotheses with a cost boost at frame t'. If we use a single criterion, the decoding can successfully find the best hypothesis for (a) but could fail for (b) because the hypotheses expanded for the boosted hypothesis can easily dominate the beam. The proposed algorithm keeps the top-N and M hypotheses for each case to deal with the problem. (Color figure online)

The more important latency value is for initialization - given a configuration file containing the vocabulary and associated parameters, and a running baseline OCR service, how long does it take until the service is ready to use the model? Constructing the appropriate state machines from a textual representation is straightforward, and times in the range 2 - 10 ms. on common desktop hardware are typical. This initialization time would be amortized over as many images as are used with the custom model.

3   Experimental Results

We explore these algorithms with three data sets, with different characteristics and different levels of information available. None of these data sets is perfect. The PubMed set has the most complete and accurate ground truth information, and is large enough to use better quality statistical tests in validating the algorithm. The other data sets don't have enough information for real statistical rigor, but give at least a qualitative sense of how the algorithm performs in other situations.

3.1   PubMed Research Papers

The PubMed Data Set. This data set is synthetic, based on a set of related biomedical research papers from PubMed, https://pubmed.ncbi.nlm.nih.gov [28].

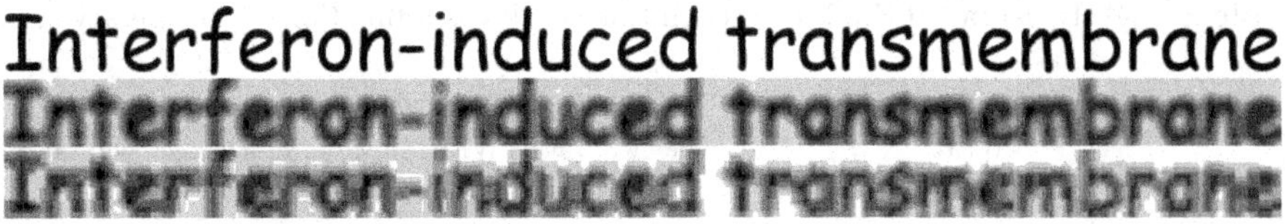

Interferon-induced transmembrane

Fig. 3. Sample images from the PubMed dataset

IFITM3 (Interferon induced transmembrane protein 3, https://www.ncbi.nlm.
nih.gov/gene/10410) [21] is implicated in the immune response to influenza, Sars-
Cov-2, and other viruses, and is an active current research topic. We took 19
research papers from PubMed related to this topic and used the Pango (https://
pango.gnome.org/) [25] typesetter program to generate text line images with random fonts and styles from their text. We then used two different levels of random
degradation on the initial images to generate images more challenging for OCR.
Figure 3 shows typical images generated in this manner.

This synthetic data set is not an exact model for real images one might see
as OCR input, but it provides an excellent experimental platform for judging
the relative efficacy of different algorithms. The original images are too easy for
this test, with word error rates under 0.5% for the baseline OCR system, but
the other two versions were used for much of the development and tuning of the
system. At each degradation level the data set contains 40898 lines and 183619
words.

Table 1. Heavily degraded PubMed recognition results as a function of vocabulary
size

Vocab size	coverage	All text				In vocabulary			
		WER		change	win ratio	WER		change	win ratio
		Baseline	Custom			Baseline	Custom		
200	0.223	7.10	6.34	−10.7%	3.62	6.00	1.96	−67.4%	14.45
400	0.287		6.28	−11.6%	3.74	5.14	1.66	−67.7%	14.48
800	0.365		6.19	−12.9%	4.10	4.51	1.45	−67.8%	14.88
1200	0.424		6.12	−13.9%	4.44	4.29	1.39	−67.6%	14.83

Fig. 4. A win: corrected "First Line of Antivirel Defenso."

PubMed Results. Table 1 shows several figures of merit for these models as
a function of vocabulary size. These results used the more heavily degraded
version of the data set. These figures were generated using a jackknife protocol,
with one paper at a time left out, and the vocabulary automatically generated.

Table 2. Lightly degraded PubMed recognition results as a function of vocabulary size

Vocab size	coverage	All text				In vocabulary			
		WER		change	win ratio	WER		change	win ratio
		Baseline	Custom			Baseline	Custom		
200	0.223	1.65	1.47	−11.0%	4.50	1.32	0.16	−88.2%	22.64
400	0.287		1.46	−11.7%	4.72	1.08	0.14	−86.9%	17.53
800	0.365		1.46	−11.8%	4.52	0.86	0.12	−86.5%	17.10
1200	0.424		1.46	−11.6%	4.11	0.75	0.11	−84.7%	13.62

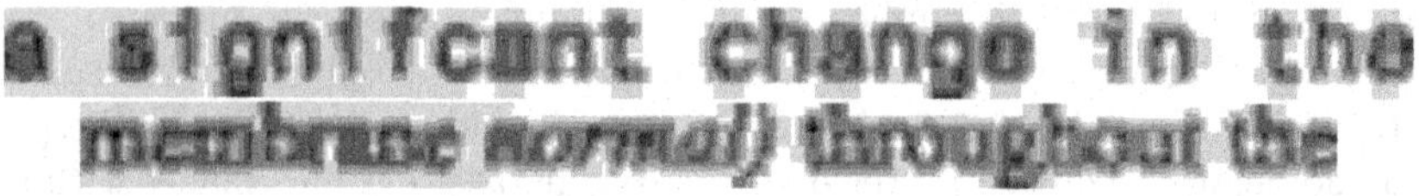

Fig. 5. Losses: changed correct "significant" to "significant" and "normal" to "formal"

At each vocabulary size we see the vocabulary coverage, the relative change in word accuracy, and the win ratio. The coverage is the portion of all the words in the document which are in vocabulary. The win ratio is the ratio of the number of errors corrected by the model to the number of words which were correct in the base model but changed to an incorrect value in the custom vocabulary model. The figure includes these values for the entire documents, and just for the in vocabulary words, as in some applications the accuracy on these items may be important to the user.

Note that the in vocabulary baseline accuracy changes with different vocabulary sizes. This is because the set of in vocabulary words changes from line to line. Each word will get exactly the same baseline results on each line, but the set of included words changes.

The model configuration anchored the vocabulary items to word start, and a word is considered in vocabulary if it benefits from the algorithm. For example, if "with" is in the vocabulary, then "within" is considered an in vocabulary word in the data set.

Table 2 shows the same results with the more lightly degraded version of the data set, with baseline word error rate 1.65. The baseline in vocabulary error rate varies with the vocabulary size, but is generally lower than the overall error rate.

Figures 4 and 5 show some wins and losses on this data set.

3.2   Handwritten Prescriptions

The Prescription Data Set. This data set contains full page images of handwritten prescriptions. A sample is shown in Fig. 6. The ground truth for these images contains only the list of medications mentioned in each prescription, and not the other text elements or the locations of the medication words. This limits

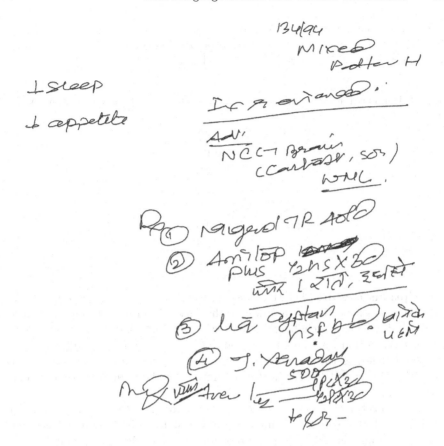

Fig. 6. This prescription contains amitop, migrol, xenadom and cyptan

the available options for analyzing OCR performance on these images, but the Jaccard Index can be used as an important measure of algorithm quality.

The Jaccard Index will be a perfect 1 if the OCR system finds all the medications in a prescription, without introducing any false positives. Given an image, if GT is the set of ground truth vocabulary words in the image, and R is the set of recognized vocabulary words in the image, then the index is

$$J(R, GT) = \frac{|R \cap GT|}{|R \cup GT|} \tag{3}$$

The Jaccard Index has important limitations, but it is an appropriate metric for this application. It provides no information on location or order, but none is required in the output. This data set also has the property that each vocabulary word appears at most once in an image, which is helpful in interpreting the metric.

The vocabulary contains 41844 words, consisting of names of medications, with 6111 of them actually used in at least one image. The data set contains 9647 images, with 47712 in vocabulary words altogether.

Table 3. Handwritten prescription recognition results as a function of vocabulary size

Vocabulary size	Jaccard Index
0	0.193
500	0.201
1000	0.209
5000	0.271
10000	0.257
41844	0.178

Prescriptions Results. For this data set we have for each image a list of the medicine names in the image, and we use the Jaccard Index as the figure of merit for the model. This is a challenging data set, containing images of messy hand-written prescriptions. As noted, this data set contains full page images rather than line images, so the error rates may include segmentation errors as well as recognition errors, but the segmentation material is the same in all experiments so we will ignore it.

This data set allows us to further explore the results of vocabulary size. The initial size is 41844 words. This includes some synonyms and abbreviations, but is larger than any list of commonly prescribed medications - the mobile version of the Physician's Desk Reference, for example, has about 2500 medications. The images actually use 6111 of the medication names.

Table 3 shows the Jaccard Index achieved by this model as a function of vocabulary size. For this experiment, to test vocabulary size V we split the input images into 5 parts. We used a word list generated by the first four parts to create a model for evaluating the fifth part. The list contains vocabulary words actually used in the first 4 parts, filled out to size V by adding random words from the master list. If more than V words are actually used in the first 4 parts, we select them by the weight as discussed above. The values in the table are pooled from evaluating all 5 data slices in this manner.

In this data set we see that including too many words is counterproductive, and better results are obtained by focusing on the common words. We might achieve better results if we had truly accurate frequency information for the whole vocabulary.

3.3 Retail Price Tags

The Price Tags Data. The next data set contains real images of price tags attached to items in a store. The vocabulary for this data is somewhat limited, containing a lot of brand names and a number of common phrases ("for a limited time only"). There are also many elements of the images well described by regular expressions, like monetary amounts or quantities ("32 oz."). The data set contains 101 images with 385 annotated lines.

Price Tags Results. The price tag task is different from the previous ones in several respects.

- The vocabulary covers essentially the whole document.
- Regular expressions are an important part of the vocabulary.
- Because of the small data set size, the training text for vocabulary statistics is the same as the test set text.

Because of the last item, the results are optimistic, but are suggestive of results one could obtain in the field.

The vocabulary included a few hand selected regular expressions; future work may allow us to generate them automatically from sample text, as we do the literal word lists. They are:

- \$ \d+
- \d+\.\d
- \d+ ?(CT | LB | OZ | EA | ML | MG)

Table 4 shows the results. We see that almost all the changes are corrections. This table also shows the effects of the dual criterion beam search. The two beam search columns use the same custom vocabulary language model, but the first uses the baseline single criterion beam search, and the second the dual criterion beam search.

Table 4. Recognition results on price tags

WER			Win ratio
Baseline	Single Beam Search	Dual Beam Search	
15.16	10.93	10.45	24.5

4   Conclusions

We observed that in a variety of data sets from different domains, language models incorporating custom domain specific vocabularies may be leveraged to substantially improve the accuracy of optical character recognition models. The language models support both literal words and regular expressions, and have a number of configuration options which enhance flexibility.

We introduced algorithms which permit these custom models to be automatically derived from a body of text in the target domain. Users may also partially or completely design their own models in special cases, for example if there is a list of key words which the user is particularly concerned to recognize correctly.

We also introduced a modified CTC decoder to support these models which in effect provides in-vocabulary lookahead in order to use information about partially completed as well as complete words to improve accuracy.

The models discussed here introduce no significant overhead to the recognition process, and they may be added to an OCR service at run time with low latency.

Future work will aim to further improve the algorithms for combining the models. It would also be useful to better address low information situations, in effect adapting to specialized input streams rather than designing a model based on prior knowledge about a domain.

References

1. Agbemenu, A.S., Yankey, J., Addo, E.O.: An automatic number plate recognition system using OPENCV and tesseract OCR engine. Int. J. Comput. Appl. **180**(43), 1–5 (2018)
2. Aho, A.V., Corasick, M.J.: Efficient string matching: an aid to bibliographic search. Commun. ACM **18**(6), 333–340 (1975)
3. Allauzen, C., Riley, M., Schalkwyk, J., Skut, W., Mohri, M.: OpenFst: A General and Efficient Weighted Finite-State Transducer Library. In: Holub, J., Žďárek, J. (eds.) CIAA 2007. LNCS, vol. 4783, pp. 11–23. Springer, Heidelberg (2007). https://doi.org/10.1007/978-3-540-76336-9_3
4. Bissacco, A., Cummins, M., Netzer, Y., Neven, H.: Photoocr: reading text in uncontrolled conditions. In: Proceedings of the IEEE International Conference on Computer Vision, pp. 785–792 (2013)
5. Bokser, M.: Omni document technologies. Proc. IEEE **80**(7), 1066–1078 (1992)
6. Caseiro, D., Trancoso, I.: A specialized on-the-fly algorithm for lexicon and language model composition. IEEE Trans. Audio Speech Lang. Process. **14**(4), 1281–1291 (2006)
7. Chen, X., et al.: Recurrent neural network language model adaptation for multi-genre broadcast speech recognition. In: Sixteenth Annual Conference of the International Speech Communication Association (2015)
8. Chin, F., Wu, F.: A microprocessor-based optical character recognition check reader. In: Proceedings of 3rd International Conference on Document Analysis and Recognition, vol. 2, pp. 982–985. IEEE (1995)
9. Diaz, D.H., Qin, S., Ingle, R., Fujii, Y., Bissacco, A.: Rethinking text line recognition models. arXiv preprint arXiv:2104.07787 (2021)
10. Dodge, J., Ilharco, G., Schwartz, R., Farhadi, A., Hajishirzi, H., Smith, N.: Fine-tuning pretrained language models: Weight initializations, data orders, and early stopping. arXiv preprint arXiv:2002.06305 (2020)
11. Du, S., Ibrahim, M., Shehata, M., Badawy, W.: Automatic license plate recognition (ALPR): a state-of-the-art review. IEEE Trans. Circuits Syst. Video Technol. **23**(2), 311–325 (2013). https://doi.org/10.1109/TCSVT.2012.2203741
12. Fujii, Y., Genzel, D., Popat, A.C., Teunen, R.: Label transition and selection pruning and automatic decoding parameter optimization for time-synchronous Viterbi decoding. In: Proceedings of the 13th International Conference on Document Analysis and Recognition, pp. 756–760. IEEE, August 2015

13. Golovin, D., Solnik, B., Moitra, S., Kochanski, G., Karro, J.E., Sculley, D. (eds.): Google Vizier: A Service for Black-Box Optimization (2017). http://www.kdd.org/kdd2017/papers/view/google-vizier-a-service-for-black-box-optimization
14. Graves, A., Fernández, S., Gomez, F., Schmidhuber, J.: Connectionist temporal classification: labelling unsegmented sequence data with recurrent neural networks. In: ICML (2006)
15. Jackel, L.D., Sharman, D., Stenard, C.E., Strom, B.I., Zuckert, D.: Optical character recognition for self-service banking. AT&T Techn. J. **74**(4), 16–24 (1995). https://doi.org/10.1002/j.1538-7305.1995.tb00189.x
16. Jelinek, F., Merialdo, B., Roukos, S., Strauss, M.: A dynamic language model for speech recognition. In: Speech and Natural Language: Proceedings of a Workshop Held at Pacific Grove, California, February 19–22, 1991 (1991)
17. Lee, T.H.: Generalized aho-corasick algorithm for signature based anti-virus applications. In: 16th International Conference on Computer Communications and Networks, pp. 792–797. IEEE (2007)
18. Nagy, G.: Twenty years of document image analysis in pami. IEEE Trans. Pattern Anal. Mach. Intell. **22**(1), 38–62 (2000)
19. Neat, L., Peng, R., Qin, S., Manduchi, R.: Scene text access: a comparison of mobile OCR modalities for blind users. In: Proceedings of the 24th International Conference on Intelligent User Interfaces, pp. 197–207 (2019)
20. Sabir, E., Rawls, S., Natarajan, P.: Implicit language model in LSTM for OCR. In: 2017 14th IAPR International Conference on Document Analysis and Recognition (ICDAR), vol. 7, pp. 27–31. IEEE (2017)
21. Sayers, E.W., et al.: Database resources of the national center for biotechnology information. Nucleic Acids Res. **49**(D1), D10 (2021)
22. Shen, H., Coughlan, J.M.: Towards a real-time system for finding and reading signs for visually impaired users. ICCHP **2**(7383), 41–47 (2012)
23. Smith, R.: An overview of the tesseract OCR engine. In: Ninth International Conference on Document Analysis and Recognition (ICDAR 2007), vol. 2, pp. 629–633. IEEE (2007)
24. Smith, R.: Limits on the application of frequency-based language models to OCR. In: 2011 International Conference on Document Analysis and Recognition, pp. 538–542. IEEE (2011)
25. Taylor, O.: Pango, an open-source unicode text layout engine (2004)
26. Thakare, S., Kamble, A., Thengne, V., Kamble, U.: Document segmentation and language translation using tesseract-OCR. In: 2018 IEEE 13th International Conference on Industrial and Information Systems (ICIIS), pp. 148–151. IEEE (2018)
27. Walker, J., Fujii, Y., Popat, A.C.: A web-based OCR service for documents. In: Proceedings of the 13th IAPR International Workshop on Document Analysis Systems (DAS), Vienna, Austria, vol. 1 (2018)
28. White, J.: Pubmed 2.0. Medical reference services quarterly **39**(4), 382–387 (2020)
29. Yan, A., McAuley, J., Lu, X., Du, J., Chang, E.Y., Gentili, A., Hsu, C.N.: Radbert: adapting transformer-based language models to radiology. Radiol. Artif. Intell. **4**(4), e210258 (2022)
30. Zhu, Y., Yao, C., Bai, X.: Scene text detection and recognition: recent advances and future trends. Front. Comp. Sci. **10**, 19–36 (2016)

A Unified Architecture for Urdu Printed and Handwritten Text Recognition

Arooba Maqsood[1,2]([✉]), Nauman Riaz[1]([✉]), Adnan Ul-Hasan[1],
and Faisal Shafait[1,2]([✉])

[1] National Center of Artificial Intelligence (NCAI), National University of Sciences
and Technology (NUST), Islamabad, Pakistan
{amaqsood.mscs20seecs,nriaz.mscs20seecs,adnan.ulhassan}@seecs.edu.pk
[2] School of Electrical Engineering and Computer Science (SEECS), National
University of Sciences and Technology (NUST), Islamabad, Pakistan
faisal.shafait@seecs.edu.pk

Abstract. Urdu text recognition (handwritten or printed) remains a challenging task due to its diverse writing styles and fonts. State-of-the-art Transformer-based OCR systems are computationally expensive because they rely on computationally expensive pretraining over text images. To address this challenge, we propose a robust architecture that utilizes a custom CNN block with a Transformer encoder for image understanding and a pre-trained Transformer decoder on Urdu language modeling. The presented model generalized well even for scarce training data without the need for pre-training on synthetic text images. Experiments show that our proposed architecture outperforms the state-of-the-art methods for Urdu printed and handwritten text recognition on several publicly available datasets including UPTI, NUST-UHWR, and MMU-OCR-21. We also combined printed and handwriting datasets to train our architecture and propose a single unified model; capable of recognizing both printed and handwritten text for maximum variations of fonts and writing styles with state-of-the-art results.

Keywords: OCR · Urdu Text Recognition · Handwriting Recognition

1 Introduction

Optical Character Recognition (OCR) technology has a rich history and is widely used by different industries and organizations to digitize their data in order to perform data analysis, streamline day-to-day operations and automate their processes [1]. Additionally, this technology is being used by the immigration department to recognize passports, the city traffic police to recognize license plates, and banking institutions to process demand drafts and checks automatically and to preserve and digitize historical writing [15]. The OCR systems can also be used to improve assistive technologies for blind and visually impaired people [6].

Urdu OCR has attracted tremendous research interest over the past ten years. According to the Ethnologue[1], Urdu is among the 10 most spoken languages in the world with over 230 million native speakers. Even though there

---

[1] https://www.ethnologue.com/.

G. A. Fink et al. (Eds.): ICDAR 2023, LNCS 14190, pp. 116–130, 2023.
https://doi.org/10.1007/978-3-031-41685-9_8

is a sizable global audience for the Urdu language and it is written and spoken in many countries, there has been little to no advancement in having its script recognized [10].

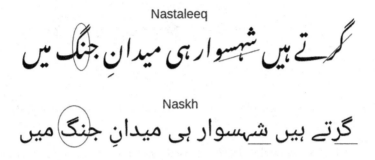

Fig. 1. An example of the Naskh and Nastaleeq scripts depicting the difference in their writing style and alignment. The beginning characters of the Naskh script are aligned along a straight baseline, whereas the characters of the Nastaleeq script have a baseline that is diagonal and shifts from right to left.

The Urdu language has a complex writing script with over 24,000 unique ligatures and different joining rules [13]. One possible reason for the complexity of Urdu script is its writing styles and scripts [15]. Devani, Kofi, Naskh, Nastaleeq, Riqa, and Taluth are only a few of the writing styles used in Urdu. The two most well-known of these are Naskh and Nastaleeq [4]. The majority of the printed material currently in circulation is in the Nastaleeq script, whereas the more prevalent script for digital content is Naskh [6]. Figure 1 illustrates an example sentence to show how these two scripts differ from one another.

It is evident from Fig. 1 that there are notable changes between the two scripts in terms of ligature formation and character shape variation. Due to diverse variations, a system trained on the Nastaleeq script performs poorly in recognizing text in the Naskh [1]. In order to have a larger range of applications, it would be useful to create an Urdu OCR system that can recognize text written in any script. Practically all of the current Urdu OCRs were created for the Nastaleeq scripts and are, therefore, useless when the source images have text in other scripts [6].

Additionally, the handwritten text also poses challenges in its recognition [15]. When it comes to handwriting, humans are extremely inventive, which results in a wide variety of writing styles, character formations, etc. Every person has a unique writing style (refer to Fig. 2), so teaching a model to recognize an unseen handwriting style is a difficult task. Therefore, while addressing text recognition, it is important to take into account several features of handwritten text, including writing styles, the type of paper used, the thickness of strokes, human mistakes, and a number of other issues.

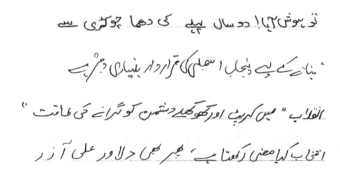

Fig. 2. Sample images of different writing styles for handwritten text.

To the best of our knowledge, there exists no system that can recognize both handwritten and printed real-world Urdu text samples. Many of the current challenges, including speech recognition, machine translation, text summarization, and image captioning, have a single end-to-end solution in deep learning [6]. When developing an end-to-end OCR system, all of the OCR's substitute tasks (i.e. printed text recognition, offline handwriting recognition, etc.) can be contained within a single model.

Printed text is easier to recognize as it is usually well-formed and printed with a consistent font. But on the other hand, handwriting text recognition is a challenging task due to the diverse variability in writing styles. Both forms of text are commonly encountered together in real-world scenarios. This motivates us to develop a unified model that can recognize both forms of text and can improve accuracy and efficiency in text recognition tasks.

Developing a unified model is valuable for digitizing Urdu documents that contain a mix of printed and handwritten text, such as application forms, invoices, receipts, and affidavits. One such system can facilitate a seamless transcription of different Urdu text forms; improving the usability of Urdu language technology for everyday tasks and helping preserve manuscripts and documents, making them easily accessible to a wider audience.

Hence, we aim to propose an end-to-end OCR system optimized as a single, unified entity that will be capable of recognizing both printed and handwritten text for a maximum variation of script and writing styles. The major contributions of our work are:

1. We propose a Transformer based text recognition model that employs a Convolutional Neural Network to extract image embeddings. The whole architecture is trained using CTC loss at the transformer encoder for image understanding and cross-entropy loss at the transformer decoder for language modeling.
2. We propose a unified architecture that gives state-of-the-art results both for printed and handwritten text recognition for the Urdu language.

The paper is mainly divided into the following sections. Section 2 gives a summary of the related work in the field of Urdu text recognition. Section 3 discusses the proposed technique for the task at hand. Section 4 describes the experimental setup including preprocessing steps and implementation details. Section 5 provides the findings and their interpretation. Lastly, Sect. 6 concludes the study and provides future research directions.

2   Related Work

In this section, we present the previous approaches for both Urdu printed and handwritten text recognition.

2.1   Printed Text Recognition

The approaches discussed in [2,3] rely on Recurrent Neural Network (RNN) and Convolutional Neural Network (CNN) blocks for the identification of Urdu ligatures. Naz et al. [2] used the stack of Multi-Dimensional Long Short Term Memory (MDLSTM) layers and feedforward neural networks followed by an output layer for sequence labeling using Connectionist Temporal Classification (CTC) loss for printed Urdu text recognition. Experimentally it was shown that the proposed model achieved an accuracy of 98% on the UPTI dataset. RNNs are widely used in situations when there is a temporal relationship between the inputs, but the nature of this interaction is less clear when it comes to visuals. To address this problem, in another study by Naz et al. [3], the authors proposed a hybrid model that comprises a CNN block followed by the MDLSTM block. The CNN block is added for implicit feature extraction from the images that help learn refined representations from the input image. On the publicly available Urdu Printed Text-line Image (UPTI) dataset, the proposed model had an accuracy of 98.12% for the classification of Urdu ligatures. Despite these changes, the architecture fails to generalize on varying scripts of Urdu. The proposed model lacks an implicit or explicit language model and relies on image signals only.

The authors in [7,9,12,16] developed a combined framework for the detection and recognition of text in video frames and natural scenes. In Mirza et al.'s study [16], text detection in video frames is carried out by fine-tuning Faster R-CNN [5] (model for object detection). Whereas, for text recognition, the authors proposed the UrduNet model (combination of CNN and LSTM blocks). An extensive series of experiments resulted in an 88.3% F1-score for text detection and an 87% recognition rate for text recognition on their custom dataset. In another study for text recognition in video frames, Rehman et al. [9] proposed a simple model that comprises a CNN block that acts as a feature extractor, a bi-directional GRU block for sequence recognition, and finally a classification layer that classifies feature vectors from prior layers into characters. The authors used the AcTiVComp20 and NUST-Urdu Ticker Text (NUST-UTT) [9] datasets for testing the proposed model and achieving encouraging results. Narwani et al. [12]

also focused on text recognition in natural scenes and also proposed an Urdu Scene Text Dataset (USTD) that contains images from real scenes like roads and streets. To prove the validity of their dataset, the authors provided an extensive comparison with baselines for both text detection (including ResNet-50, EAST, and Seglink) and recognition (including variants of CRNN). The end-to-end combination of ResNet-50 with CRNN outperformed with an F1-score of 0.66. Since Urdu Nastaleeq text uses a modified version of Arabic script, there is still a challenge in localizing, detecting, and recognizing it. To further improve the baselines, the authors in [7], enhanced approaches for both text detection and recognition. The Connected Component Analysis (CCA) and Long Short-Term Memory (LSTM) units are used in the initial stage to detect text. The detected text is recognized in the second phase using a hybrid Convolution Neural Network and Recurrent Neural Network (CNN-RNN) architecture. The proposed method performs better than the ones currently in use, with an overall accuracy of 97.47% due to the use of CCA, which has the capability to process higher dimensional data as well.

The different scripts of Urdu have variations in cursive writing styles which pose a challenge in text recognition and most of the systems for Urdu OCR are developed that do not cater to these variations of writing styles. To further aid the research in this direction, the authors of [10] focused on developing a framework that can recognize the text irrespective of its script and writing style. The authors not only created a large-scale multi-font printed Urdu text recognition data set but also presented extensive experimentation using the CNN-based ResNet-18 model with an accuracy of 85 percent. Further addressing the scarcity of multi-font and multi-lines datasets for the Urdu language, the authors of [6] presented a very large 'Multi-level and Multi-script Urdu (MMU-OCR-21)' corpus. The corpus is made up of over 602,472 images in total, including ground truth for text-line and word images in three well-known fonts. Additionally, the authors provided extensive experimentation with text-line and word-level images using a variety of cutting-edge deep learning baselines with encouraging results.

The previous approaches discussed for printed text recognition so far rely only on information from the text image and do not incorporate language modeling. Moreover, to capture the temporal information the authors relied only on networks like RNNs, GRUs, and LSTMs, which suffer from vanishing gradients for very large sequences which is the case in text line recognition. The lack of language understanding and extensive use of RNNs hinders the generalization of these models on varying Urdu text fonts and thus a generalized OCR system cannot be proposed for the Urdu language.

2.2 Handwritten Text Recognition

The authors in [8,11,14] presented analytical approaches based on implicit character segmentation. Similar to the approach in [3], the authors [8] employed a CNN block that works as a feature extractor and an LSTM block for sequence classification of the Urdu characters. The experimentation on custom data of

6,000 unique handwritten text lines gave a character recognition accuracy of 83.69%. The authors proposed to extend their experimentation to other publically available datasets for better generalization of their approach. Similar to the idea of Naz et al. [2], the authors of [14] used a similar approach based on CNN block as a feature extractor and an MDLSTM block as the classifier. The authors provided an extensive comparison of several baseline datasets and achieved satisfying results. Mushtaq et al. [11] also used CNNs as they produce effective results compared to traditional handmade feature extraction methods and do not require explicit feature engineering. The authors achieved a recognition rate of 98.82% on their custom dataset.

Zia et al. [13] demonstrates how convolutional-recursive architecture can be utilized to recognize recursive text effectively. The papers aimed to address the challenges of recognizing the complex ligatures in the Urdu Language by developing a robust architecture based on CNN-RNN blocks aided by an n-gram language model (LM). In this proposed model, the implicit character level segmentation is done using CNN, and RNN acts as the classifier. On top of these blocks, the n-gram language model acts as a spelling corrector. The reported character error rate (CER) for this approach is 5.82%. Additionally, to address the scarcity of Urdu language datasets, the authors also presented a dataset named 'NUST-UHWR'. The authors used a character-level deep learning model on the output of which a word-level n-gram model acts as a spelling corrector. This architecture fails to recognize out-of-vocab words. In order to address these issues, Riaz et al. [15] mapped the problem of handwritten text recognition as a Seq2Seq problem. The authors proposed an encoder-decoder Conv. Transformer model that not only leverages the task at hand by capturing the inter-language dependencies and caters to diverse alignments of characters at the embedding level. Due to the inherent property of how transformers work, the proposed model also learns a language model for language understanding hence eliminating the need for an explicit language model as proposed by Zia et al. [13] and also effectively handles the out-of-vocab words. Riaz et al's [15] approach gave the CER of 5.31% on the publicly available NUST-UHWR [13] dataset. The authors trained the Conv-Transformer architecture from scratch and thus the language model learning for the decoder of the transformer is restricted to text image datasets. The decoder can be pre-trained on an Urdu language modeling task on a large Urdu text dataset like Urdu News Dataset 1M [17].

3   Methodology

Taking our inspiration from [15,20] and addressing the lack of language modeling in current Urdu OCR systems, we propose a composition of CNN along with a transformer architecture as shown in Fig. 3.

In [20], the authors proposed TrOCR, which is a transformer architecture that uses vision transformers as encoders [22,26]. These vision transformers rely on heavy pretraining over synthetic text images before fine-tuning over the respective task of text recognition. The scalability of vision transformers over huge

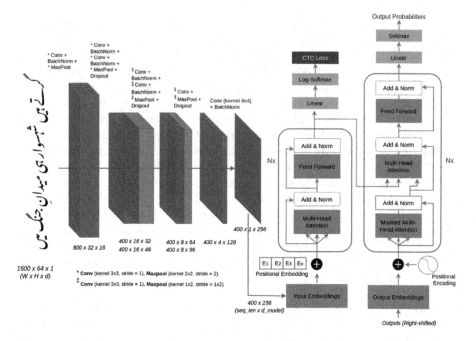

Fig. 3. Overview of the Proposed Architecture. The proposed architecture comprises a custom Convolutinal block coupled with a Transformer encoder for image understanding and a pre-trained Transformer decoder on Urdu language modeling.

datasets gives state-of-the-art results but on the other hand, they struggle in cases where data is not abundant and overfit very quickly lacking generalization. Moreover, due to the n^2 computational complexity of the transformers [19], the image is resized to reduce the spatial resolution leading to reduced training times but on the other hand information loss as well.

To address these issues, we propose a CNN block (as shown in Fig. 3) before the transformer which uses max pooling layers to reduce the spatial resolution of the feature maps. Essentially, the max-pooling layers capture the important features and eradicate the problem of information loss due to resizing of original images. Moreover, the presence of attention layers in the transformer encoder after the convolutional block is used to capture the global context of the text images, and then a transformer decoder is used for language understanding. We use character-level tokenization since the publicly available Urdu text image datasets are smaller in size.

3.1 Convolutional Encoder

We use simple convolutional blocks with max pooling layers to reduce the spatial resolution of feature maps and rely mostly on a transformer encoder for text image understanding. The configuration of CNN is shown in Fig. 3. We follow the standard practices of building convolution blocks using batch normalization and dropout layers. The dimension of the feature map after the convolution block is (W/4, 1, 256) where W is the original image width. This is reshaped to (W/4, 256) and fed into the transformer encoder. The height is reduced to 1 whereas the depth of the feature map is treated as embedding vectors. So we have a sequence length of W/4 and an embedding dimension of 256. These configurations of convolutional blocks give the best results for Urdu text recognition. The sparsity of connections in CNNs leads to better generalization when trained on smaller datasets of text images.

3.2 Transformer Encoder-Decoder

The features extracted from CNN are used as input to the transformer encoder. We use trainable vectors for positional embeddings that are fused with the feature maps at the transformer encoder end. The embeddings generated after the transformer encoder are fed to the decoder as keys and values for cross attention [21] and also through a linear layer that changes the embedding dimension to *vocab* size (refer to Fig. 3). Log Softmax probabilities are generated over the *vocab* dimension and then CTC loss is calculated using the ground truth during training. We use 3 stacks of transformer encoder layers with the embedding dimension of 256 and 8 attention heads for multi-head attention.

We use a pre-trained vanilla transformer decoder on the Urdu language modeling task over the 'Urdu News Corpus 1M Dataset' [17] with sinusoidal position encodings [21]. The configuration of the transformer decoder is the same as the transformer encoder. Softmax probabilities are generated over the vocab dimension and cross-entropy loss is calculated at the decoder end during training. The decoder is trained in a teacher-forcing manner and both the CTC loss and cross-entropy loss are used for backpropagation and training the model. The use of both CTC and cross-entropy Loss leads to effective training of the architecture.

3.3 Data Augmentations

To add diversity and enhance the generalization of the decoder, we use Tiling and Corruption (TACo) technique [24] for data augmentation. As per this technique, tiling is the process of dividing an input image into many, tiny, equal-sized tiles. As part of the corruption stage after tiling, a portion of the tiles is swapped out for corrupted ones. The enhanced image is then created by stitching the tiles back together in the same sequence. A sample instance of this process can be seen in Fig. 4).

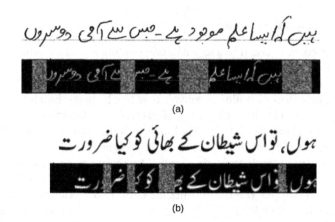

(a)

(b)

Fig. 4. The figure shows samples of input images after applying the TACo augmentations. **(a)** shows a respective example from the NUST-UHWR handwriting dataset. **(b)** shows input sample from UPTI2.0 dataset for printed text.

4   Experimental Setup

Different experiments were carried out for Urdu Text Recognition to provide a comparison with the baselines and the current state-of-the-art. The details of the datasets, implementation and hyper-parameters are discussed below.

4.1   Datasets Used

To evaluate the performance of our proposed model, we utilized both printed and handwritten open-source datasets for the Urdu language. The different datasets used for experimentation in this study include:

Urdu News Dataset 1M [17]: This dataset provides a text corpus of more than 1 million Urdu news articles for four different subject areas: business and economics, science and technology, entertainment, and sports. The dataset is used for pretraining the decoder for URDU language modeling. We use this pre-trained decoder in our proposed architecture for all experiments.

UPTI2.0 [18]: The dataset was formed by collecting sample images from the web, new articles, and books. It contains around 120,000 unique text lines in four different scripts namely Alvi Nastaleeq, Jameel Nori Nastaleeq, Pak Nastaleeq, and Nafees Nastaleeq. This dataset is primarily used for training our architecture and showing generalization capabilities on other datasets like URTI.

URTI [31]: The dataset comprises text lines that have been scanned from printed and calligraphic Urdu magazines, newspapers, poems, and novels. There are 694 text lines from novels, 971 text lines from periodicals, 233 text lines from books, and 282 text lines from poetry. Due to the diversity of fonts in this dataset, it is not included in training and is kept only for testing and inference purposes. It serves well to test the generalization capability of our architecture.

NUST-UHWR [13]: This dataset is obtained from a variety of websites, including social networking and news websites; and contains a total of 10,606 samples of handwritten Urdu text recognition.

MMU-OCR-21 [6]: It is the largest collection of printed Urdu text. The corpus is made up of over 602,472 images in total, including ground truth for text-line and word images in three well-known scripts.

4.2   Implementation Details and Hyperparameters

We evaluate our model separately for printed and handwriting text recognition with state-of-the-art OCR systems at first and then propose a unified training approach to train a single model for inference on both tasks.

For printed text recognition, different configurations of printed text datasets for training and testing are used. At first UPTI 2.0 is utilized for training and URTI for inference. The same dataset splits are used as in [18]. This configuration tests the generalization of different state-of-the-art architectures with ours. Furthermore, our architecture is evaluated on datasets MMU-OCR-2021 with training and testing configurations as in [6]. For unconstrained offline Urdu handwriting recognition, we utilize the NUST-UHWR dataset. The same training and testing splits are employed as presented in [13,15].

Our proposed architecture gives state-of-the-art results for all configurations which inspire us to further test its generalization potential. For this purpose, we combine printed and handwritten text images from UPTI 2.0 and NUST-UHWR respectively for training the architecture. The training is converged against a validation set of equal printed and handwritten text samples from the same datasets. The model is tested against the URTI dataset for printed and NUST-UHWR testing split for handwriting text images. The results are presented and discussed in Sect. 5.

The proposed architecture is trained on RTX 3080 GPU with a batch size of 8. All the text images are resized to (1600×64), keeping the aspect ratio. 'GELU' [32] is used as an activation function throughout the architecture with dropout layers having a probability of 0.1 (refer to Fig. 3). We utilize AdamW [23] optimizer with a learning rate of 3×10^3 for updating the weights. During training, we calculate the CTC and cross-entropy loss on the transformer encoder and decoder output respectively. For inference, the decoder part of the transformer gives the best results after utilizing beam search compared to the transformer encoder.

Table 1. Comparison of CERs of various architectures on URTI [31] to show generalization after training on multi-font subset of UPTI2.0 [18].

Subset	Character Error Rate (CER%)				
	BDLSTM [18]	MDLSTM [3]	CLE Nastaliq [28]	TrOCR [20]	**Proposed Printed**
Magazine	50.69	47.44	50.3	18.4	**11.85**
Book	58.10	56.73	50.12	23.76	**19.23**
Poetry	59.40	58.38	64.55	22.47	**17.17**
Novel	57.70	58.99	38.65	28.75	**21.68**

Table 2. Comparison of CERs of CNN+LSTM based architectures with our proposed model on MMU-OCR-21 dataset [6].

Models	Character Error Rate (CER%)		
	train	val	test
CNN+BLSTM+CTC [6]	0.1	7.4	7.2
VGG-16+BLSTM+CTC [6]	35.5	49.0	49.0
Encoder-Decoder [27]	0.1	7.4	7.3
Proposed Printed	2.0	**6.6**	**6.7**

5   Results and Analysis

We perform extensive experimentation and provide results that establish a new state-of-the-art in Urdu printed and handwritten text recognition. To first test the generalization of our model in comparison with other baselines for printed text (referred as proposed printed), we carry out training on 80,000 images containing each script in equal proportion from UPTI 2.0 that cover 18,000 ligatures [18]. The training converged against a validation set of 10,000 images. Then we benchmarked our model against various other architectures on the URTI dataset.

The results are shown in Table 1. BDLSTM, MDLSTM, and CLE Nastaliq did not generalize well when tested on variations of scripts. The absence of CNN and language modeling in all these models leads to reducing the generalizability and yielding unsatisfactory results. TrOCR is trained from scratch and it quickly overfits during training. Transformers in general require heavy pretraining before fine-tuning over a specific task. This is quite evident from our results.

Our proposed printed model gives the best CER on different scripts compared to other architectures with a significant margin (as given in Table 1). The use of Convolution before the Transformer proves to work best for small datasets, improving generalization.

Next, we benchmark our printed model against various CNN-based architectures (refer to Table 2) on the MMU-OCR-21 dataset [6]. The CNN+BLSTM [6] and VGG16+BLSTM [6] use CTC loss for transcription without any language modeling. Encoder-Decoder [27] architecture comprises a CNN + LSTM encoder

Table 3. Comparison of CER between baselines and our proposed HWR model for handwritten text recognition on NUST-UHWR test split [13].

Models	(CER%)	
	Val	Test
BLSTM [25]	27.39	27.05
Modified CRNN [30]	18.57	19.34
MDLSTM [33]	14.11	19.15
CNN-RNN [29]	13.25	14.12
BiGRU [34]	13.50	13.28
TrOCR [20]	20.12	21.34
Conv. Recursive [13]	7.25	7.35
Conv. Transformer [15]	6.0	6.4
Proposed HWR	**5.9**	**6.2**

Table 4. Performance of our proposed model on URTI dataset [31] for printed text recognition and NUST-UHWR test split [13] for handwriting text recognition after unified training on single dataset.

Dataset		CER(%)
URTI dataset [31]	Magazine	8.11
	Book	13.32
	Poetry	12.39
	Novel	22.34
NUST-UHWR test split [13]		6.6

and an LSTM language modeling decoder. Our proposed printed architecture gives superior results on validation and testing sets in comparison. Higher CER on the training set is due to the TACo augmentations we use during training. The printed model achieves state of the art for Urdu printed text recognition.

For offline handwriting text recognition, we propose to use the same proposed architecture (referred as proposed HWR) using the NUST-UHWR [13] dataset. The same splits were used as in [13,15]. The results are shown in Table 3. Our proposed HWR model performs superior in terms of CER compared to other architectures.

The proposed printed and HWR model gives better generalization when trained on small datasets for both Urdu printed and handwriting recognition. This encourages us to train this model on a unified dataset of printed and hand-written text to define a single model (namely proposed unified) for both tasks. We take the same 80,000 multi-font printed text images from UPTI 2.0 as described in previous experiments and append the UHWR train split of 8483 handwriting text images to create the unified dataset. We train our architecture

Table 5. CER of our proposed unified model on URTI keeping fixed samples from UPTI 2.0 and gradually increasing NUST UHWR samples on each stage for training.

URTI dataset [31]	Character Error Rate (CER%)		
	25% handwriting sample	50% handwriting sample	75% handwriting sample
Magazine	10.12	10.43	9.73
Book	16.73	14.24	13.87
Poetry	16.74	16.32	14.34
Novel	21.69	23.54	23.43

on this dataset and test it against the URTI and UHWR test split. The results are shown in Table 4. The unified training on a single architecture gives state-of-the-art results for both printed and handwriting recognition. A marginal decrease in CER over the URTI dataset can be seen. Additionally, the URTI dataset comprises text lines that have been scanned from machine-printed magazines, newspapers and novels, and calligraphic handwritten Urdu poems, this indicates that the training and testing data distributions are not entirely the same. This leads to significant variation in CERs of different splits of the URTI dataset.

We perform analysis on the unified architecture by keeping the printed text samples fixed and training it on fewer handwriting samples. The handwriting samples are gradually increased for each stage. Each stage is then tested over the URTI dataset as shown in Table 5. URTI consists of a diverse range of fonts including calligraphy. The results in the table show that with an increase in the handwriting dataset for training the model consistently performs better. The diversity of different handwriting styles aids the model in generalizing better to diverse sets of fonts in URTI.

We demonstrate that utilizing both handwriting and printed text for unified training leads to better results for both tasks despite the different distributions in URTI and diverse writing styles in UHWR, our architecture was able to generalize substantially better than current state-of-the-art Urdu OCR models. These findings suggest that a real-world application using such an architecture would produce correct results due to its superior generalization.

6   Conclusion

In this paper, we present an end-to-end Transformer based OCR model for text recognition that utilizes a CNN architecture to extract image embeddings and a pre-trained transformer decoder for language modeling. To the best of our knowledge, we are the first ones to propose a single end-to-end framework that recognizes both printed and handwritten Urdu text images. Vision transformers despite being the new advancement in computer vision suffer from overfitting when it comes to small datasets. They rely on computationally expensive pre-training over synthetic text images and fail if trained from scratch on datasets with moderate sizes. This effect is greatly reduced when we utilize the generalization capabilities of a CNN to extract image embeddings and then pass them through the transformer network to benefit from the attention mechanism. The hyperparameters of our model are tuned for efficient results on smaller datasets. The scalability of our model on larger datasets is yet to be tested which is proposed as a future direction.

References

1. Khan, N.H., Adnan, A.: Urdu optical character recognition systems: present contributions and future directions. IEEE Access **6**, 46019–46046 (2018)
2. Naz, S., Umar, A.I., Ahmed, R., Razzak, M.I., Rashid, S.F., Shafait, F.: Urdu Nasta'liq text recognition using implicit segmentation based on multi-dimensional long short term memory neural networks. In: SpringerPlus, 5(1), pp. 1–16 (2016)
3. Naz, S., et al.: Urdu nastaliq recognition using convolutional-recursive deep learning. Neurocomputing **243**, 80–87 (2017)
4. Naz, S., Hayat, K., Razzak, M.I., Anwar, M.W., Madani, S.A., Khan, S.U.: The optical character recognition of urdu-like cursive scripts. Pattern Recogn. **47**(3), 1229–1248 (2014)
5. Ren, S., He, K., Girshick, R., Sun, J.: Faster R-CNN: towards real-time object detection with region proposal networks. In: Advances in Neural Information Processing Systems, 28 (2016)
6. Nasir, T., Malik, M.K., Shahzad, K.: MMU-OCR-21: towards end-to-end urdu text recognition using deep learning. IEEE Access **9**, 124945–124962 (2021)
7. Umair, M., et al.: A multi-layer holistic approach for cursive text recognition. Appl. Sci. **12**(24), 12652 (2022)
8. Hassan, S., Irfan, A., Mirza, A., Siddiqi, I.: Cursive handwritten text recognition using Bi-directional LSTMs: a case study on urdu handwriting. In: 2019 International Conference on Deep Learning and Machine Learning in Emerging Applications (Deep-ML), pp. 67–72. IEEE (2019)
9. Rehman, A., Ul-Hasan, A., Shafait, F.: High performance Urdu and Arabic video text recognition using convolutional recurrent neural networks. In: Barney Smith, E.H., Pal, U. (eds.) ICDAR 2021. LNCS, vol. 12916, pp. 336–352. Springer, Cham (2021). https://doi.org/10.1007/978-3-030-86198-8_24
10. Rehman, A.U., Hussain, S.U.: Large scale font independent Urdu text recognition system. In: arXiv, preprint: arXiv:2005.06752 (2020)
11. Mushtaq, F., Misgar, M.M., Kumar, M., Khurana, S.S.: UrduDeepNet: offline handwritten Urdu character recognition using deep neural network. Neural Comput. Appl. **33**(22), 15229–15252 (2021)
12. Narwani, K., Lin, H., Pirbhulal, S., Hassan, M.: Towards AI-enabled approach for Urdu text recognition: a legacy for Urdu image apprehension. In: IEEE Access (2022)
13. Zia, N., Naeem, M.F., Raza, S.M.K., Khan, M.M., Ul-Hasan, A., Shafait, F.: A convolutional recursive deep architecture for unconstrained Urdu handwriting recognition. Neural Comput. Appl. **34**(2), 1635–1648 (2022)
14. Husnain, M., et al.: Recognition of Urdu handwritten characters using convolutional neural network. Appl. Sci. **9**(13), 2758 (2019)
15. Riaz, N., Arbab, H., Maqsood, A., Nasir, K.B., Ul-Hasan, A., Shafait, F.: Conv-transformer architecture for unconstrained Off-Line Urdu handwriting recognition. Int. J. Document Anal. Recogn. (IJDAR) **25**, 373–384 (2022)

16. Mirza, A., Zeshan, O., Atif, M., Siddiqi, I.: Detection and Recognition of Cursive Text from Video Frames. In: EURASIP J. Image Video Process. **2020**(1), 1–19 (2020)
17. Hussain, K., Mughal, N., Ali, I., Hassan, S., Daudpota, S.M.: Urdu News Dataset 1M. In: Mendeley Data, 3 (2021)
18. Naeem, M.F., Awan, A.A., Shafait, F., ul-Hasan, A.: Impact of ligature coverage on training practical Urdu OCR systems. In: 2017 14th IAPR International Conference on Document Analysis and Recognition (ICDAR), vol. 1, pp. 131–136. IEEE (2017)
19. Riaz, N., Latif, S., Latif, R.: From transformers to reformers. In: 2021 International Conference on Digital Futures and Transformative Technologies (ICoDT2), pp. 1–6. IEEE (2021)
20. Li, M., et al.: TrOCR: transformer-based optical character recognition with pre-trained models. In: arXiv, preprint arXiv:2109.10282 (2021)
21. Vaswani, A., et al.: Attention is All You Need. In: Advances in Neural Information Processing Systems, 30 (2017)
22. Dosovitskiy, A., et al.: An Image is Worth 16x16 words: Transformers for Image Recognition at Scale. In: arXiv preprint arXiv:2010.11929 (2020)
23. Loshchilov, I., & Hutter, F.: Decoupled Weight Decay Regularization. In: arXiv preprint arXiv:1711.05101 (2017)
24. Chaudhary, K., Bali, R.: Easter2. 0: Improving convolutional models for handwritten text recognition. In: arXiv preprint arXiv:2205.14879 (2022)
25. Ul-Hasan, A., Ahmed, S.B., Rashid, F., Shafait, F., Breuel, T.M.: Offline printed Urdu nastaleeq script recognition with bidirectional LSTM networks. In: 2013 12th International Conference on Document Analysis and Recognition, pp. 1061–1065. IEEE (2013)
26. Bao, H., Dong, L., Piao, S., Wei, F.: Beit: BERT pre-training of Image Transformers. In: arXiv preprint arXiv:2106.08254 (2021)
27. Sutskever, I., Vinyals, O., Le, Q.V.: Sequence-to-sequence learning with neural networks. In: Advances in Neural Information Processing Systems, vol. 27 (2014)
28. Hussain, S., Niazi, A., Anjum, U., Irfan, F.: Adapting tesseract for complex scripts: an example for Urdu nastalique. In: 2014 11th IAPR International Workshop on Document Analysis Systems, pp. 191–195. IEEE (2014)
29. Safarzadeh, V.M., Jafarzadeh, P.: Offline Persian handwriting recognition with CNN and RNN-CTC. In: 2020 25th International Computer Conference, Computer Society of Iran (CSICC), pp. 1–10. IEEE (2020)
30. Shi, B., Bai, X., Yao, C.: An end-to-end trainable neural network for image-based sequence recognition and its application to scene text recognition. IEEE Trans. Pattern Anal. Mach. Intell. **39**(11), 2298–2304 (2016)
31. Shafait, F., Keysers, D., Breuel, T.M.: Layout analysis of Urdu document images. In: 006 IEEE International Multitopic Conference, pp. 293–298. IEEE (2006)
32. Hendrycks, D., Gimpel, K.: Gaussian Error Linear Units (GELUs). In: arXiv preprint arXiv:1606.08415 (2016)
33. Graves, A., Schmidhuber, J.: Offline handwriting recognition with multidimensional recurrent neural networks. In: Advances in Neural Information Processing Systems, vol. 21 (2008)
34. Chen, L., Yan, R., Peng, L., Furuhata, A., Ding, X.: Multi-layer recurrent neural network based offline Arabic handwriting recognition. In: 2017 1st International Workshop on Arabic Script Analysis and Recognition (ASAR), pp. 6–10. IEEE (2017)

Sampling and Ranking for Digital Ink Generation on a Tight Computational Budget

Andrei Afonin[1], Andrii Maksai[2(✉)], Aleksandr Timofeev[1], and Claudiu Musat[2]

[1] EPFL, Lausanne, Switzerland
[2] Google Research, Zürich, Switzerland
amaksai@google.com

Abstract. Digital ink (online handwriting) generation has a number of potential applications for creating user-visible content, such as handwriting autocompletion, spelling correction, and beautification. Writing is personal and usually the processing is done on-device. Ink generative models thus need to produce high quality content quickly, in a resource constrained environment.

In this work, we study ways to maximize the quality of the output of a trained digital ink generative model, while staying within an inference time budget. We use and compare the effect of multiple sampling and ranking techniques, in the first ablation study of its kind in the digital ink domain.

We confirm our findings on multiple datasets - writing in English and Vietnamese, as well as mathematical formulas - using two model types and two common ink data representations. In all combinations, we report a meaningful improvement in the recognizability of the synthetic inks, in some cases more than halving the character error rate metric, and describe a way to select the optimal combination of sampling and ranking techniques for any given computational budget.

1 Introduction

Digital ink (online handwriting) offers users of digital surfaces a way of expression similar to pen and paper. This mode of expression is gaining popularity with the increasing adoption of styluses and digital pens for tablets. In its digital form, ink is a medium that offers rich possibilities for personalized intelligent assistance for creativity and productivity. One direct way of offering the assistance is via ink synthesis, enabling user-facing features such as handwriting autocompletion, spelling correction, beautification, assisted diagramming and sketching.

Making these assistance experiences convenient and comfortable requires maximizing the output quality of the models, while respecting privacy and latency constraints. The same is true of other types of generated content, but standards might be higher in the case of digital ink generation, for example:

– Since assistive handwriting content appears in the same space as the content generated by the user, it's vital that the generated content is readable and not look "out-of-place". The users of generative image models for content creation purposes might

---

A. Afonin—Work done as a student researcher at Google Research, Zürich, Switzerland.

A. Afonin and A. Maksai—These authors contributed equally to this work and share first authorship.

G. A. Fink et al. (Eds.): ICDAR 2023, LNCS 14190, pp. 131–146, 2023.
https://doi.org/10.1007/978-3-031-41685-9_9

be more forgiving to model mistakes, because there the model assists in the creative process where the users don't necessarily know what exactly they are looking for.

- Personalized assistive handwriting often requires the models to observe the user's handwriting and transfer that style to the generated output. Unlike other modalities, handwriting is a personally-identifiable data. Therefore, it is important for the models to run on-device, rather than server-side.
- Generating suggestions (for example when doing autocompletion in handwriting) requires the models to be fast enough to produce their suggestions before the user has moved on or decided to add new content themselves. When the content is produced too slowly, it gets in the way of the user's flow rather than helping. This problem is further exacerbated by the constraint that the models run on-device.

In this work, we aim, given a trained generative model of digital ink and a computation budget, to produce readable outputs as often as possible, under the assumption that the model is going to be run on-device. To achieve this goal, we consider two classes of approaches that work well together.

Sampling. This constrained ink modelling problem resembles text and audio generation. Following the work that has been done there [3,6,19,22,36], we first concentrate on using perturbed probability distributions for sampling from autoregressive models. This improves the quality within a single inference call, by picking a sampling technique that minimizes the number of repetitive or incoherent samples. Examples of generated digital ink can be found in Fig. 3.

Ranking. We additionally train ranking models to predict the recognizability of an ink. We employ these models by first generating a diverse set of candidates and then ranking them to select the best output. This improves the quality if the time budget allows for multiple inference calls.

Our proposed ranking approach would actually work for any binary quality measure (like thresholded L_2 distance in the style embedding space for style transfer [9] or edit-aware Chamfer distance for spelling correction [26]), but we focus on recognizability, since likely for any application of digital ink synthesis, the output should be recognizable.

Our contributions are as follows[1]:

- We use sampling and ranking techniques for digital ink generation, and perform an ablation study on the ranking model objective, training, and tuning. To our knowledge, ours is the first work on this topic in the digital ink space.
- We show that selecting appropriate sampling parameters improves the quality of the output significantly compared to the typically used baselines, across multiple datasets, model types, and data representations.
- We show that ranking further improves the quality, and discover that depending on the computational budget, the highest quality ranking models may not lead to optimal quality. We provide practical way of selecting the ranking model.

---

[1] A notebook accompanying this submission that can run inference on example models for each dataset, data representation, and model type, and includes test label sets, is available here: https://colab.research.google.com/drive/1AkwmDOkEIkifbOYEBdcB9PrR_Ll-fcmz.

2   Related Work

Errors in Autoregressive Generative Models. Autoregressive generative models often generate samples with artifacts [19]. Artifacts appear when the generation process gets stuck in either high- or low-probability regions of the sampling space, and results in two types of errors, overconfidence (usually manifested as repeated tokens) [4] and incoherence errors, respectively. We show examples of such errors during Digital Ink generation process in Fig. 3. This is also known as the likelihood trap [32] and stems from exposure bias [18], which is difference between training done with 'teacher forcing' and inference [5].

Sampling. One common way of finding the trade-off between overconfidence and incoherence errors, often used in Text-to-Speech (TTS) and Natural Language Processing (NLP), is sampling [4], which modifies the distribution from which the points in the autoregressive model are sampled. Sampling from original distribution is called ancestral sampling; popular sampling techniques that extend it include Top-K [13] and Top-P, or nucleus [19] sampling. Originally introduced for text generation, they propose picking a word from the distribution of the top most likely next words, limited by either number (in Top-K) or cumulative probability (in Top-P). Variations of the sampling techniques above include Typical sampling [27], which selects components closest to a dynamically selected probability, Mirostat sampling [4], which select K in Top-K sampling adaptively, and Beam search [30].

Ranking Models. Another way to improve the generation quality is to generate several samples and choosing the best one among them. This is frequently done in information retrieval domains such as question answering [23], text summarization [29], and code generation [36]. Approaches most similar to ours are the ones that use ranking models for conditional generative modeling. In [22], the ranking model is trained to predict the best text continuation, with positive samples coming from real text and negative samples coming from different parts of the text and model-generated continuations. In [6], two ranking models are trained to predict the match between the generated audio and the target label, as well as between the generated audio and the source audio used for style extraction. They are combined with weights specified by the user, to rank audio generated with specific style.

Handwriting Synthesis. Two of the most popular models for digital ink generation are multi-layer LSTMs with monotonic attention over the label [15] (also known in TTS as Tacotron [35]) and the encoder-decoder Transformer architecture [34]. Other architectures include VRNN [11] used in [2], Neural ODEs [12], and Diffusion models [25].

These architectures underpin applications such as sketch generation [17] and completion [31], style transfer [21], beautification [2], spelling correction [26], and assisted diagramming [1].

Metrics for evaluating the quality of digital ink generative models of text typically include Character Error Rate for text generation readability [2,9,21], writer identification for style transfer [21], and human evaluation [2,7,21].

Most digital ink generation approaches use either ancestral sampling or greedy sampling, with exception of [10], which uses biased sampling [15] for the task of generating the synthetic training data.

To our knowledge, no studies on the effects of sampling and ranking for digital ink generation have been performed. Similarly, no studies have looked at the relationship between the generation speed and quality.

3 Method

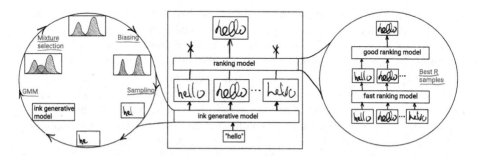

Fig. 1. The diagram of the proposed solution. The input to the model is a single text label. The generative model is run to produce B candidates. The highest scoring one according to the ranking model is returned. In the generative model, we use different sampling modes to modify the output distribution of the model. The ranking model consists of two blocks, first taking B generated inks and scoring them, then taking the R inks with the highest scores and re-ranking them.

Given an autoregressive generative model of digital ink that takes a text label as input and produces a sequence representing digital ink as output, we are interested in maximizing the average quality $M_{\Theta_S, \Theta_R}(S, B, R)$ of the model output, while guaranteeing that the maximum inference time does not exceed a certain threshold $T_{\max}$. Here, S is the sampling method used by the generative model, B is the size of the batch for generation, and R is an inference-time parameter of the ranking model, Θ_S are fixed trained weights of the model, Θ_R are the trainable parameters of the ranking model, which we will describe below.

During inference, given a label, the generative model will use sampling method S to produce a batch of B digital inks, which will be scored according to the ranking model Θ_R. The highest-ranking sample will be returned as the output; if $B = 1$, the ranking model is bypassed. Figure 1 illustrates the approach.

Our main results concern the trade-off between the inference time and model output quality, and are presented in Sect. 4. The rest of this section is organized as follows: we describe our approach to measuring quality and inference time in Sect. 3.1; Sect. 3.2 outlines the data representation for digital ink and sampling methods S that can be used with it; Sect. 3.3 describes the ranking models we use and how to train them.

3.1 Evaluation

We propose an evaluation method linked to the system's usability. Similar to other works [2,9,10,21], as quality measure M we use the Character Error Rate (CER) of a trained handwriting recognition model on the generated samples. This stems from the assumption that the generated text is not useful if it is not readable, regardless of other attributes like style and beauty.

A second axis of interest for usability is the inference time. We report the **worst case** inference time **per character**. We measure the worst case latency, with the assumption that exceeding the budget makes the functionality unusable for users. We measure time per character since processing time is expected to scale linearly with the sequence length.

3.2 Data Representation and Sampling

Two frequently used representations of the digital ink data are **raw** and **curve** representation, which both encode the ink as a sequence of input tokens in $\mathbb{R}^d \times \{0,1\}^2$, with first d values describing the shape of the stroke between two points, and the last 2 binary values indicating whether (i) a particular token is at the end of the stroke, and whether (ii) it is the last token in the sequence (end of ink). For the **raw** representation, $d = 2$ and describes the offset between two adjacent points, and for the **curve** representation, $d = 6$ and describes the parameters of Bezier curve fit to a segment of the stroke [33].

Following the approach of [15] and most of the later literature on the topic, we parameterize the output distribution of every step of the autoregressive generative model by a set of parameters $(\pi, \mu, \Sigma, e_s, e_i)$, where π, μ, Σ describe weights, means, and covariances of a mixture of Gaussians, from which $\mathbb{R}^d$ stroke parameters are sampled, and e_s and e_i describe the parameters of Bernoulli distributions from which the pen-up (end-of-stroke) and end-of-sequence events are sampled. Σ is full-covariance matrix for raw features ($d = 2$) and diagonal otherwise. We provide more details in Sect. 4.2.

Sampling. We consider two types of distortions for the output distribution: distortion of the mixture weights π and distortion of the diagonal components of the covariance matrix Σ. To distort the mixture weights, we consider several standard approaches: Top-K (parameterized by the value of K), and Top-P and Typical sampling (both parameterized by the value of P). To distort the covariance matrix, we subtract a *sampling bias* value b from the diagonal elements of the covariance matrix, before applying the softplus [14] function to it to ensure positive values. This reduces the variance after the model has been trained, to avoid sampling in low-confidence regions. The sampling parameters $S = (s, m, b)$ are therefore the sampling method $s \in \{\text{Top-K}, \text{Top-P}, \text{Typical}\}$, the mixture parameter m, and the sampling bias value b.

3.3 Ranking Models

Running a ranking model to order the generated samples may be computationally costly. For this reason, we differentiate between a process to rank all candidates and one that ranks only the most promising ones. Following the approach commonly used in information retrieval [23,29], our ranking approach is two-staged, with a "fast" ranker $\mathcal{R}_1$ that runs on all B generated outputs simultaneously, and a slower, more trustworthy "good" ranker $\mathcal{R}_2$, which is used to re-rank the samples ranked highest by $\mathcal{R}_1$. The inference time parameter R of the ranking model, introduced at the beginning of this section, is the number of top samples according to $\mathcal{R}_1$ that are re-ranked by $\mathcal{R}_2$. When $R = B$, this corresponds to using only $\mathcal{R}_2$, and when $R = 1$, only $\mathcal{R}_1$ is used. We describe both rankers below, and provide more details about them in Sect. 4.1.

"Good" Ranker $\mathcal{R}_2$. Since our goal is to generate samples with lowest possible Character Error Rate, an obvious choice for $\mathcal{R}_2$ to use the recognizer model that measures CER as the ranking model - that is, select the sample that is perfectly recognizable or has the lowest character error rate. However, running the recognizer on-device can be slow depending on the implementation, and we will see that having a faster first stage is beneficial.

"Fast" Ranker $\mathcal{R}_1$. Following the approach of [6], our $\mathcal{R}_1$ ranker is a model learned to predict whether the generated sample is recognizable or not, that is, whether the recognizer would return the target label given the generated ink. In other words, this ranker is an approximation of the "good" ranker and tries to predict its output. Since inference time is one of the main focuses of our work, we consider a much simpler ranking model than the one described in [6]. Instead of looking at both the generated ink and target label, our ranker just uses the generated ink. It consists of two convolutional layers followed by global average pooling. We study this choice of ranking model in terms of inference speed and the types of errors that it can address in Sect. 4.

Training Dataset for $\mathcal{R}_1$. As described above, $\mathcal{R}_1$ ranker is trained to be a fast approximation of the $\mathcal{R}_2$ ranker, and it predicts whether synthesized ink is even close to being recognizable. To train $\mathcal{R}_1$, we don't use real data: we use the synthesizer for generating a sample for a given text label, and $\mathcal{R}_2$ ranker for generating a binary label of whether the sample is recognizable (recognition result matches the text label) or not. The pair of generated ink and binary label is the training data for $\mathcal{R}_1$ (more details in Sect. 4.2).

We first train the ranking model, and then, select the sampling method S that performs best on the $\mathcal{D}_{\text{tune}}$ dataset. Doing the reverse would require training a ranking model for each possible sampling parameter setting, which would be prohibitively expensive. This means that during training of $\mathcal{R}_1$, the sampling method is yet unknown. To accommodate this, we create the training dataset for $\mathcal{R}_1$ by generating samples with (s, m, b) selected at random, for each sample. This allows $\mathcal{R}_1$ to be robust to any future selection of S, so that the sampling parameters can be chosen after the ranker is trained. We evaluate this method of training dataset creation in Sect. 4.

4 Results

4.1 Setup

To show that both sampling and ranking bring forth significant improvements in generation quality, and show the robustness of the proposed approach, we will evaluate it on

4 datasets across 3 different languages, with two frequently used model types, and two data representations.

We consider 4 digital ink datasets for text generation: English **Deepwriting** [2] and **IAMonDB** [24], Vietnamese **VNonDB** [28], and an internal **Math** dataset of mathematical expressions. We use two data representations described in Sect. 3.2, **raw** and **curve**, and evaluate two different model types, **Tacotron** [15,35] and **Transformer** [34].

4.2 Implementation Details

For both **Tacotron** and **Transformer**, we use 10-component Gaussian mixtures in the model output. For **Tacotron**, we use one-hot encoding of labels and 3 layers of size 256 in the decoder. For **Transformer**, we use 2 layers with 4 attention heads and embedding size 64 in the label encoder, and 6 layers with 4 attention heads and embedding size 128 in the decoder. We use the Pre-LN implementation [3]. We train models with Adam with global clipnorm of 0.1, and learning rate of 1e-3 for **Tacotron** and learning rate schedule described in [34] for **Transformer**. Models are trained for 2×10^6 steps with batch size 256. For training the $\mathcal{R}_1$ ranker, we generate 10^5 samples with labels from the generator training data as the training set, and 1000 samples with labels from the generator validation data as the validation set. As described in Sect. 3, for each sample, we select a sampling method at random to generate it. The pool of sampling methods includes Top-P, Typical samplings with $m \in \{0.0, 0.1, \ldots, 1.0\}$ and Top-K sampling with $m \in \{1, 2, \ldots, 10\}$, and sampling biases $b \in \{0, 1, 5, 25, 100, \infty\}$. The $\mathcal{R}_2$ ranker is a state-of-the-art recognizer that has been trained on internal data not related to public datasets and is an LSTM-CTC model with 6 layers of size 216 [8], which is combined with word and character language models during beam search decoding, similar to [20].

For **IAMonDB**, we use *testset_v* for validation, *testset_f* for tuning sampling parameters (via grid search over all possible samplings), and *testset_t* for testing. For **VNonDB**, we use the version of the dataset split by individual words. Since this dataset does not have the tuning subset, we use validation data labels for tuning sampling parameters. For **Deepwriting**, since this dataset does not have tuning or testing subset, we extracted 1500 labels whose lengths have the same mean and variance as the **Deepwriting** validation data, from the labels present in the **IAMonDO** dataset (we include these labels with the submission for clarity). Models were implemented in Tensorflow and the time measurements were done after conversion to TFLite on a Samsung Galaxy Tab S7+ tablet.

4.3 Baselines

Sampling Model Baseline. We compare the model with tuned sampling parameters, with a model with fixed sampling method. Since different works in the literature consider different sampling methods, to have a fair comparison to them, as to a baseline, we report the best result with $S = (\text{Top-P}, m, b), m \in \{0.0, 1.0\}, b \in \{0.0, \infty\}$, that is, greedy or ancestral sampling of component with infinite or zero bias for the offset parameters. We will refer to the optimal sampling method as S_{opt}, and to baseline as S_{base}.

Ranking Model Baseline. We compare the $\mathcal{R}_1$ ranker that predicts the recognizability of the generated ink, described in Sect. 3, with an approach described in [22], which trains a model to distinguish between real and synthesized samples, with the goal of selecting the most "real-looking" samples. We will refer to it as $\mathcal{R}_{base}$.

4.4 Quantitative Analysis

Table 1. CER for different sampling and ranking strategies. For S_{base} and S_{opt}, we use $B = 1$, meaning that no ranker is used. For $\mathcal{R}_1$ and $\mathcal{R}_{base}$, we use $B = 5$ and $R = 1$, meaning that "good" $\mathcal{R}_2$ ranker is not used. For $\mathcal{R}_2$, we use $B = 5$ and $R = 5$, meaning that the samples are ranked according to the "good" ranker only. This number is also a bound on the quality achievable with a "fast" ranker $\mathcal{R}_1$.

Dataset	Data	Model	S_{base}	S_{opt}	$\mathcal{R}_{base}$	$\mathcal{R}_1$	$\mathcal{R}_2$
Deepwriting	raw	**Tacotron**	$4.6_{\pm0.6}$	$2.6_{\pm0.2}$	$2.3_{\pm0.3}$	$1.7_{\pm0.2}$	$0.7_{\pm0.1}$
		Transformer	$8.1_{\pm2.9}$	$6.7_{\pm1.8}$	$5.8_{\pm1.3}$	$4.9_{\pm1.1}$	$1.8_{\pm0.5}$
	curve	**Tacotron**	$5.9_{\pm0.5}$	$5.6_{\pm0.7}$	$4.5_{\pm0.7}$	$2.1_{\pm0.2}$	$0.9_{\pm0.1}$
		Transformer	$8.9_{\pm1.5}$	$6.6_{\pm0.9}$	$4.7_{\pm0.5}$	$2.8_{\pm0.3}$	$1.0_{\pm0.1}$
IAMonDB	raw	**Tacotron**	$5.8_{\pm3.1}$	$3.8_{\pm0.7}$	$3.7_{\pm0.9}$	$2.6_{\pm0.4}$	$1.3_{\pm0.1}$
		Transformer	$13.3_{\pm2.9}$	$12.3_{\pm2.0}$	$10.9_{\pm0.2}$	$9.3_{\pm1.2}$	$5.3_{\pm1.2}$
	curve	**Tacotron**	$14.9_{\pm1.2}$	$9.1_{\pm0.9}$	$9.1_{\pm0.6}$	$3.8_{\pm0.0}$	$2.1_{\pm0.1}$
		Transformer	$16.8_{\pm1.4}$	$12.0_{\pm1.6}$	$11.7_{\pm1.0}$	$8.2_{\pm0.4}$	$3.9_{\pm0.7}$
VNonDB	raw	**Tacotron**	$4.0_{\pm0.5}$	$3.2_{\pm0.6}$	$3.2_{\pm0.5}$	$2.1_{\pm0.2}$	$0.7_{\pm0.1}$
		Transformer	$4.3_{\pm0.9}$	$3.7_{\pm0.6}$	$3.0_{\pm0.4}$	$2.6_{\pm0.4}$	$0.8_{\pm0.1}$
	curve	**Tacotron**	$2.1_{\pm0.1}$	$2.2_{\pm0.2}$	$2.2_{\pm0.2}$	$1.8_{\pm0.2}$	$0.7_{\pm0.1}$
		Transformer	$2.0_{\pm0.2}$	$2.0_{\pm0.2}$	$2.0_{\pm0.2}$	$1.8_{\pm0.3}$	$0.7_{\pm0.0}$
Math	raw	**Tacotron**	$28.5_{\pm1.0}$	$23.1_{\pm1.1}$	$22.3_{\pm0.4}$	$18.5_{\pm0.6}$	$8.3_{\pm0.5}$
		Transformer	$28.1_{\pm4.0}$	$22.8_{\pm2.5}$	$20.3_{\pm3.0}$	$19.7_{\pm2.9}$	$8.3_{\pm1.1}$
	curve	**Tacotron**	$9.4_{\pm0.5}$	$9.4_{\pm0.6}$	$9.0_{\pm0.1}$	$9.0_{\pm0.1}$	$3.1_{\pm0.1}$
		Transformer	$13.6_{\pm1.8}$	$10.8_{\pm0.7}$	$9.6_{\pm0.6}$	$9.2_{\pm0.4}$	$4.0_{\pm0.1}$

Effect of Sampling and Ranking. In Table 1, we compare the results of applying different sampling and ranking techniques for all datasets, model types, and data types.

A first major finding of our study is that **tuning the sampling technique helps in almost all cases** - in 13 cases out of 16, with the remaining ones being ties.

The second conclusion is that using a ranking model helps **in all cases**.

There is still a significant gap between the performance when using $\mathcal{R}_1$ and the quality-optimal $\mathcal{R}_2$. However, as we show in the next paragraph, achieving such quality comes with penalties for inference time.

Finally, we can conclude that using ranker that predicts whether the ink is recognizable or not is superior to using a baseline ranker [22] that predicts whether a given ink is real or synthetic. However the latter ranker also helps in most cases, as compared to not using ranking at all.

Comparison Under a Time Budget. The inference time for the model consists of 3 separate parts: *(i)* generating a batch of B samples; *(ii)* ranking them with the $\mathcal{R}_1$ ranker (unless $B = R$, in which case we can use just $\mathcal{R}_2$); *(iii)* Re-ranking the top R candidates with $\mathcal{R}_2$ (unless $B = 1$ in which case the generated sample can be returned directly). We show how these values scale with the input batch size for the model (that is, B for generative model and $\mathcal{R}_1$, and R for $\mathcal{R}_2$), in Table 2, and the trade-off between CER and inference time in Fig. 2.

Table 2. Model inference time per character, in milliseconds, for generative model, ranking model $\mathcal{R}_1$, and recognizer $\mathcal{R}_2$. Average across 1000 labels, **Tacotron** model on **Deepwriting** data with **curve** data representation. The generation process can be efficiently vectorized and scales sublinearly. The inference time of $\mathcal{R}_1$ is almost negligible, and the inference time of $\mathcal{R}_2$ scales linearly.

Batch size	Generation	$\mathcal{R}_1$	$\mathcal{R}_2$
1	15.5	0.05	2.79
2	20.6	0.05	5.19
4	26.6	0.09	11.40
8	35.0	0.15	23.04
16	45.0	0.24	41.39
32	66.3	0.45	76.97
64	128.6	0.91	163.47

Here we present the comparison of model quality vs inference time budget, by varying the values of B and R.

To connect the input sequence length to inference time, we fix the maximum number of decoding steps the model is allowed to make per input sequence symbol. In other words, our inference time is measured as time needed for one decoding step times the maximum allowed number of tokens per input symbol. The generation is always run until the maximum number of frames. In the models we used for this evaluation, 99% of the samples generated less than 5 frames per output character, which is the ratio that we fixed.

Table 2 shows the inference time for synthesis model, $\mathcal{R}_1$, and $\mathcal{R}_2$, in ms per character as a function of the input batch size. Notice that both the autoregressive generative model and the convolution-based ranker are able to take advantage of vectorization and are 7.5 and 3.2 times faster for large batch sizes than if run individually. The recognizer, used as $\mathcal{R}_2$, however, does not parallelize well due to CTC [16] decoding and combination with language models, thus scaling linearly with the batch size.

Based on the data in Table 2, we plot the numbers for model quality and worst-case inference time for different values of B and R in Fig. 2. Points with $(B = 4, R = 2)$, $(B = 8, R = 4)$, and $(B = 16, R = 8)$ are on the Pareto frontier, verifying our earlier statement that there are scenarios where the best performance can be achieved by combining the two rankers. Points $(B = 2, R = 1)$ and $(B = 4, R = 1)$ are also on the frontier, verifying our statement that there are cases where the best performance can be achieved without using the recognizer part of the ranking model at all.

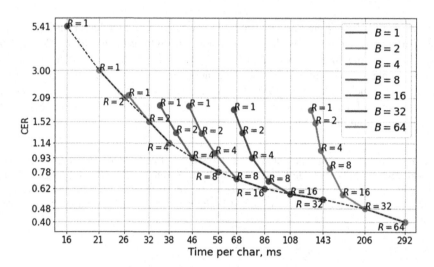

Fig. 2. Model inference time (upper bound, per char) vs CER for various values of B and R. For each values of B, we report results for values of R in $\{1, 2, 4, 8, 16, 32, 64\}$ s.t. $R \leq B$. The gray dotted line shows a Pareto-optimal frontier. Both axes on the log-scale. As visible, there are points on the Pareto frontier that include the use of both $\mathcal{R}_1$ and $\mathcal{R}_2$, justifying our claim that there are scenarios where optimal performance for a given computational budget can be achieved by a combination of both.

Discussion and Limitations. We note that the findings we present here are not universal, and the exact inference time depends on a multitude of factors such as specific generative model type and size, hardware, length of the sequence to be generated (processor caching makes longer sequences faster on a per-character basis), ranking model type and size (for the recognizer ranker, we rely on a model using CTC decoding which is hard to vectorize, whereas Seq2Seq models may parallelize better, although usually have worse accuracy). Furthermore, the average/median inference time might differ from the worst case significantly: The generative model produces an average 3.7 output frames per input character, compared to 5 which we used for the worst case analysis. Also when using the recognizer as a ranker, we need not recognize all of the candidates as we can stop at the first candidate that is perfectly recognizable, which may happen sooner or later depending on the exact sampling type and model quality. However, we believe that this does not invalidate our findings: depending on the time budget, better performance may be achieved by using a fast learned ranking model or combining it with a recognizer.

Ablation Study. In Table 3 we evaluate our choice of the construction of the ranker training dataset, and tuning of the sampling parameters for every setup (generation model type and feature type).

Firstly, we compare our approach of generating training data for the ranker by using random sampling parameters for every label to two other baseline approaches: *(i)* using a fixed ancestral sampling when generating the training data; this intuitively makes sense as sampling from "widest" possible distribution should cover all the whole diver-

Table 3. Ablation study for the ranker. The first column contains the results obtained when using $\mathcal{R}_1$ as the ranker. The next group of columns ablates the way of constructing the training dataset - by always generating samples using ancestral sampling, or by always generating samples using the sampling that yields the optimal performance when using $\mathcal{R}_2$ as the ranker. The last column shows that the optimal sampling parameters are different for each setup, ablating our choice of always tuning the sampling parameters.

Dataset	Data	Model	$\mathcal{R}_1$	Ranker training data		Opt. sampling
				Anc.	Rec.	
Deepwriting	**raw**	**Tacotron**	$1.7_{\pm 0.2}$	$1.9_{\pm 0.2}$	$2.0_{\pm 0.2}$	Top-P, 0.9, 5.0
		Transformer	$4.9_{\pm 1.1}$	$5.4_{\pm 1.0}$	$5.0_{\pm 0.9}$	Top-K, 9, ∞
	curve	**Tacotron**	$2.1_{\pm 0.2}$	$2.0_{\pm 0.4}$	$2.0_{\pm 0.4}$	Top-K, 3, ∞
		Transformer	$2.8_{\pm 0.3}$	$2.7_{\pm 0.3}$	$2.8_{\pm 0.3}$	Top-K, 5, ∞
IAMonDB	**raw**	**Tacotron**	$2.6_{\pm 0.4}$	$2.8_{\pm 0.5}$	$2.6_{\pm 0.4}$	Top-P, 0.9, 100.0
		Transformer	$9.3_{\pm 1.2}$	$9.1_{\pm 1.3}$	$9.3_{\pm 1.5}$	Top-K, 6, ∞
	curve	**Tacotron**	$3.8_{\pm 0.0}$	$3.8_{\pm 0.1}$	$4.3_{\pm 0.3}$	Top-K, 2, ∞
		Transformer	$8.2_{\pm 0.4}$	$8.6_{\pm 0.8}$	$8.2_{\pm 0.8}$	Top-K, 4, ∞
VNonDB	**raw**	**Tacotron**	$2.1_{\pm 0.2}$	$2.5_{\pm 0.2}$	$2.4_{\pm 0.2}$	Top-P, 0.9, 100.0
		Transformer	$2.6_{\pm 0.4}$	$2.8_{\pm 0.4}$	$2.9_{\pm 0.4}$	Top-P, 0.9, 5.0
	curve	**Tacotron**	$1.8_{\pm 0.2}$	$2.0_{\pm 0.1}$	$1.7_{\pm 0.1}$	Top-P, 0.4, ∞
		Transformer	$1.8_{\pm 0.3}$	$2.8_{\pm 0.4}$	$2.9_{\pm 0.4}$	Top-P, 0.3, ∞
Math	**raw**	**Tacotron**	$18.5_{\pm 0.6}$	$19.4_{\pm 0.6}$	$19.0_{\pm 0.6}$	Top-P, 0.9, 5.0
		Transformer	$19.7_{\pm 2.9}$	$20.5_{\pm 2.7}$	$20.0_{\pm 2.1}$	Top-K, 8, ∞
	curve	**Tacotron**	$7.7_{\pm 0.3}$	$8.4_{\pm 0.1}$	$7.7_{\pm 0.2}$	Top-P, 0.3, ∞
		Transformer	$9.2_{\pm 0.4}$	$10.2_{\pm 0.5}$	$9.3_{\pm 0.1}$	Top-P, 0.3, ∞

sity of the generated data. *(ii)* for each setup, using the sampling parameters that yield the lowest CER if $\mathcal{R}_2$ is used as ranker; this makes sense as $\mathcal{R}_1$ tries to approximate $\mathcal{R}_2$, and it is reasonable to assume that their optimal sampling parameters should be similar. We observe that on average our proposed way of constructing a training dataset is optimal, never being more than one decimal point worse than other approaches, but at times significantly outperforming them.

Secondly, we show that the optimal sampling parameters differ a lot between the setups, so it is important to tune them for each setup. The only reliable signals we observed was that for the **curve** representation, it is often preferable to sample more "greedily" (lower value of K in Top-K or P in Top-P sampling) than for the **raw** representation, and that the optimal samplings seem to be somewhat close between the two model types.

4.5 Qualitative Analysis

In this section, we first attempt to confirm that: *(i)* the two types of errors, overconfidence and incoherence, actually happen when generating digital ink samples, and *(ii)*

Table 4. Number of overconfidence and incoherence errors for various values of p in Top-P sampling, for a model with and without $\mathcal{R}_1$ ranker.

P	No ranking		Ranking with $\mathcal{R}_1$	
	Overconf	Incoher	Overconf	Incoher
0.1	81	120	42	157
0.2	75	115	37	114
0.3	69	140	23	111
0.4	59	170	16	109
0.5	41	180	9	109
0.6	33	216	3	121
0.7	30	246	2	137
0.8	22	281	1	149
0.9	14	375	1	197
1.0	7	466	1	282

both the choice of sampling and ranking has effect on these errors. Results are presented with the **Tacotron** model on **Deepwriting** dataset with **curve** representation, but we have observed largely similar trends for other cases. Afterwards, we present examples of model output on various datasets.

Figure 3 shows examples of generated ink with various samplings - with both incoherence and overconfidence examples visible. As we can observe, overconfidence errors typically result in very long ink, that can not be recognized as the label, with repeating pattern inside. Given this observation, we attempt to quantify the number of errors of

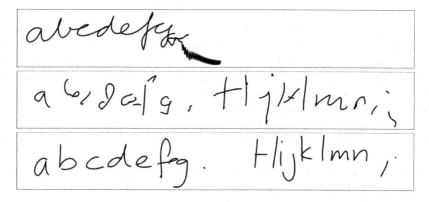

Fig. 3. Examples of model outputs for different sampling parameters. Input label is "*abcdefg. Hijklmn,*". Sampling parameters used are: Top - (Top-P, 0.0, ∞); Middle - (Top-P, 1.0, 0.0); Bottom - (Top-P, 0.5, 5.0). The overconfidence error is clearly visible in the top example, while the middle example is incoherent and hard to recognize. The bottom row shows the importance of carefully selecting sampling for optimal performance.

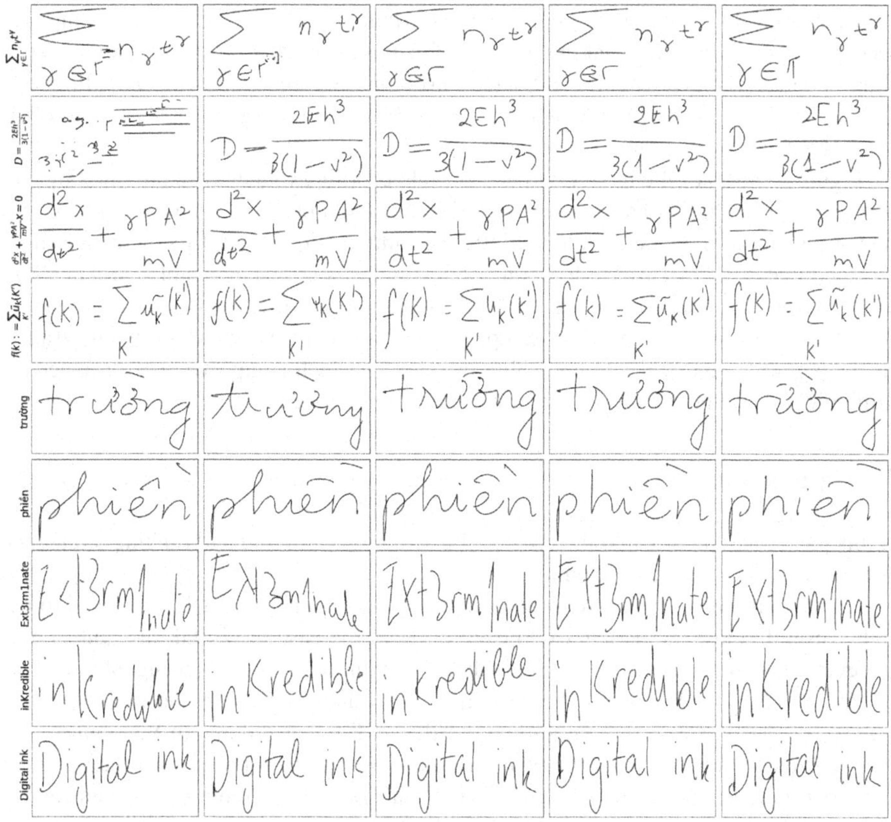

Fig. 4. Examples of model outputs. **Transformer** with **curve** representation for **Math** data, **Tacotron** with **curve** representation for **VNonDB** data, **Tacotron** with **raw** representation for **Deepwriting** data. In each case, 5 samples were generated, and sorted left-to-right according to the score provided by the $\mathcal{R}_1$ ranker model, with the rightmost image being the most recognizable according to the ranker. The first column shows some examples of samples that are not recognizable and are scored low by the ranker, ex. stray strokes (first row), overconfident generation of repeated lines (second row), misplaced tilde sign over u (fourth row), one extra diacritic (fifth row), missing dash over t (seventh row). More examples can be obtained in demo colab: https://colab.research.google.com/drive/1AkwmDOkEIkifbOYEBdcB9PrR_Ll-fcmz.

each type by looking at *samples that can not be recognized* (meaning the label returned by the recognizer differs from the input label to the generative model), and within those samples, whether the generation process reached the maximum number of steps (implying overconfidence) or not (implying incoherence). Table 4 shows the number of errors, estimated by this approach, as a function of sampling parameters (value of p in Top-P sampling), and it confirms the intuition about how it should behave. We can see that as the sampling parameters go from greedy sampling closer to ancestral sampling, the number of overconfidence errors goes down, while the number of incoherence errors goes up. When we use the ranking model, we see that the number of incoherence samples first goes down, and then goes up. We attribute this to the fact that as sampling becomes more diverse, the ranking model is able to select better candidates, but as sampling becomes too diverse, all candidates start being less recognizable. Overall, using ranking seems to reduce the number of overconfidence errors by 50–90%, and number of incoherence errors by up to 50%.

Figure 4 shows of the model outputs, sorted according to the score provided by the ranker, left-to-right. As can be seen, the rightmost sample in every row is recognizable and matches the label, while the leftmost sample is mostly not recognizable. It is expected that in many cases at least one of 5 samples is not recognizable - if that were not the case, that would mean that the selected sampling method is too conservative and should be relaxed to produce samples with higher diversity (which would trade-off having all 5 candidates recognizable in "easy" cases for improved performance in "difficult" cases where all 5 samples were not recognizable).

5   Conclusion

In this paper, we investigated the effects of combining sampling and ranking strategies to improve digital ink generation.

These methods, used before in other domains such as NLG and TTS, proved to be highly useful, and complementary to each other in the case of digital ink. Until now, however, they were not explored in this domain, with most methods using ancestral or greedy sampling, and no candidate ranking. We evaluate sampling and ranking techniques, on four datasets - two containing writing in English and one in Vietnamese, as well as a fourth one with mathematical formulas. We test the robustness of the findings using two model types (Tacotron and Transformer) and two common ink data representations (**raw** and **curve**). In all the combinations, we report significant improvements in the recognizability of the synthetic inks: taken together, a well-chosen sampling method, followed by fast ranking consistently improve recognizability, in many cases halving the character error rates.

An important factor in the perceived quality of ink synthesis is speed. Potential applications, such as handwriting autocompletion, spelling correction, and beautification usually process user inputs on-device, so ink generative models need to be fast. We thus report the findings with respect to a given computational budget.

References

1. Aksan, E., Deselaers, T., Tagliasacchi, A., Hilliges, O.: Cose: compositional stroke embeddings. arXiv preprint arXiv:2006.09930 (2020)
2. Aksan, E., Pece, F., Hilliges, O.: Deepwriting: making digital ink editable via deep generative modeling. In: Proceedings of the 2018 CHI Conference on Human Factors in Computing Systems (2018)
3. Baevski, A., Auli, M.: Adaptive input representations for neural language modeling. arXiv preprint arXiv:1809.10853 (2018)
4. Basu, S., Ramachandran, G.S., Keskar, N.S., Varshney, L.R.: Mirostat: a neural text decoding algorithm that directly controls perplexity (2020). https://doi.org/10.48550/ARXIV.2007.14966. https://arxiv.org/abs/2007.14966

5. Bengio, S., Vinyals, O., Jaitly, N., Shazeer, N.: Scheduled sampling for sequence prediction with recurrent neural networks (2015). https://doi.org/10.48550/ARXIV.1506.03099.https://arxiv.org/abs/1506.03099
6. Betker, J.: TorToiSe text-to-speech, April 2022. https://github.com/neonbjb/tortoise-tts
7. Cao, N., Yan, X., Shi, Y., Chen, C.: AI-sketcher: a deep generative model for producing high-quality sketches. In: Proceedings of the AAAI Conference on Artificial Intelligence (2019)
8. Carbune, V., et al.: Fast multi-language LSTM-based online handwriting recognition (2020)
9. Chang, J., Shrivastava, A., Koppula, H., Zhang, X., Tuzel, O.: Style equalization: Unsupervised learning of controllable generative sequence models. arXiv preprint arXiv:2110.02891 (2021)
10. Chang, J.H.R., et al.: Data incubation-synthesizing missing data for handwriting recognition. In: ICASSP 2022–2022 IEEE International Conference on Acoustics, Speech and Signal Processing (ICASSP), pp. 4188–4192. IEEE (2022)
11. Chung, J., Kastner, K., Dinh, L., Goel, K., Courville, A., Bengio, Y.: A recurrent latent variable model for sequential data (2015). https://doi.org/10.48550/ARXIV.1506.02216. https://arxiv.org/abs/1506.02216
12. Das, A., Yang, Y., Hospedales, T., Xiang, T., Song, Y.Z.: Sketchode: learning neural sketch representation in continuous time. In: International Conference on Learning Representations (2021)
13. Fan, A., Lewis, M., Dauphin, Y.: Hierarchical neural story generation (2018). https://doi.org/10.48550/ARXIV.1805.04833, https://arxiv.org/abs/1805.04833
14. Glorot, X., Bordes, A., Bengio, Y.: Deep sparse rectifier neural networks. In: Proceedings of the Fourteenth International Conference on Artificial Intelligence and Statistics, pp. 315–323. JMLR Workshop and Conference Proceedings (2011)
15. Graves, A.: Generating sequences with recurrent neural networks. arXiv preprint arXiv:1308.0850 (2013)
16. Graves, A., Fernández, S., Gomez, F., Schmidhuber, J.: Connectionist temporal classification: labelling unsegmented sequence data with recurrent neural networks. In: Proceedings of the 23rd International Conference on Machine Learning, pp. 369–376 (2006)
17. Ha, D., Eck, D.: A neural representation of sketch drawings. arXiv preprint arXiv:1704.03477 (2017)
18. He, T., Zhang, J., Zhou, Z., Glass, J.R.: Quantifying exposure bias for open-ended language generation (2020)
19. Holtzman, A., Buys, J., Du, L., Forbes, M., Choi, Y.: The curious case of neural text degeneration (2019). https://doi.org/10.48550/ARXIV.1904.09751. https://arxiv.org/abs/1904.09751
20. Keysers, D., Deselaers, T., Rowley, H., Wang, L., Carbune, V.: Multi-language online handwriting recognition. IEEE Trans. Pattern Anal. Mach. Intell. (2017)
21. Kotani, A., Tellex, S., Tompkin, J.: Generating handwriting via decoupled style descriptors. In: Vedaldi, A., Bischof, H., Brox, T., Frahm, J.-M. (eds.) ECCV 2020. LNCS, vol. 12357, pp. 764–780. Springer, Cham (2020). https://doi.org/10.1007/978-3-030-58610-2_45
22. Krishna, K., Chang, Y., Wieting, J., Iyyer, M.: Rankgen: improving text generation with large ranking models (2022). https://doi.org/10.48550/ARXIV.2205.09726. https://arxiv.org/abs/2205.09726
23. Liu, B., Wei, H., Niu, D., Chen, H., He, Y.: Asking questions the human way: scalable question-answer generation from text corpus. In: Proceedings of the Web Conference 2020, pp. 2032–2043 (2020)
24. Liwicki, M., Bunke, H.: IAM-OnDB-an on-line English sentence database acquired from handwritten text on a whiteboard. In: ICDAR 2005. IEEE (2005)
25. Luhman, T., Luhman, E.: Diffusion models for handwriting generation (2020). https://doi.org/10.48550/ARXIV.2011.06704, https://arxiv.org/abs/2011.06704

26. Maksai, A., Rowley, H., Berent, J., Musat, C.: Inkorrect: online handwriting spelling correction (2022). https://doi.org/10.48550/ARXIV.2202.13794. https://arxiv.org/abs/2202.13794

27. Meister, C., Pimentel, T., Wiher, G., Cotterell, R.: Typical decoding for natural language generation (2022). https://doi.org/10.48550/ARXIV.2202.00666. https://arxiv.org/abs/2202.00666

28. Nguyen, H., Nguyen, C., Bao, P., Nakagawa, M.: A database of unconstrained Vietnamese online handwriting and recognition experiments by recurrent neural networks. Pattern Recognition (2018)

29. Ravaut, M., Joty, S., Chen, N.F.: Summareranker: a multi-task mixture-of-experts re-ranking framework for abstractive summarization. arXiv preprint arXiv:2203.06569 (2022)

30. Reddy, R.: Speech understanding systems: A summary of results of the five-year research effort at carnegie mellon university. Tech. rep. (1977)

31. Ribeiro, L., Bui, T., Collomosse, J., Ponti, M.: Sketchformer: transformer-based representation for sketched structure. In: Proceedings of the IEEE/CVF Conference on Computer Vision and Pattern Recognition (2020)

32. See, A., Pappu, A., Saxena, R., Yerukola, A., Manning, C.D.: Do massively pretrained language models make better storytellers? (2019). https://doi.org/10.48550/ARXIV.1909.10705, https://arxiv.org/abs/1909.10705

33. Song, Y.: Béziersketch: A generative model for scalable vector sketches. University of Surrey, Tech. rep. (2020)

34. Vaswani, A., et al.: Attention is all you need. In: Advances in neural information processing systems (2017)

35. Wang, Y., et al.: Tacotron: towards end-to-end speech synthesis. arXiv preprint arXiv:1703.10135 (2017)

36. Zhang, T., et al.: Coder reviewer reranking for code generation. arXiv preprint arXiv:2211.16490 (2022)

Linguistic Knowledge Within Handwritten Text Recognition Models: A Real-World Case Study

Samuel Londner[1]([✉]), Yoav Phillips[2], Hadar Miller[3], Nachum Dershowitz[4], Tsvi Kuflik[3], and Moshe Lavee[2]

[1] School of Engineering, Tel Aviv University, Tel Aviv, Israel
`samuell@mail.tau.ac.il`
[2] Department of Jewish History and Thought, University of Haifa, Haifa, Israel
[3] Department of Information Systems, University of Haifa, Haifa, Israel
[4] School of Computer Science, Tel Aviv University, Tel Aviv, Israel

Abstract. State-of-the-art handwritten text recognition models make frequent use of deep neural networks, with recurrent and connectionist temporal classification layers, which perform recognition over sequences of characters. This architecture may lead to the model learning statistical linguistic features of the training corpus, over and above graphic features. This in turn could lead to degraded performance if the evaluation dataset language differs from the training corpus language.

We present a fundamental study aiming to understand the inner workings of OCR models and further our understanding of the use of RNNs as decoders. We examine a real-world example of two graphically similar medieval documents but in different languages: rabbinical Hebrew and Judeo-Arabic. We analyze, computationally and linguistically, the cross-language performance of the models over these documents, so as to gain some insight into the implicit language knowledge the models may have acquired. We find that the implicit language model impacts the final word error by around 10%. A combined qualitative and quantitative analysis allow us to isolate manifest linguistic hallucinations. However, we show that leveraging a pretrained (Hebrew, in our case) model allows one to boost the OCR accuracy for a resource-scarce language (such as Judeo-Arabic).

All our data, code, and models are openly available at https://github. com/anutkk/ilmja.

Keywords: Optical character recognition · Handwritten text recognition · Transfer learning · Language model · Hebrew manuscripts

1 Introduction

Modern optical character recognition (OCR) algorithms have come a long way in their ability to accurately recognize handwritten text. However, it remains an

Supported in part by the Deutsch Foundation, the Israeli Ministry of Science and Technology (grant number 3-17516) and a grant from Tel Aviv University Center for AI and Data Science (TAD) in collaboration with Google, as part of the initiative "AI and DS for Social Good".

G. A. Fink et al. (Eds.): ICDAR 2023, LNCS 14190, pp. 147–164, 2023.
https://doi.org/10.1007/978-3-031-41685-9_10

open question whether these algorithms are able to capture linguistic features of the text in addition to graphical features. These algorithms use neural networks, specifically ending with recurrent layers and connectionist temporal classification (CTC) layers [16,27]. This architecture may lead to the model learning statistical linguistic features of the training corpus, over and above graphic features. And this would lead to sensitivity of the model towards the document language and to degraded performance if the evaluation dataset language differs from the training corpus language.

This paper investigates this question by examining the performance of OCR algorithms on two manuscripts written by the same scribe, one in medieval Judeo-Arabic and the other in medieval rabbinic Hebrew, but both in the same Hebrew script. Manuscripts in Hebrew script demonstrate high variability due to the wide dispersion of Jewish communities across different geo-cultural milieus. The use of manuscripts written by the selfsame person allows us to control for graphical features and focus on the rôle of linguistic knowledge in OCR performance. Our hypothesis is that OCR algorithms that are able to capture linguistic features will show higher accuracy in recognizing the handwritten text. By analyzing this real-world experimental design, we aim to shed light on the extent to which linguistic knowledge is incorporated in modern OCR algorithms.

If our hypothesis is supported, it would have important practical implications for the development and deployment of OCR algorithms. Specifically, it would suggest that it may not be possible to use a single OCR model for multiple languages with comparable accuracies, but rather a separate model for each language would be required. Accordingly, building multilingual OCR systems and making them more cost-effective so as to support a wider range of languages requires additional research and engineering.

One potential application of this idea is to use a model trained on a relatively data-rich language as a starting point for recognizing other, poorer languages resource-wise. For example, a model trained on Hebrew could be fine-tuned on Judeo-Arabic, a related Semitic language with relatively little available data. This approach would allow us to leverage the larger amount of available training data for Hebrew to improve OCR performance for Judeo-Arabic. The Ktiv database of the National Library of Israel[1] lists 61,096 known, extant manuscripts and fragments in Judeo-Arabic.

Overall, the results of this study have the potential to inform the design and implementation of OCR systems for multiple languages, with implications for a range of applications including historical document preservation, digital humanities, and language learning.

2   Related Work

2.1   Handwritten Text Recognition

We use off-the-shelf methods for automatic page segmentation, layout analysis, and line segmentation. Machine-learning based systems have seen wide use

---

[1] https://www.nli.org.il/en/discover/manuscripts/hebrew-manuscripts.

recently for these tasks [2, 6, 9, 10, 12, 17, 31, 35, 45], the majority using combinations of CNNs and LSTMs. Traditional computer-vision methods have advantages for some types of manuscripts [33, 35]. State-of-the-art methods have been implemented in kraken [22] and eScriptorium [23] for mixed models in various scripts, including Hebrew, and for a wide range of manuscript types.

The current best transcription results for such manuscripts are achieved by combinations of CNNs and BLSTMs [11, 19, 22]. OCR efforts working with medieval Hebrew manuscripts include [23–25]. The Sofer Mahir project (https://sofermahir.hypotheses.org) applied kraken's OCR to 20 large manuscripts of early rabbinic compositions. In the Tikkoun Sofrim project [24, 44], crowdsourcing and machine learning have been used to correct the errors of the automatic transcriptions of several large manuscripts of medieval exegetical literature. Character error rates (CER) of 2–3% were attained usually for manuscripts with homogeneous layout and script but only around 9% when there were complications. Modern end-to-end systems (segmentation plus OCR) include [5, 20].

2.2   Implicit Linguistic Knowledge in OCR Models

Previous works have employed synthetic data to show OCR models' sensitivity to language, and thus that they implicitly learn linguistic features. The authors of [43] test the performance of an LSTM-based OCR trained on one language and tested on other languages. The difference in performance is indicative of the model's reliance on an implicit language model (LM). However, no explanation or linguistic analysis is provided. Moreover, no attention is paid to the fact that the languages being compared, English and French, share linguistic features and even complete lexemes to a significant degree. The authors of [32] established and characterized the strength of the implicit LM in LSTM-based OCR systems by synthesizing printed English text and shuffling the characters in each sentence. This approach, although proving the existence of the implicit internal language model, is not applicable to evaluate the cross-lingual generalization capability (or lack thereof) of pretrained language models. Furthermore, in the experiment described in [32], shuffling characters does not affect the distribution of characters, thus leaving some linguistic hints to the (hypothesized) LM.

In this work, we combine an in-depth linguistic qualitative analysis and a quantitative approach to examine the degree of reliance of OCR models on linguistic features. We account for similarities and differences between the languages and present specific examples of seemingly linguistic "hallucinations" in OCR models. To the best of our knowledge, this is the first time such a hybrid approach is applied in the field of digital humanities with the goal to isolate and quantify the influence of language on the OCR model's performance. Taking into account the fact that synthesizing data is less relevant for historical manuscripts, we leverage a real-world case of two manuscripts, in two different languages, which share the same graphical features, having been written by the same scribe.

2.3   Transfer Learning

Manuscript handwriting styles are highly dependent on time, place, training, and individual predilections. Improving over state-of-the-art models by leveraging transfer learning is an obvious choice. Models pretrained over a large corpus are fine-tuned on the first few annotated pages of a manuscript in order to help decipher the rest of the manuscript. In this way, the representation learned over a *source* dataset can be refined to solve the *target* task, namely transcribing documents of a smaller, disjoint dataset [14]. Recent research [1,18] shows that the optimal method to improve accuracy is to fine-tune the parameters of the whole recognition model, while the first layer can be frozen without any meaningful performance degradation. In [15], the authors successfully apply this concept for Latin-alphabet handwriting to historical handwritten Italian titles of plays. The technique also allows one to transfer the representation from Arabic printed text to genuine handwriting [29]. Transductive methods, using purely synthetic data with data rendering and augmentation, along with domain adaptation, cycle-consistent adversarial networks, and a combination of a domain-adversarial neural network approach with a convolutional recurrent neural network architecture, have been used to advantage in [20] for Tibetan Buddhist historical texts in a variety of scripts.

3   Linguistic Background

Judeo-Arabic is a general term describing an Arabic-based Jewish language or ethnolect, with a wide variety of regional dialects, which gradually developed in Jewish communities across Arabic-speaking Islamic regions, from the 8th century until the mid-20th century. Although these dialects were influenced by local variants of Arabic, they had their own distinct characteristics that distinguished them as a unique communal dialect. On the other hand, most Judeo-Arabic dialects shared common features forming a distinctive Jewish ethnolect. The most common distinctive feature is the Hebrew orthography that was common to all Judeo-Arabic dialects (apart of some Karaite writings that used Arabic characters). The implementation of Hebrew orthography was mostly phonetic; therefore, it may have differed from one Jewish community to another due to different local pronunciation tendencies. Another common feature was the grammatical and syntactical integration of Hebrew roots, words, and phrases into the Arabic. Most manuscripts written in the Middle Ages, roughly between the 10th and the 13th centuries, as is the case for the manuscripts with which we will be working, were written in a relatively high register defined as Classical Judeo-Arabic (CJA). Simply put, this means that the core Arabic elements of the text are similar to its literary Arabic counterpart, while the differences between the various dialects within CJA are relatively mild [21,36].

For the purpose of this investigation, words were classified into four different linguistic categories:

1,2. The two basic groups are Hebrew and Judeo-Arabic. Under Hebrew we included the odd Aramaic words that are frequent in Hebrew medieval works and hence are assumed to be part of the linguistic knowledge of a model trained on Hebrew manuscripts. Each word in the manuscript is classified either as Hebrew (including Aramaic) or Judeo-Arabic.

3. A third category comprises homographs (distinct words that are written in the same manner): Since our manuscripts, like most Hebrew and Judeo-Arabic texts, lack vowels (the Hebrew and Arabic alphabets are partial abjads), many of them can be read both as a Hebrew word and as an Arabic word with divergent meanings.

4. The fourth group classified consists of abbreviated words. Our manuscripts, like most Hebrew and Judeo-Arabic texts, have a tendency to abbreviate words by dropping one or more letters at the end and adding an apostrophe or dot on top of the last letter of the shortened word, as in 'גו ($g\bar{o}$') for גומר ($g\bar{o}mer$). Shortening, which is not common in Arabic texts, is also applied to Arabic words in Judeo-Arabic texts, as for instance, 'ק (q') for קאל ($q\bar{a}la$). Thus, we have a Hebrew textual convention applied to both Hebrew and Arabic words. For completeness of the comparison of the model's performance between both languages, we group these potentially ambiguous strings in a separate category.

As in many Judeo-Arabic manuscripts, our scribe tended to separate the Arabic definitive article from the rest of the word. The abundant use of the definitive article in Arabic with the graphical effect of this Judeo-Arabic phenomenon was analysed separately. It should be noted that the definitive article [Arabic ٵ ($^{\circ}al$), which in our manuscript may be signified by the ال ($^{\circ}al$) ligature], stripped of its context, was usually classified as a homograph since it can be read as Hebrew or Arabic, although the adjacent word to which it refers was not necessarily Arabic. In a case like החמים ٵ ($^{\circ}al\ \d{h}a\d{h}am\bar{\imath}m$), the definitive form may be classified as a homograph and the noun החמים ($\d{h}a\d{h}am\bar{\imath}m$) as Hebrew [7].

4 Data

For our experiments, we use the manuscripts, MS Genève Comites Latentes 146 [3] and Oxford Bodleian Library MS Huntington 115 [30]. See Fig. 1. MS Genève 146 contains a rabbinic homiletic work from late antiquity, *Midrash Tanḥuma*. MS Huntingtion 115 contains *Kitab al-Tuffāḥa*, an unpublished Judeo-Arabic homiletic work by Shamariah Hacohen (d. between 1124–1137) [13, 26, 28]. The majority of MS Huntington 115 (from p. 103r on) was copied by the same scribe who copied MS Genève 146, in an Oriental Hebrew Script of the 14th century.

The main evaluation set is composed of 5 pages from MS Huntington 115. It amounts to 1559 words, or 5818 characters. The manuscript was first transcribed using a base OCR system. The transcription was then manually corrected by two experts of the language and the relevant literature. The resulting ground-truth text is not corrected, that is, it includes "typos" that actually appear in the data. Labeling was performed using eScriptorium [23].

A character k-gram, also known as a "k-mer", is a sequence of k consecutive letters of the alphabet or other characters (spaces and punctuation). As detailed below, for advanced analysis, we compare k-mer distributions of our

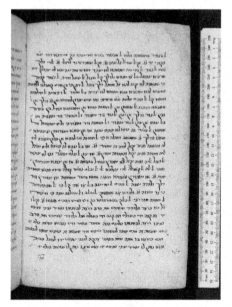
(a) MS Huntington 115

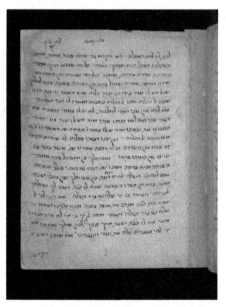
(b) MS Genève 146

(c) MS Huntington 115 – zoom in

(d) MS Genève 146 – zoom in

Fig. 1. Sample pages of the manuscripts used.

texts with the distributions within two larger literary bases: one for rabbinic Tanḥumic Hebrew (which parallels the language of MS Genève 146) and another for Judeo-Arabic (parallel to MS Huntington 115). The Tanḥumic Hebrew corpus is a subset of Sefaria's dataset [34],[2] and the Judeo-Arabic corpus is from the Friedberg Judeo-Arabic Corpus [42].[3]

5   Methodology

5.1   Training

We fine-tune a pretrained model over Hebrew and test it over Judeo-Arabic and conversely. Transfer learning is an efficient approach to attain state-of-the-art OCR performance over a specific data distribution with a limited amount of data. The pretrained model, which is composed of four convolutional layers, three LSTM layers and a CTC layer, has been trained over a heterogeneous batch of generic medieval manuscripts [41]. We fine-tune it to get optimal performance over a specific manuscript, using the Adam optimizer (constant learning rate: 0.001, momentum: 0.9). Fine-tuning is performed using the kraken package.[4]

Fine-tuning the models' parameters [41] over the first few pages of the manuscript (whose ground-truth text is known) indeed improves performance dramatically. Preliminary results show that character accuracy can be boosted by around 18% by fine-tuning the recognition models over only three labeled pages (see Fig. 2). It appears that the maximum achievable accuracy with the current architecture and limited data scope is approximately 96–98%, as evidenced by state-of-the-art results for pretrained models in larger datasets [41]. When fine-tuning a model on a manuscript that exhibits a similar graphical and linguistic distribution to the pretraining dataset, only a minimal quantity of data is necessary to optimize the model's weights for the new manuscript, which accounts for the observed "saturation" phenomenon. As such, the particular choice of the source model does not seem to impact performance, nor does adding more labeled data. We note that the same technique can be applied to segmentation models.

We use a model pretrained on a corpus of biblical and rabbinical Hebrew [41]. The same base model is used for fine-tuning over Hebrew as well as Judeo-Arabic.

n.b. The original models were taken from [41] and are available from kraken's Zenodo archive [37–40].[5]

5.2   Inference

OCR is generally composed of two steps: segmentation of the image into lines and recognition of the identified segments as text. The model is applied to images and their corresponding ground-truth segmentation, generating output through

---

[2] See https://www.sefaria.org/texts. We selected all the available texts from books that belong to the Tanḥumic Hebrew corpus: *Tanḥuma, Pesikta Rabbati, Shemot Rabbah, Bemidbar Rabbah*, and *Devarim Rabbah*.

[3] See https://ja.genizah.org/Home.aspx.

[4] https://kraken.re/, https://github.com/mittagessen/kraken.

[5] https://zenodo.org/communities/ocr_models.

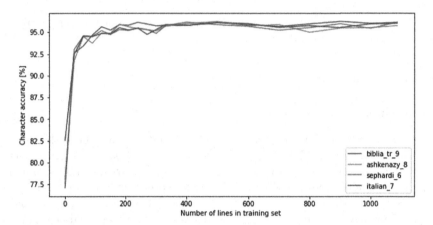

Fig. 2. Character accuracy achieved by transfer learning, as a function of additional labeled lines used for fine-tuning. Models courtesy [41].

CNN, RNN, and CTC layers. These outputs are exported to files and subsequently evaluated against ground truth, as elaborated next.

To neutralize the impact of incorrect segmentation as much as possible, we use manual ground-truth segmentation and focus only on the recognition network.

5.3   Evaluation

Some characters in Judeo-Arabic do not exist in Hebrew, mainly diacritics. We ignored these signs in the comparison, since a model trained on Hebrew material cannot generate Judeo-Arabic–specific symbols.

We compare character error rate (CER) in Table 1 and the word error rate (WER) in Table 2 over the complete evaluation sets. Although previous work [32, 43] dealt only with CER, we include WER in our analysis, since we expect the hypothesised implicit language model to affect WER more significantly than CER.

We present results for four subsets of words in the Judeo-Arabic evaluation set: (a) all words; (b) Hebrew words; (c) homographs (Judeo-Arabic spelled like other Hebrew words); (d) words in Judeo-Arabic that do not exist in Hebrew. This classification was performed manually by experts. It allows us to infer the level – if any – of the linguistic features the model may have learned: character level, part-of-word level, or word level. For example, if the model learned features related to k-mer distributions, but not features related to word n-gram distributions, we would except the homograph group error rate to be similar to the Hebrew error rate. On the other hand, if the model learned language modeling features related to context, we may expect the homograph group to have a higher error rate, since the inter-word context in the evaluation set is very

Table 1. CER [percent].

Set	Hebrew model	Judeo-Arabic model
Hebrew MS	6.7	9.1
Judeo-Arabic MS – All	8.2	6.3
Judeo-Arabic MS – Hebrew	5.2	–
Judeo-Arabic MS – Hebrew homographs	5.8	–
Judeo-Arabic MS – Arabic	8.0	–

Table 2. WER [percent].

Set	Hebrew model	Judeo-Arabic model
Hebrew MS	13.9	24.6
Judeo-Arabic MS – All	17.1	14.0
Judeo-Arabic MS – Hebrew	12.7	–
Judeo-Arabic MS – Hebrew homographs	10.0	–
Judeo-Arabic MS – Arabic	21.2	–

dissimilar from the training-set context. To facilitate the manual comparison, we used Dicta's Synopsis Builder [4,8].

For reference, we include the resulting error rates of the reciprocate Judeo-Arabic model over the whole Hebrew and Judeo-Arabic datasets. Note that diacritics are ignored in the evaluation.

We also compare distributions of errors of the model trained over MS Genève 146 over the MS Genève 146 holdout test set and the MS Huntington 115 dataset. See Fig. 4 for confusion matrices. Moreover, to account for the different distribution of characters in the two languages, we normalize each column in the confusion matrix by the number of respective characters in the ground truth; see Fig. 5. We also report the actual error rate distribution by character in Fig. 6.

To see if the model reproduces statistical patterns from Tanḥumic Hebrew, we compare the distribution of 1,2,3-mers in the transcription in Hebrew and Judeo-Arabic. For this specific comparison, we ignore differences of ligature; specifically, ﭏ (ʾal) is considered identical to אל (ʾal). The numerical scores are cosine metrics between the (sorted) distributions. Results are detailed in Fig. 3.

6   Results and Analysis

6.1   Error Rates

The main result leading our analysis is the difference in the error rates between the Hebrew model's transcriptions over Hebrew and Arabic words (the first and last rows in Tables 1 and 2). We note that the CER difference, although existent (around 2% – consistent with previous results [32,43]), is modest compared to

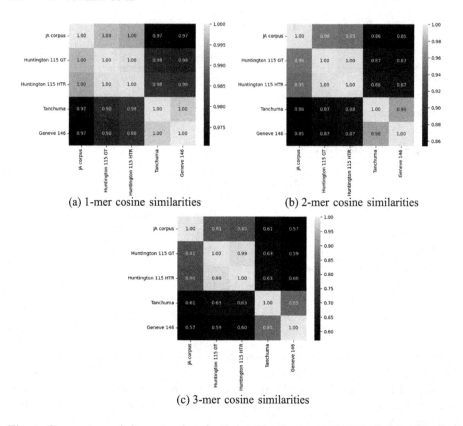

(a) 1-mer cosine similarities

(b) 2-mer cosine similarities

(c) 3-mer cosine similarities

Fig. 3. Comparison of character distributions. "Huntington 115 GT" denotes the distribution of ground-truth text in the Judeo-Arabic manuscript MS Huntington 115, whereas "Huntington 115 HTR" denotes the distribution of the transcription performed by the model trained on MS Genève 146.

the WER gap of more than 8%. This gap is preserved in the overall error rates, without distinction between subsets of words (second row in Tables 1 and 2). This is a strong indication that an implicit language model exists and is sensitive to the specific language of the transcribed text. The fact that the error rates for the Hebrew and homograph words (third and fourth rows in Tables 1 and 2) are similar to the pure Hebrew error rate indicates that the learned linguistic features are intraword and not interword, that is, they are on the k-mer level.

An additional finding is the difference between the normalized error rates per character between the holdout Hebrew text and the whole Judeo-Arabic dataset (Fig. 6). The modest but significant gaps may be explained by the sensitivity of the model to language.

This conclusion is further reinforced by the converse finding that the Judeo-Arabic model performs much better on the Judeo-Arabic holdout test set that on the Hebrew text, by a margin of more than 10%. Incidentally, since the base pretrained model was trained on Hebrew data only, and fine-tuned on a limited

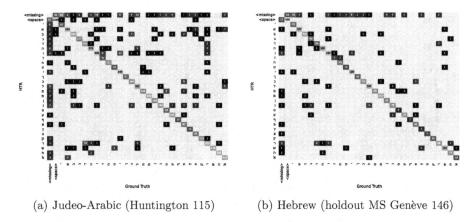

(a) Judeo-Arabic (Huntington 115) (b) Hebrew (holdout MS Genève 146)

Fig. 4. Confusion matrices of the Hebrew model, evaluated over Judeo-Arabic and Hebrew.

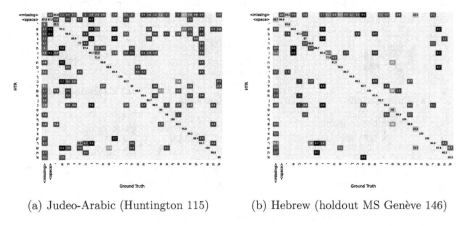

(a) Judeo-Arabic (Huntington 115) (b) Hebrew (holdout MS Genève 146)

Fig. 5. Normalized confusion matrices of the Hebrew model, evaluated over Judeo-Arabic and Hebrew. Units are in percent of corresponding characters in GT.

amount of Judeo-Arabic data, this shows that the implicit language model can be relatively easily updated. This means that – provided the graphemes are close enough – transferring the graphical knowledge and updating the language model by transfer learning may allow the leveraging of pretrained models for the benefit of data-scarce languages. Further research may analyze the influence of fine-tuning only the part of the model that is suspected to act as an implicit language model, namely the recurrent layers.

On the other hand, the k-mer distribution of the transcribed text (see Fig. 3) is significantly more similar to the corresponding distribution of the ground-truth Judeo-Arabic text than to Hebrew distributions (Tanḥuma or MS Genève 146).

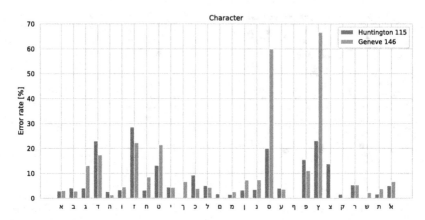

Fig. 6. Comparison of error rate per character. The outlier frequencies of ס and ץ are due to the low number of these characters in the holdout MS Genève 146 test set. Kolmogorov-Smirnov test after normalization: *statistic* = 0.129, *p* = 0.963.

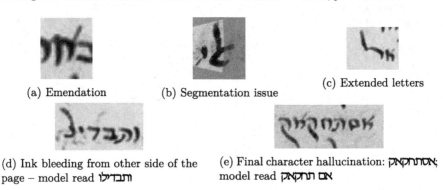

(a) Emendation   (b) Segmentation issue   (c) Extended letters

(d) Ink bleeding from other side of the page – model read וחבד׳ל׳

(e) Final character hallucination: אמתחתקאן; model read אם תחתק אן

Fig. 7. Examples of erroneous readings.

We theorize that although the model's output mainly depends on purely graphical features, in case of ambiguous readings linguistic features "tip the scale". A potential explanation for the observed phenomenon could be the following: The RNN, positioned at the conclusion of the model, likely functions as a self-supervised conditional language model, primarily utilizing the target text during the training process. To validate this theory, we performed a semi-qualitative analysis of the identified errors.

6.2   Graphical Errors

In many cases, a graphic issue can explain mistakes: the ink bleeding from the reverse side of the page causes confusions (Fig. 7d), or the scribe made a slight emendation or wrote the letter in a manner that resembles another letter, and so forth. For example, in Fig. 7a it seems that the scribe wrote the letter ח (*ḥ*) by

Table 3. Errors excluding graphical issues.

Language	Words	Error count	Error rate [%]
Hebrew	524	58	11.07
Arabic	499	90	18.04
Homographs	405	27	6.67
Other	54	11	20.37
Total	1482	186	–

mistake, and then emended it to the correct letter א (ʾ) by adding the upper right stroke, which is hardly seen. Indeed the model read ח. Another type of mistake that is not related to acquisition of the language is due to segmentation and hence should not be counted as evidence for the question of semantic knowledge. For instance, in Fig. 7b, the segmentation missed the exact beginning of the line, and hence the ligature ﭏ (ʾl) was read as the last letter in it, ל (l).

The scribe tends to write wide letters to fill the space at the end of line, so it comes out adjusted (see Fig. 7c). The model did not "learn" this feature, frequently failing for such stretched letters. However, a human reader who knows the language has the ability to overcome graphic issues, and our assumption is that since we taught the model full lines, we should expect some knowledge about frequencies of letters, that would help the model to overcome graphic issues.

6.3 Evidence for Linguistically Triggered Errors

To validate our assumption, we excluded mistakes obviously caused by a graphical reason (i.e. segmentation, ink bleed). Indeed, the ratio between the mistakes in both language did not change. See Table 3.

Subject to the danger of the rule of small numbers, the following report suggests an analysis based on a qualitative review of the material and some quantitative related analyses. We examined all mistakes and noted the following phenomena.

The most frequent mistake is the replacement of ר (*resh*) in place of ד (*dalet*). The shapes of these two letters are very similar, and our scribe writes them in an extremely inconsistent manner. In many cases, without the semantic context, a human reader will be unable to distinguish between them. The directionality of the mistakes is very clear. Only one time ר (*r*) was read as ד (*d*), while ד (*r*) was read as ר (*d*) 35 times (19% of its appearances in the examined pages).

There are 20 mistakes in Judeo-Arabic words (out of 89 words containing the letter and 499 words in total) and 15 mistakes are in Hebrew words or homographs (out of 84 words containing the letter and 954 words in total). This clearly shows a typical cause for the larger proportions of mistakes in Judeo-Arabic. At least in 11 cases the mistake created a valid word in Hebrew, so if there is any accumulation of linguistic knowledge it could not support the model decision making in those cases. Of special importance are two cases in which

the model also split the word wrongly, so that a valid Hebrew word is created:
וגדנאתה (wagadnāhā) becomes וגר נאה (ve-ger nāhā), אדדאר (ʾd-dār) becomes
א ראה (ʾel raʾah). Note that וגר (ve-ger) and ראה (raʾah) are valid Hebrew words.

In both Rabbinic and Biblical Hebrew, the frequency of ר resh is double
that of ד dalet. This explains why the model mistakes dalet for resh and not
vice versa. In Judeo-Arabic the ratio changes significantly, possibly because the
Hebrew letter ד dalet represents, in Arabic, both د (dāl) and ذ (dāl – usually
written with a diacritic, דֿ).[6] In this case, which is the most glaring one, we can
clearly see that the frequency of single letter is the cause of the different ratio
of mistakes in the two languages.

Another frequent mistake is the reading of ס (samekh) as ם (final mem).
These characters are graphically similar. This case is important because final
mem always comes at the end of a word. Indeed, 11 out the 12 errors are
ones in which the model read final mem as the last letter rather than the
actual samekh. Another case of reading final mem mistakenly was also at
the end of the word. Out of these mistakes, there are three striking cases in
which the samekh was in a middle of the word, but the model both read it
as final mem and split the word wrongly after the final mem, clearly demonstrating an inclination to represent the frequency of appearance of a space
after final mem. For instance, in Fig. 7e, the OCR model erroneously mis-
took a ס for a ם, and hallucinated a space after the mistaken ם, turning
אסתחקאק (ʾistiḥqāq) into אם תחקאק (ʾimtiḥqāq). A model trained on Judeo-
Arabic would probably be familiar with the sequence אסת (ʾist) which is part
of the conjugation אסתפעל (استفعال = ʾistifʿāl). Additional errors of this genre
include קי׳ סיידנא (qaw' sayīdnā) being mistaken for קם יירנא (waqaʾm yīrnā)
, and מע סעאדה (maʿa saʿādah) for מננים עאדיר (mananīm ʿādir).

Except once, all final mem mistakes are in Judeo-Arabic words. More telling
is the following observation: Out of a total nine cases of samekh at the end of a
Judeo-Arabic word, seven were wrongly read! Samekh appears in the middle of
a word in Judeo-Arabic 36 times, and only one of them was read as final mem in
the middle of word. In the three other cases, samekh in the middle of the word
was read as final mem and followed by an imaginary space (presented above).

In Hebrew, the corpus has only 2 words ending with samekh, one read as final
mem, and none of the total 24 cases of samekh in the middle of a Hebrew word
was read as final mem. Indeed the frequency of final mem versus samekh at the
end of the word in the Hebrew MS Genève 146 gold transcription is about 30:1.
It seems obvious that the model "learned" that samekh hardly ever appears at
the end of a word and that a final mem is final.

Another hint for a certain acquisition of knowledge concerns the frequency
of letters and sequences is the reading of zayin (ז) as vav (ו), as shown in Table 4.
Once again, these are two similar letters, though much more distinguishable to
the human eye in the hand of this specific scribe. Zayin is very rare, whereas

---

[6] As mentioned, since diacritics are not used in standard Hebrew, and do not appear
in the Hebrew model's training data, we should ignore them in our analysis and
error rate computation.

Table 4. *Zayin/vav* confusions.

	Hebrew	Judeo-Arabic
Total (mistakes/words) *zayin*	5/25	4/18
Beginning of a word	5/16	1/2
Middle of a word	0/9	3/16

vav is very frequent, and hence the clear directionality of the mistakes (as in the case of *resh* and *dalet*). The model read *zayin* as *vav* 9 times, 5 of them in the beginning of a word. This is related to the frequent function of *vav* as a conjunction, which appears at the beginning of a word.

Zayin is much more frequent in Hebrew than in Judeo-Arabic, especially at the beginning of a word. As a result, this is a rare case where the OCR model's character error rate is significantly higher in Hebrew than in Judeo-Arabic.

7   Conclusion

This paper presents a fundamental study aiming to understand the inner workings of OCR models and further our understanding of the use of RNN as decoders. We find that a network concluded by a RNN, trained to recognize words in one language, suffers a bias for that language, and therefore performs less well on texts in another natural (not artificial) language with the same alphabet and distribution of letters. Specifically, our combined quantitative and qualitative analysis shows that although OCR models mainly base their output on graphical features, linguistic features play a significant rôle in the transcription process and affect the final word accuracy by around 10%. By combining a qualitative approach to the linguistic features of the transcription and a quantitative analysis of the error distributions, we were able to isolate specific cases of seemingly linguistic hallucinations. We surmise that the decoder functions as a self-supervised conditional language model, primarily utilizing the target text during the training process.

The results demonstrate the need to train specific models for languages other than Hebrew in Hebrew script. Our conclusions are probably relevant to other Jewish languages in Hebrew script, such as Yiddish and Ladino (Judeo-Español), to Aramaic, and perhaps to the different languages written in Arabic characters.

Moreover, the existence of a low-level internal language model in OCR models suggests that post-OCR correction using a character-level or *k*-mer language model may be less likely to be helpful than using a semantic language model.

It may be feasible to moderate the extent of learning, such as by training on multilingual datasets or randomized synthetic data, although this may result in reduced accuracy for the original target language due to the implicit language model's capacity to "pre-correct" errors. An alternative approach involving training on a data-rich language and subsequently fine-tuning all or part of the network on a closely related data-poor language may yield superior outcomes. In fact, the similarities between the languages leave the door open for fine-tuning pretrained models over less data-rich datasets, although special attention needs to be given to language-specific glyphs such as diacritics.

References

1. Aradillas, J.C., Murillo-Fuentes, J.J., Olmos, P.M.: Boosting offline handwritten text recognition in historical documents with few labeled lines. IEEE Access, pp. 76674–76688 (2021). https://ieeexplore.ieee.org/stamp/stamp.jsp?arnumber=9438636
2. Barakat, B., Droby, A., Kassis, M., El-Sana, J.: Text line segmentation for challenging handwritten document images using fully convolutional network. In: Proceedings of the 16th International Conference on Frontiers in Handwriting Recognition (ICFHR), pp. 374–379. IEEE (2018)
3. Bibliothèque de Genève: Comites Latentes 146: Midrash Tanhuma (Leviticus-Numbers-Deuteronomy) (2015). https://www.e-codices.unifr.ch/en/list/one/bge/cl0146
4. Brill, O., Koppel, M., Shmidman, A.: FAST: Fast and accurate synoptic texts. Digital Scholarship in the Humanities **35**(2), 254–264 (2020)
5. Carbonell, M., Mas, J., Villegas, M., Fornés, A., Lladós, J.: End-to-end handwritten text detection and transcription in full pages. In: Proceedings of the International Conference on Document Analysis and Recognition Workshops (ICDARW), vol. 5, pp. 29–34. IEEE (2019)
6. Chen, K., Seuret, M., Liwicki, M., Hennebert, J., Ingold, R.: Page segmentation of historical document images with convolutional autoencoders. In: Proceedings of the 13th International Conference on Document Analysis and Recognition (ICDAR), pp. 1011–1015. IEEE (2015)
7. Connolly, M.M.: Splitting definitives: the separation of the definite article in medieval and pre-modern written Judeo-Arabic. J. Jewish Lang. **9**(1), 32–76 (2021)
8. Dicta: Synopsis Builder. https://synoptic.dicta.org.il
9. Diem, M., Kleber, F., Fiel, S., Grüning, T., Gatos, B.: cBAD: ICDAR2017 competition on baseline detection. In: Proceedings of the 14th IAPR International Conference on Document Analysis and Recognition (ICDAR), vol. 1, pp. 1355–1360. IEEE (2017)
10. Droby, A., Kurar Barakat, B., Madi, B., Alaasam, R., El-Sana, J.: Unsupervised deep learning for handwritten page segmentation. In: Proceedings of the 17th International Conference on Frontiers in Handwriting Recognition (ICFHR), pp. 240–245. IEEE (2020)
11. Dutta, K., Krishnan, P., Mathew, M., Jawahar, C.V.: Improving CNN-RNN hybrid networks for handwriting recognition. In: Proceedings of the 16th International Conference on Frontiers in Handwriting Recognition (ICFHR), pp. 80–85. IEEE (2018)
12. Fink, M., Layer, T., Mackenbrock, G., Sprinzl, M.: Baseline detection in historical documents using convolutional U-nets. In: Proceedings of the 13th IAPR International Workshop on Document Analysis Systems (DAS), pp. 37–42. IEEE (2018)
13. Gan-Zvi, M.: Parashat Pinchas in Kitáb-al-Tuffaha and the Early Judeo-Arabic Homiletics. Master's thesis, The University of Haifa (2018)
14. Goodfellow, I., Bengio, Y., Courville, A.: Deep Learning. MIT Press (2016). http://www.deeplearningbook.org

15. Granet, A., Morin, E., Mouchère, H., Quiniou, S., Viard-Gaudin, C.: Transfer learning for handwriting recognition on historical documents. In: Proceedings of the 7th International Conference on Pattern Recognition Applications and Methods (ICPRAM) (2018)
16. Graves, A., Fernández, S., Gomez, F., Schmidhuber, J.: Connectionist temporal classification: labelling unsegmented sequence data with recurrent neural networks. In: Proceedings of the 23rd International Conference on Machine Learning (ICML), pp. 369–376 (2006)
17. Grüning, T., Leifert, G., Strauß, T., Michael, J., Labahn, R.: A two-stage method for text line detection in historical documents. Int. J. Document Anal. Recogn. (IJDAR) **22**(3), 285–302 (2019). https://doi.org/10.1007/s10032-019-00332-1
18. Jaramillo, J.C.A., Murillo-Fuentes, J.J., Olmos, P.M.: Boosting handwriting text recognition in small databases with transfer learning. In: Proceedings of the 16th International Conference on Frontiers in Handwriting Recognition (ICFHR), pp. 429–434. IEEE (2018)
19. Kahle, P., Colutto, S., Hackl, G., Mühlberger, G.: Transkribus-a service platform for transcription, recognition and retrieval of historical documents. In: Proceedings of the 14th IAPR International Conference on Document Analysis and Recognition (ICDAR), vol. 4, pp. 19–24. IEEE (2017)
20. Keret, S., Wolf, L., Dershowitz, N., Werner, E., Almogi, O., Wangchuk, D.: Transductive learning for reading handwritten Tibetan manuscripts. In: Proceedings of the International Conference on Document Analysis and Recognition (ICDAR), pp. 214–221. IEEE (2019)
21. Khan, G.: Judeo-Arabic. In: Handbook of Jewish Languages, pp. 22–63. Brill (2016)
22. Kiessling, B.: Kraken – An universal text recognizer for the humanities. In: Digital Humanities (DH2019) (2019)
23. Kiessling, B., Tissot, R., Stokes, P., Stökl Ben Ezra, D.: eScriptorium: an open source platform for historical document analysis. In: Proceedings of the International Conference on Document Analysis and Recognition Workshops (ICDARW), vol. 2, pp. 19–19. IEEE (2019)
24. Kuflik, T., et al.: Tikkoun Sofrim combining HTR and crowdsourcing for automated transcription of Hebrew medieval manuscripts. In: Digital Humanities (DH2019) (2019)
25. Kurar Barakat, B., El-Sana, J., Rabaev, I.: The Pinkas dataset. In: Proceedings of the International Conference on Document Analysis and Recognition (ICDAR), pp. 732–737. IEEE (2019)
26. Lavee, M.: Literary canonization at work: the authority of aggadic midrash and the evolution of havdalah poetry in the Genizah. AJS Rev. **37**(2), 285–313 (2013)
27. Liwicki, M., Graves, A., Fernàndez, S., Bunke, H., Schmidhuber, J.: A novel approach to on-line handwriting recognition based on bidirectional long short-term memory networks. In: Proceedings of the 9th International Conference on Document Analysis and Recognition (ICDAR) (2007)
28. Nahra, R.: Kitab al-Tuffāḥa: A Collection of Judaeo-Arabic Homilies on the Torah, from the End of the 11th or the Beginning of the 12th Century. Introduction with an Edition of the Homilies on the Book of Bereshit. Ph.D. thesis, Hebrew University of Jerusalem (2016), [Hebrew]
29. Noubigh, Z., Mezghani, A., Kherallah, M.: Transfer learning to improve Arabic handwriting text recognition. In: Proceedings of the 21st International Arab Conference on Information Technology (ACIT), pp. 1–6. IEEE (2020)
30. Oxford University, Bodleian Library: MS. Huntington 115 (2015). https://www.e-codices.unifr.ch/en/list/one/bge/cl0146

31. Reul, C., et al.: OCR4all-An open-source tool providing a (semi-) automatic OCR workflow for historical printings. Appl. Sci. **9**(22), 4853 (2019)
32. Sabir, E., Rawls, S., Natarajan, P.: Implicit language model in LSTM for OCR. In: Proceedings of the 14th IAPR International Conference on Document Analysis and Recognition (ICDAR), vol. 7, pp. 27–31. IEEE (2017)
33. Sadeh, G., Wolf, L., Hassner, T., Dershowitz, N., Stökl Ben Ezra, D.: Viral transcript alignment. In: Proceedings of the 13th International Conference on Document Analysis and Recognition (ICDAR), pp. 711–715. IEEE (2015)
34. Sefaria Inc: A living library of Torah texts online, December 2021. https://github.com/Sefaria/Sefaria-Export
35. Seuret, M., Stökl Ben Ezra, D., Liwicki, M.: Robust heartbeat-based line segmentation methods for regular texts and paratextual elements. In: Proceedings of the 4th International Workshop on Historical Document Imaging and Processing, pp. 71–76 (2017)
36. Stillman, N.A.: The Judeo-Arabic heritage. In: Zion, Z. (ed.) Sephardic & Mizrahi Jewry: From the Golden Age of Spain to Modern Times, pp. 40–54. NYU Press (2005)
37. Stökl Ben Ezra, D.: Medieval Hebrew manuscripts in Ashkenazi bookhand (2021). https://zenodo.org/record/5468478. Accessed 31 Jan 22
38. Stökl Ben Ezra, D.: Medieval Hebrew manuscripts in Italian bookhand, version 1.0 (2012). https://zenodo.org/record/5468573. Accessed 31 Jan 22
39. Stökl Ben Ezra, D.: Medieval Hebrew manuscripts in Sephardi bookhand, version 1.0 (2021). https://zenodo.org/record/5468665. Accessed 31 Jan 22
40. Stökl Ben Ezra, D.: Medieval Hebrew manuscripts, version 1.0 (2021). https://zenodo.org/record/5468286. Accessed 31 Jan 22
41. Stökl Ben Ezra, D., Brown-DeVost, B., Jablonski, P., Lapin, H., Kiessling, B., Lolli, E.: BiblIA-a general model for medieval Hebrew manuscripts and an open annotated dataset. In: Proceedings of the 6th International Workshop on Historical Document Imaging and Processing (HIP), pp. 61–66 (2021)
42. The Friedberg Jewish Manuscript Society: The Friedberg Judeo-Arabic Project (2014). https://ja.genizah.org/. Accessed 2022 01 08
43. Ul-Hasan, A., Breuel, T.M.: Can we build language-independent OCR using LSTM networks? In: Proceedings of the 4th International Workshop on Multilingual OCR, pp. 1–5 (2013)
44. Wecker, A.J., et al.: Tikkoun Sofrim: Making ancient manuscripts digitally accessible: The case of Midrash Tanhuma. ACM J. Comput. Cultural Heritage (JOCCH) **15**(2), 1–20 (2022)
45. Xu, Y., He, W., Yin, F., Liu, C.L.: Page segmentation for historical handwritten documents using fully convolutional networks. In: Proceedings of the 14th IAPR International Conference on Document Analysis and Recognition (ICDAR), vol. 1, pp. 541–546. IEEE (2017)

Decoupled Learning for Long-Tailed Oracle Character Recognition

Jing Li[1], Bin Dong[2], Qiu-Feng Wang[1(✉)], Lei Ding[2], Rui Zhang[3], and Kaizhu Huang[4]

[1] School of Advanced Technology, Xi'an Jiaotong-Liverpool University, Suzhou, China
Jing.Li19@student.xjtlu.edu.cn, Qiufeng.Wang@xjtlu.edu.cn
[2] Ricoh Software Research Center(Beijing) Co., Ltd., Beijing, China
{Bin.Dong,Lei.Ding}@cn.ricoh.com
[3] School of Mathematics and Physics, Xi'an Jiaotong-Liverpool University, Suzhou, China
Rui.Zhang02@xjtlu.edu.cn
[4] Data Science Research Center, Duke Kunshan University, Suzhou, China
Kaizhu.Huang@dukekunshan.edu.cn

Abstract. Oracle character recognition has recently made significant progress with the success of deep neural networks (DNNs), but it is far from being solved. Most works do not consider the long-tailed distribution issue in oracle character recognition, resulting in a biased DNN towards head classes. To overcome this issue, we propose a two-stage decoupled learning method to train an unbiased DNN model for long-tailed oracle character recognition. In the first stage, we optimize the DNN under instance-balanced sampling, obtaining a robust backbone but biased classifier. In the second stage, we propose two strategies to refine the classifier under class-balanced sampling. Specifically, we add a learnable weight scaling module which can adjust the classifier to respect tail classes; meanwhile, we integrate the KL-divergence loss to maintain attention to head classes through knowledge distillation from the first stage. Coupling these two designs enables us to train an unbiased DNN model in oracle character recognition. Our proposed method achieves new state-of-the-art performance on three benchmark datasets, including OBC306, Oracle-AYNU and Oracle-20K.

Keywords: Oracle character recognition · Long tail · Decoupled learning · Knowledge distillation

1 Introduction

As the earliest known writing system in China, the oracle bone script plays a significant role in archaeology, palaeography, and history. These characters were carved on turtle nails and animal bones for divination during the Shang Dynasty. Recognizing them usually requires a high level of expertise, which is both time-consuming and costly. Recently, much attention has been paid to investigating automatic recognition technologies for oracle characters. Such research has

G. A. Fink et al. (Eds.): ICDAR 2023, LNCS 14190, pp. 165–181, 2023.
https://doi.org/10.1007/978-3-031-41685-9_11

made great progress [1–3], but its performance still needs improvement so as to meet the requirements of practical applications. With the burst of deep neural networks (DNNs) and their successful applications in computer vision, natural language processing, etc [4], researchers have recently explored DNNs in oracle character recognition [2,3,5]. The success of DNN models usually needs a large size of labelled training samples. However, obtaining sufficient oracle character data is challenging due to the scarcity of sources and difficult labelling.

Thanks to the great efforts from the research community, there are some available oracle datasets, which can be divided into two categories: real rubbing character images and hand-copied character images. Some examples are shown in Fig. 1. As we can see, real images scanned from turtle nails and animal bones contain various noises, e.g., partially missing, dense white regions, and bone fractures. One public representative dataset is OBC306, collected by [2], which contains 309,511 character-level instances belonging to 306 classes. In contrast, hand-copied images are high-resolution images without noise, but it needs to invite experts to transcribe them. Two additional available datasets are Oracle-20K [1] and Oracle-AYNU [3]. Oracle-20K contains 19,491 character images and 249 classes, while Oracle-AYNU has 2,584 classes with 39,072 instances.

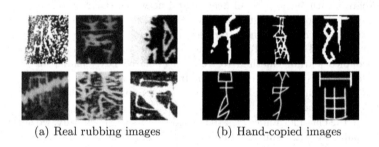

(a) Real rubbing images   (b) Hand-copied images

Fig. 1. Examples of oracle character images.

Unfortunately, all these current oracle datasets suffer from the common issue arising in the long-tailed distribution, as shown in Fig. 2 for the OBC306 dataset. It is apparent that the number of samples varies significantly among classes. Specifically, in the training set of OBC306, the top five majority classes have over 10,000 instances while many classes have just one or two instances; in the test set of OBC306, around 17% of classes contain fewer than ten samples, while the largest class contains 6,474 samples. Such distribution leads the training of DNNs to suffer from a strong bias towards head classes; consequently, the learned model cannot learn robust classification for tail classes[1]. To address this issue, Zhang et al. [3] proposed a nearest neighbour classifier with metric learning for imbalanced oracle character recognition, and successfully improved the accuracy

---

[1] We divide the oracle data into three categories: the classes with many samples as head classes, the classes with few samples as tail classes, and the remainder are the medium classes described in Sect. 4.3.

on Oracle-AYNU and Oracle-20K. Furthermore, Li et al. [5] designed a mix-up strategy by combining softmax loss and triplet loss, and demonstrated the state-of-the-art performance on oracle datasets, including Oracle-20K, Oracle-AYNU, and OBC306. Albeit these advances, the issue of long-tailed distribution is still far from being solved in oracle character recognition.

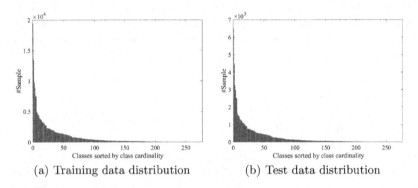

(a) Training data distribution   (b) Test data distribution

Fig. 2. Data distributions of OBC306.

Although the long-tailed issue has not yet received extensive attention in oracle character recognition, it has been intensively studied in general visual recognition tasks. For example, re-sampling training samples [6,7] or adjusting the loss value for each class [8,9] has been widely used to balance the class distribution during training. In addition, some methods [10,11] utilize label frequencies to shift output logits of models during training or post-processing. Some recent efforts have also been made to re-balance DNN models. In particular, studies [12–15] have shown that it is effective to decouple the one-stage training process into representation learning and classifier learning for imbalanced data. In general, such works adjust the classifier to focus on tail classes while sacrificing the performance of head classes during the training. They then validate the effectiveness on the test data with a balanced distribution across all classes. However, the test data in the oracle datasets is also heavily long-tailed as shown in Fig. 2. Performance sacrifice on head classes may unfortunately lead to the degradation of the total accuracy, though the average accuracy is still improved as presented in Table 2 of Sect. 4.4.

Motivated from the aforementioned analysis, we propose a two-stage-based decoupled learning method for long-tailed oracle character recognition, where the DNN model is split into a ViT [16] as the backbone network and a single fully connected layer network as the classifier. In the first stage, we train the DNN model with the standard cross-entropy loss under instance-balanced sampling. As the oracle data is limited, we utilize mixup augmentation [17] to exploit current oracle samples fully. Although a robust backbone model is learned in the first stage, the long-tailed oracle data distribution makes the classifier strongly biased towards the head classes. Therefore, we further propose two strategies to refine the classifier under class-balanced sampling in the second stage while freezing the backbone. First, we add a learnable weight scaling (LWS) module to adjust the classifier to respect tail classes. Second, we integrate the KL-divergence loss to keep noticing head classes through knowledge distillation from the first stage. Coupling these two designs enables us to train an unbiased DNN model on both tail and head classes in oracle character recognition, thus offering the strong potential to improve both the average and total accuracies. We evaluate the proposed method on benchmarks including OBC306, Oracle-AYNU, and Oracle-20K. Experimental results show that our novel design attains new state-of-the-art performance.

2   Related Work

2.1   Oracle Character Recognition

Identifying characters from hand-copied or scanned oracle bone images has long been considered as a challenging problem. It has attracted much attention and achieved tremendous advances [1,2,18]. Earlier studies often adopted traditional pattern recognition techniques on oracle character recognition. For example, the work in [18] treated oracle bone inscriptions as undirected graphs and applied graph isomorphism for identification. Guo et al. [1] proposed a hierarchical representation for oracle characters, consisting of a Gabor-related low-level representation and a sparse-encoder-related mid-level representation. Liu et al. [19] recognized oracle characters by extracting block histogram-based features and employing support vector machines.

Recently, DNN-based methods have also been applied in oracle character recognition. In the early stage, researchers combined Convolutional Neural Network (CNN) models with traditional feature representation [1]. Next, Huang et al. [2] evaluated several popular CNNs (e.g., ResNet, InceptionNet) in their established OBC306. As DNN models usually require a large number of labelled samples for training, researchers have to make significant efforts on the oracle data collection. To this end, Guo et al. [1] first collected about 20,000 legible oracle character images called Oracle-20k. Then, Anyang Normal University constructed another hand-copied dataset Oracle-AYNU [1], and Huang et al. [2] released a large-scale scanned oracle character dataset called OBC306.

Owing to the difficulty in obtaining oracle characters, current oracle data is both rare and seriously long-tailed, making the DNN-based recognition of oracle characters challenging. Zhang et al. [3] first investigated the seriousness of this issue and proposed a nearest neighbour classifier with metric learning. Following that, Li et al. [5] integrated mix-up augmentation and triplet loss to improve the recognition performance. However, such long-tailed distribution issue is far from being solved. In this paper, we aim to train an unbiased DNN model for long-tailed oracle character recognition via the proposed two-stage decoupled learning method.

2.2   Long-Tailed Visual Recognition

It is crucial to obtain an unbiased model for all classes in long-tailed visual recognition. Most existing works can be divided into two categories: re-sampling and re-weighting. Re-sampling-based techniques typically obtain a more balanced data distribution by over-sampling tail classes or under-sampling head classes [6,7]. Re-weighting methods assign appropriate weights to the loss of each class to re-balance classes [8,9]. In addition, some methods adjust model logits during training or post-processing based on label frequencies to achieve relatively large margins between classes, which can also strengthen the classification of tail classes and mitigate the long-tailed distribution issue [10,11].

Recently, researchers have started studying two-stage-based decoupled learning in DNN models for imbalanced recognition instead of end-to-end learning. Kang et al. [12] proposed to decouple the one-stage training of the DNN model into feature representation (i.e., backbone network) and classifier learning. This work demonstrated that it could learn well-generalized representation under the normal instance-balanced sampling in the first training stage, and merely adjusting the classifier in the second stage is effective for imbalanced recognition. Based on this study, researchers innovated the decoupled learning scheme from different aspects. For example, KCL [13] developed a k-positive contrastive loss to learn a more balanced and discriminative feature representation. MiSLAS [14] proposed to adopt mixup augmentation in the first stage to enhance the representation learning and applied a label-aware smoothing strategy in the second stage. The work [15], following the weight re-balancing direction, proposed to tune weight decay in the first stage and utilized class-balanced loss with tuning weight decay in the second stage.

Despite the effectiveness, these present approaches conduct their evaluation on the test data, which usually follows a balanced distribution across all classes. In case of imbalanced test data, these methods would largely degrade their performance as they aim to obtain a uniform distribution during the training. Several works have noticed such an issue which is still less explored [20,21]. In this paper, we focus on decoupled learning for long-tailed oracle character recognition with imbalanced test data.

2.3   Knowledge Distillation

Knowledge distillation (KD) is proposed to achieve better generalization performance by transferring knowledge from pre-trained models (i.e., teachers) to target networks (i.e., students) [22]. The concept of KD was proposed by Hinton et al. [23], which transfers the knowledge via minimizing the KL-Divergence loss between predicted logits of teachers and students. Several recent works have explored KD for imbalanced visual recognition. Xiang et al. [24] divided the whole dataset into subsets and trained multiple teacher models for each subset. Meanwhile, a unified student model was trained by using adaptive KD in an easy-to-hard curriculum instance selection approach. Following this multi-expert framework, Wang et al. [20] introduced one special KD approach to simplify the multi-expert model by training a student network with multiple experts. In this paper, we propose to integrate the idea of knowledge distillation in our two-stage-based decoupled learning framework for long-tailed oracle recognition, aiming to make the classifier not ignore the head classes by transferring the knowledge from the first stage.

3 Main Methodology

In this paper, we develop a two-stage decoupled learning to train an unbiased DNN model for both tail and head classes in long-tailed oracle character recognition. The overview of our work is illustrated in Fig. 3. In the first stage, the DNN model is trained by the cross-entropy loss under instance-balanced sampling. To be noted, we divide the DNN model into two components, i.e., backbone and classifier. Due to the insufficient oracle data, we adopt mixup augmentation to explore the training samples in the first stage fully. Although a well-generalized representation can be obtained from the backbone in the first stage [12], the classifier is usually biased to head classes because of the long-tailed distribution. To overcome this issue, we train a learnable weight scaling (LWS) module to adjust the classifier with the frozen backbone under class-balanced sampling in the second stage, thus paying more attention to the tail classes. In addition, we integrate the KL-Divergence loss in the second stage to maintain attention on the head classes by the knowledge distillation from the first stage. In the inference stage, the oracle character images pass through the backbone and classifier with LWS to output the final recognition result.

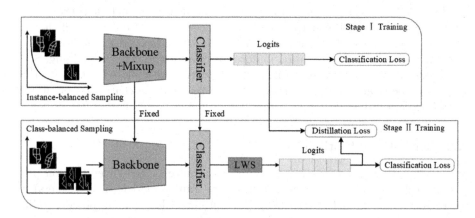

Fig. 3. Overall structure of the proposed scheme. LWS represents the learnable weight scaling, aiming to adjust the classifier to respect the tail classes.

3.1   Backbone Architecture

In terms of the excellent performance of Transformer in natural language processing and computer vision areas [16,25,26], we employ ViT [16] as the backbone network for oracle character recognition in this paper.

ViT mainly consists of patch embedding, position embedding, class token, and Transformer encoder [27]. First, we convert an image into a sequence of 1-D patch embeddings. Given an image $\boldsymbol{X} \in \mathbb{R}^{H \times W \times M}$ (H, W, M denote the height, width, and the number of channels, respectively), we divide $\boldsymbol{X}$ into a series of patches denoted as $\boldsymbol{X}_p \in \mathbb{R}^{S \times P \times P \times M}$, where $P \times P$ is the patch size and S is the number of patches (i.e., $S = HW/P^2$). Then, we flatten these cropped patches to be $\boldsymbol{X}_{p'} \in \mathbb{R}^{S \times (P^2 \cdot M)}$ and utilize a trainable linear projection $\boldsymbol{E} \in \mathbb{R}^{(P^2 \cdot M) \times D}$ to generate the sequence of 1-D embedding of all patches $\boldsymbol{X}_{p''} \in \mathbb{R}^{S \times D}$. Second, motivated from BERT [25], we add a learnable class token $\boldsymbol{x}_{class}$ to the beginning of the patch embeddings, which can be regarded as the representation of the input image and fed into the latter classifier. Third, we integrate the learnable position embedding $\boldsymbol{E}_{pos} \in \mathbb{R}^{(S+1) \times D}$ to retain the position information in the sequence as the Transformer encoder is permutation-invariant. Therefore, the overall input to the Transformer encoder can be defined as:

$$Z_0 = [\boldsymbol{x}_{class}; \boldsymbol{X}_{p''}] + \boldsymbol{E}_{pos}. \tag{1}$$

The Transformer encoder is composed of L layers. Every layer contains one multi-headed self-attention (MSA) block and one multi-layer perceptron (MLP) block. MLP includes two linear layers with a GELU. LayerNorm (LN) is applied before each block, while residual connections are applied after each block. More details can be seen in [27]. The process can be simply formulated as follows:

$$\begin{cases} \hat{\boldsymbol{Z}}_l = MSA(LN(\boldsymbol{Z}_{l-1})) + \boldsymbol{Z}_{l-1}, l = 1 \ldots L \\ \tilde{\boldsymbol{Z}}_l = MLP(LN(\hat{\boldsymbol{Z}}_l)) + \hat{\boldsymbol{Z}}_l, \quad l = 1 \ldots L \\ \boldsymbol{z} = LN(\tilde{\boldsymbol{Z}}_L[0]). \end{cases} \tag{2}$$

3.2   Mixup Augmentation

Mixup [17] and its variants [28,29] have been widely adopted as data augmentation strategies in long-tailed tasks [5,14,30], which enable to improve the generalization of DNN models. The basic concept of Mixup is to generate new samples by interpolating two randomly sampled input images with their labels $(\boldsymbol{X}, \boldsymbol{y})$ and $(\boldsymbol{X}', \boldsymbol{y}')$ as follows:

$$\tilde{\boldsymbol{X}} = \lambda \boldsymbol{X} + (1 - \lambda)\boldsymbol{X}', \tag{3}$$

$$\tilde{\boldsymbol{y}} = \lambda \boldsymbol{y} + (1 - \lambda)\boldsymbol{y}', \tag{4}$$

where $\tilde{\boldsymbol{X}}$ denotes the mixed new sample and its label is $\tilde{\boldsymbol{y}}$, λ is a mixing factor drawn from a Beta distribution $Beta(\alpha, \beta)$ and $\alpha = \beta = 1$ in our experiments. We integrate this original Mixup augmentation method in the first training stage for better representation learning of the backbone. Since the LWS module trained in the second stage is lightweight, we remove Mixup to reduce the complexity. Due to the limited space, we do not compare other variants of Mixup [28,29] in this paper, and more comparisons can be found in [5].

3.3 Decoupled Learning

In the paper, we decouple the DNN model into backbone and classifier in the two-stage training framework, where the backbone is learned in the first stage and the classifier is adjusted with the LWS module in the second stage for the long-tailed oracle character recognition.

In the first stage, we adopt the ViT model as the backbone to learn the representation z from the input X, then obtain the classification logit $\hat{y}$ by feeding z into the linear classifier. Finally, the predicted class could be given by argmax $\hat{y}$ as follows:

$$\hat{y} = C(z) = W^T z + b, \tag{5}$$

where $C(\cdot)$ represents the classifier, W denotes the weight matrix and b denotes the bias. In this stage, the backbone and classifier are learned jointly by minimizing the standard cross-entropy loss between ground truth y and argmax $\hat{y}$ under instance-balanced sampling. Here, the probability of sampling data from the class j is proportional to its cardinality n_j. Therefore, the long-tailed data distribution makes the learned model biased to head classes.

In the second stage, LWS aims to rectify the imbalanced decision boundaries between head and tail classes via re-scaling the magnitude of weights in W for each class. To this end, we utilize a scaling factor f_j for the j-th class to adjust the weights of the classifier:

$$\tilde{w}_j = f_j * w_j, \tag{6}$$

where $w_j \in \mathbb{R}^d$ represents the weight vector of class j. Then the whole weight matrix $W = \{w_j\} \in \mathbb{R}^{d \times Y}$ is re-scaled to $\tilde{W}$ so that $\hat{y}$ becomes $\tilde{W}^T z + b$, where Y is the number of classes. In this way, merely the LWS block (denoted as $f = \{f_j\} \in \mathbb{R}^Y$) is learned by class-balanced sampling in the second stage, while the backbone and classifier (i.e., W and b) are fixed. Under class-balanced sampling, each class shares an equal probability of being selected. Once a class is selected, an instance is sampled uniformly from the chosen class, so it is unbiased sampling. To be noted, LWS is very lightweight, thereby its learning can converge quickly in the second stage of training.

3.4 Logit-Based Knowledge Distillation

Although LWS in the second stage can promote the learned DNN model to highlight tail classes, it will lose the importance of the head classes. However, the oracle test data is also a long-tailed distribution, and the recognition of head classes plays a crucial role in the overall performance. Since the model learned in the first stage better understands the head classes, we can utilize this model to guide LWS in the second stage to keep attention on head classes. Motivated by this, we propose a knowledge distillation strategy in the second training stage.

Logit-based knowledge distillation aims to transfer the knowledge from a teacher model to a student model by aligning their logit predictions [22]. In this paper, we leverage the first stage model as the teacher model to guide the second stage model, where the popular soft targets [23] are adopted as our logit-based knowledge. Specifically, given classification logits $\hat{y}$ of the classes, the soft targets can be obtained by the softmax function as follows:

$$p_j = \frac{\exp(\hat{y}_j/T)}{\sum_{t=1}^{Y} \exp(\hat{y}_t/T)}, \tag{7}$$

where T denotes the temperature that controls the importance of each soft target and is set to 1 in our experiments. The classical KD adopts KL-Divergence as the distillation loss for soft targets, which can be rewritten as:

$$\mathrm{KL}(\boldsymbol{p}^T||\boldsymbol{p}^S) = \sum_{i=1}^{Y} p_i^T \log(\frac{p_i^T}{p_i^S}). \tag{8}$$

Here, p^T and p^S represent the teacher and student model output logits, respectively. By integrating the distillation loss in the second stage, the decision boundaries of head classes can be protected to some extent.

3.5 Overall Training

Our work follows a two-stage training scheme, which can be summarized as (1) representation learning by training the backbone and classifier under the cross-entropy loss with instance-balance sampling, and (2) classifier learning by training the integrated LWS under the cross-entropy and KL losses with class-balanced sampling.

In the first training stage, we aim to obtain well-generalized representations and achieve higher performance in the head classes for subsequent knowledge transfer. According to [12], instance-balanced sampling with the cross-entropy loss yields a more general representation than other re-sampling methods. Therefore, we adopt this training strategy in the paper. In addition, we integrate Mixup [17] to improve further the representation ability and recognition performance motivated by the previous works [5,14]. The loss function in the first stage is the cross-entropy loss:

$$L_{s1} = \sum_{n=1}^{N} -\tilde{\boldsymbol{y}}_n \log(C(B(\tilde{\boldsymbol{X}}_n))), \tag{9}$$

where N is the number of training samples. $B(\cdot)$ represents the backbone network.

In the second stage, Zhong et al. [14] indicate that Mixup has negligible or even negative effects on classifier learning. Moreover, our trainable LWS is lightweight. Therefore, we remove Mixup at this stage. Furthermore, to keep noticing head classes, the distillation loss is adopted for classes with larger cardinalities. The overall loss function of the second stage is defined as:

$$L_{s2} = \sum_{n=1}^{N} -\boldsymbol{y}_n \log(C(B(\boldsymbol{X}_n))) + \sum_{i \in \mathbb{N}} \mathrm{KL}_i(\boldsymbol{p}^T || \boldsymbol{p}^S). \tag{10}$$

Here, $\mathbb{N}$ represents the subset of classes whose cardinality is larger than γ, indicating those classes with knowledge distillation. In our experiments, we set γ as 100, 150 and 160 in OBC306, Oracle-AYNU and Oralce-20k, respectively, which are tuned empirically.

4 Experiments

4.1 Datasets

OBC306. OBC306 [2] is currently the largest public dataset of oracle bone scripts to our knowledge, which contains 309,551 character samples with 306 classes in total. As shown in Fig. 2, this dataset suffers from a typical long-tailed distribution, i.e., 70 classes cover 83.82% of total samples while 52 classes have fewer than ten samples. We remove 29 classes with only one sample since we do not consider the out-of-vocabulary (OOV) performance. Then, the remaining dataset is divided into training and test sets following a 3:1 ratio. Finally, OBC306 used in the paper has 277 classes with 309,522 samples. The imbalance ratios (i.e., the size of the largest class: the smallest class) of the training set and test set are 19,424:1 and 6,474:1, respectively. To be noted, all samples in OBC306 are oracle bone rubbing images with various noises.

Oracle-AYNU. Oracle-AYNU [3] consists of 39,072 hand-copied oracle character samples with 2,584 classes. Specifically, the cardinality of each class varies from 2 to 287, and about 68% of classes have fewer than ten samples. We divide the dataset into the training and test sets following a 9:1 ratio, and then the imbalance ratios of the training set and test are 259:1 and 28:1, respectively. We can see that this dataset also suffers a long-tailed distribution issue, but not severe as that in OBC306.

Oracle-20K. Oracle-20K [1] contains 19,491 hand-copied samples with 249 classes, where class cardinalities range from 25 to 291. We split the dataset into training and test sets following a ratio of 2:1, and then the imbalance ratio of the training set and test set are 194:17 and 97:8, respectively. We can see that its imbalanced issue is not very severe compared to the other two benchmark datasets.

4.2 Implementation Details

We implement our model by Pytorch. In all experiments, we adopt ViT-Base [16] pre-trained on ImageNet as the backbone model, and utilize the SGD optimizer with momentum 0.9, batch size 64, image size 256 × 256. In the first stage, the initial learning rate is 0.01 and deceased by 0.1 at the m_1-th and m_2-th epochs (i.e., $m_1 = 15$ and $m_2 = 20$ in OBC306, $m_1 = 100$ and $m_2 = 150$ in Oracle-AYNU, $m_1 = 150$ and $m_2 = 200$ in Oracle-20K). In the second stage, we restart the learning rate to train the LWS module with 0.2 for OBC306 and 0.01 for both Oracle-AYNU and Oracle-20K. The learning rate is decreased by 0.1 at m_1'-th and m_2'-th epochs (i.e., $m_1' = 2$ and $m_2' = 4$ in both OBC306 and Oracle-AYNU, $m_1' = 5$ and $m_2' = 10$ in Oracle-20K).

4.3 Evaluation Metrics

Most previous papers adopt the total accuracy as defined in Eq. (11) to evaluate the recognition performance. However, if the test data also follows long-tailed data distribution, this metric will be dominated by those head classes. To reflect the effectiveness on tail classes as well, we exploit another metric additionally to evaluate average accuracy as defined in Eq. (12). Following [5], such two metrics are formulated by

$$Total = \frac{\sum_{j=1}^{Y} r_j}{\sum_{j=1}^{Y} n_j}, \tag{11}$$

$$Average = \frac{1}{Y} \sum_{j=1}^{Y} \frac{r_j}{n_j}, \tag{12}$$

where r_j and n_j denote the number of correctly classified samples and total samples in the j-th class, respectively. In addition, to better demonstrate the performance on the long-tailed distribution, we split each dataset into three subsets, namely *Head* (more than 100 samples), *Medium* (20~100 samples), and *Tail* (fewer than 20 samples).

4.4 Ablation Study

Effects of different patch sizes of ViT. Similar to ViT [16], we also compare the effectiveness of different patch sizes in oracle character recognition. As shown in Fig. 4, if the patch size is too small, there will be too many less expressive patches input to the Transformer encoder, e.g., background patches or similar patches; if the patch size is too large, some non-trivial spatial and local information will be lost, leading to low discriminability. We try three sizes without pretraining, including 16 × 16, 32 × 32 and 64 × 64. The results are reported in Table 1. We can see that 32 × 32 provides the best performance. Furthermore, ViT usually requires pre-training on a large amount of data and then transferring to small datasets to obtain good results. Therefore, we choose the "Base" variant of ViT with 32 × 32 patch size pre-trained on ImageNet as our backbone, and fine-tune it together with the linear classifier on oracle datasets.

Effects of Different Components. In this part, we evaluate the effectiveness of the proposed LWS and KD. For clarity, we also report the performance of

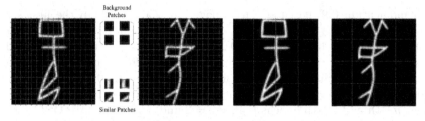

(a) Small input patches   (b) Large input patches

Fig. 4. Visualization of different patch sizes.

Table 1. Comparison on OBC306, Oracle-AYNU and Oracle-20K in terms of the average and total accuracies (%) of ViT-Base with different patch sizes.

Patch Size	OBC306		Oracle-AYNU		Oracle-20K	
	Average	Total	Average	Total	Average	Total
16 × 16	65.36	87.81	61.54	73.84	82.78	86.79
32 × 32	**68.08**	**88.97**	**66.65**	**77.88**	**87.80**	**90.76**
64 × 64	60.38	85.88	65.18	76.68	86.30	90.20

Mixup. The results are summarized in Table 2, and its first row shows the baseline model where the backbone and classifier are jointly trained in one stage without any proposed component. Compared with the baseline model, we can see that Mixup (the second row in Table 2) improves the performance significantly in terms of both the average and total accuracy, demonstrating the effectiveness of the Mixup strategy.

To facilitate the recognition of tail classes, we learn LWS in the second stage of training to adjust the decision boundaries from the baseline model as shown in the third row of Table 2. We observe that LWS improves the average accuracy, especially on OBC306 and Oracle-AYNU, with significant gains of 5.08% and 4.79%, respectively, since both datasets have many tail classes. LWS adjusts the weights of the classifier under class-balanced sampling, which favors tail classes while weakening the dominance of head classes. Therefore, we suppose that the learned LWS will reduce the performance of the head classes, resulting in a reduction of total accuracy on OBC306. As OBC306 has an extremely large imbalance ratio (i.e., 9,424:1 and 6,474:1 in the training and test set, respectively) and the head classes dominate the test set, a slight suppression of the head classes significantly impacts the total accuracy. In contrast, the total accuracy is improved on Oracle-AYNU as the tail classes account for a large proportion in this dataset; however, the improvement is fairly smaller than the average accuracy. Combining Mixup and LWS (the fifth row of Table 2), recognition performance can be lifted up further, but the total accuracy on OBC306 is still reduced significantly.

To mitigate the issue of LWS on head classes, we plug the distillation loss in the second training stage to preserve the decision boundaries of head classes. The

Table 2. Ablation study for three proposed components on OBC306, Oracle-AYNU, and Oracle-20K in terms of the average and total accuracies (%). The first row is the baseline model. Mixup: adding Mixup to the baseline model. LWS: training LWS in the second stage. KD: adding the distillation loss in the second stage.

Mixup	LWS	KD	OBC306		Oracle-AYNU		Oracle-20K	
			Average	Total	Average	Total	Average	Total
✗	✗	✗	79.50	93.02	77.07	84.42	93.78	95.37
✔	✗	✗	77.07	**93.12**	79.75	86.23	94.57	96.03
✗	✔	✗	84.58	90.05	81.86	86.68	93.92	95.32
✔	✔	✗	84.38	90.55	**83.53**	87.97	94.87	96.03
✔	✔	✔	**85.01**	92.40	83.27	**88.02**	94.9	**96.06**

results are listed in the last row of Table 2. On the one hand, adding KD boosts the total accuracy, especially on OBC306, from 90.55% to 92.40%; on the other hand, it makes little impact on the average accuracy, with a slight increase on OBC306 and Oracle-20K due to its effectiveness on the head classes. In summary, by combining all the proposed components, our proposed method can achieve a good trade-off between total and average accuracies, finally improving the overall recognition performance significantly.

To demonstrate the effectiveness of each component in detail, we show the total accuracy for the three subsets of OBC306 and Oracle-AYNU in Fig. 5. In OBC306, we can see that Mixup benefits the head subset more due to the severe long-tailed problem; after adding LWS, the accuracies of the tail and medium classes boost while head accuracy drops significantly by 2.76%. To keep noticing head classes, we adopt knowledge distillation to keep the decision boundaries of head classes while facilitating the tail classes. The proposed method (Mixup+LWS+KD) obtains a fair trade-off between the head and tail classes. Finally, it increases the tail accuracy by 3.61% compared to Mixup+LWS, without much loss in head performance compared to Mixup. In Oracle-AYNU, Mixup generates positive influences on all the three subsets. For LWS, it improves the accuracy of tail classes while reducing the accuracy of head classes. By combining it with KD, the accuracies of all the three subsets increase compared with the baseline model, from 94.14% to 94.68%, 93.12% to 93.19% and 74.41% to 81.69%, respectively. In summary, these results demonstrate the effectiveness of our proposed method apparently.

4.5 Comparison to Previous Methods

In Table 3, we compare the proposed method with the previous competitive oracle character recognition methods regarding average and total accuracy on three benchmark datasets. From the table, we can see that our approach surpasses the other methods on both average and total accuracy in all datasets. Specifically, the proposed method achieves superior performance on OBC306 with the serious long-tailed data issue, improving the average and total accuracies over

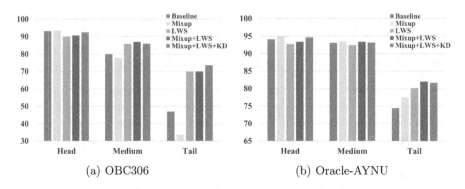

(a) OBC306   (b) Oracle-AYNU

Fig. 5. Total accuracy of each component on the head, medium and tail classes of OBC306 and Oracle-AYNU.

Table 3. Comparison to previous methods on Oracle-AYNU, OBC306, and Oracle-20K. † denotes the accuracy excluding 29 classes which do not have any training samples (The original average accuracy is reported as 70.28% on all the 306 classes [2]).

Method	OBC306		Oracle-AYNU		Oracle-20K	
	Average	Total	Average	Total	Average	Total
Zhang et al. [3]	–	–	–	83.37	–	92.43
Huang et al. [2]†	77.64	–	–	–	–	–
Li et al. [5]	80.16	91.74	79.96	83.65	93.50	94.64
Ours	**85.01**	**92.4**	**83.27**	**88.02**	**94.9**	**96.06**

the previous best oracle character recognition method [5] by 4.85% and 0.66%, respectively. For the Oracle-AYNU dataset with more than 2,000 classes, the proposed method achieves 83.27% in average accuracy and 88.02% in total accuracy, which outperforms the prior methods by a significant margin of 3.31% and 4.37%, respectively. In Oracle-20K, the long-tailed issue is not evident; however, our proposed method can also yield better average and total results. In summary, the proposed method achieves new state-of-the-art performance on three benchmark datasets.

4.6   Error Analyses

In Fig. 6, we present some error recognition examples of the proposed approach, which are roughly summarized into four categories: long tail, similar looking, intra-class variance, and severe noises. First, the long-tailed issue has not been fully resolved, and some tail characters are mis-recognized as shown in Fig. 6(a). Second, it is sometimes confusing for our model to identify similar characters. For example, the characters in Fig. 6(b) look very similar between the ground truth and prediction (e.g., in the second row, the character '1035' is mis-recognized to '100003'). Third, since there was no standardization of the oracle bone script during historical periods, some characters may have a high degree of intra-class

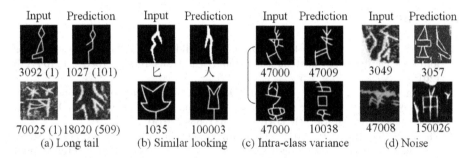

Input	Prediction	Input	Prediction	Input	Prediction	Input	Prediction
3092 (1)	1027 (101)	と	人	47000	47009	3049	3057
70025 (1)	18020 (509)	1035	100003	47000	10038	47008	150026
(a) Long tail		(b) Similar looking		(c) Intra-class variance		(d) Noise	

Fig. 6. Error examples of our model. Input columns present the input images with their ground-truth labels underneath (characters or digital codes). Prediction columns present samples from the corresponding incorrect predicted classes. Note that Oracle-AYNU and OBC306 are annotated by digital codes, not by real characters like Oracle-20K. In (a), the number in parentheses represents the number of training samples for that class. In (c), round brackets represent they belong to the same class.

variation in shapes, structures, and number of strokes in Fig. 6(c). Last, due to the long history, the real oracle characters on bones are polluted by serious noises and abrasions as shown in Fig. 6(d), making them difficult to be recognized even by humans.

5   Conclusion

In this paper, we propose a two-stage decoupled learning method for long-tailed oracle character recognition, aiming to train an unbiased DNN model on both tail and head classes. In the first stage, we train a ViT backbone model and a linear classifier under the instance-balanced sampling, where mixup augmentation is utilized to exploit current oracle samples fully. In the second stage, we propose a learnable weight scaling module to refine the classifier to respect tail classes. Meanwhile, the KL-divergence loss is also integrated to maintain attention on head classes by the knowledge distillation from the first stage. Extensive experiments on three oracle benchmark datasets demonstrate the effectiveness of both LWS and KD components, finally achieving new state-of-the-art recognition performance on average and total accuracy. However, there is still room to improve further the recognition performance, especially on tail classes. In the future, we will explore how to enhance the representation ability of the backbone on tail classes so as to further promote the decoupled learning framework.

Acknowledgements. This research was funded by National Natural Science Foundation of China (NSFC) no.62276258, Jiangsu Science and Technology Programme (Natural Science Foundation of Jiangsu Province) no. BE2020006-4, and Xi'an Jiaotong-Liverpool University's Key Program Special Fund no. KSF-T-06.

References

1. Guo, J., Wang, C., Roman-Rangel, E., et al.: Building hierarchical representations for oracle character and sketch recognition. IEEE Trans. Image Process. **1**, 104–118 (2016)
2. Huang, S., Wang, H., Liu, Y., et al.: OBC306: a large-scale oracle bone character recognition dataset. In: Proceedings of International Conference on Document Analysis and Recognition (ICDAR), pp. 681–688 (2019)
3. Zhang, Y.-K., Zhang, H., Liu, Y.-G., et al.: Oracle character recognition by nearest neighbor classification with deep metric learning. In: Proceedings of International Conference on Document Analysis and Recognition (ICDAR), pp. 309–314 (2019)
4. Huang, K., Hussain, A., Wang, Q.-F., Zhang, R.: Deep Learning: Fundamentals, Theory and Applications, vol. 2 (2019)
5. Li, J., Wang, Q.-F., Zhang, R., Huang, K.: Mix-up augmentation for oracle character recognition with imbalanced data distribution. In: Proceedings of International Conference on Document Analysis and Recognition (ICDAR), pp. 237–251 (2021)
6. Chawla, N.V., Bowyer, K.W., Hall, L.O., Kegelmeyer, W.P.: SMOTE: synthetic minority over-sampling technique. J. Artif. Intell. Res. **16**, 321–357 (2002)
7. Estabrooks, A., Jo, T., Japkowicz, N.: A multiple resampling method for learning from imbalanced data sets. Comput. Intell. **20**(1), 18–36 (2004)
8. Cui, Y., Jia, M., Lin, T.-Y., et al.: Class-balanced loss based on effective number of samples. In: Proceedings of Conference on Computer Vision and Pattern Recognition (CVPR), pp. 9268–9277 (2019)
9. Cao, K., Wei, C., Gaidon, A., et al.: Learning imbalanced datasets with label-distribution-aware margin loss. In: Proceedings of Advances in Neural Information Processing Systems (NeurIPS), pp. 1565–1576 (2019)
10. Menon, A.K., Jayasumana, S., Rawat, A.S., et al.: Long-tail learning via logit adjustment. In: Proceedings of International Conference on Learning Representations (ICLR) (2021)
11. Wu, T., Liu, Z., Huang, Q., et al.: Adversarial robustness under long-tailed distribution. In: Proceedings of Conference on Computer Vision and Pattern Recognition (CVPR), pp. 8659–8668 (2021)
12. Kang, B., Xie, S., Rohrbach,, M., et al.: Decoupling representation and classifier for long-tailed recognition. In: Proceedings of International Conference on Learning Representations (ICLR) (2020)
13. Kang, B., Li, Y., Xie, S., et al.: Exploring balanced feature spaces for representation learning. In: Proceedings of International Conference on Learning Representations (ICLR) (2021)
14. Zhong, Z., Cui, J., Liu, S., Jia, J.: Improving calibration for long-tailed recognition. In: Proceedings of Conference on Computer Vision and Pattern Recognition (CVPR), pp. 16489–16498 (2021)
15. Alshammari, S., Wang, Y.-X., Ramanan, D., Kong S.: Long-tailed recognition via weight balancing. In: Proceedings of Conference on Computer Vision and Pattern Recognition (CVPR), pp. 6887–6897 (2022)
16. Dosovitskiy, A., Beyer, L., Kolesnikov, A., et al.: An image is worth 16x16 words: transformers for image recognition at scale. In: Proceedings of International Conference on Learning Representations (ICLR) (2021)
17. Zhang, H., Cissé, M., Dauphin, Y.N., Lopez-Paz, D.: mixup: beyond empirical risk minimization. In: Proceedings of International Conference on Learning Representations (ICLR) (2018)

18. Li, Q., Yang, Y., Wang, A.: Recognition of inscriptions on bones or tortoise shells based on graph isomorphism. Jisuanji Gongcheng yu Yingyong (Comput. Eng. Appl.) **47**(8), 112–114 (2011)

19. Liu, Y., Liu, G.: Oracle bone inscription recognition based on SVM. J. Anyang Normal Univ. **2**, 54–56 (2017)

20. Wang, X., Lian, L., Miao, Z., et al.: Long-tailed recognition by routing diverse distribution-aware experts. In: Proceedings of International Conference on Learning Representations (ICLR) (2021)

21. Zhang, Y., Hooi, B., Hong, L., Feng, J.: Self-supervised aggregation of diverse experts for test-agnostic long-tailed recognition. In: Advances in Neural Information Processing Systems (NeurIPS) (2022)

22. Gou, J., Baosheng, Yu., Maybank, S.J., Tao, D.: Knowledge distillation: a survey. Int. J. Comput. Vision **129**(6), 1789–1819 (2021)

23. Hinton, G.E., Vinyals, O., Dean, J.: Distilling the knowledge in a neural network (2015)

24. Xiang, L., Ding, G., Han, J.: Learning from multiple experts: self-paced knowledge distillation for long-tailed classification. In: Vedaldi, A., Bischof, H., Brox, T., Frahm, J.-M. (eds.) ECCV 2020. LNCS, vol. 12350, pp. 247–263. Springer, Cham (2020). https://doi.org/10.1007/978-3-030-58558-7_15

25. Devlin, J., Chang, M.-W., Lee, K., Toutanova, K.: BERT: pre-training of deep bidirectional transformers for language understanding. In: Proceedings of the North American Chapter of the Association for Computational Linguistics: Human Language Technologies (NAACL-HLT), pp. 4171–4186 (2019)

26. He, K., Chen, X., Xie, S., et al.: Masked autoencoders are scalable vision learners. In: Proceedings of Conference on Computer Vision and Pattern Recognition (CVPR), pp. 15979–15988 (2022)

27. Vaswani, A., Shazeer, N., Parmar, N., et al.: Attention is all you need. In: Proceedings of Advances in Neural Information Processing Systems (NeurIPS), pp. 5998–6008 (2017)

28. Verma, V., Lamb, A., Beckham, C., et al.: Manifold mixup: better representations by interpolating hidden states. In: Proceedings of International Conference on Machine Learning (ICML), pp. 6438–6447 (2019)

29. Yun, S., Han, D., Chun, S., et al.: Cutmix: regularization strategy to train strong classifiers with localizable features. In: Proceedings of International Conference on Computer Vision (ICCV), pp. 6022–6031 (2019)

30. Zhang, Y., Wei, X.-S., Zhou, B., Wu, J.: Bag of tricks for long-tailed visual recognition with deep convolutional neural networks. In: Proceedings of Association for the Advancement of Artificial Intelligence (AAAI), pp. 3447–3455 (2021)

Faster DAN: Multi-target Queries with Document Positional Encoding for End-to-End Handwritten Document Recognition

Denis Coquenet[1]([⊠]) [iD], Clément Chatelain[2,3] [iD], and Thierry Paquet[2,4] [iD]

[1] Conservatoire National des Arts et Métiers, CEDRIC, Paris, France
`denis.coquenet@lecnam.net`
[2] LITIS Laboratory, EA 4108, Rouen Cedex, France
`{clement.chatelain,thierry.paquet}@litislab.eu`
[3] Rouen University, Mont-Saint-Aignan, France
[4] INSA of Rouen, Saint-Etienne-du-Rouvray, France

Abstract. Recent advances in handwritten text recognition enabled to recognize whole documents in an end-to-end way: the Document Attention Network (DAN) [9] recognizes the characters one after the other through an attention-based prediction process until reaching the end of the document. However, this autoregressive process leads to inference that cannot benefit from any parallelization optimization. In this paper, we propose Faster DAN, a two-step strategy to speed up the recognition process at prediction time: the model predicts the first character of each text line in the document, and then completes all the text lines in parallel through multi-target queries and a specific document positional encoding scheme. Faster DAN reaches competitive results compared to standard DAN, while being at least 4 times faster on whole single-page and double-page images of the RIMES 2009, READ 2016 and MAUR-DOR datasets. Source code and trained model weights are available at https://github.com/FactoDeepLearning/FasterDAN.

Keywords: Handwritten Document Recognition · Document Layout Analysis · Handwritten Text Recognition · Transformer

1 Introduction

Unconstrained offline handwritten text recognition has been studied for decades now. Until recently, all the proposed approaches were focused on recognizing the text from cropped parts (text regions) of the original document, leading to a sequential multistep approach, namely text region segmentation, ordering and recognition. Numerous advances enabled to extend the recognition stage to handle increasingly complex inputs. In the 90's, the use of Hidden Markov Model (HMM) enabled to go from isolated character recognition [19] to multi-character (word or line) recognition [12,14]. Thereafter, the democratization of

G. A. Fink et al. (Eds.): ICDAR 2023, LNCS 14190, pp. 182–199, 2023.
https://doi.org/10.1007/978-3-031-41685-9_12

deep neural networks, combined with the Connectionist Temporal Classification (CTC) loss [15], made the line-level approach the standard framework to handle handwritten document recognition [7,11,16,22,23,30,33].

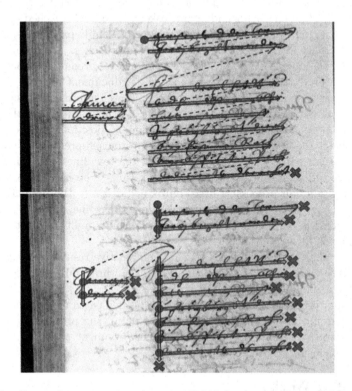

Fig. 1. Reading order comparison between DAN (top) and Faster DAN (bottom). Circles and crosses represent the start and the end of a pass, respectively. The first pass is showed in red, and the second one in blue. The DAN (top) sequentially predicts the characters of the whole documents in a single pass. The Faster DAN first predicts the first character of each line (as well as the layout tokens), and then predicts the remaining of all the text lines in parallel, in a second pass. (Color figure online)

More recently, few advanced works focused on text recognition at paragraph level [1,2,8,32], reaching similar performance compared to line-level recognition [10]. However, whether it is at character, word, line or paragraph level, this three-step paradigm has many drawbacks: the errors accumulate from one step to another, additional physical segmentation annotations are required to train the segmentation step, the use of a rule-based ordering stage is limited for documents with a complex layout, and the stages are performed independently, so they cannot benefit from one another.

Based on these observations, we proposed in [9] a new end-to-end paradigm named Handwritten Document Recognition (HDR). It aims at serializing documents in an XML-way, combining both Handwritten Text Recognition (HTR) and Document Layout Analysis (DLA), through layout XML-markups. We proposed the Document Attention Network (DAN) [9] to tackle HDR.

It is made up of a Fully Convolutional Network (FCN) encoder to extract features from the input image, and a transformer [29] decoder to recurrently predict the different character and layout tokens. The DAN reached competitive results, recognizing both text and layout at page or double-page levels, compared to state-of-the-art line-level or paragraph-level approaches. The main drawback of the DAN is about its autoregressive character-level prediction process, which leads to high prediction times (a few seconds per document image).

In this paper, we propose Faster DAN, a novel approach to significantly reduce the prediction time of end-to-end HDR, without impacting the training time. This approach is based on a new document positional encoding whose aim is to inject the line membership information to each predicted character. In this way, we can parallelize the recognition of the text lines still using a single model, while reducing the total number of iterations. The Faster DAN relies on a two-step prediction process: a first step is dedicated to the prediction of the layout tokens, as well as the first character of each text line; all the text lines are then recognized in parallel in the second stage through multi-target transformer queries. This is illustrated in Fig. 1.

We show that the Faster DAN reaches competitive results compared to the original DAN, while being at least 4 times faster on three public datasets: READ 2016, RIMES 2009 and MAURDOR.

This paper is organized as follows. Section 2 is dedicated to the related works. DAN background is presented in Sect. 3. We detail the proposed approach in Sect. 4. Section 5 presents the experimental environment and the results. We draw the conclusion in Sect. 6.

2   Related Works

Nowadays, the most popular HTR framework is made up of three stages: the input document image is segmented into text line crops, which are then ordered and recognized. Indeed, the concept of text line is widely used as a building block in many works, and has been studied from different angles.

The text line has mostly been studied from the physical point of view: the majority of the works focused on predicting text line bounding boxes, either through a pixel-by-pixel classification task [3,21,24] or through an object-detection approach [5,6]. Detecting the start-of-line information was also studied as part of the segmentation stage. In [20], a model is trained to predict the coordinates of the bottom-left corner of each text line, as well as their height. Similarly, in [28,31], the authors considered the prediction of the start-of-line coordinates as an object detection task, using a region proposal network. Scale and rotation values are also associated to each text line to handle monotonic slanted lines. Contrary to these works, the Faster DAN strategy we propose only relies on language supervision: we do not need any additional physical annotations.

Recent works proposed to perform the recognition step at paragraph level [1,2,8,33]. Although not relying on raw physical text line annotations, most paragraph-level text recognition works take advantage of the physical properties of text lines in single-column layout: the whole horizontal axis is associated to a text line, no matter its length. The authors of [32] and [8] concatenate the representation of the different text lines, or the text line predictions, respectively, to get back to a one-dimensional alignment problem. In [1,10], the text lines are processed recurrently through a line-level attention mechanism.

Another approach to deal with multi-line images consists in relying on an autoregressive character-level prediction process [2,9,25,27]. This time, the notion of text line is limited to the use of a dedicated line break token, used as any other character token. This way, this approach is no longer limited to single-column document. This strategy is also used in [18] for visual question-answering, information extraction or classification of documents, the OCR task being reduced to pretraining. In [9], we proposed the Document Attention Network to tackle Handwritten Document Recognition, by predicting opening and closing layout markup tokens in an XML way: all the character and layout tokens are sequentially and indifferently predicted, leading to hundreds or even thousands of iterations for single-page or double-page document images. It results in long prediction times: approximately one second for 100 characters on a single GPU V100.

In this paper, we propose to speed up the prediction of this latter approach by reading text lines in parallel. This way, we take the best of both worlds: we can deal with documents with complex layout through this character-level attention, and we use the concept of text line more directly through the prediction of the first character of each line and by using a dedicated document positional encoding scheme, but without using any additional physical annotations.

3 DAN Background

We proposed the Document Attention Network (DAN) in [9] for the task of Handwritten Document Recognition. It takes an input image of a whole document $X \in \mathbb{R}^{H_i \times W_i \times C_i}$, where H_i, W_i, C_i are the height, the width and the number of channels, respectively. It outputs the associated XML-like serialized representation $\hat{y}$, $i.e.$, a sequence of tokens, each token $\hat{y}_i$ representing either a layout markup or a character among an alphabet $\mathcal{A}^*$. For an input document represented by N tokens, we can note the expected output sequence as $y^* \in \mathcal{A}^{*N}$. For instance, a three-line document, split into two paragraphs, could be represented as:

$$\text{<D><P>Line 1}\backslash \text{ nLine 2</P><P>Line 3</P></D>}$$

where <D> and <P> corresponds to document and paragraph markups, respectively.

The DAN is made up of two main components. An FCN encoder is used to extract 2D features $f^{2D} \in \mathbb{R}^{H \times W \times d}$ from the input image X, with $H = \frac{H_i}{32}$, $W = \frac{W_i}{8}$ and $d = 256$. A Transformer decoder is used to iteratively predict the tokens $\hat{y}_i$. To this aim, a special start-of-transcription token is used to initialize the prediction ($\hat{\boldsymbol{y}}_0$ = <sot>) and a special end-of-transcription token is added to the ground truth to stop it. This way, the new target sequence is $\boldsymbol{y} \in \mathcal{A}^{N+1}$ with $\boldsymbol{y}_{N+1}$ = <eot> and $\mathcal{A} = \mathcal{A}^* \cup \{<eot>\}$. During inference, a maximum number of iterations $N_{\max} = 3,000$ is fixed in case of the <eot> token is not predicted.

The transformer attention mechanism being invariant to the order of its input sequences, positional encoding is added to inject the positional information: 2D positional encoding $\boldsymbol{P}^{2D} \in \mathbb{R}^{H \times W \times d}$ for the 2D features of the image, and 1D positional encoding $\boldsymbol{P}^{1D} \in \mathbb{R}^{N_{\max} \times d}$ for the previously predicted tokens. Both positional encodings are defined as a fixed encoding based on sine and cosine functions with different frequencies, as proposed in the original Transformer paper [29]. The image features are flattened afterward, for transformer needs:

$$f^{1D} = \text{flatten}(f^{2D} + P^{2D}). \tag{1}$$

The DAN can be seen under the prism of the question-answering paradigm. At iteration t, the question corresponds to the previously predicted tokens $\hat{\boldsymbol{y}}^t = [\hat{\boldsymbol{y}}_0, ..., \hat{\boldsymbol{y}}_{t-1}]$, referred to as *context* in this work, and the answer is the next token $\hat{\boldsymbol{y}}_t$. Formally, the tokens are first embedded through a learnable matrix $\boldsymbol{E} \in \mathbb{R}^{(|\mathcal{A}|+1) \times d}$ (+1 for the <sot> token), leading to $\boldsymbol{e}^t = [\boldsymbol{e}_0, ..., \boldsymbol{e}_{t-1}]$, with $\boldsymbol{e}_i = \boldsymbol{E}_{\hat{\boldsymbol{y}}_i}$ ($\in \mathbb{R}^d$). Positional embedding is then added to get the transformer input query $\boldsymbol{q}^t = [\boldsymbol{q}_0, ..., \boldsymbol{q}_{t-1}]$ with $\boldsymbol{q}_i = \boldsymbol{e}_i + \boldsymbol{P}_i^{1D}$.

The transformer's self-attention and cross-attention mechanisms compute an output $\boldsymbol{o}_i \in \mathbb{R}^d$ for each query input $\boldsymbol{q}_i$ by comparing them with the other query tokens, and with the image features $\boldsymbol{f}^{1D}$, respectively. Formally,

$$o^t = [o_0, ..., o_{t-1}] = \text{decoder}(q^t, f^{1D}), \tag{2}$$

where the decoder corresponds to a stack of 8 standard transformer decoder layers [29]. This process being autoregressive, the query at position i can only attend to positions from 0 to i. In addition, the intermediate computations are preserved for each layer from one iteration to another in order to avoid computing the same output multiple times.

A score $\boldsymbol{s}_i^t$ is computed for each token i of the alphabet $\mathcal{A}$ using a single densely-connected layer of weights $\boldsymbol{W}_p$ ($\boldsymbol{s}^t \in \mathbb{R}^{|\mathcal{A}|}$):

$$s^t = W_p \cdot o_{t-1}. \tag{3}$$

Probabilities are obtained through softmax activation: $\boldsymbol{p}_i^t = \frac{\exp s_i^t}{\sum_j \exp s_j^t}$. The predicted token at iteration t is the one whose probability is maximum:

$$\hat{y}_t = \arg\max(p^t). \tag{4}$$

The model is trained in an end-to-end fashion using the cross-entropy loss over the sequence of tokens:

$$\mathcal{L}_{\text{DAN}} = \sum_{t=1}^{N+1} \mathcal{L}_{\text{CE}}(y_t, p^t). \tag{5}$$

This autoregressive process can be parallelized during training through teacher forcing, but this is not possible during inference. That is why we propose the Faster DAN strategy.

4 Faster DAN

The standard character-level attention-based approach for HTR is to sequentially recognize all the characters y_i of the whole input image X. This way the number of iterations, thus the prediction time, grows linearly with the number of characters in the document. This may be negligible for isolated text line images, for which the image feature extraction stage is predominant, but this becomes significant for whole page images (around one second for 100 characters on a GPU V100).

We propose the Faster DAN, a novel approach for Handwritten Document Recognition, to noticeably reduce the prediction time. The goal is to take advantage of the line-based structure of documents to parallelize the recognition of the text lines. Considering the layout markups and the <eot> tokens as lines by themselves (of unit length), we can rewrite the target sequence as $y = \text{concatenate}(y^1, ..., y^L)$ where L is the number of lines in the document and $y^j \in \mathcal{A}^{n_j}$ represent the different text lines (y_i^j is the character i of line j).

Using one model per line is prohibitive in terms of GPU memory consumption. Instead, the parallelization is carried out among a single model which processes multi-target queries through masking in the second pass. This is feasible thanks to the dedicated document positional encoding scheme we propose. It is important to note that the proposed approach is not specific to the DAN architecture. It could be used with any attention-based HDR model. However, to our knowledge, the only available end-to-end HDR model is the DAN.

Reading Lines in Parallel. Parallelizing the recognition faces two main challenges: detecting all the text lines, and recognizing them in parallel through transformer queries without mixing them. Moreover, since our goal is to perform HDR, and not only HTR, we also need to recognize the layout entities.

To tackle these issues, we opted for a two-pass process, as illustrated in Fig. 2b. In a first pass, the model sequentially predicts the layout tokens as well as the first character of each text lines, solving both layout recognition and text line detection. Then, the different text lines are completed in parallel based on their previously predicted first character. To this end, it is crucial to determine which token belongs to which line.

Document Positional Encoding. To parallelize the recognition of the text lines, we propose a new positional encoding scheme, as shown in Fig. 2. We associate to each predicted token $\hat{y}_i^j$ (with $\hat{y}_0^0 = \text{<sot>}$) two 1D positional embedding: one for the index of the line, and the other one for the index of the token in the line, leading to the global positional embedding $P^{\text{doc}} \in \mathbb{R}^{l_{\max} \times n_{\max} \times d}$,

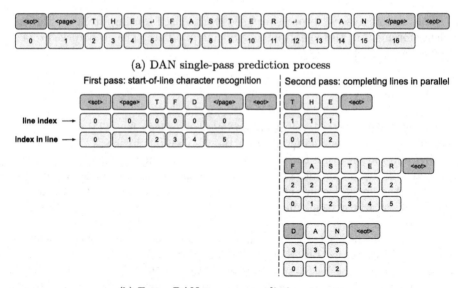

(a) DAN single-pass prediction process

(b) Faster DAN two-pass prediction process

Fig. 2. Comparison of the prediction process and positional encoding scheme between DAN and Faster DAN. This illustrates the example of a document input with three one-word text lines. The DAN associates a unique positional value for each token, which continues from one text line to the next. The Faster DAN uses two positional values: the index of the text line and the position of the token in this text line. Special (start and end) tokens are in blue and layouts tokens are in green. (Color figure online)

where l_{max} is the maximum number of line and n_{max} is the maximum number of characters per line. y_i^j is associated to:

$$P_{j,i}^{\text{doc}} = \text{concatenate}(P_j^{\text{1D}'}, P_i^{\text{1D}'}), \tag{6}$$

where $P^{\text{1D}'}$ is equivalent to P^{1D} but encoded on half channels ($P_i^{\text{1D}'} \in \mathbb{R}^{d/2}$). The transformer input queries become $q_i^j = E_{\hat{y}_i^j} + P_{j,i}^{\text{doc}}$. The idea of injecting the line information was already used in [27], but it was computed as a ratio with an arbitrary maximum number of lines, and concatenated to the token embedding directly. In addition, the position of the tokens was absolute, and not relative to the current text line, as for the standard DAN.

First Pass. The Faster DAN follows the standard autoregressive process to predict the first token $\hat{y}_0^j$ of each line j based on Eqs. 2 to 4. At iteration t, $q^t = [q_0^0, ..., q_0^{t-1}]$.

Second Pass. The standard Transformer decoding process is to give a sequence of query tokens q^t as input and keep the output corresponding to the last token only (o_{t-1}), as single output for iteration t. Instead, the output of the last token of each line o_{t-1}^j are kept in this second pass. We refer to this as multi-target queries. $\hat{y}_0^j$ are duplicated into $\hat{y}_1^j$ to initiate the second pass; the modification of the associated position in line (from 0 to 1) indicates to the model a change of expected behavior: from the prediction of the first token of the next line to the prediction of the next token of the current line. By setting $q^t = [q_0^0, ..., q_{t-1}^0, ..., q_0^L, ..., q_{t-1}^L]$ (the t first tokens of all the lines), we obtain $o^t = [o_0^0, ..., o_{t-1}^0, ..., o_0^L, ..., o_{t-1}^L]$ through Eq. 2. In this way, the tth tokens of each line j are computed in a single iteration t:

$$\hat{y}_t^j = \arg\max(W_p \cdot o_{t-1}^j). \tag{7}$$

Extra tokens ($\hat{y}_i^j$ with $i > n_j$) are discarded through masking.

Context Exploitation. The naive approach to recognize the text lines in parallel would be to recognize them independently, by applying a mask to discard the tokens from all the other text lines. It means that q_i^j could only attend to line j (itself) and position 0 to i in that line. However, this would lead to an important loss of context. Instead, we propose to take advantage of all the partially predicted text lines: q_i^j can attend to all lines, from 0 to L, and from position 0 to i in those lines, this is illustrated in Fig. 3

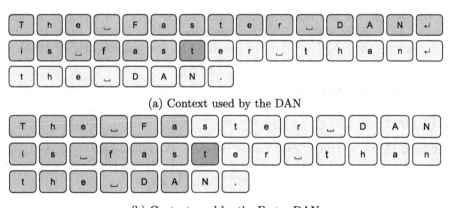

(a) Context used by the DAN

(b) Context used by the Faster DAN

Fig. 3. Context comparison between DAN and Faster DAN. The colored cells represent the current character to predict (in purple), the previously predicted tokens *i.e.* the context (in blue and green), the token used for the prediction (in green), and the remaining characters to recognize (in gray). (Color figure online)

The major drawback of parallelizing the line recognition, compared to purely sequential recognition, is the loss of context. Indeed, the standard DAN benefits from all the past context during prediction: this is partially available for the Faster DAN since the past context is limited to the beginning of the text lines.

In this way, it becomes harder for the model to focus on the correct text part, especially for very short contexts. Indeed, a sequence of characters may appear several times in a document, especially if this sequence is short, *e.g.*, at the beginning of the recognition process. We counterbalance the loss of context from past by combining partial context from both past and future. We show the impact of this approach in Sect. 5.6.

Training and Inference. The model is trained over the target sequence using the cross-entropy loss:

$$\mathcal{L} = \sum_{j=1}^{L} \sum_{\substack{i=0 \\ i \neq 1}}^{n_j} \mathcal{L}_{\text{CE}}(\boldsymbol{y}_i^j, \boldsymbol{p}_i^j). \tag{8}$$

It has to be noted that the training time is not impacted by this two-step decoding strategy since the whole expected sequence prediction (from both passes) is trained in parallel through teacher forcing, with appropriate masks.

During inference, the Faster DAN reduces the number of iterations I from

$$I_{\text{DAN}} = \sum_{j=1}^{L} n_j = N + 1 \quad \text{to} \quad I_{\text{FasterDAN}} = L + \max_j(n_j),$$

by considering the line breaks as belonging to the lines. For example, 25 text lines of 50 characters, structured according to 3 layout entities, leads to 1,251 iterations for the DAN, and only 76 iterations for the proposed Faster DAN.

5   Experimental Study

5.1   Datasets

We used three document-level public datasets to evaluate the proposed approach: RIMES 2009 [17], READ 2016 [26] and MAURDOR [4]. Document image examples from these three datasets are showed in Fig. 4.

RIMES 2009. The RIMES 2009 dataset corresponds to French grayscale handwritten page images. These pages are letters produced in the context of writing mail scenarios. Text regions are classified among one of the following 7 classes: sender coordinates, recipient coordinates, object, date & location, opening, body and PS & attachment. We used these classes as layout tokens.

READ 2016. The READ 2016 dataset corresponds to Early Modern German handwritten pages from the Ratsprotokolle collection. Images are RGB encoded. We used two versions of this dataset: single-page images and double-page images. The layout classes are as follows: page, section, margin annotation and body.

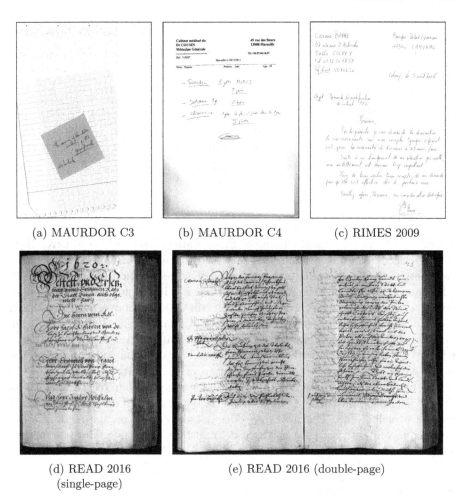

(a) MAURDOR C3   (b) MAURDOR C4   (c) RIMES 2009

(d) READ 2016   (e) READ 2016 (double-page)
(single-page)

Fig. 4. Document image examples from the RIMES 2009, READ 2016 and MAURDOR datasets.

MAURDOR. The MAURDOR dataset consists in a heterogeneous collection of documents. We used the same configuration as in [9] *i.e.* we only use the English and French documents, and we focus on the C3 and C4 subsets of this dataset, which corresponds to private or professional correspondences. The documents are either handwritten, printed, or a mix of both. There is no sufficient annotation to produce the layout tokens, so we only evaluate the HTR task on this dataset.

Table 1 details the splits in training, validation and test, as well as the number of characters in the alphabet and the number of layout tokens (2 by class, for opening and closing markups) for each dataset.

Table 1. Splits and number of character and layout tokens for each dataset.

Dataset	Training	Validation	Test	# char tokens	# layout tokens
RIMES 2009	1,050	100	100	108	14
READ 2016 (single-page)	350	50	50	89	10
READ 2016 (double-page)	169	24	24	89	10
MAURDOR (C3)	1,006	148	166	134	✗
MAURDOR (C4)	721	111	114	127	✗

5.2 Metrics

In addition to the standard Character Error Rate (CER) and Word Error Rate (WER) metrics used to evaluate the text recognition performance, we proposed two metrics in [9] to evaluate the specific layout recognition of the HDR task. The Layout Ordering Error Rate (LOER) consists in considering the document layout as a graph and computing the graph edit distance between the prediction and the ground truth. The LOER aims at evaluating the layout recognition only, considering the reading order between layout entities. Since LOER and CER/WER only evaluate the layout and text recognition independently, the mAP_{CER} is used to evaluate the recognition of the layout with respect to the text content. It is computed as the area under the precision/recall curve, as in object detection approaches [13] for instance, but it is based on a CER threshold instead of a IoU one. The mAP_{CER} does not dependent on the reading order between layout entities. That is why it is important to consider all these metrics altogether to evaluate the HDR task.

5.3 Training Details

In [9], we used some pretraining and curriculum training strategies to speed up the convergence of the DAN, and to not use any physical segmentation annotation during training. To be fairly comparable with this work, we follow the exact same training configuration, whose major points are as follows:

- The encoder is pretrained on synthetic isolated text line images using the CTC loss and a dedicated FCN line-level OCR model.
- The Faster DAN is trained on a mixture of real and synthetic documents. Using a curriculum strategy, the Faster DAN is trained on increasingly complex synthetic documents through the epochs. The complexity varies from two aspects: the number of lines contained in the document image, and the size of this image. The ratio between synthetic and real document also evolves during training, from 90%/10% to 20%/80%.
- A rule-based post-processing is used to make sure that the layout markups have the correct format (no unpaired markup, for instance).

– Whether it is for pretraining or training, input images are downsized to 150 dpi, normalized and data augmentation is performed 90% of the time.

We carried out 2-day pretraining and 4-day training on a single GPU V100 (32 Go), using automatic mixed-precision. We used the Adam optimizer with an initial learning rate of 10^{-4}. We do not use any external data, external language model nor lexicon constraints.

5.4 Comparison with the State of the Art

To our knowledge, the only work performing HDR is the DAN [9]. Tables 2, 3 and 4 provide an evaluation of the Faster DAN on the READ 2016, RIMES 2009 and MAURDOR datasets, respectively, as well as a comparison with the state of the art.

Table 2. Evaluation of the Faster DAN on the test set of the READ 2016 dataset and comparison with the state of the art. Metrics are expressed in percentages.

Architecture	READ 2016 (single-page)				READ 2016 (double-page)			
	CER ↓	WER ↓	LOER ↓	mAP$_{CER}$ ↑	CER ↓	WER ↓	LOER ↓	mAP$_{CER}$ ↑
DAN [9]	**3.43**	**13.05**	5.17	93.32	**3.70**	**14.15**	4.98	93.09
Faster DAN	3.95	14.06	**3.82**	**94.20**	3.88	14.97	**3.08**	**94.54**

Table 3. Evaluation of the Faster DAN on the test set of the RIMES 2009 dataset and comparison with the state of the art. Metrics are expressed in percentages.

Architecture	RIMES 2009			
	CER ↓	WER ↓	LOER ↓	mAP$_{CER}$ ↑
DAN [9]	**4.54**	**11.85**	**3.82**	**93.74**
Faster DAN	6.38	13.69	4.48	91.00

Table 4. Evaluation of the Faster DAN on the test set of the MAURDOR dataset and comparison with the state of the art. Metrics are expressed in percentages.

Architecture	C3		C4		C3 & C4	
	CER ↓	WER ↓	CER ↓	WER ↓	CER ↓	WER ↓
DAN [9]	**8.62**	**18.94**	**8.02**	**14.57**	11.59	27.68
Faster DAN	8.93	19.00	9.88	16.52	**10.50**	**19.64**

The Faster DAN reaches competitive results compared to the DAN on the three datasets. For the READ 2016 dataset, it even reaches state-of-the-art results in terms of LOER and mAP$_{CER}$ for both single-page and double-page versions, involving a better recognition of the layout. Results are not as good for the RIMES 2009 dataset, which includes more variability in terms of layout. We assume that this higher variation makes the first pass of the Faster DAN more difficult. This is confirmed when measuring the CER for the first pass only: it is of 4.72% and 5.34% for READ 2016 at single-page and double-page levels, and of 9.10% for RIMES 2009. Concerning the MAURDOR dataset, the Faster DAN reaches competitive results on the C3 and C4 categories, taken separately, and it reaches new state-of-the art results when mixing both categories with 10.50% of CER, compared to 11.59% for the standard DAN.

Discussion. It has to be noted that it is more difficult to compare the text recognition performance at document level than at line level. Indeed, the reading order is far more complex for documents, to go from one paragraph to another, and to one line to the next, than for isolated lines. This way, even perfectly recognized, the CER can be severely impacted if the paragraphs are recognized in the wrong order. On the contrary, the mAP_{CER} is invariant to the order of the layout entities, but it is dependent to the well recognition of the layout.

Another point to emphasize is about the severity of the errors made. There are two types of errors to be distinguished. The first corresponds to standard character addition, removal, or substitution cases. During the first pass of the Faster DAN, this kind of error may have a great impact because a whole text line may be duplicated or discarded. However, during the second pass, we assume that the impact of such errors is rather equivalent for both DAN and Faster DAN. The second kind of errors is related to the end-of-transcription token prediction. Indeed, although rare, the model may not predict the end of the transcription and loop on the same text region again and again until reaching an arbitrary chosen iteration limit. For this later issue, the standard DAN is more impacted than the Faster DAN. Indeed, the DAN only have one iteration limit, which corresponds to the global number of tokens to predict for the whole document. For the Faster DAN, we used two iteration limits: one for the number of lines, and one for the number of characters per line. Given that the range of values for a line length is smaller than for the whole document, the impact is less important for the Faster DAN.

Prediction Time. Table 5 shows a comparison of the Faster DAN with the DAN in terms of prediction times for the three datasets: RIMES 2009, READ 2016 and MAURDOR. To be fairly comparable, these times account for the whole prediction process, including the time dedicated to the encoder part and to formatting instructions. Additional details are given for each dataset such as the image sizes, the number of characters, lines, and layout tokens per image, and the number of characters per line. The values are given as average for the test set of each dataset. As one can note, the Faster DAN is significantly faster than the DAN for all the datasets, speeding up the prediction process by at least 4.

We showed that the Faster DAN reaches competitive results on three document-level datasets while being at least 4 times faster than the standard DAN at prediction time. We now evaluate the performance on heterogeneous documents, by mixing both RIMES 2009 and READ 2016 datasets.

Table 5. Prediction time comparison between the DAN and the Faster DAN. Times (in seconds) are averaged on the test set for a single document image, using a single GPU V100.

	RIMES 2009	READ 2016		MAURDOR		
		single-page	double-page	C3	C4	C3 & C4
Dataset details (averaged for a document on the test set)						
width (px)	1,235	1,190	2,380	1,336	1,240	1,297
height (px)	1,751	1,755	1,755	1,658	1,754	1,697
# chars	578	528	1,062	481	706	575
# lines	18	23	47	16	22	18
# chars/line	31	22	22	30	31	30
# layout tokens	11	15	30	0	0	0
Prediction times (in seconds)						
DAN [9]	5.6	4.6	8.5	5.8	7.7	6.6
Faster DAN	**1.4**	**0.9**	**1.9**	**1.0**	**1.6**	**1.3**
Speed factor	x4	x5.1	x4.5	x5.8	x4.8	x5.1

5.5   Evaluation on Heterogeneous Documents

In this experiment, we mixed both RIMES 2009 and READ 2016 datasets at single page level, for both training and evaluation. Examples from both datasets are balanced at training time, *i.e.*, the models have been trained on the same number of documents for both datasets. These are the first results for such an experiment; we also train the standard DAN for comparison purposes. Results are shown in Table 6. As one can note, results are rather similar when training on datasets separately or altogether, except for the DAN on the RIMES dataset whose CER increases from 4.54% up to 7.96%.

Table 6. Evaluation of the Faster DAN on heterogeneous data (mixing READ 2016 and RIMES 2009 for both training and evaluation) and comparison with the state of the art.

Architecture	RIMES 2009 (page)				READ 2016 (single-page)			
	CER ↓	WER ↓	LOER ↓	mAP$_{CER}$ ↑	CER ↓	WER ↓	LOER ↓	mAP$_{CER}$ ↑
DAN [9]	7.96	15.76	8.72	**91.59**	**3.50**	**13.36**	**3.86**	**94.23**
Faster DAN	**6.73**	**15.22**	**5.56**	90.10	3.81	14.30	4.32	93.57

Table 7. Ablation study of the Faster DAN and DAN. Results (in percentages) are given for the test set of the RIMES 2009 and READ 2016 datasets.

Architecture	RIMES 2009 (page)			READ 2016 (single-page)			READ 2016 (double-page)		
	CER	LOER	mAP$_{CER}$	CER	LOER	mAP$_{CER}$	CER	LOER	mAP$_{CER}$
Faster DAN	**6.38**	4.48	91.00	3.95	**3.82**	**94.20**	**3.88**	**3.08**	**94.54**
(1) No line encoding	79.39	6.21	0.00	75.08	11.81	0.29	75.01	10.79	5.44
(2) Single-line context	94.73	**4.30**	0.00	91.23	4.61	0.00	91.22	4.03	0.00
(3) First-pass context	8.27	4.90	90.73	6.68	4.50	88.37	6.87	5.22	87.93
(4) Sum PE	6.88	4.90	**91.06**	3.82	4.27	94.08	4.55	4.39	92.76

5.6   Ablation Study

In Table 7, we propose an ablation study of the proposed approach on the RIMES 2009 and READ 2016 datasets. The first line corresponds to the Faster DAN baseline. In experiment (1), the document positional encoding is replaced by standard 1d positional encoding, *i.e.*, a unique index is associated to each token. The model does not succeed to recognize the text, showing the necessity of injecting line positional information to parallelize the recognition. The model can only access to tokens of the text line to recognize in (2), also preventing the text recognition. Indeed, it is nearly impossible to predict the next character with only a one-character query (beginning of the second pass) since characters are not unique in a document. For both experiments, one can note that the LOER is nearly not impacted, this is because the layout recognition takes place in the first pass, before the parallelization.

In experiment (3), in addition to the tokens of the text line to recognize, the first character of all the text lines, as well as the layout markup tokens, are available. This leads to an increase of the CER of at least 1.89 points for RIMES 2009, and up to 2.99 points for READ 2016 at double-page level, compared to the baseline. This shows the efficiency of the text line detection performed in the first pass, since the text recognition is parallelized, but it also demonstrates that gathering the context from past and future lines helps to improve the performance. In experiment (4), the positional encoding of the line and of the index in the line are summed instead of being concatenated. As one can note, results are slightly in favor of the concatenation.

6   Conclusion

In this paper, we proposed the Faster DAN, a novel approach for end-to-end Handwritten Document Recognition. We evaluate this approach with the current state-of-the-art architecture and showed that this approach reaches competitive results on three document-level datasets while being at least 4 times faster. This way, we preserved the advantages of using a single end-to-end approach, while greatly mitigating the major drawback of prediction time. In this work, we focused on line-level multi-target queries to show the gain in prediction time.

However, it would also be possible to perform this parallelization at paragraph level in order to have a more important language modeling of the past: this would represent an in-between in terms of prediction time.

Acknowledgments. This work was granted access to the HPC resources of IDRIS under the allocation 2020-AD011012155.

References

1. Bluche, T.: Joint line segmentation and transcription for end-to-end handwritten paragraph recognition. In: Advances in Neural Information Processing Systems (NIPS), vol. 29, pp. 838–846 (2016)
2. Bluche, T., Louradour, J., Messina, R.O.: Scan, attend and read: end-to-end handwritten paragraph recognition with MDLSTM attention. In: International Conference on Document Analysis and Recognition (ICDAR), pp. 1050–1055 (2017)
3. Boillet, M., Kermorvant, C., Paquet, T.: Robust text line detection in historical documents: learning and evaluation methods. Int. J. Doc. Anal. Recogn. (IJDAR) **25**, 95–114 (2022)
4. Brunessaux, S., et al.: The Maurdor project: improving automatic processing of digital documents. In: International Workshop on Document Analysis Systems (DAS), pp. 349–354 (2014)
5. Carbonell, M., Fornés, A., Villegas, M., Lladós, J.: A neural model for text localization, transcription and named entity recognition in full pages. Pattern Recogn. Lett. **136**, 219–227 (2020)
6. Chung, J., Delteil, T.: A computationally efficient pipeline approach to full page offline handwritten text recognition. In: Workshop on Machine Learning (WML@ICDAR), pp. 35–40 (2019)
7. Coquenet, D., Chatelain, C., Paquet, T.: Recurrence-free unconstrained handwritten text recognition using gated fully convolutional network. In: 17th International Conference on Frontiers in Handwriting Recognition (ICFHR), pp. 19–24 (2020)
8. Coquenet, D., Chatelain, C., Paquet, T.: SPAN: a simple predict & align network for handwritten paragraph recognition. In: Lladós, J., Lopresti, D., Uchida, S. (eds.) ICDAR 2021. LNCS, vol. 12823, pp. 70–84. Springer, Cham (2021). https://doi.org/10.1007/978-3-030-86334-0_5
9. Coquenet, D., Chatelain, C., Paquet, T.: Dan: a segmentation-free document attention network for handwritten document recognition. IEEE Trans. Pattern Anal. Mach. Intell. (2023). https://doi.org/10.1109/TPAMI.2023.3235826
10. Coquenet, D., Chatelain, C., Paquet, T.: End-to-end handwritten paragraph text recognition using a vertical attention network. Trans. Pattern Anal. Mach. Intell. (TPAMI) **45**(1), 508–524 (2023)
11. Coquenet, D., Soullard, Y., Chatelain, C., Paquet, T.: Have convolutions already made recurrence obsolete for unconstrained handwritten text recognition ? In: Workshop on Machine Learning (WML@ICDAR), pp. 65–70 (2019)
12. El-Yacoubi, M.A., Gilloux, M., Sabourin, R., Suen, C.Y.: An hmm-based approach for off-line unconstrained handwritten word modeling and recognition. Trans. Pattern Anal. Mach. Intell. (TPAMI) **21**(8), 752–760 (1999)
13. Everingham, M., Gool, L.V., Williams, C.K.I., Winn, J.M., Zisserman, A.: The pascal visual object classes (VOC) challenge. Int. J. Comput. Vis. **88**(2), 303–338 (2010)

14. Gilloux, M., Lemarié, B., Leroux, M.: A hybrid radial basis function network/hidden markov model handwritten word recognition system. In: Third International Conference on Document Analysis and Recognition (ICDAR), pp. 394–397 (1995)

15. Graves, A., Fernández, S., Gomez, F.J., Schmidhuber, J.: Connectionist temporal classification: labelling unsegmented sequence data with recurrent neural networks. In: International Conference on Machine Learning (ICML), vol. 148, pp. 369–376 (2006)

16. Graves, A., Schmidhuber, J.: Offline handwriting recognition with multidimensional recurrent neural networks. In: Advances in Neural Information Processing Systems (NIPS), vol. 21, pp. 545–552 (2008)

17. Grosicki, E., Carré, M., Brodin, J., Geoffrois, E.: Results of the RIMES evaluation campaign for handwritten mail processing. In: 10th International Conference on Document Analysis and Recognition (ICDAR), pp. 941–945 (2009)

18. Kim, G., et al.: Ocr-free document understanding transformer. In: Avidan, S., Brostow, G., Cissé, M., Farinella, G.M., Hassner, T. (eds.) ECCV 2022. LNCS, vol. 13688, pp. 498–517. Springer, Cham (2022). https://doi.org/10.1007/978-3-031-19815-1_29

19. LeCun, Y., et al.: Handwritten digit recognition with a back-propagation network. In: Advances in Neural Information Processing Systems, vol. 2 (1989)

20. Moysset, B., Kermorvant, C., Wolf, C.: Full-page text recognition: learning where to start and when to stop. In: International Conference on Document Analysis and Recognition (ICDAR), pp. 871–876 (2017)

21. Oliveira, S.A., Seguin, B., Kaplan, F.: dhSegment: a generic deep-learning approach for document segmentation. In: 16th International Conference on Frontiers in Handwriting Recognition (ICFHR), pp. 7–12 (2018)

22. Ptucha, R.W., Such, F.P., Pillai, S., Brockler, F., Singh, V., Hutkowski, P.: Intelligent character recognition using fully convolutional neural networks. Pattern Recogn. **88**, 604–613 (2019)

23. Puigcerver, J.: Are multidimensional recurrent layers really necessary for handwritten text recognition? In: 14th IAPR International Conference on Document Analysis and Recognition (ICDAR), pp. 67–72 (2017)

24. Renton, G., Soullard, Y., Chatelain, C., Adam, S., Kermorvant, C., Paquet, T.: Fully convolutional network with dilated convolutions for handwritten text line segmentation. Int. J. Doc. Anal. Recogn. (IJDAR) **21**(3), 177–186 (2018)

25. Rouhou, A.C., Dhiaf, M., Kessentini, Y., Salem, S.B.: Transformer-based approach for joint handwriting and named entity recognition in historical documents. Pattern Recogn. Lett. **155**, 128–134 (2022)

26. Sánchez, J., Romero, V., Toselli, A.H., Vidal, E.: ICFHR2016 competition on handwritten text recognition on the READ dataset. In: 15th International Conference on Frontiers in Handwriting Recognition (ICFHR), pp. 630–635 (2016)

27. Singh, S.S., Karayev, S.: Full page handwriting recognition via image to sequence extraction. In: Lladós, J., Lopresti, D., Uchida, S. (eds.) ICDAR 2021. LNCS, vol. 12823, pp. 55–69. Springer, Cham (2021). https://doi.org/10.1007/978-3-030-86334-0_4

28. Tensmeyer, C., Wigington, C.: Training full-page handwritten text recognition models without annotated line breaks. In: International Conference on Document Analysis and Recognition (ICDAR), pp. 1–8 (2019)

29. Vaswani, A., et al.: Attention is all you need. In: Advances in Neural Information Processing Systems (NIPS), vol. 30, pp. 5998–6008 (2017)

30. Voigtlaender, P., Doetsch, P., Ney, H.: Handwriting recognition with large multi-dimensional long short-term memory recurrent neural networks. In: International Conference on Frontiers in Handwriting Recognition (ICFHR), pp. 228–233 (2016)
31. Wigington, C., Tensmeyer, C., Davis, B., Barrett, W., Price, B., Cohen, S.: Start, follow, read: end-to-end full-page handwriting recognition. In: Ferrari, V., Hebert, M., Sminchisescu, C., Weiss, Y. (eds.) ECCV 2018. LNCS, vol. 11210, pp. 372–388. Springer, Cham (2018). https://doi.org/10.1007/978-3-030-01231-1_23
32. Yousef, M., Bishop, T.E.: OrigamiNet: weakly-supervised, segmentation-free, one-step, full page text recognition by learning to unfold. In: Conference on Computer Vision and Pattern Recognition (CVPR), pp. 14698–14707 (2020)
33. Yousef, M., Hussain, K.F., Mohammed, U.S.: Accurate, data-efficient, unconstrained text recognition with convolutional neural networks. Pattern Recogn. **108**, 107482 (2020)

Shared-Operation Hypercomplex Networks for Handwritten Text Recognition

Giorgos Sfikas[1]([✉]), George Retsinas[2], Panagiotis Dimitrakopoulos[3], Basilis Gatos[4], and Christophoros Nikou[3]

[1] Department of Surveying and Geoinformatics Engineering, School of Engineering, University of West Attica, Athens, Greece
gsfikas@uniwa.gr

[2] School of Electrical and Computer Engineering, National Technical University of Athens, Athens, Greece
gretsinas@central.ntua.gr

[3] Department of Computer Science and Engineering, University of Ioannina, Ioannina, Greece
p.dimitrakopoulos@uoi.gr, cnikou@cse.uoi.gr

[4] Computational Intelligence Laboratory, Institute of Informatics and Telecommunications, National Center for Scientific Research "Demokritos", Athens, Greece
bgat@iit.demokritos.gr

Abstract. Parameterized hypercomplex layers have recently emerged as very useful alternatives of standard neural network layers. They allow for the construction of extremely lightweight architectures, with little to no sacrifice of accuracy. We propose networks of Shared-Operation Parameterized Hypercomplex layers, where the operation parameterization is co-learned by all layers in tandem. In this manner, we mitigate the computational burden of operation parameterization, which grows cubically with respect to the hypercomplex dimension. We attain good word and character error rate at only a small fraction of the memory footprint of non-hypercomplex models as well as previous non-shared operation hypercomplex ones (up to -96.8% size reduction).

Keywords: Parameterized Hypercomplex Layers · Hypercomplex Algebra · Handwritten Text Recognition · Low memory footprint

1 Introduction and Related Work

Handwritten Text Recognition (HTR) is one of the major pattern recognition tasks in the field of document image processing. The most typical use-case involves segmented lines of text images as inputs, which are to be automatically converted to a string of characters. While the task itself is similar to that of Optical Character Recognition (OCR), which conventionally involves recognition of printed text, handwritten text offers a significantly greater challenge.

G. A. Fink et al. (Eds.): ICDAR 2023, LNCS 14190, pp. 200–216, 2023.
https://doi.org/10.1007/978-3-031-41685-9_13

The variability of handwriting style, which may be important not only between writers, but within the output of the same writer is one of the major difficulties. Indeed, it is not an uncommon occurence to have a learning system that fails, when trained on one writer or group of writers and tested on another [24,25].

Current state-of-the-art systems for HTR include different variants of Deep Neural Networks. Due to the sequential nature of text, Recurrent Neural Networks (RNN) have been a popular choice of neural network for HTR. Combined with Connectionist Temporal Classification (CTC)-based objectives, which allow for a loss value to be computed during training and without requiring exact alignment between prediction and target, RNNs have been the basis of various excellent-performing systems [6,21]. Sequence-to-sequence models constistute an alternative approach to HTR, which is based on decoupling encoding to a feature vector and decoding to the target string as two separate network components. Compared to Convolutional Neural Networks (CNNs), RNNs have the advantage of capturing information dependencies in a sequential manner, and usually have led to better HTR models w.r.t. to the former. Encoding prior knowledge about the nature of our inputs is primarily enforced through the inherent inductive bias that is represented by each model. Specifically, convolutional layers model statistical dependence of some form for each time frame w.r.t. neighbouring frames, or spatial dependence of line image cues w.r.t. spatially close pixels. The dependence range is hard-coded in the form of the characteristics of each convolution, and primarily as the size of each kernel, along with other hyperparameters (dilation, stride). On this note, there has been important recent work on flexible, adaptive convolutional operations, e.g. [9,27]. With recurrent layers on the other hand, we encode our belief that our data are inherently sequential. Forward and backward dependencies are captured easily through bidirectional recurrent variants. Compared to CNNs however, RNNs are known to be difficult to train and converge to an acceptable solution. To this end, architectures that combine convolutional and recurrent components have been proposed [23,26]. A very much used recipe involves a convolutional backbone that is charged with transforming the input segmented image into a useful feature map. This feature map is then pooled or reshaped into a sequence of features that is fed into a recurrent component [13,23,26]. Regarding convolutional-recurrent model architecture, a technique that has also worked well involves supplying the main recurrent network with an auxiliary CTC-based component [26,36]. This can be understood as a penalty on cross-entropy loss over a recurrent decoder. In practice, this CTC shortcut can be fully convolutional, including 2D and 1D (temporal) convolutions, which translates to less recurrent components and faster convergence.

Transformers have emerged as an antagonist to both convolutional and recurrent architectures, in the sense of the aspiration to replace either one for most important vision tasks. Initially proposed in a Natural Language Processing setting [34], they have been tailored for tasks that can be cast as sequential processing, and can in principle capture complex, far-reaching input interdependencies [20]. The main ingredient in transformers is the self-attention operation. Input sequence vectors are transformed into a set of "keys", "queries" and "values" through learnable, shared transformations which create a dictionary of features that are subsequently recombined as a softmax-weighted average and transformed through a fully-connected layer into an output sequence. Multiple transformations for the same inputs have shown to work well in practice, which corresponds to the so-called multi-head variant of self-attention, or simply multi-head self-attention. In turn, cascades of multi-head self-attention can be grouped together to form so-called transformer layers [20]. Use of transformers has been explored in the field of HTR by recent work, where excellent results are reported [8]; interestingly, transformers are however related to inherent shortcomings such as poorly handling text repetitions [35], which in turn are rooted to the indirect manner that they handle sequence positional dependence. Another important disadvantage is that models that fully depend on a transformer structure tend to be orders of magnitude larger than their non-transformer counterparts [36].

A direction of research that is orthogonal to optimizing HTR accuracy w.r.t. to NN architectural components and structure, involves creating a network that is as resource-demanding as possible. Network size in terms of numbers of parameters is one such metric. The general trend in learning is to have ever-larger models, with NN sizes reaching billions of parameters in some tasks [20]. The largest models in HTR are Transformer-based and comprise hundreds of millions of parameters [36]. In a real application setting, where budget constraints may be very tight (e.g. on embedded devices), large models are unfortunately inapplicable. To this end, a host of works have explored sparsity in neural networks [4,12,22,39], where, in broad terms, the goal is to train NNs with as many zero-ed connections as possible, at as less of an accuracy loss as possible. A family of techniques involves augmenting the training objective with terms that will encourage model sparsity. In [12], a L_0 term is added to the total objective. In [39], a variational inference model is proposed, where model weights follow a zero-mean prior, pushing the posterior towards small magnitudes; values that are under a threshold are pruned. Feature pyramid components are connected with group-wise factors in [4], and in this context variational inference amounts to a neural architecture search scheme. Hypercomplex networks are another group of techniques that follow a very different philosophy in achieving network sparsity. They lead to models which have alternate layers of standard NN layers (fully-connected, convolutional, etc.) but which enforce extensive parameter sharing. Quaternion neural networks were the first type of hypercomplex networks [7,16]. By treating model neurons and weights as quaternionic, which are inherently 4-dimensional, an impressive 75% economy in model size is easily attained. After quaternionic versions of convolutional networks [17,40], proposals for other types of quaternionic layer have followed suit, like recurrent layers, transformers, or extensions to graph neural networks [15,19,32]. Generalizing this paradigm, parameterized hypercomplex networks have recently been proposed as an alternative to quaternionic networks, where the level of parameter sharing is defined as a model hyperparameter. Crucially, in parameterized hypercomplex networks the manner in which weight tuples are multiplied is learnable. This type of multiplication has been named Parameterized Hypercomplex Multiplication (PHM) [38], and has lead to models that were (in practice) downscaled up to 16×.

In this work, we argue that the marginal benefit with respect to increasing network size, structure and training complexity is a critical factor when it comes to choosing an architecture for our HTR model. Is using a model that is $10x$ or $100x$ larger than the previous baseline worth it, only to obtain a decrease in Character Error Rate (CER) by 1–2% as an end result? Also, it is hard to tell whether small accuracy differences generalize well and whether they lead to *any* improvement on out-of-distribution test data. We believe that a trade-off of benchmark accuracy and resource requirements should be considered. In light of the aforementioned considerations, we propose a HTR model that uses a new, even more compact model of Parameterized Hypercomplex Layers that builds on a Convolutional-Recurrent architecture with a CTC shortcut [23, 26]. Our model achieves good HTR results at a model size as small as $\sim 500,000$ parameters, which amounts to up to 32-fold compression or 3.1% the size of our baseline model.

The paper is structured as follows. We begin with a brief outline of the prerequisites for hypercomplex algebra and define hypercomplex layers in Sect. 2. In Sect. 3 we present the proposed Shared Hypercomplex architecture for Handwritten Text Recognition. We evaluate the discussed models in Sect. 4, where we show that the proposed model using shared-operation hypercomplex layers performs very adequately on an HTR task while being significantly smaller than competing models; also, we test the model against non-hypercomplex architectures given a tight resource budget. We close with concluding remarks on our contribution and future work in Sect. 5.

2 Hypercomplex Numbers and Hypercomplex Layers

2.1 Quaternions

Hypercomplex algebras are mathematical structures of numbers of "intrinsically high" dimensionality. Historically, quaternions (the set of which is denoted as $\mathbb{H}$ in this text) were the first type of hypercomplex numbers, discovered by Hamilton in the 19th century [10]. Motivated by extending complex numbers $\mathbb{C}$, which are made up of a real and an imaginary part, to an algebra of a multitude of parts, quaternions were defined as numbers q:

$$q = a + b\boldsymbol{i} + c\boldsymbol{j} + d\boldsymbol{k}, \tag{1}$$

where $a, b, c, d \in \mathbb{R}$ and $\boldsymbol{i}, \boldsymbol{j}, \boldsymbol{k}$ are imaginary units. The three imaginary units, along with the real unit, are deemed independent and perpendicular to one another. They can also be rewritten as a sum of a scalar part $S(q) = a$ and a vector part $V(q) = b\boldsymbol{i} + c\boldsymbol{j} + d\boldsymbol{k}$, isomorphic to $\mathbb{R}$ and $\mathbb{R}^3$ respectively. All imaginary units admit to being square roots of negative unity:

$$\boldsymbol{i}^2 = \boldsymbol{j}^2 = \boldsymbol{k}^2 = -1, \tag{2}$$

a property which actually extends to infinite elements in $\mathbb{H}$. Additive and multiplication rules are necessary to define an algebra, of which the first is quite straightforward; corresponding real or imaginary parts are added together, with no operations acting between different number real or imaginary parts. Formally,

$$p + q = (a_p + a_q) + (b_p + b_q)\boldsymbol{i} + (c_p + c_q)\boldsymbol{j} + (d_p + d_q)\boldsymbol{k}, \tag{3}$$

where $p = a_p + b_p\boldsymbol{i} + c_p\boldsymbol{j} + d_p\boldsymbol{k}$ and $q = a_q + b_q\boldsymbol{i} + c_q\boldsymbol{j} + d_q\boldsymbol{k}$. Multiplication of quaternions requires first defining a way to multiply imaginary units. Except Eq. 2, we have

$$\boldsymbol{ij} = \boldsymbol{k}, \boldsymbol{jk} = \boldsymbol{i}, \boldsymbol{ki} = \boldsymbol{j}, \boldsymbol{ji} = -\boldsymbol{k}, \boldsymbol{jk} = -\boldsymbol{i}, \boldsymbol{ki} = -\boldsymbol{j}, \tag{4}$$

and the real unity acts as a multiplicative identity as in $\mathbb{R}$. Note that from Eq. 4 we have the corollary that in general $pq \neq qp$, so multiplication is not commutative in $\mathbb{H}$ (but it still is associative). Combined with a transitive property over our definition of addition, we have the rule of multiplication (also referred to as a "Hamilton" product):

$$pq = S(p)S(q) - V(p) \cdot V(q) + S(p)V(q) + S(q)V(p) + V(p) \times V(q), \tag{5}$$

where $\cdot$ is the inner product and $\times$ is the cross product. Note from the above, that for "pure" quaternions ($S(p) = S(q) = 0$) that are also perpendicular, the multiplication rule becomes simply a cross product $V(p) \times V(q)$. The multiplication rule can be written in an expanded form as:

$$pq = (a_p a_q - b_p b_q - c_p c_q - d_p d_q) + \tag{6}$$
$$(a_p b_q + b_p a_q + c_p d_q - d_p c_q)\boldsymbol{i} + \tag{7}$$
$$(a_p c_q - b_p d_q + c_p a_q + d_p b_q)\boldsymbol{j} + \tag{8}$$
$$(a_p d_q + b_p c_q - c_p b_q + d_p a_q)\boldsymbol{k}, \tag{9}$$

where we replaced $S(\cdot)$ and $V(\cdot)$ with their definitions. Especially interestingly regarding the proposed model in this work, we can rewrite the above in a matrix-vector form, as:

$$\begin{bmatrix} a_{pq} \\ b_{pq} \\ c_{pq} \\ d_{pq} \end{bmatrix} = Pq = \begin{bmatrix} a_p & -b_p & -c_p & -d_p \\ b_p & a_p & -d_p & c_p \\ c_p & d_p & a_p & -b_p \\ d_p & -c_p & b_p & a_p \end{bmatrix} \begin{bmatrix} a_q \\ b_q \\ c_q \\ d_q \end{bmatrix}, \tag{10}$$

where for the resulting quaternion pq we write $pq = a_{pq} + b_{pq}\boldsymbol{i} + c_{pq}\boldsymbol{j} + d_{pq}\boldsymbol{k}$. With a slight abuse of notation we write q also to denote the vector $\in \mathbb{R}^4$ that includes the coefficients of quaternion q, and we write P for the matrix that corresponds to quaternion p. Matrices structured as P form a 4-dimensional subspace of $\mathbb{R}^{4 \times 4}$ that is isomorphic to $\mathbb{H}$, to which we shall refer to as S_4. It is straightforward to see that a basis for S_4 is formed by the following matrices:

$$A_1^4 = I_4, A_2^4 = \begin{bmatrix} 0 & -1 & 0 & 0 \\ 1 & 0 & 0 & 0 \\ 0 & 0 & 0 & -1 \\ 0 & 0 & 1 & 0 \end{bmatrix}, A_3^4 = \begin{bmatrix} 0 & 0 & -1 & 0 \\ 0 & 0 & 0 & 1 \\ 1 & 0 & 0 & 0 \\ 0 & -1 & 0 & 0 \end{bmatrix}, A_4^4 = \begin{bmatrix} 0 & 0 & 0 & -1 \\ 0 & 0 & -1 & 0 \\ 0 & 1 & 0 & 0 \\ 1 & 0 & 0 & 0 \end{bmatrix},$$
$$\tag{11}$$

i.e. any matrix in S_4 can be written as a linear combination of $\{A_i\}_{i=1}^4$, and $\{A_i\}_{i=1}^4$ are linearly independent. Hence, we can rewrite Eq. 10 as:

$$
\begin{bmatrix} a_{pq} \\ b_{pq} \\ c_{pq} \\ d_{pq} \end{bmatrix} = (a_p A_1^4 + b_p A_2^4 + c_p A_3^4 + d_p A_4^4) \begin{bmatrix} a_q \\ b_q \\ c_q \\ d_q \end{bmatrix}. \tag{12}
$$

Now, supposing that we need to multiply N quaternions $p_1, p_2, \cdots, p_N$ with N quaternions $q_1, q_2, \cdots, q_N$, we can rewrite this process again as a matrix-vector product $q^N = P^N q^N$, where our input and resulting vectors are now of dimensions equal to $4N$, and the transformation matrix is $\in \mathbb{R}^{4N \times 4N}$. Depending on whether we want to multiply only N pairs of quaternions $p_i q_i$ for $\forall i \in [1, N]$ or all possible pairs of p_i $\forall i \in [1, N]$ and q_j $\forall j \in [1, N]$, we'll have a dimensionality equal to $4N$ or $4N^2$ respectively. This is still much less than the containing space of $4N \times 4N$, for which $dim\{\mathbb{R}^{4N \times 4N}\} = 16N^2$. Furthermore, the matrix P^N can be written as a sum of Kronecker products of matrices A_1, A_2, A_3, A_4 with matrices that contain the elements of $p_1, p_2, \cdots, p_N$ [38]. Recall that the Kronecker product of $A \in \mathbb{R}^{k \times l}$ and $B \in \mathbb{R}^{m \times n}$ is defined as:

$$
A \otimes B = \begin{bmatrix} a_{11}B \ \ldots \ a_{1l}B \\ \vdots \ \ddots \ \ \vdots \\ a_{k1}B \ \ldots \ a_{kl}B \end{bmatrix}, \tag{13}
$$

where $A \otimes B \in \mathbb{R}^{km \times ln}$. Note that in general $A \otimes B \neq B \otimes A$, but the two results are related through a "perfect shuffle" permutation [33]. This means that, depending on whether we stack quaternion elements in input and resulting vectors as $a_p^1, a_p^2, a_p^3, \cdots, d_p^2, d_p^3, d_p^4$ or $a_p^1, b_p^1, c_p^1, \cdots, b_p^N, c_p^N, d_p^N$, we can either use a sum of factors with A_i matrices multiplying from the left or from the right. Writing a set of quaternion products as a single Kronecker-factored product is important with respect to the generalization of the Hamilton product to Parameterized Hypercomplex Multiplication, which we discuss in Subsect. 2.2.

Higher-order Hypercomplex Numbers. Quaternions aside, there exist other types of hypercomplex numbers of dimensionality higher than 4. As a rule of the thumb, the more we progress to higher dimensions, the less "easy-to-use" each hypercomplex algebra becomes; for example, quaternions have a non-commutative but associative multiplication rule but octonions and sedenions, of dimensionalities equal to 8 and 16 respectively, are neither commutative nor associative.

2.2 Quaternion and Parameterized Hypercomplex Layers

The restatement of the Hamilton product as a matrix-vector product (Eq. 10) has in fact provided the basis for the definition of quaternion layers as part of quaternionic extensions of standard layers in neural networks. As a multiplication by a matrix corresponds to a linear transformation, this construction has been used as a replacement of standard linear transformation components. Furthermore, a linear transformation matrix sized $M \times N$, which would normally have a dimension equal to MN, when replaced with its quaternionic counterpart only has a dimension equal to four times less, $MN/4$, as we saw in the previous subsection. In practical terms, this means that the layer uses four times less parameters for each quaternionic component. As linear components are ubiquitous in neural networks, whole networks can be revamped to their quaternionic versions [18].

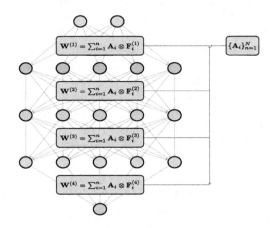

Fig. 1. An illustration of the proposed idea. Standard Parameterized Hypercomplex Networks use multiple layers that are parameterized using two sets of matrices: $\{A_i\}_{i=1}^n$ and $\{F_i\}_{i=1}^n$, of which the former can be thought of a learnable generalization of the Hamilton product rules. In this work, with "Shared-Operation" Hypercomplex Networks, we argue that learning only one set of $\{A_i\}_{i=1}^n$ for the whole network, is enough to construct a useful and very light-weight model.

The major constraint related to quaternionic neural networks is that dimensionality is reduced according to a fixed factor of *four*. Zhang et al. [38] have proposed a generalization of the aforementioned construction from quaternions and four-fold economy to arbitrary-dimension hypercomplex constructions. The matrix W in all cases is defined as a decomposition into matrices $\{A_i\}_{n=1}^N$ and $\{F_i\}_{n=1}^N$. This is based on writing the transformation matrix W for a given linear component as a sum of Kronecker factors:

$$W = \sum_{i=1}^{n} A_i \otimes F_i, \tag{14}$$

where $A_i \in \mathbb{R}^{n \times n}$ and $F_i \in \mathbb{R}^{f/n \times g/n}$. The resulting matrix W is of size $f \times g$, but the total number of independent parameters is significantly less. Indeed, we have a total of $n^3 + fg/n$ parameters, where supposing that n is much less than either input and output dimensionalities f or g, the second parameter should be dominant. For a total of L independent linear components in the network, we have $Ln^3 + Lfg/n$ free parameters.

3   Proposed Model for Handwritten Text Recognition

3.1   Shared-Operation Parameterized Hypercomplex Layer

In this work, we propose the use of a new variant of Parameterized Hypercomplex Multiplication Networks (PHM), to which we refer to as *Shared-Operation Hypercomplex Network* (SOHN). "Operation sharing" in SOHN refers to having a shared component that is learned by all hypercomplex layers jointly. Our motivation is related to the three following points:

a) In standard PHM, layer linear components are decomposed as sums of Kronecker products of the form $\sum_i A_i \otimes F_i$. The two sets of matrices A_i, F_i are related to a different intuitive use. Matrices A_i are related to how the shared parameter groups interact with one another. Note for example, that we can understand the quaternionic case as a special case of parameterized hypercomplex multiplication where $n = 4$ and A_i are as in Eq. 11. These matrices control the structure of the resulting Kronecker product, and they are a direct consequence of the definition of the Hamilton product, which in turn stems from the multiplication rules that were set so that $\mathbb{H}$ can form an algebra. In the parameterized hypercomplex case, we no longer have these constraints, and in a sense the equivalent of the Hamilton product is re-learnt for arbitrary n.

b) The number of what we called "independent" linear layers in our network can in effect be significantly larger than the number of layers themselves. For example, for a Hypercomplex LSTM layer, we have 4 or 8 linear transformations assuming unidirectional or bidirectional recurrence; these correspond to each of the related gates (input, output, forget, cell input). For the convolutional case, we can also write a convolution in a matrix-vector form $Wx + b$ with W as a circulant matrix, but it will suffice to deal with a form where parameters are packed in an order-4 tensor as in [5,31]). Nevertheless, the complexity factor related to the size of the A_i matrices can quickly (w.r.t. increasing n) become significant. (In our HTR experiments, we show that choosing $n = 32$ accounts for almost half the network parameter complexity).

c) Also importantly, in practice we have observed that results for $PHM/n = 4$ and quaternionic variants are similar in terms of network accuracy/efficiency (e.g. [31]). This may hint that *learning a separate set of multiplication rules for each linear component is unnecessary*.

For reasons of clarity of presentation, we consider a case where we have only feed-forward connections and all linear operations are hypercomplex; this case can easily be extended to recurrent connections and NNs that include non-hypercomplex layers. Formally, we write our network as a cascade of L layers, i.e.:

$$SOHN(x; \{A_i\}_{i=1}^n, \{F_i^{(1)}\}_{i=1}^n, \{F_i^{(2)}\}_{i=1}^n, \cdots, \{F_i^{(L)}\}_{i=1}^n) =$$

$$l^{(L)}(\{A_i\}_{i=1}^n, \{F_i^{(L)}\}_{i=1}^n) \circ \cdots \circ l^{(1)}(\{A_i\}_{i=1}^n, \{F_i^{(1)}\}_{i=1}^n)(x), \qquad (15)$$

each layer uses parameterized hypercomplex operations:

$$y^{(j)} = PHM(y^{(j-1)}; \{A_i\}_{i=1}^n, \{F_i^{(j)}\}_{i=1}^n), \qquad (16)$$

where y^j is the output of layer j. In practical terms, a SOHN can be implemented by adding skip connections starting from a single learnable tensor of size $n \times n \times n$ which would represent the A_i matrices, towards all hypercomplex layers. An illustration of this idea can be seen in Fig. 1.

3.2 Model Architecture

As a baseline architecture, in the sense of choosing the layout, number of blocks, channels, dimensionality and other component features, we have followed the convolutional-recurrent architecture proposed in [23]. Written as a Python-style data structure –this will come in handy for comparing architectures in Sect. 4– we denote architecture as $[(2, 64), mpool, (4, 128), mpool, (4, 256)], (256, 3)$. A list of tuples corresponds to convolutional backbone of ResNet blocks, and the final tuple corresponds to the setup of the recurrent component before the softmax output. Tuples for the convolutional backbone signify (number of ResNet blocks, number of channels). "mpool" signifies a 2×2 max-pooling operation. Regarding the tuple corresponding to the recurrent component, it signifies (hidden feature dimensionality, number of LSTM layers). After the final convolutional block, a column-wise max-pooling follows, which branches out to two routes. The first, "main" route is fed to the LSTM recurrent component, followed by a fully-connected layer which maps the output number of channels to the number of output classes, so we get a sequence of softmax-activated probability vectors as a result. This result is fed to a CTC loss. The second route that branches after the column-wise max-pooling operation, is transformed to softmax-activated probability vectors directly, i.e. bypassing the recurrent component. This again is fed to a CTC loss that is weighed down $(0.1\times)$ with respect to the loss that corresponds to the first, recurrent route; this architectural choice has shown to aid convergence in recurrent architectures [23,26]. In all cases, we have opted for using channel size of convolutional and recurrent layers that are multiples of the highest hypercomplex parameter that we have used, $n = 32$. This choice was made so as to ease comparisons between different hyperparameter choices. Setting channel sizes according to this rule was not possible in a number of cases, namely regarding input and output size. In these cases, we use 1×1 convolution to transform a tensor from or to the desired channel size. Dropout with probability 20% is used between recurrent layers (Fig. 2).

4   Experiments

4.1   Datasets

We have tested our models on three document image datasets: IAM [14], Rimes [1], and Memoirs [28]. Excerpts from the three datasets can be compared in Fig. 3.

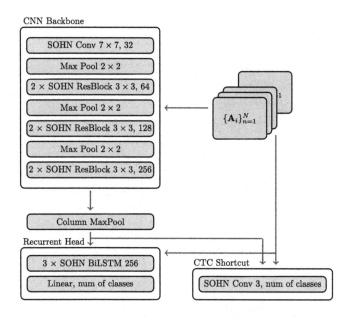

Fig. 2. The proposed shared-operation architecture for HTR.

IAM. The IAM dataset encompasses text produced by 657 different writers, split into writer-independent partitions (we use the same training/validation/set partition as in [21]). There is a total of 6482 lines for the training set, 976 lines for the validation set, and 2915 lines for the test set. IAM is written in a Latin script and in the English language.

RIMES. RIMES encompasses 11333 lines in the training set and 778 lines in the test set. It is written in a Latin script and in the French language.

Memoirs. The Memoirs dataset [28] comprises 46 manuscripts, written in the 19th as a personal diary of Sophia Trikoupi, sister of a contemporary Greek prime minister. We have a total of 4, 941 words, which correspond to 385 lines for the training set, 129 lines for the validation set and 179 lines of text for the test set. We use the training and test partitions defined in [28].[1] All experiments follow the setting of line-level recognition, with a lexicon-free unconstrained greedy CTC decoding scheme.

4.2 Varying the Hypercomplex Dimension and PHM vs SOHN

In Table 1 we report results of HTR on the three aforementioned datasets. We have run different variants of the proposed hypercomplex model, using either standard Parameterized Hypercomplex Multiplication [38] (PHM), or the proposed Shared-Operation Hypercomplex variant (SOHN). In both cases we report

---

[1] The dataset is publicly available at https://github.com/sfikas/sophia-trikoupi-handwritten-dataset/.

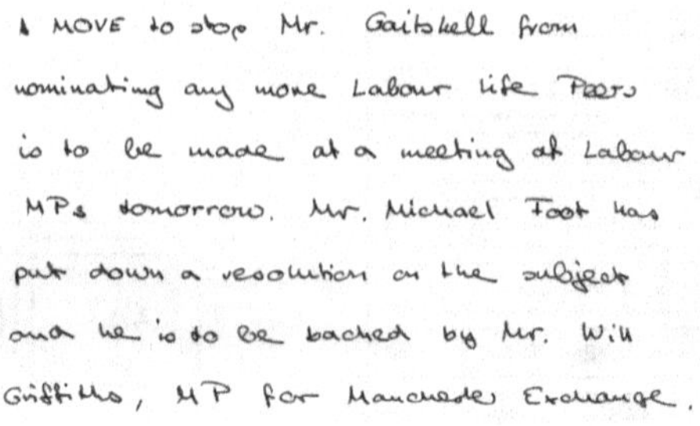
A MOVE to stop Mr. Gaitskell from nominating any more Labour life Peers is to be made at a meeting of Labour MPs tomorrow. Mr. Michael Foot has put down a resolution on the subject and he is to be backed by Mr. Will Griffiths, MP for Manchester Exchange.

(a) IAM

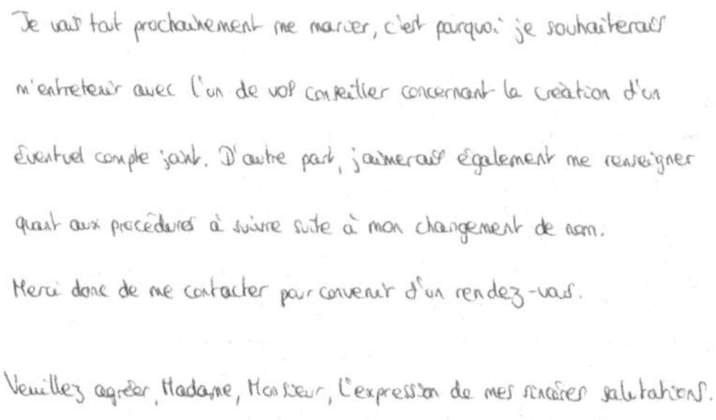
Je vais tout prochainement me marier, c'est pourquoi je souhaiterais m'entretenir avec l'un de vos conseiller concernant la création d'un éventuel compte joint. D'autre part, j'aimerais également me renseigner quant aux procédures à suivre suite à mon changement de nom.
Merci donc de me contacter pour convenir d'un rendez-vous.

Veuillez agréer, Madame, Monsieur, l'expression de mes sincères salutations.

(b) RIMES

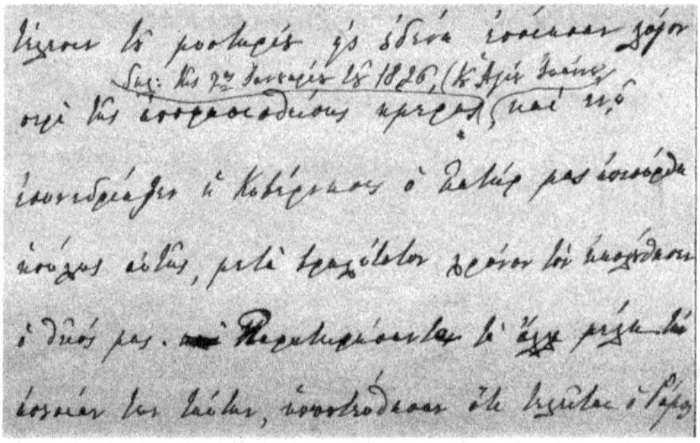

(c) Memoirs

Fig. 3. Excerpts from the datasets used for our experiments.

the employed hypercomplex parameter (n). Different values for n correspond to different structure of the A_i and F_i Kronecker factors, and in general a higher value for n leads to more compression. The exception to this rule comes when the cubic factor of complexity $O(n^3)$ for the A_i matrices surpasses that of the F_i matrices, i.e. $O(fg/n)$. In the SOHN variant the former is significantly mitigated, as we only require a single n^3-sized tensor for the whole network, hence the savings in parameter size. (Note that the difference in accuracy between [23] stems from the custom implementation of LSTM that was written to extend to the hypercomplex variants – the architecture of the "Retsinas et al." and

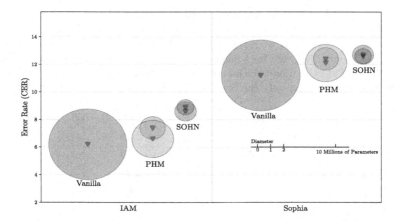

Fig. 4. Comparison of HTR models in terms of model size and test set error rate. The area of each circle is proportional to the number of trainable parameters of a variant in a model family. Different circle sizes for SOHN and PHM correpond to different hyperparameter values ($n = 16$ and $n = 32$ for the larger and smaller radii respectively).

"Real-valued" models as referenced in Table 1 is otherwise identical). A graphical illustration can also be examined in Fig. 4, where we present a comparison of both error rate and model size.

4.3   PHM Model vs Real-Valued Model on a Resource Budget

We have compared against a variants of CNN-RNN with a constrained, fixed budget of trainable parameters. We have tested models on two different budgets, namely 500 thousand and 750 thousand parameters, which we can examine in Tables 2 and 3 respectively. In these cases, we used the proposed SOHN models for hypercomplex dimensions $n = 16$ and $n = 32$, and compared against non-hypercomplex models that use approximately the same number of parameters. As there is naturally more than one way to build a model given a set budget, we built different non-hypercomplex models in order to have as much an unbiased and fair comparison as possible. The setups for these model variants are:

1. $[(4, 64)], (64, 2)$ (500k parameters)
2. $[(1, 64)], (100, 2)$ (513k parameters)
3. $[(1, 64)], mpool, (1, 128)], (64, 2)$ (576k parameters)
4. $[(2, 32), mpool, (2, 64), mpool, (2, 64)], (96, 1)$ (521k parameters)
5. $[(2, 8), mpool, (4, 8), mpool, (4, 8)], (112, 2)$ (494k parameters)
6. $[(1, 64), mpool, (2, 112)], (64, 2)$ (746k parameters)
7. $[(2, 96)], (100, 2)$ (768k parameters)
8. $[(2, 16), mpool, (4, 18), mpool, (4, 32)], (128, 2)$ (744k parameters)

where we have used the same convention to represent architectural choices as in Sect. 3.

Table 1. Comparison of HTR models in terms of model size and test set error rate. Model size is evaluated in terms of model number of parameters, and error rate is evaluated in terms of CER over the test set of three different datasets (IAM [14], RIMES [1], Memoirs [28]). For both considered metrics, a lower value is more desirable. Both works of Li et al. [11] and Wick et al. (+syn) [36] use additional synthetic data in training, while for the rest only the predefined training set is used to train the model.

	#Millions of Params	IAM	RIMES	Memoirs
Li et al. (+syn) [11]	334	3.42	–	–
Li et al. (+syn) [11]	558	2.89	–	–
Wick et al. (+syn) [36]	4.8	3.96	–	–
Wick et al. [36]	4.8	5.09	–	–
Yousef et al. [37]	3.4	4.9	–	–
Diaz et al. [3]	12	2.75	–	–
Retsinas et al. [23]	10	4.62	2.75	–
Real-valued model	10	6.2	3.9	11.2
PHM/$n = 2$ model	5.1	6.5	6.5	11.4
Quaternion model	2.6	6.9	4.3	11.8
PHM/$n = 4$ model	2.6	6.9	7.0	11.4
PHM/$n = 8$ model	1.4	7.5	4.7	11.8
PHM/$n = 16$ model	1.03	7.4	4.5	12.4
SOHN/$n = 16$ model	0.74	8.6	5.5	12.7
PHM/$n = 32$ model	2.8	6.6	3.9	12.1
SOHN/$n = 32$ model	0.46	8.9	5.7	12.6

As a remark on the results of Tables 2, 3, we can state that it seems that trimming down a network to less parameters leads to error rates that vary widely; some of the compared variants are much worse than the proposed SOHN models, while others come quite close. We can however deduce that while applying a Shared-Operation architecture invariably leads to a slight detriment of accuracy, attempting to enforce a constrained budget on a real-valued architecture is very much dependent on the way the baseline network is "pruned-down".

Table 2. Peformance comparison of HTR models on a budget of 500k parameters. Loss, CER and WER over the IAM test set are reported. See text for details.

Model type	Number of Parameters	Test Loss	Best CER	Best WER
Ours (SOHN/n=32)	459k	25.3	8.9	29.6
Non-PHM variant #1	500k	35.4	16.6	50.9
Non-PHM variant #2	513k	72.4	37.2	82.7
Non-PHM variant #3	576k	44.8	17.3	52.9
Non-PHM variant #4	521k	27.3	9.0	30.6
Non-PHM variant #5	494k	61.7	26.3	65.4

Table 3. Peformance comparison of HTR models on a budget of 750k parameters. Loss, CER and WER over the IAM test set are reported. See text for details.

Model type	Number of Parameters	Test Loss	Best CER	Best WER
Ours (SOHN/n=16)	742k	26.4	8.6	28.9
Non-PHM variant #6	746k	34.2	13.1	42.2
Non-PHM variant #7	768k	53.7	23.6	64.5
Non-PHM variant #8	744k	31.2	10.9	34.9

5 Conclusion and Future Work

We have presented a new type of hypercomplex architecture called Shared-Operation Hypercomplex Networks. This scheme is based on the idea that the multiplication operation matrices used in hypercomplex layers can be shared across the network. As more parameters are shared, the computational complexity alongside the trainable parameters is significantly reduced. Our claims are *to an extent* corroborated by the reported experimental results, which test SOHN based networks and PHM variants of our method against the current state of the art in handwritten text recognition. While cutting down the number of matrices to a single one does not throw off the model mechanics and adequate error rates are still achieved, our experiments suggest that learning multiple A_i matrices (as in standard PHM) is still beneficial, and by no means the difference in accuracy between PHM and SOHN is insignificant. In general however, integration of the proposed method results in high network compression with small accuracy drop. For future work, we plan to extend the model with other ways of constraining the expenses related to the two Kronecker factors, like low-rank approximations or other types of factorizations. Another direction of research could involve imposing contraints in the form of a probabilistic prior over the A_i matrices, which could in this context help regulate between choosing a global, shared representation versus a local representation [29,30]. In terms of applications, interesting use cases include segmentation-free HTR frameworks [2] or word-level recognition systems [26].

Acknowledgments. This research has been partially co - financed by the EU and Greek national funds through the Operational Program Competitiveness, Entrepreneurship and Innovation, under the call "OPEN INNOVATION IN CULTURE", project *Bessarion* (T6YBΠ - 00214).

References

1. Augustin, E., Carré, M., Grosicki, E., Brodin, J.M., Geoffrois, E., Prêteux, F.: Rimes evaluation campaign for handwritten mail processing. In: International Workshop on Frontiers in Handwriting Recognition (IWFHR 2006), pp. 231–235 (2006)
2. Coquenet, D., Chatelain, C., Paquet, T.: DAN: a segmentation-free document attention network for handwritten document recognition. IEEE Trans. Pattern Anal. Mach. Intell. **45**, 8227–8243 (2023)
3. Diaz, D.H., Qin, S., Ingle, R., Fujii, Y., Bissacco, A.: Rethinking text line recognition models. arXiv preprint arXiv:2104.07787 (2021)
4. Dimitrakopoulos, P., Sfikas, G., Nikou, C.: Variational feature pyramid networks. In: International Conference on Machine Learning, pp. 5142–5152. PMLR (2022)
5. Grassucci, E., Zhang, A., Comminiello, D.: Lightweight convolutional neural networks by hypercomplex parameterization. arXiv preprint arXiv:2110.04176 (2021)
6. Graves, A., Fernández, S., Gomez, F., Schmidhuber, J.: Connectionist temporal classification: labelling unsegmented sequence data with recurrent neural networks. In: Proceedings of the 23rd International Conference on Machine Learning, pp. 369–376. ACM (2006)
7. Isokawa, T., Kusakabe, T., Matsui, N., Peper, F.: Quaternion neural network and its application. In: Palade, V., Howlett, R.J., Jain, L. (eds.) KES 2003. LNCS (LNAI), vol. 2774, pp. 318–324. Springer, Heidelberg (2003). https://doi.org/10.1007/978-3-540-45226-3_44
8. Kang, L., Riba, P., Rusiñol, M., Fornés, A., Villegas, M.: Pay attention to what you read: non-recurrent handwritten text-line recognition. Pattern Recogn. **129**, 108766 (2022)
9. Knigge, D.M., et al.: Modelling long range dependencies in ND: from task-specific to a general purpose CNN. arXiv preprint arXiv:2301.10540 (2023)
10. Kuipers, J.B.: Quaternions and Rotation Sequences: A Primer with Application to Orbits, Aerospace and Virtual Reality. Princeton University Press, Princeton (1999)
11. Li, M., et al.: TROCR: transformer-based optical character recognition with pre-trained models. arXiv preprint arXiv:2109.10282 (2021)
12. Louizos, C., Welling, M., Kingma, D.P.: Learning sparse neural networks through l_0 regularization. arXiv preprint arXiv:1712.01312 (2017)
13. Markou, K., et al.: A convolutional recurrent neural network for the handwritten text recognition of historical greek manuscripts. In: Del Bimbo, A., et al. (eds.) ICPR 2021. LNCS, vol. 12667, pp. 249–262. Springer, Cham (2021). https://doi.org/10.1007/978-3-030-68787-8_18
14. Marti, U.V., Bunke, H.: The IAM-database: an English sentence database for offline handwriting recognition. Int. J. Doc. Anal. Recogn. **5**(1), 39–46 (2002)
15. Nguyen, T.D., Phung, D., et al.: Quaternion graph neural networks. In: Asian Conference on Machine Learning, pp. 236–251. PMLR (2021)
16. Nitta, T.: A quaternary version of the backpropagation algorithm. In: Proceedings of ICNN 1995 - International Conference on Neural Networks, pp. 2753–2756 (1995)

17. Parcollet, T., Morchid, M., Linarès, G.: Quaternion convolutional neural networks for heterogeneous image processing. In: ICASSP 2019–2019 IEEE International Conference on Acoustics, Speech and Signal Processing (ICASSP), pp. 8514–8518. IEEE (2019)
18. Parcollet, T., Morchid, M., Linarès, G.: A survey of quaternion neural networks. Artif. Intell. Rev. **53**(4), 2957–2982 (2020)
19. Parcollet, T., et al.: Quaternion recurrent neural networks. arXiv preprint arXiv:1806.04418 (2018)
20. Prince, S.J.: Understanding Deep Learning. MIT Press (2023). https://udlbook.github.io/udlbook/
21. Puigcerver, J.: Are multidimensional recurrent layers really necessary for handwritten text recognition? In: 2017 14th IAPR International Conference on Document Analysis and Recognition (ICDAR), vol. 1, pp. 67–72. IEEE (2017)
22. Retsinas, G., Elafrou, A., Goumas, G., Maragos, P.: Online weight pruning via adaptive sparsity loss. In: 2021 IEEE International Conference on Image Processing (ICIP), pp. 3517–3521. IEEE (2021)
23. Retsinas, G., Sfikas, G., Gatos, B., Nikou, C.: Best practices for a handwritten text recognition system. In: Uchida, S., Barney, E., Eglin, V. (eds.) DAS 2022. LNCS, vol. 13237, pp. 247–259. Springer, Cham (2022). https://doi.org/10.1007/978-3-031-06555-2_17
24. Retsinas, G., Sfikas, G., Louloudis, G., Stamatopoulos, N., Gatos, B.: Compact deep descriptors for keyword spotting. In: 2018 16th International Conference on Frontiers in Handwriting Recognition (ICFHR), pp. 315–320. IEEE (2018)
25. Retsinas, G., Sfikas, G., Nikou, C.: Iterative weighted transductive learning for handwriting recognition. In: Lladós, J., Lopresti, D., Uchida, S. (eds.) ICDAR 2021. LNCS, vol. 12824, pp. 587–601. Springer, Cham (2021). https://doi.org/10.1007/978-3-030-86337-1_39
26. Retsinas, G., Sfikas, G., Nikou, C., Maragos, P.: From Seq2Seq to handwritten word embeddings. In: British Machine Vision Conference (BMVC) (2021)
27. Romero, D.W., Bruintjes, R.J., Tomczak, J.M., Bekkers, E.J., Hoogendoorn, M., van Gemert, J.C.: Flexconv: continuous kernel convolutions with differentiable kernel sizes. arXiv preprint arXiv:2110.08059 (2021)
28. Sfikas, G., Giotis, A.P., Louloudis, G., Gatos, B.: Using attributes for word spotting and recognition in polytonic greek documents. In: 2015 13th International Conference on Document Analysis and Recognition (ICDAR), pp. 686–690. IEEE (2015)
29. Sfikas, G., Nikou, C., Galatsanos, N., Heinrich, C.: MR brain tissue classification using an edge-preserving spatially variant Bayesian mixture model. In: Metaxas, D., Axel, L., Fichtinger, G., Székely, G. (eds.) MICCAI 2008. LNCS, vol. 5241, pp. 43–50. Springer, Heidelberg (2008). https://doi.org/10.1007/978-3-540-85988-8_6
30. Sfikas, G., Nikou, C., Galatsanos, N., Heinrich, C.: Majorization-minimization mixture model determination in image segmentation. In: CVPR 2011, pp. 2169–2176. IEEE (2011)
31. Sfikas, G., Retsinas, G., Gatos, B., Nikou, C.: Hypercomplex generative adversarial networks for lightweight semantic labeling. In: El Yacoubi, M., Granger, E., Yuen, P.C., Pal, U., Vincent, N. (eds.) ICPRAI 2022, Part I. LNCS, vol. 13363, pp. 251–262. Springer, Cham (2022)
32. Tay, Y., et al.: Lightweight and efficient neural natural language processing with quaternion networks. arXiv preprint arXiv:1906.04393 (2019)
33. Van Loan, C.F.: The ubiquitous kronecker product. J. Comput. Appl. Math. **123**(1–2), 85–100 (2000)

34. Vaswani, A., et al.: Attention is all you need. In: Advances in Neural Information Processing Systems, vol. 30 (2017)

35. Wick, C., Zöllner, J., Grüning, T.: Transformer for handwritten text recognition using bidirectional post-decoding. In: Lladós, J., Lopresti, D., Uchida, S. (eds.) ICDAR 2021. LNCS, vol. 12823, pp. 112–126. Springer, Cham (2021). https://doi.org/10.1007/978-3-030-86334-0_8

36. Wick, C., Zöllner, J., Grüning, T.: Rescoring sequence-to-sequence models for text line recognition with CTC-prefixes. In: Uchida, S., Barney, E., Eglin, V. (eds.) DAS 2022. LNCS, vol. 13237, pp. 260–274. Springer, Cham (2022). https://doi.org/10.1007/978-3-031-06555-2_18

37. Yousef, M., Hussain, K.F., Mohammed, U.S.: Accurate, data-efficient, unconstrained text recognition with convolutional neural networks. Pattern Recogn. **108**, 107482 (2020)

38. Zhang, A., et al.: Beyond fully-connected layers with quaternions: Parameterization of hypercomplex multiplications with $1/n$ parameters. In: International Conference on Learning Representations (ICLR 2021) (2021)

39. Zhao, C., Ni, B., Zhang, J., Zhao, Q., Zhang, W., Tian, Q.: Variational convolutional neural network pruning. In: Proceedings of the IEEE Conference on Computer Vision and Pattern Recognition, pp. 2780–2789 (2019)

40. Zhu, X., Xu, Y., Xu, H., Chen, C.: Quaternion convolutional neural networks. In: Ferrari, V., Hebert, M., Sminchisescu, C., Weiss, Y. (eds.) ECCV 2018. LNCS, vol. 11212, pp. 645–661. Springer, Cham (2018). https://doi.org/10.1007/978-3-030-01237-3_39

DSS: Synthesizing Long Digital Ink Using Data Augmentation, Style Encoding and Split Generation

Aleksandr Timofeev[1(✉)], Anastasiia Fadeeva[2(✉)], Andrei Afonin[1(✉)], Claudiu Musat[2(✉)], and Andrii Maksai[2(✉)]

[1] EPFL, Lausanne, Switzerland
aleksandr.timofeev.m@gmail.com, afonin.ad@phystech.edu
[2] Google Research, Zürich, Switzerland
fadeich@google.com, cmusat@google.com, amaksai@google.com

Abstract. As text generative models can give increasingly long answers, we tackle the problem of synthesizing long text in digital ink. We show that the commonly used models for this task fail to generalize to long-form data and how this problem can be solved by augmenting the training data, changing the model architecture and the inference procedure. These methods use contrastive learning technique and are tailored specifically for the handwriting domain. They can be applied to any encoder-decoder model that works with digital ink. We demonstrate that our method reduces the character error rate on long-form English data by half compared to baseline RNN and by 16% compared to the previous approach that aims at addressing the same problem. We show that all three parts of the method improve recognizability of generated inks. In addition, we evaluate synthesized data in a human study and find that people perceive most of generated data as real.

Keywords: digital ink · online handwriting · generative models · length generalization

1 Introduction

With the growing usage of tablets and styluses, handwriting is an increasingly used human computer interaction (HCI) method. Recent developments in natural language processing make the interaction increasingly bidirectional. In the past handwriting was mostly used to facilitate the input of information for the human user, making its recognition the primary digital ink task [6,24]. With the advent of highly capable digital assistants [40] however, the interactions are becoming ever more natural. One way to add to this immersive HCI experience is to have the digital assistant respond in the same modality as the user input—handwriting. Handwriting synthesis is the process of converting printed text labels into handwriting. Traditionally, this was proposed for

---

A. Timofeev and A. Afonin—Work done as a student researcher at Google Research, Zürich, Switzerland.

A. Timofeev and A. Fadeeva—These authors contributed equally to this work and share first authorship.

G. A. Fink et al. (Eds.): ICDAR 2023, LNCS 14190, pp. 217–235, 2023.
https://doi.org/10.1007/978-3-031-41685-9_14

user-facing features like autocompletion and error correction [1]. These tasks all need to deal only with short textual sequences—operating mostly on a word level. For instance, in the case of error correction, only the misspelled word needs to be replaced. Whereas, responses from recent text generative models can contain multiple paragraphs of text. We propose methods to synthesize long inks to accommodate those responses.

The primary trait of synthesized ink is that it needs to be readable in order to be useful to a user. This observation leads to character error rate being the evaluation of choice in the past work on synthesis [7, 30]. For applications like autocompletion, similarity between the writer's style and generated ink is critical [1, 30]. In the current work we adopt recognisability metric (CER) but the evaluation of stylistic similarity is beyond the scope of this paper.

Some concerns about the synthesis quality are directly linked to the length of the ink. Ensuring the stability of the generated ink style is one of them, as it is needed to construct an immersive experience, where the generated ink is consistent and looks real to people. The presence of artifacts is another, as the number of artifacts is linked to the ink size. We tackle evaluations in both a qualitative and quantitative way, to see which methods of generating long inks are the best and whether they are good enough by human standards.

We propose three distinct methods of improving long generation. To extend the ink generation to long inks we propose data augmentation to bridge the gap between training and test conditions. To improve style consistency for long ink generation we utilize contrastive learning for style transfer. To ensure synthesis generalization to any input length we propose split generation. We show their impact using two different synthesis model architectures—LSTM and Transformer-based, on two ink representations—points and Bézier curves.

We compare ink synthesizer using the three improvements with both internal and external baselines and we observe recognizability improvements ranging from 16% to 50%, depending on the architecture. To sum up,

- we create a system that successfully synthesizes long digital ink[1]
- we blend three different approaches: Data augmentation, Style encoder and Split generation (we refer to it as **DSS**), resulting in large improvements in ink recognizability across multiple ink representations and architectures
- we perform an ablation study to quantify the additive individual impact of the three proposed components
- we run a user study to strengthen the quantitative analysis with a qualitative one, showing that most of the synthetic ink is perceived as real

2   Related Work

Handwriting synthesis has been a topic of interest for many years [13]. Digital ink can be represented in images which carry information about stroke style, color, and width, as well as in the background they are drawn on [2, 15] or sequence of coordinates with time information [16]. Both of those representations have their benefits. In this study we will focus on sequence-based approaches.

---

[1] A notebook to test model inference is available here: https://colab.research.google.com/drive/1SB_vyDcsdSq1CtE9IOD9opBR9IDgG0ly.

Digital ink synthesis is a generation task where a model receives text as input and outputs a sequence of coordinates that represents handwriting. Methods in this field have evolved from parametric models with a set of handmade style features [26] and sigma lognormal modeling [12] to deep neural networks such as LSTM [16] and Transformers [42]. Sequence to sequence models are frequently used for machine translation [39], speech recognition [19], speech synthesis [43] and abstractive text summarization [31]. These models can be applied to ink generation as well [16]. Mixture density network [5] are commonly used as the last layer to generate coordinates or Bézier curves [38] to capture a variety of possible next strokes.

Generation of personalized handwriting is of particular research interest due to many appealing applications like spelling correction [1] and completion [37]. In this setting a model has two inputs - text to generate and style ink. It outputs an ink with the text in the given style. Style determines the appearance of the output ink but doesn't change the content that is written. In the existing approaches, style is usually represented by a vector [1] or sequence of embeddings [7]. In this work we adopt style transfer to guarantee style consistency of long digital inks.

Previous studies in other domains suggest that deep neural networks tend to degrade in quality on long-form inputs [3,41]. For recurrent models LSTM [17] block and attention mechanisms [42] were proposed to facilitate long sequence modeling. However, even with those methods the gap in quality between training and out-of-distribution lengths is still present for encoder-decoder RNN models [9,22].

For transformer models incorporating positional information together with long data in training is a key to modelling long sequences [21,33,36]. In tasks where long-form training data is unavailable, self-supervised pre-training can help achieve state of the art performance across all lengths [10]. However, this pre-training requires a vast amount of unlabeled data that is not available in the handwriting domain.

While the previous methods focused on improvements in various architectures, they all rely on the existence of these long sequences in the training data. For the tasks where long training data is unavailable, data augmentation is one way to improve quality for long sequences [32,41]. In the case of speech recognition, simulating long-form data improved the quality by 27% [32] and for machine translation data augmentation boosted the performance by 30% [23].

In this paper we will evaluate common architectures for digital ink synthesis on long-form data. We will also investigate the affect of possible improvements like data augmentation and style conditioning on the out-of-domain length quality.

3   Method

In this section we propose the methods that focus on improving generalization of encoder-decoder to long sequences models in the digital ink domain. It is especially important because handwriting data collection is complex as it needs special equipment such as electronic whiteboard [27] or a tablet [1]. Thus, methods that address the issue without additional data collection are of particular interest.

3.1   Data Augmentation

One of possible approaches to improve performance on longer sequences is to create synthetic long training examples. This idea was explored for speech recognition where speech fragments from the same person were concatenated together [28]. Our approach also concatenates inks but instead of samples from the same writer we use stylistically similar inks. Algorithm 1 shows our general method of long ink construction from the original training dataset D. Thanks to uniform sampling in the algorithm 1 (denoted as U) we can get a set of long inks by simply calling the function multiple times. Algorithm 1 receives the following inputs:

- Style model f: receives two inks and outputs a score from 0 to 1 which represents the similarity between the writing styles.
- Concatenation function: receives two inks and outputs one ink which consists of two parts (example in Fig. 1).
- batch size bs controls the diversity of the result dataset, with higher values the probability of overlap between two long inks increases.
- Similarity threshold t controls the style consistency, with lower values the chance to concatenate inks with different styles increases.

In the following two subsections we describe the style model and concatenation function in more detail.

Algorithm 1: algorithm for generating a long sample

Input : training dataset D, target length l, batch size bs, threshold t, style model f, concat function
Output: long ink r
1 $r \sim U(D)$;
2 **while** $|r| < l$ **do**
3 $\quad$ $ink_0, \ldots, ink_{bs} \sim U(D)$;
4 $\quad$ $\forall\ ink \in$ inks: similarity$(ink) = f(r, ink)$;
5 $\quad$ candidates $= \{i \in$ inks $\mid$ similarities$(i) \geq t\ or\ i = \arg\max$ similarity$\}$;
6 $\quad$ ink $\sim U$(candidates);
7 $\quad$ r=concat$(r,$ ink$)$;
8 **end**
9 **return** r;

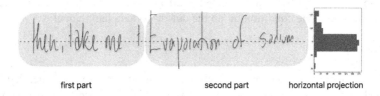

first part   second part   horizontal projection

Fig. 1. Example of two inks concatenation. In order to put inks on one line, we match the medians of their histograms and scale the second ink to have the same height.

Style Model. We train a model to have similar representations of different texts written in the same style. We use the fact that most inks are stylistically consistent, so we can use different parts of one ink as examples with different text but similar style. To achieve this we use contrastive learning technique. We train a classifier to distinguish between pieces of one ink versus pieces from different inks. We adapt the SimCLR approach from [8] where two inputs x_i and x_j are mapped with the same function g into two representations h_i and h_j of a fixed size and then the similarity of these vectors is computed:

$$\text{similarity}(i, j) = \cos(h_i, h_j)$$

This similarity represents how close in style inks x_i and x_j are. To generate training data for this task we randomly split existing inks into three pieces - beginning part, random gap and final part (we randomly choose two cut points in an ink). The random gap is used to generate multiple training examples from one ink. During training we want a similarity of parts from one ink to be close to 1 and for parts from different inks - to be around 0. During inference we use a full ink to compute the style representation and subsequently in similarity computation.

Ink Concatenation. In order to obtain long samples we need a method to merge two inks into one line. In case of misalignment, a training set may lose style consistency, which can lead to model performance degradation. Concatenation can be achieved using a *baseline* - imaginary line on which handwriting is written. We estimate baselines for both inks and apply an rigid transformation to one of them. There are many language specific baseline estimation methods [35], but for the purposes of this study we use a more general method of horizontal projection, shown in Fig. 1. During concatenation we add spaces in between inks which can be adjusted for the languages without space separation like Chinese. Hereby, we generate synthetic long samples from training examples with a similar style.

3.2   Generation with Style Conditioning

Autoregressive models are known to suffer from accumulation of mistakes during inference [4,25]. This problem becomes even more pressing for generation of long sequences. In order to overcome this, we explicitly add a style conditioning into the ink generation model. Conditioning on style is commonly used for style transfer [1,7,18], but we use it as a tool to mitigate past mistakes in generation.

We propose to train an encoder-decoder model with style conditioning end-to-end, as shown in Fig. 2. Encoder inputs the text label and the style model receives an example of ink to generate a style representation vector which is used in the decoding process. Training style model weights at the same time as encoder-decoder parameters helps to adapt style information to a given encoder-decoder model. For model optimisation we combine the negative log likelihood which is typically used to train generative models with SimCLR loss [8]:

$$\mathscr{L}(x) = \mathscr{L}_{\text{NLL}} + \mathscr{L}_{\text{SimCLR}} = -\sum_{i \leq n} \log \mathrm{P}(x_i | x_{<i}) + \mathscr{L}_{\text{SimCLR}}(x)$$

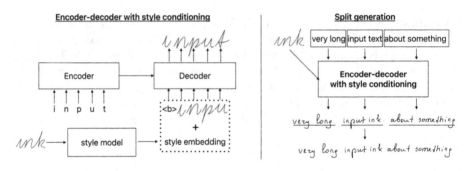

Fig. 2. On the left - training of generative model with style conditioning. On the right - split generation procedure with two words.

The first part of this loss is a likelihood to correctly generate an ink x. This part depends on encoder, decoder and style function f weights, as we use style embedding in the decoder's input. The second part of this loss ensures that function f outputs style representation and doesn't contain information about written text. Similar to the previous section we use the SimCLR approach from [8] where two inputs x_i and x_j are mapped with the same function f into two representations h_i and h_j of a fixed size and then the similarity of these vectors is computed as $\cos(h_i, h_j)$. In $\mathscr{L}_{\text{SimCLR}}$ we generate from ink x two pieces $x_{:k}$ and $x_{l:}$ (for some random k and l, $k < l$). These pieces are supposed to have high similarity, whereas similarity between $x_{:k}$ and the ink pieces from other inks should be low.

During training we use a piece of target ink as a style input, which can cause target leaking as $\mathscr{L}_{\text{NLL}}$ uses the full target including the style input. However, the use of Sim-CLR loss promotes independence of style and the text content, therefore no additional steps are needed to prevent target leaking.

During inference we use a separate set of inks for style extraction. Additionally, we use a full ink in the style model rather than just a piece and provide this embedding on each step of decoding as a reminder to the model. This helps avoid the style drift, where the model slowly changes the style of the writing until it becomes completely unrecognizable. Thus, we incorporated a style model into the long ink synthesizer to improve style consistency of long generation.

3.3 Split Generation

We propose an inference technique for long sequences inspired by dynamic overlapping inference in recognition [9]. The idea is to split a long input into separate pieces and calculate their results separately. We split a label at the word boundaries and the maximum number of words in one piece is a hyperparameter of this method, shown in Fig. 2. This way we retain the capability to generate cursive handwriting. The main idea of this method is to find the input length that is optimal for model's performance and use it in the inference. However, the main drawback of generating ink pieces separately is the fact that the result parts don't share any information. An example is shown in Fig. 4 random arrangement.

In case of a style mismatch between pieces, people can easily recognize that the ink is synthetic. We propose to use the model with style conditioning from the previous subsection in order to ensure style consistency of the individual generated pieces, that we combine together to obtain the result ink. To merge pieces together, we match the medians of their horizontal projections, as in the data augmentation step Fig. 1. We don't need to match their scale as they are already similar due to style conditioning.

Mismatch in slope between pieces is also easily recognizable by humans (see Fig. 3). To fix this problem we can sample multiple candidates and choose the most horizontal among them using linear regression on the y trajectory. This problem occurs because training inks are not always written horizontally and similar model behavior in split generation leads to mismatched slopes seen in Fig. 3.

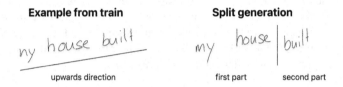

Fig. 3. Reason to generate multiple samples and choose the most straight ink in each part of Split Generation.

3.4 Combining Proposed Methods

The proposed methods are complementary to each other as they focus on different parts of an encoder-decoder model. In the training phase, we propose the following order of actions:

1. train a style model with SimCLR loss on dataset D
2. generate an augmented dataset $\hat{D}$ with longer inks from D
3. train an encoder-decoder model with style conditioning on $D \cup \hat{D}$

The final result is the trained encoder-decoder model with style conditioning. In the inference, we suggest the following sequence of steps:

1. choose an ink for style extraction
2. split input text by n words
3. run inference of the encoder-decoder model on each piece with the same style input
4. combine pieces into the result ink

4 Experiments

4.1 Setup

In this section we will apply our modifications to two architectures - **RNN** [16] and **Transformer** [42]. A classic approach to ink synthesis [16] proposed a multi-layer LSTM model with monotonic attention over the labels and gaussian mixture model output for ink synthesis. In our RNN implementation the label encoder is simply one-hot encoding of input characters with dictionary size 70. The decoder consists of an LSTM with 128 units, GMM monotonic attention layer (size 128, one attention head and 10 components), two LSTM blocks with 256 units, a dense layer and an output GMM model with 10 components [16].

In the transformer model we use a standard encoder-decoder architecture with an output GMM head similar to the RNN model. In both encoder and decoder we use sinusoidal positional embeddings [42]. The transformer label encoder has 4 self-attention blocks (8 heads, 16 units per head, ReLU activation and drop-out rate 0.1). The decoder has 6 standard cross-attention blocks (4 heads, 32 units per head, ReLU activation and drop-out rate 0.1) followed by two dense layers with ReLU activation.

We train both models with batch size of 128 and the Adam optimizer. For the RNN model we use learning rate 1e-4 and for transformer learning rate is 1e-3 with the schedule from [42]. We train each of these models with two different ink representations - raw [16] and curve [11]. At inference time, all transformers and the raw RNN models use random sampling of the GMM with bias ∞ similar to [16]. For the curve RNN we use greedy sampling due to better validation performance.

Style Encoder. For style encoding we use a model that consists of one bidirectional LSTM layer with 256 units, followed by a unidirectional LSTM layer with 256 units that outputs the last state and three dense layers on top with sizes 256, 256 and 16. In order to condition a model on a given style we sum a style vector with the decoder input.

Datasets. We train our models on the **DeepWriting** dataset which contains more than 34,000 handwritten samples in English [1]. Examples in the training dataset on average consist of 12 characters and 2.8 words. To evaluate the quality of long generation quantitatively and qualitatively we need a suitable source of handwriting in English. For this purpose, we use **IAMonDo** which consists of 1,000 pages with English handwriting, diagrams and drawings in them [20]. We extracted handwriting from those pages and split them into three categories - *long* - more than 7 words, *medium* - 4–7 words and *short* - less than 4 words. For short samples we chose a threshold to match the mean length in the training set (ss shown in Table 1). We then split the rest of the inks into two sets to see progression of quality from in-distribution to increasingly out-of-distribution lengths. For clarity, we include the label sets with our submission[2].

Evaluation Metric. Following [29] we measure the recognizability of generated inks with Character Error Rate (CER) on a set of test labels. Our recognizer is an RNN model trained on a private dataset described in [6]. It performs well on all lengths present in the test sets, see the results on the original data in Table 9.

---

[2] A notebook with test sets and model inference https://colab.research.google.com/drive/1SB_vyDcsdSq1CtE9IOD9opBR9IDgG0ly.

Table 1. Dataset statistics for train, validation and test datasets.

dataset		words	mean characters (std)	samples	characters
DeepWriting	train	2.8 (1.2)	12.41 (5.08)	34K	421K
	valid	2.8 (1.2)	12.30 (4.97)	680	8K
IAMonDO	long	> 7	48.22 (9.06)	310	15K
	medium	4–7	30.52 (7.98)	864	26K
	short	< 4	10.91 (5.63)	811	9K

Baseline Model. We compare our method to the VRNN model [1] which generates one character at a time and propagates RNN states between characters to control the style. This model easily generates long input text (see Table 2) as very limited information is shared between the characters. Another result of this procedure is that characters are rarely connected in the final handwriting. Thus, this model struggles to generate cursive writing.

4.2 Quantitative Results

In this section we apply proposed changes to two architectures with two different ink representations - raw [16] and curve [11]. We train each model 3 times in order to compute the variance between runs. In Table 2 we present the mean CER and standard deviation of the models without any changes and with proposed improvements. Comparing baseline models we see that RNN models degrade on long inputs less than transformers, which matches previous results in the text domain [33]. It is expected as positional embeddings for remote positions are unknown to the transformer model, hurting accuracy, while RNN has monotonic attention which helps with longer sequences. However, even in case of RNN models CER gets 2 times worse for curve setup and 5 times for raw between short and long validation datasets. Thus, all four models can benefit from better out-of-domain length generalisation.

Table 2. Recognizability comparison of our method with baseline models and the VRNN model [1]. Datasets: long > 7 words, medium 4–7 words, short < 4 words.

model		CER long	CER medium	CER short	CER avg
VRNN [1]		5.94 (0.22)	5.05 (0.04)	4.8 (0.1)	5.26
transformer curve	DSS	**3.96 (0.23)**	**4.0 (0.37)**	5.3 (0.57)	**4.42**
	base	60.62 (3.05)	34.91 (3.36)	10.35 (0.47)	35.29
transformer raw	DSS	12.73 (1.23)	9.89 (1.61)	8.83 (2.06)	10.48
	base	64.34 (1.11)	41.24 (2.15)	11.4 (1.4)	38.99
RNN curve	DSS	6.6 (0.39)	6.03 (0.95)	7.08 (0.95)	6.57
	base	12.25 (0.78)	9.27 (0.93)	5.29 (0.28)	8.94
RNN raw	DSS	4.42 (0.55)	4.22 (0.51)	4.56 (0.48)	**4.4**
	base	21.48 (5.18)	11.74 (3.18)	**4.11 (0.43)**	12.44

In this section we provide details on the hyperparameters utilized for our results and the criteria for their selection. For each of four models we added *10K long samples* built from train dataset and picked the optimal target length l based on validation quality presented in Table 11 (in the appendix). In style condition during inference we used random examples from *DeepWriting validation dataset* as a source of style. This ensured that we did not introduce any stylistic bias compared to unconditional generation, as the distribution of validation dataset matches the training dataset. In order to determine the best number of words in the split generation we applied data augmentation method to validation dataset. We matched the target length of this dataset to the long test dataset. Then, we evaluated models with additional data and style conditioning on the said dataset. Results are presented in Table 12 (in the appendix).

In conclusion, our method significantly outperforms the baseline in all four cases, with curve transformer as well as raw RNN showing the best overall quality. This quality is also better than VRNN model [1] which requires character segmentation in the training data. This information is absent in most open source handwriting datasets [34]. In our results we show that without any additional annotation, superior quality is attainable across all three target lengths.

4.3   Ablation Study

In this section we remove each of the three changes that we proposed from the combination and measure the quality to show that all three changes contribute to the optimal model performance. In addition, we do a more detailed ablation study on the curve transformer setup, as it provides the best overall quality.

Impact of Data Augmentation. In order to measure the impact of synthetic long data, we remove it from training. All of the other parameters stay the same and we train a new set of models only on the original data. This experiment is especially important because split generation significantly reduces the input text length during inference. However, data augmentation improves model performance on short samples as well (see Table 11 in the appendix) possibly due to the moderate size of the training set.

Table 3. Effect of data augmentation.

models	data	CER long	CER medium	CER short	CER avg
transformer curve	yes	3.96 (0.23)	4.0 (0.37)	5.3 (0.57)	4.42
	no	6.61 (0.5)	6.89 (0.29)	7.71 (0.95)	7.07
transformer raw	yes	12.73 (1.23)	9.89 (1.61)	8.83 (2.06)	10.48
	no	13.35 (0.59)	11.42 (1.08)	11.21 (1.23)	11.99
RNN curve	yes	6.6 (0.39)	6.03 (0.95)	7.08 (0.95)	6.57
	no	6.78 (0.62)	6.05 (0.64)	6.85 (0.64)	6.56
RNN raw	yes	4.42 (0.55)	4.22 (0.51)	4.56 (0.48)	4.4
	no	4.85 (0.6)	4.66 (0.83)	4.94 (1.04)	4.82

In Table 3 we can see that for transformers augmented data plays a very important role. The decrease in quality is almost 60% for curve transformer and around 16% for raw transformer. As mentioned in the main results, transformer models fail to generalize to long-form sequences due to constant positional embeddings unseen at training time. The restriction is overcome through the use of synthetic long data. However, for raw RNN model the loss is also quite significant - around 10%, which shows that augmented data is beneficial in most cases.

Fig. 4. Style, random and same ink arrangements.

In DSS we propose to concatenate existing training examples based on their style similarity. We compare this approach to repetition of the same ink many times and concatenation of random samples into one ink. In Fig. 4 we can see that random arrangement doesn't preserve style consistency which is an important quality of digital inks. On the other hand, repetition of the same ink returns stylistically consistent inks but lacks diversity in text. These strategies decrease recognizability compared to style-based arrangement in the curve transformer setup, as shown in Table 4.

Table 4. Different data augmentation strategies for curve transformer setup.

data arrangement	CER long	CER medium	CER short	CER avg
style	3.96 (0.23)	4.0 (0.37)	5.3 (0.57)	**4.42**
random	5.01 (0.52)	4.96 (0.33)	5.58 (0.49)	5.18
repetition	7.15 (0.34)	7.02 (0.6)	8.56 (0.57)	7.58

Impact of Style Encoding. Next, we remove style encoder from encoder-decoder architecture, leaving everything else unchanged. In Table 5 we show results for split generation where we generate label pieces independently and don't share any information between them (rows with no style). In this case split generation returns stylistically inconsistent inks similar to random data augmentation as shown in Fig. 4. In a qualitative evaluation, we show that stylistically inconsistent inks are 33% less likely to be recognized as real. For RNN models character error rate is better without style than with style conditioning, but it is not the case for transformer models where CER gets significantly higher without any style - 110% for the curves setup.

We also compare these results to a different method of ensuring style consistency, proposed in [16]. In split generation we can prime a model on the style ink before generating each piece by teacher-forcing a style input and asking a model to complete it with input text. This way we get result ink with similar style to the style ink. We implicitly determine consistency of the result by using the same style ink in each piece of split generation. The main disadvantage of this approach is that by doing a completion of

Table 5. Effect of style encoding. No style rows show metrics for stylistically inconsistent inks (see random arrangement in Fig. 4). In priming we prompt a model with the style ink before target ink generation.

models	style	CER long	CER medium	CER short	CER avg
transformer curve	yes	3.96 (0.23)	4.0 (0.37)	5.3 (0.57)	4.42
	prime	6.34 (0.65)	6.19 (0.7)	7.39 (1.69)	6.64
	no	10.02 (1.77)	10.32 (1.04)	7.73 (0.74)	9.36
transformer raw	yes	12.73 (1.23)	9.89 (1.61)	8.83 (2.06)	10.48
	prime	17.92 (1.99)	23.06 (2.31)	18.74 (2.95)	19.91
	no	13.15 (2.55)	10.16 (2.56)	9.81 (2.21)	11.04
RNN curve	yes	6.6 (0.39)	6.03 (0.95)	7.08 (0.95)	6.57
	prime	7.49 (0.63)	6.85 (0.3)	7.69 (0.69)	7.34
	no	5.14 (0.95)	4.45 (0.96)	4.42 (0.95)	4.67
RNN raw	yes	4.42 (0.55)	4.22 (0.51)	4.56 (0.48)	4.4
	prime	4.49 (1.46)	4.25 (1.28)	4.08 (1.13)	4.27
	no	3.66 (0.2)	3.59 (0.86)	2.16 (0.45)	3.14

style ink we increase the target ink length which may result in subpar quality. In Table 5 we can see that for curve transformer completion decreases CER by 30% compared to no style and for RNN raw the quality of completion is similar to style encoding.

Impact of Split Generation. Finally, we evaluate models quality without split generation. Table 6 shows that models with fully autoregressive inference perform by 30–100% worse than with split generation. The gap in quality is especially pronounced for long evaluation sets as we split the labels there into many pieces. Whereas, the quality on short data stays almost the same as in many cases we don't split targets there. However, raw RNN even without split generation still performs similarly to VRNN (see Table 2).

Table 6. Effect of split generation.

model	split	CER long	CER medium	CER short	CER avg
transformer curve	yes	3.96 (0.23)	4.0 (0.37)	5.3 (0.57)	4.42
	no	10.6 (1.42)	10.6 (1.42)	5.33 (0.53)	8.84
transformer raw	yes	12.73 (1.23)	9.89 (1.61)	8.83 (2.06)	10.48
	no	22.44 (4.37)	22.4 (2.04)	9.88 (1.6)	18.24
RNN curve	yes	6.6 (0.39)	6.03 (0.95)	7.08 (0.95)	6.57
	no	14.82 (2.87)	10.28 (1.43)	7.43 (0.97)	10.84
RNN raw	yes	4.42 (0.55)	4.22 (0.51)	4.56 (0.48)	4.4
	no	7.32 (1.95)	4.86 (0.95)	4.54 (0.5)	5.57

In Table 7 we show that quality can vary quite a bit depending on the number of words in one split. It may seem that generating only one word at a time is the most simple task and that would lead to the best performance. However, we noticed that curve transformer fails to generate one word with punctuation at the end probably due to lack of similar data in training. Thus, it is important to choose number of words that is most suited for each model. This value is consistent for curve transformer between different datasets: long synthetic validation in Table 12, long and medium validation in Table 7.

Table 7. Different number of words in split generation for curve transformer setup.

n words	CER long	CER medium	CER short	CER avg
1	6.3 (0.67)	5.96 (0.75)	6.96 (0.97)	6.41
2	4.66 (0.5)	4.56 (0.33)	5.62 (0.23)	4.95
3	3.96 (0.23)	4.0 (0.37)	5.3 (0.57)	**4.42**
5	4.53 (0.01)	4.83 (0.2)	5.3 (0.57)	4.89

4.4   Qualitative Evaluation

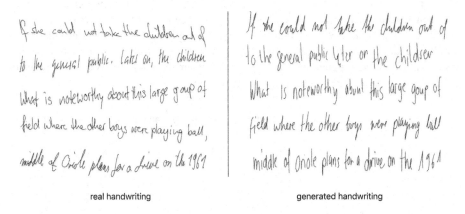

real handwriting   generated handwriting

Fig. 5. Examples of inks from human study.

To ensure that the generated samples are not only recognizable but also look realistic, we performed a human evaluation. In it, we asked the participants to evaluate a mix of generated and real inks - 100 each with long texts from IAMonDO dataset, see examples in Fig. 5. Our model's performance was evaluated on long texts because baseline models struggled the most with generating long samples (see Table 2). We used a curve transformer model with style encoding, split generation of two words as the direction pattern (see Fig. 3) is less pronounced there compared to three words. We also use the sampling of 5 candidates to pick the most horizontal ink. In Table 10 in the appendix we show that picking the most horizontal pieces in split generation of two words doesn't significantly decrease recognition quality.

We study whether people can differentiate between real and generated data by showing one ink at a time with a question "Does this ink look real?". We collected 600 responses from 12 people who has worked with digital ink before (3 answers per ink). The results are presented in Table 8. The majority of generated inks were labeled as real - **79%**, as a result the F1 score of human answers is only **0.34**. To check whether original inks are labeled real with higher probability we use Fisher's exact test [14]. We get the p-value equal to **0.04**. Thus, we can reject the null hypothesis and conclude that original inks are more likely to be labeled real than generated ones. To sum up, generated inks are frequently perceived as real but there is still a statistically significant difference between the two datasets.

Table 8. Human study results.

data	look real	consistent style	no artifacts	readable
real	**0.89**	**0.96**	**0.77**	**0.8**
generated	0.79	0.85	0.7	0.77

Moreover, we asked about the style consistency, presence of artifacts like additional lines or dots and readability of the ink see in Fig. 6. In the latter two cases the gap between real and original quality is not as pronounced as in the first two and is not statistically significant. We see that according to people real and generated inks have similar readability which matches our quantitative results (see Tables 9 and 10). It is also important to note that generated inks with artifacts are 42% less likely to be perceived as real and for style inconsistency this number is 33%.

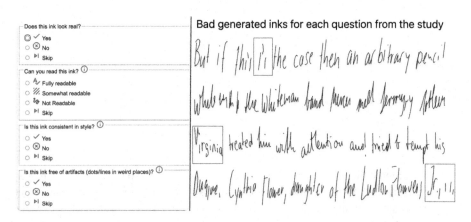

Fig. 6. Human study questionnaire on the left and examples of synthesizer's mistakes on the right. The mistakes are marked with rectangles.

5   Conclusion

We have presented three improvements to encoder-decoder models for digital ink generation - data augmentation, style conditioning and split generation. We have been able to decrease character error rate compared to baseline RNN model by **50%** (baseline RNN curve 8.94 vs DSS transformer curve 4.42) and by **16%** compared to the previous approach. We've shown that all three proposed methods play an important part in optimal model performance: data augmentation is very important for transformer models, style conditioning has higher recognisability than priming in split generation, split generation has the biggest impact on quality among the three parts of our approach. In addition, we have conducted a qualitative evaluation where we have verified the quality of generated long inks. People perceive most of synthesized inks as real, but there is still a statistically significant discrepancy between answers in real and generated buckets. We believe that our findings regarding long ink synthesis can be used in real world applications and as a stepping stone for future research in this field.

6   Appendix

Table 9. Recognizer CER on test data, caused by data noise and the model's mistakes.

metric	value
CER long	4.8
CER medium	4.41
CER short	5.5
CER avg	4.9

Table 10. CER of the curve transformer on long test with and without the most horizontal ink.

n words	1 attempt	5 attempts
1	6.29 (0.67)	7.43 (0.53)
2	4.64 (0.48)	5.0 (0.17)
3	3.98 (0.26)	4.26 (0.32)
5	4.53 (0.01)	4.71 (0.36)

Table 11. CER for different lengths in data augmentation procedure with candidate pool of 15K and threshold 0.5. We compare sets with average lengths of 26, 37, 48, 65, 79, 100. For RNN models, training with lengths > 26 results in poor performance.

model	mean length	CER valid
transformer curves	–	8.0 (0.69)
	26	7.46 (1.02)
	37	7.58 (1.56)
	48	6.19 (0.42)
	65	6.98 (0.23)
	79	6.08 (0.66)
	100	6.62 (0.36)
transformer raw	–	9.72 (1.85)
	48	9.97 (1.1)
	65	9.08 (3.68)
	79	11.01 (2.6)
RNN curves	–	5.94 (0.74)
	26	5.23 (0.5)
RNN raw	-	3.4 (0.58)
	26	2.06 (0.64)

Table 12. CER for different number of words in split generation on synthetic long validation dataset.

model	n words	CER long valid
transformer curves	1	8.0 (0.64)
	2	6.03 (0.27)
	3	**5.57 (0.18)**
	5	5.78 (0.2)
transformer raw	1	17.83 (0.92)
	2	**14.21 (1.17)**
	3	14.63 (0.23)
	5	20.17 (3.48)
RNN curves	**1**	**7.15 (0.87)**
	2	7.47 (0.76)
	3	7.92 (0.95)
	5	9.28 (1.03)
RNN raw	1	7.4 (0.34)
	2	7.17 (0.37)
	3	**6.66 (0.31)**
	5	7.03 (0.7)

References

1. Aksan, E., Pece, F., Hilliges, O.: DeepWriting: Making Digital Ink Editable via Deep Generative Modeling. In: SIGCHI Conference on Human Factors in Computing Systems. CHI 2018, New York, NY, USA. ACM (2018)
2. Alonso, E., Moysset, B., Messina, R.O.: Adversarial generation of handwritten text images conditioned on sequences. In: 2019 International Conference on Document Analysis and Recognition (ICDAR), pp. 481–486 (2019)
3. Anil, C., et al.: Exploring length generalization in large language models. In: Oh, A.H., Agarwal, A., Belgrave, D., Cho, K. (eds.) Advances in Neural Information Processing Systems (2022). https://openreview.net/forum?id=zSkYVeX7bC4
4. Bengio, S., Vinyals, O., Jaitly, N., Shazeer, N.M.: Scheduled sampling for sequence prediction with recurrent neural networks. arXiv:abs/1506.03099 (2015)
5. Bishop, C.M.: Mixture density networks (1994)
6. Carbune, V., et al.: Fast multi-language LSTM-based online handwriting recognition. Int. J. Doc. Anal. Recogn. (IJDAR) **23**, 89–102 (2020)
7. Chang, J.H.R., Shrivastava, A., Koppula, H.S., Zhang, X., Tuzel, O.: Style equalization: Unsupervised learning of controllable generative sequence models (2022). https://arxiv.org/abs/2110.02891
8. Chen, T., Kornblith, S., Norouzi, M., Hinton, G.: A simple framework for contrastive learning of visual representations (2020)

9. Chiu, C.C., et al.: RNN-T models fail to generalize to out-of-domain audio: causes and solutions, pp. 873–880 (2021). https://doi.org/10.1109/SLT48900.2021.9383518

10. Dai, Z., Yang, Z., Yang, Y., Carbonell, J., Le, Q.V., Salakhutdinov, R.: Transformer-xl: attentive language models beyond a fixed-length context (2019). https://doi.org/10.48550/ARXIV.1901.02860, https://arxiv.org/abs/1901.02860

11. Das, A., Yang, Y., Hospedales, T., Xiang, T., Song, Y.-Z.: BézierSketch: a generative model for scalable vector sketches. In: Vedaldi, A., Bischof, H., Brox, T., Frahm, J.-M. (eds.) ECCV 2020. LNCS, vol. 12371, pp. 632–647. Springer, Cham (2020). https://doi.org/10.1007/978-3-030-58574-7_38

12. Djioua, M., Plamondon, R.: An interactive system for the automatic generation of huge handwriting databases from a few specimens. In: 2008 19th International Conference on Pattern Recognition, pp. 1–4 (2008)

13. Elarian, Y., Abdel-Aal, R., Ahmad, I., Parvez, M.T., Zidouri, A.: Handwriting synthesis: classifications and techniques. Int. J. Document Anal. Recogn. (IJDAR) 17(4), 455–469 (2014). https://doi.org/10.1007/s10032-014-0231-x

14. Fisher, R.A.: On the interpretation of from contingency tables, and the calculation of p. J. Roy. Stat. Soc. 85 (1922). http://www.medicine.mcgill.ca

15. Gan, J., Wang, W.: HIGAN: handwriting imitation conditioned on arbitrary-length texts and disentangled styles. In: AAAI Conference on Artificial Intelligence (2021)

16. Graves, A.: Generating sequences with recurrent neural networks. arXiv:abs/1308.0850 (2013)

17. Hochreiter, S., Schmidhuber, J.: Long short-term memory. Neural Comput. 9, 1735–80 (1997). https://doi.org/10.1162/neco.1997.9.8.1735

18. Hsu, W.N., et al.: Disentangling correlated speaker and noise for speech synthesis via data augmentation and adversarial factorization. In: ICASSP 2019–2019 IEEE International Conference on Acoustics, Speech and Signal Processing (ICASSP), pp. 5901–5905 (2019). https://doi.org/10.1109/ICASSP.2019.8683561

19. Huber, C., Hussain, J., Stüker, S., Waibel, A.H.: Instant one-shot word-learning for context-specific neural sequence-to-sequence speech recognition. In: 2021 IEEE Automatic Speech Recognition and Understanding Workshop (ASRU), pp. 1–7 (2021)

20. Indermühle, E., Liwicki, M., Bunke, H.: Iamondo-database: an online handwritten document database with non-uniform contents. In: International Workshop on Document Analysis Systems (2010)

21. Ke, G., He, D., Liu, T.Y.: Rethinking the positional encoding in language pre-training (2020)

22. Koehn, P., Knowles, R.: Six challenges for neural machine translation. In: Proceedings of the First Workshop on Neural Machine Translation, Vancouver, August 2017, pp. 28–39. Association for Computational Linguistics (2017). https://doi.org/10.18653/v1/W17-3204, https://aclanthology.org/W17-3204

23. Kondo, S., Hotate, K., Hirasawa, T., Kaneko, M., Komachi, M.: Sentence concatenation approach to data augmentation for neural machine translation, pp. 143–149 (2021). https://doi.org/10.18653/v1/2021.naacl-srw.18

24. Krishnan, P., Jawahar, C.: HWNET v2: an efficient word image representation for handwritten documents. Int. J. Doc. Anal. Recogn. (IJDAR) 22, 387–405 (2019)

25. Lamb, A., Goyal, A., Zhang, Y., Zhang, S., Courville, A., Bengio, Y.: Professor forcing: a new algorithm for training recurrent networks (2016)

26. Lin, Z., Wan, L.: Style-preserving English handwriting synthesis. Pattern Recogn. 40(7), 2097–2109 (2007). https://doi.org/10.1016/j.patcog.2006.11.024, https://www.sciencedirect.com/science/article/pii/S0031320306004985

27. Liwicki, M., Bunke, H.: Iam-ondb - an on-line English sentence database acquired from handwritten text on a whiteboard. In: Eighth International Conference on Document Analysis and Recognition (ICDAR 2005), vol. 2, pp. 956–961 (2005). https://doi.org/10.1109/ICDAR.2005.132

28. Lu, Z., et al.: Input length matters: improving RNN-T and MWER training for long-form telephony speech recognition (2021)

29. Luo, C., Zhu, Y., Jin, L., Li, Z., Peng, D.: Slogan: Handwriting style synthesis for arbitrary-length and out-of-vocabulary text (2022). https://doi.org/10.48550/ARXIV.2202.11456, https://arxiv.org/abs/2202.11456

30. Maksai, A., Rowley, H., Berent, J., Musat, C.: INKORRECT: online handwriting spelling correction (2022)

31. Nallapati, R., Zhou, B., dos Santos, C., Gulcehre, C., Xiang, B.: Abstractive text summarization using sequence-to-sequence RNNs and beyond. In: Proceedings of the 20th SIGNLL Conference on Computational Natural Language Learning, Berlin, Germany, August 2016, pp. 280–290. Association for Computational Linguistics (2016). https://doi.org/10.18653/v1/K16-1028, https://aclanthology.org/K16-1028

32. Narayanan, A., Prabhavalkar, R., Chiu, C.C., Rybach, D., Sainath, T.N., Strohman, T.: Recognizing long-form speech using streaming end-to-end models. In: 2019 IEEE Automatic Speech Recognition and Understanding Workshop (ASRU), pp. 920–927 (2019). https://doi.org/10.1109/ASRU46091.2019.9003913

33. Neishi, M., Yoshinaga, N.: On the relation between position information and sentence length in neural machine translation. In: Proceedings of the 23rd Conference on Computational Natural Language Learning (CoNLL), Hong Kong, China, November 2019, pp. 328–338. Association for Computational Linguistics (2019). https://doi.org/10.18653/v1/K19-1031, https://aclanthology.org/K19-1031

34. Nguyen, H., Nguyen, C., Bao, P., Nakagawa, M.: A database of unconstrained vietnamese online handwriting and recognition experiments by recurrent neural networks. Pattern Recogn. 78, 291–306 (2018)

35. Pechwitz, M., Margner, V.: Baseline estimation for Arabic handwritten words. In: Proceedings Eighth International Workshop on Frontiers in Handwriting Recognition, pp. 479–484 (2002). https://doi.org/10.1109/IWFHR.2002.1030956

36. Raffel, C., et al.: Exploring the limits of transfer learning with a unified text-to-text transformer (2019). https://doi.org/10.48550/ARXIV.1910.10683, https://arxiv.org/abs/1910.10683

37. Ribeiro, L.S.F., Bui, T., Collomosse, J., Ponti, M.: Sketchformer: Transformer-based representation for sketched structure (2020). https://doi.org/10.48550/ARXIV.2002.10381, https://arxiv.org/abs/2002.10381

38. Schaldenbrand, P., Liu, Z., Oh, J.: Styleclipdraw: Coupling content and style in text-to-drawing translation (2022). https://doi.org/10.48550/ARXIV.2202.12362, https://arxiv.org/abs/2202.12362

39. Sutskever, I., Vinyals, O., Le, Q.V.: Sequence to sequence learning with neural networks. In: Proceedings of the 27th International Conference on Neural Information Processing Systems. NIPS 2014, Cambridge, MA, USA, vol. 2, pp. 3104–3112. MIT Press (2014)

40. Thoppilan, R., et al.: LAMDA: language models for dialog applications (2022)

41. Varis, D., Bojar, O.: Sequence length is a domain: length-based overfitting in transformer models, pp. 8246–8257 (2021). https://doi.org/10.18653/v1/2021.emnlp-main.650

42. Vaswani, A., et al.: Attention is all you need. In: Guyon, I., et al. (eds.) Advances in Neural Information Processing Systems, vol. 30. Curran Associates, Inc. (2017), https:// proceedings.neurips.cc/paper/2017/file/3f5ee243547dee91fbd053c1c4a845aa-Paper.pdf
43. Zhang, J.X., Ling, Z.H., Dai, L.R.: Forward attention in sequence- to-sequence acoustic modeling for speech synthesis. In: 2018 IEEE International Conference on Acoustics, Speech and Signal Processing (ICASSP), pp. 4789–4793 (2018). https://doi.org/10.1109/ICASSP.2018. 8462020

Precise Segmentation for Children Handwriting Analysis by Combining Multiple Deep Models with Online Knowledge

Simon Corbillé[1]($\boxtimes$)(iD), Éric Anquetil[2](iD), and Élisa Fromont[1](iD)

[1] Univ Rennes, IRISA, 35000 Rennes, France
{simon.corbille,elisa.fromont}@irisa.fr
[2] INSA Rennes, IRISA, 35000 Rennes, France
eric.anquetil@irisa.fr

Abstract. We present a strategy, called Seq2Seg, to reach both precise and accurate recognition and segmentation for children handwritten words. Reaching such high performance for both tasks is necessary to give personalized feedback to children who are learning how to write. The first contribution is to combine the predictions of an accurate Seq2Seq model with the predictions of a R-CNN object detector. The second one is to refine the bounding box predictions provided by the detector with a segmentation lattice computed from the online signal. An ablation study shows that both contributions are relevant, and their combination is efficient enough for immediate feedback and achieves state of the art results even compared to more informed systems.

Keywords: Handwriting Recognition and Segmentation · R-CNN object detector · Seq2Seq network · French Children Handwriting

1 Introduction

The paradox of Sayre [1] is a famous problem in the handwriting recognition domain. This dilemma claims that a handwritten word cannot be recognized without being segmented in letters and at the same time cannot be segmented in letters without the word being recognized. To tackle the handwriting recognition task, the systems use an analytic or a holistic approach. The analytic approach segments the handwriting and tries to recognize letters, while the holistic approach tries to recognize the whole word without explicit segmentation. State-of-the-art methods use holistic approaches based on deep learning model. They are designed for recognition only and are efficient in solving this task. However, in a context of learning spelling, the letter segmentation provided by these approaches is not precise enough to provide a useful spatial feedback on spelling mistake to a user.

We aim at designing a support system for **learning cursive handwriting at school** and more particularly in a dictation context. Tackling both the challenges of **recognition and segmentation of children handwriting** may

G. A. Fink et al. (Eds.): ICDAR 2023, LNCS 14190, pp. 236–252, 2023.
https://doi.org/10.1007/978-3-031-41685-9_15

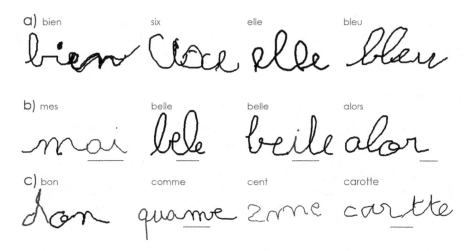

Fig. 1. Examples of cursive children handwriting. The oral French instruction given to the children is provided in *orange* and examples of feedback are drawn in *red*. Line *a* shows some degraded handwriting, line *b*, phonetic errors and line *c* shows other types of errors in a context of learning spelling. (Color figure online)

allow a system to provide a fine-grained analysis on the handwritten words and to **deliver immediate spelling feedback** to children. The children are in a learning process and therefore, their handwriting is degraded and contains misspelling errors. Line *a* of Fig. 1 illustrates several examples of degraded handwriting. We can see that a distortion of the letter "e" can be interpreted as a letter "l" and vice versa for the word "elle" in the third position of this line. Line *b* shows several examples of phonetic errors. In the first example, where the instruction is "mes" [mɛ], the child writes "mai" [me], which sounds very similar in French. These homophonic errors can be anticipated in automatic systems using a language model that would take into account the contextual information. However, other types of errors in a context of learning spelling illustrated in line *c* such as dyslexia and out-of-vocabulary words cannot. The first example of line *c* shows a common mistake in French where the child confuses the letters "b" and "d" which are phonetically close.

To provide an accurate recognition and segmentation of the children handwritten words, we propose the two following contributions included in the Seq2Seg system:

- We present an original combination strategy using a model dedicated to recognition and an object detector dedicated to segmentation. The recognition model is used to recognize a word and to select the segmentation predictions of the object detector corresponding to the letters of the recognized word.
- We use an segmentation lattice [2–4] which encodes expert knowledge to refine the letter segmentation provided by the object detector and thus improve the precision of the segmentation.

This paper is organized as follows. The related works are described in Sect. 2. The contributions are detailed in Sect. 3. Section 4 presents an ablation study of our approach and compares it with the state of the art. We conclude in Sect. 5.

2 Related Work

This section presents works related to the **recognition and segmentation** of handwritten words. The first part introduces the latest methods in the handwriting recognition domain. The second part sets out the limits of these methods for a segmentation task and presents the state of the art in children handwriting recognition and segmentation. Finally, we provide a brief presentation of object detection models to show their relevance in an handwriting segmentation context.

2.1 Handwriting Recognition

The state of the art in handwriting recognition is achieved by the **Sequence-to-Sequence** (Seq2Seq) [5,6] and Transformer [7,8] networks. Seq2Seq use an encoder-decoder paradigm enhanced by an attention mechanism, while Transformers are based on a feature extractor followed by multi-head attention mechanisms. Transformers are slightly more accurate but need much more data to be optimized than Seq2Seq. This is often dealt with data generation and data augmentation techniques.

In our work, a rather small dataset is available compare to adult handwriting ones than can be found in [9,10] due to the cost associated with data collection in schools and degraded handwriting annotations. We thus decided to rely upon a Seq2Seq network for word recognition because of its good compromise accuracy/need of labeled data.

2.2 Handwriting Segmentation

To our knowledge, there is no (reasonable sized) public dataset for handwriting (semantic) segmentation, *i.e.,* handwriting words with annotations at the letter pixel level. This task is particularly tedious and time-consuming but is not necessary, nowadays, to achieve excellent recognition results for the architectures mentioned above. For this reason, handwriting letter segmentation methods are difficult to compare quantitatively. For the networks designed for handwriting recognition, the letter segmentation can be computed from the position of the receptive field associated to the letter prediction. The width and height of the receptive field being fixed, this approach, which lacks flexibility, does not provide a precise segmentation. Furthermore, most networks are trained with the connectionist temporal classification (CTC) [11] approach. CTC manages the alignment between an input data sequence and an output sequence of frames of variable size. CTC is known to have a peaky behavior [12] *i.e.,* it predict one frame per letter. This impacts the segmentation performance since a frame has a fixed size while a handwritten letter has a variable one. In [13,14], the authors modified CTC to enforce a better alignment between the frames and the real letters. However, despite these efforts, the segmentation was still lacking precision.

The authors of [4] and its extensions [15,16] use an analytic approach to reach the state of the art in children handwriting recognition and segmentation. The **letter recognition** is made from letter splitting hypotheses coming from a **segmentation lattice** [2–4]. Then, the method selects the best path of the lattice where his associated word is closest to the instruction or to a phonetically close word. However, this system uses a **language model** to guide the analysis of children handwriting using assumptions of probable phonetic errors. It is thus specific and dedicated to the French language and cannot be easily adapted to other languages. Moreover, as already shown in the Introduction in Fig. 1, some children errors cannot be prevented using a language model. In this work, we want to achieve results on par with [15] without relying on a language model.

2.3   Object Detection

We rely on an existing, very successful, two-stage deep learning-based object detector [17] to perform a precise localization of the letters in the handwritten words. **Two-stage detectors** [17–19] are known to be a little bit more precise for localization than their one-stage counterparts [20–23] even though they are usually slower. Object detectors provide a joint classification of objects into classes and a regression of the bounding boxes that best localize each object in an image (or in a video frame). In a two-stage detector, candidate regions are generated by a RPN (region-proposal network) and processed to perform the detection task.

Based on the output of the object detector (*i.e.* labeled bounding boxes), the final segmentation is obtained using all the handwriting pixels within the predicted bounding box. Note that in our work, the image of the handwriting comes from an online signal, therefore, this image is noise-free both on the background and on the handwriting pixels. This makes it possible to extract the letter segmentation from the bounding box coordinates. Also note that we could have used the semantic segmentation output of an instance-based semantic segmentation network such as Mask-RCNN [17] to directly segment the letters. However, the complexity in terms of parameters of such segmentation networks (the semantic segmentation part of the network is usually independent from the object detection part), and the limited number of realistic labeled children words to train it, made bounding boxes of traditional object detectors better candidates to tackle our segmentation problem when the segmentation target is too ambiguous.

3   Methods

We propose Seq2Seg, a method to combine two deep learning models (Seq2Seq and R-CNN) and expert online knowledge to accurately segment and recognize children handwritten words. Seq2Seg, illustrated in Fig. 2 leverages each method to provide a precise semantic segmentation of the children words. The first level (**Level A**) uses a model dedicated to the recognition task as an oracle to filter out the bounding box's predictions of the object detector. **Level B** uses an expert segmentation lattice [2–4] to refine the letter segmentation associated to the bounding boxes predicted by the object detector. The segmentation lattice use online data, while the object detector and the recognizer use online data converted to offline.

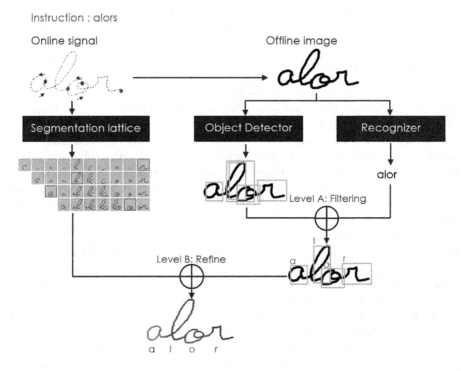

Fig. 2. Summary of levels A and B contributions.

In our work, we use the Seq2Seq architecture defined in [15] as the text recognition model and the R-CNN architecture defined in [17] as the object detector. The Seq2Seq performs well on the recognition task but provides an imprecise segmentation while R-CNN performs well on the segmentation task but is less accurate in recognition than the Seq2Seq (see Table 2 in Sect. 4 for the detailed results).

3.1   Level A: Filtering Bounding Boxes Predictions with an Accurate Recognition Model

The recognition model is trained solely on the recognition task and outputs a word. From this word, one can deduce, in particular, the number of letters to be segmented. This information makes it possible to select a fixed number of object detector segmentation predictions **during the inference** and use the more accurate recognition result of the recognition model. The process of selecting the predictions from the object detector can be difficult and ambiguous in certain cases, as illustrated in Fig. 3: *e.g.*, in a letter "m", two letters "n" can be recognized but they cannot be both true at the same time so here, a more global view is necessary to choose the right segmentation. The use of the precise recognition model, providing that it does not introduce other errors, makes it possible to remove these ambiguities.

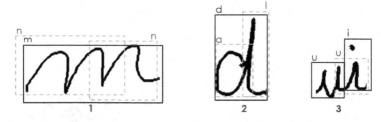

Fig. 3. Examples of ambiguity in object detector predictions: the correct prediction is in full line and the wrong ones in dash.

The object detector has several output x-ordered predictions. The goal is to select the object detector prediction corresponding to the letter segmentation. This method illustrated in Fig. 4 is broken down into three steps: (Step 1) we **compute all object detector prediction sequences**; (Step 2) we **filter the sequences according to the length of the word recognized by the recognition model**; (Step 3) we **compute the score associated with each sequence**. The final selected sequence is the one with the highest score. R-CNN natively includes two Non-Maximum Suppression (NMS) phases to filter out its predictions. The first is applied to the regions proposals to reduce the number of proposals to consider, while the second is applied to predictions (bounding boxes and labels) to keep the best prediction for the objects predictions with the same label. In a letter-in-word detection context, there is little overlap between letters unlike a classic COCO-style object detection. In order to handle the cases where several letter predictions are nested as emphasized in Fig. 3, we have added an NMS on the predictions of the model which is **independent of the class**. Our method uses the predictions **before** the last NMS to have a wide variety of prediction to filter with segmentation ambiguities. The purpose of the method is to remove these ambiguities.

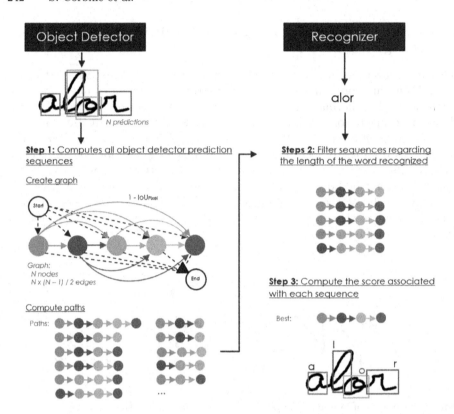

Fig. 4. Level A: Example the three steps of the process of filtering object detector predictions with the result of a recognition model.

(1) Compute All the Prediction Sequences: consider a directed graph $G(V, E)$, where V and E correspond to the sets of vertices and edges. For each prediction of the object detector ordered by x_{min} from the bounding box coordinates, a vertex is added in G as illustrated in Fig. 4. The weight of an edge $e_{ij} = (v_i, v_j) \in E$ is computed as $e_{ij} = 1 - \text{IoU}_{Pixel}$ between the predictions ordered by x_{min} associated to the vertices. IoU_{Pixel} stands for the Intersection Over Union of the handwriting pixels contained in the two bounding boxes corresponding to the two vertices: the predictions with the higher overlap have a weaker link. A sequence of predictions ordered by x_{min} corresponds to a graph path, *i.e.* a list of connected vertices in the graph.

(2) Filter the Sequences According to the Length of the Word Predicted by the Recognition Model: there are three selection scenarios (Table 3 in Sect. 4 details the result of each type of scenario):

– **Perfect matching:** The number of predicted letters of the object detector and of the recognition model is equal. In this case, we expect our filtering to only improve the recognition part of the object detector (that we do not use explicitly).

- **Matching:** The number of predicted letters of the object detector and of the recognition model is different but there is at least one possible matching in the solutions. In this case, we expect that the use of the word classifier as an oracle will help to remove some ambiguities for the object detector. This may improve both the recognition and the segmentation.

- **No matching:** The number of predicted letters of the object detector and of the recognition model is different and there is no possible matching in the predictors' solutions. In this case, both the classification and the segmentation of the object detector are used (the Seq2Seq is ignored). In practice, in this case, we noticed that the Seq2seq was either predicting an additional letter or was missing one. It is thus important for the object detector to be able to ignore the oracle prediction when there is a strong conflict between both models. This filtering might thus improve the overall recognition results since the object detector will take over the Seq2Seq but only for the most difficult predictions.

(3) Compute the Score Associated with Each Sequence: the score of a sequence of size N_a takes into account the degree of overlap between all the bounding boxes involved in the sequence. In particular, it minimizes the inter-letter overlap and also includes a coverage criterion to ensure a good coverage of the entire handwritten text. The overlapping score, $s_{overlap}$, is the product of all edge weights $weight\ v$ in the path of the graph $G(V, E)$ corresponding to a sequence:

$$s_{overlap} = \Pi_{i=1}^{N_a} weight\ v_i \tag{1}$$

The larger the overlap, the lower the score is. On the contrary to classic COCO-style object detection contexts [24], in the handwriting context, there is almost no overlap between objects to detect except for the ligature area between the letters. To compute the coverage score and to count each pixel only once, we add the number of pixels contained in each prediction and the number of pixels contained in the intersection of the two predictions is subtracted from the number of pixels contained in each prediction. Then, the predicted number of pixels is divided by the total number of pixels as follows:

$$s_{cover} = (\Sigma_{i=1}^{N_a} N_p\ pred_i - \Sigma_{i=2}^{N_a} N_p\ inter(pred_{i-1}, pred_i))/N_p\ total \tag{2}$$

The final $s_{alignment}$ score is defined as:

$$s_{alignement} = s_{overlap} + s_{cover} \tag{3}$$

The output of the Seq2Seg model is the semantic segmentation computed from the bounding boxes of the sequence with the highest $s_{alignment}$ score together with the predictions of the Seq2Seq model for each letter. We can note that the efficiency of this method in terms of computation time depends on the size of the generated graphs. In our children handwriting context, the graphs associated with the words are small because the words are smaller than 10 letters. The computation time of this method is therefore low enough to provide immediate feedback.

3.2   Level B: Use of a Segmentation Lattice Based on Online Handwriting

The **online handwriting** can be split *a priori* into different segmentation hypotheses grouped in a **segmentation lattice** using heuristics (and without letter recognition). This process is detailed in [2] and consolidated in [4]. Furthermore, the online signal makes it possible to obtain a first automatic **semantic segmentation** for each hypothesis where two classes are considered: background and handwriting. Our goal is to use this lattice to find the "nearest" hypotheses of the segmentation lattice associated to the bounding boxes predicted by the object detector as illustrated in Fig. 5.

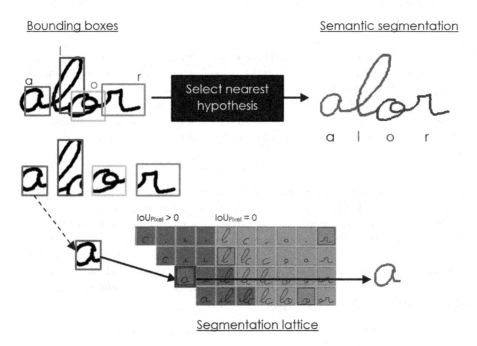

Fig. 5. Example of bounding boxes refinement with the online segmentation lattice. IoU_{Pixel} is used to select the best lattice hypothesis.

The similarity between the lattice nodes and the bounding boxes is computed with an IoU_{Pixel} *i.e.* an Intersection-over-Union between the handwriting pixels contained in a bounding box (easily accessible as explained before) and the ones in a node of the segmentation lattice. By associating the hypotheses of the lattice with the bounding boxes, this method **refines the coordinates of the**

bounding boxes and thus increase the precision of the segmentation of the object detector. Moreover, this approach also provides a better segmentation for slant handwriting than the "bounding box to segmentation" trivial correspondence proposed in Sect. 2.3. This is illustrated in Fig. 5 for the letter "l" and "o".

4 Experiments

4.1 Dataset

Children cursive handwriting is acquired on pen-based tablet at schools. The **French** handwritten words are acquired as an **online signal** encoded by a sequence of points represented by two coordinates (x, y), a pressure and a timestamp. The online signal is used to compute the segmentation lattice presented in the previous section and is converted into an image with linked points and a thickness of 2 on the links. The input images are padded (x axis) and resized (y axis) at $1\,280 \times 128$ pixels to fit the used deep learning models. Table 1 details the datasets used to train/test the deep learning models (which are all variants of the acquired children words). The original dataset is composed of 8 054 French handwritten cursive words annotated at the word level that are useful to train the Seq2Seq network. Besides, 2 126 words are annotated (*i.e.* segmented) at the letter level to train the R-CNN object detector. The number of letter-annotated data being limited, we redefined the splits compared to [15] and between the two models, to better train the object detector. The children writers are different for the training and the testing and the test set is the same for all models. Due to GDPR restrictions on children's private data, this dataset is not public.

Table 1. Details on the data used to train the deep learning models.

Models	Annotation type	Training	Validation	Test	Total
Seq2Seq	Words	6 022	1 000	1 032	8 054
R-CNN	Letters	918	176	1 032	2 126
R-CNN with synthesis	Letters	27 540	176	1 032	28 748

To better train the R-CNN model (that is data greedy), we perform data augmentation only on the training set (called "with synthesis" in the table). Among a list of usual offline deformations (stretching, slant) and more recent ones (stroke stretching, curvature [25]), each word is augmented 30 times with random parameters.

4.2   Implementation and Evaluation Metrics

Implementation: The Seq2Seq model follows the same architecture and training protocol as in [15]. The model performs poorly with only children handwriting dataset and to our knowledge, there is no other children handwriting dataset available. Therefore the model is pre-trained on an adult handwriting dataset [10] and then fine-tuned on the children handwriting dataset. The model is trained during 200 epochs with a batch size of 16. The RMS prop optimizer is used with a learning rate of 0.001. Since the test set is different, we reevaluate the method from [15] on our dataset. The R-CNN with a ResNet-FPN backbone is trained during 60 epochs with a batch size of 4. The AdamW [26] optimizer is used with a learning rate of 0.0001. The R-CNN parameters are indicated in the original article [17] except that we ignore the Mask branch, we use the Complete Intersection Over Union (CIoU) [27] criterion to match the ground truths and the predictions during the training phase. We also add an NMS filtering independent of the class on the outputs to handle nested predictions as explained in Sect. 3.1. We set all parameters of the R-CNN model on the validation set using the Mean Average Precision (MAP) performance score before evaluating the best model on the test set.

Metrics: To evaluate the performance of our Seq2Seg approach, we use the usual Character Error Rate (CER) and Word Error Rate (WER) with a Damerau Leveinshtein [28] distance for the recognition performance. We use Intersection Over Union (IoU) and IoU at pixel level to evaluate the segmentation performance. As explained before, the IoU_{Pixel} focuses on the handwriting lines and ignores the (mostly white) background.

4.3   Quantitative Results

We first perform an ablation study to measure the impact of our different contributions as well as the choice of features extractor backbone in object detector. Then, we compare our approach to the state-of-the-art models on our data. All experiments are evaluated in terms of **recognition, segmentation and computing speed** on the test set. While the networks are trained on GPU, the processing time is computed on a laptop with an Intel Core i7-8665U CPU. Indeed, education applications are run on a pen-based tablet, where an internet connection is not always available. Therefore, timing analysis is more relevant on a CPU-equipped laptop. We consider acceptable an analysis time lower than 2 s to deliver immediate feedback to the children.

Table 2 shows the results of the ablation study, where Level A corresponds to the filtering method of the object detector predictions with the result of a recognition model presented in Sect. 3.1 and Level B corresponding to the refinement of the bounding boxes coordinates of the object detector with a segmentation lattice presented in Sect. 3.2. We denote Seq2Seq as the result of the encoder part and use only the encoder result in this work, as recommended in [15]. We can see in the table that the choice of a deeper backbone (we tried 18 to 101

Table 2. Ablation study of the object detector backbone and impact of our contributions. Recognition is evaluated with Character Error Rate (CER) and Word Error Rate (WER) (lower values are better). Segmentation is evaluated with Intersection Over Union (IoU) and IoU$_{Pixel}$ (higher values are better). The *Average time* is the averaged number of seconds for a method to analyze a word.

Method	Backbone	Recognition		Segmentation		Time
		CER (%)	WER (%)	IoU (%)	IoU$_{Pixel}$ (%)	Average Time (s)
Seq2Seq [15]		**5.3**	**19.4**	48.4	59.4	0.12
R-CNN [17]	ResNet-18 FPN	12.0	36.4	78.6	80.3	1.25
	ResNet-34 FPN	12.2	37.7	79.6	81.3	1.29
	ResNet-50 FPN	11.4	34.7	80.5	81.5	1.61
	ResNet-101 FPN	10.7	34.2	81.0	82.2	2.17
Level A	ResNet-18 FPN	5.2	19.0	81.7	83.6	1.43
	ResNet-34 FPN	**5.0**	**18.6**	82.3	84.0	1.47
	ResNet-50 FPN	5.2	18.9	82.8	84.0	1.79
	ResNet-101 FPN	5.1	19.0	82.0	83.5	2.35
Level B	ResNet-18 FPN	12.0	36.4	82.6	85.0	1.40
	ResNet-34 FPN	12.2	37.7	83.3	85.9	1.44
	ResNet-50 FPN	11.4	34.7	83.8	86.3	1.77
	ResNet-101 FPN	10.7	34.2	84.4	87.1	2.32
Seq2Seg: Levels A + B	ResNet-18 FPN	5.2	19.0	85.9	**88.3**	**1.58**
	ResNet-34 FPN	5.0	18.6	86.1	**88.9**	1.62
	ResNet-50 FPN	5.2	18.9	86.3	**89.0**	1.95
	ResNet-101 FPN	5.1	19.0	85.6	**88.4**	2.50

layers) in the object detector (R-CNN) improves the performance in recognition and segmentation (-1.3% of CER from ResNet 18 to ResNet 101; +1.9% of IoU$_{Pixel}$ from 18 to 101 layers). On the other hand, the computing time increases of more than 2 s. The Seq2Seq model remains much more accurate in recognition (CER/WER) than all versions of the R-CNN. The choice of the backbone had no significant impact on our contribution (see bottom part of the table). We chose the backbone ResNet-34 FPN for the next experiments due to its speed and slightly better performance in recognition.

Level A: filtering the object detector's predictions with the results of the Seq2Seq allows us to obtain slightly better results in recognition (CER of 5%) than the Seq2Seq alone (CER of 5.3%). The reasons for this are given in the "no matching case" of the second step of the first contribution presented in Sect. 3.1. Furthermore, this method selects the bounding boxes to maximize the coverage and minimize the overlap of the handwriting and thus improves the segmentation of the object detector. Table 3 details the different scenarios of filtering and their contributions to the performance compared to the object detector and the Seq2Seq performance alone:

- In the scenario where the number of predictions of the two models is equal, the Level A improves only the recognition performance as expected. This scenario concerns most of the words.
- In the scenario where the number of predictions is different and a matching exist, the gain is the highest. Indeed, the strategy makes it possible to filter the bad predictions of the object detector.
- For a few words, nothing is filtered out and thus this contribution does not improve the object detector performance. In practice, this corresponds to words for which the recognizer makes more mistakes than the object detector.

Table 3. Number of words by scenario of filtering between R-CNN and Seq2Seq. Performance of models alone and level A contribution. R-CNN uses ResNet34-FPN backbone.

		R-CNN		Seq2Seq		Level A	
Filtering type	#Words	CER (%)	IoU (%)	CER (%)	IoU (%)	CER (%)	IoU (%)
Perfect Matching	857	8.0	**85.6**	**3.8**	50.2	**3.8**	**85.6**
Matching	164	34.7	47.9	**11.5**	40.0	**11.5**	**64.6**
No Matching	11	**1.8**	**87.2**	36.6	31.3	**1.8**	**87.2**

Level B: refining the bounding boxes coordinates by the use of a segmentation lattice improves the R-CNN segmentation performance for a small computing cost.

The results of the competitors are shown in Table 4. The best recognition and segmentation performance on our dataset are given by [15] with a small margin compared to Seq2Seg (+0.1% CER, +1.6% IoU$_{Pixel}$), a high computation cost (5,07s, +3,45s compared to Seq2Seg) and using a language model. To overcome this computation cost, the authors of [15] have proposed a pruning strategy (shown in the second line). This strategy degrades the recognition performance as well as the segmentation one which makes it **significantly lower than Seq2Seg for recognition and segmentation** (-2.6% CER, -4.3% WER, +1.3% IoU, +2.6% IoU$_{Pixel}$).

The following section presents a qualitative analysis of the results obtained and shows the limits associated with the children handwriting recognition and segmentation tasks.

4.4   Qualitative Results

This section presents a qualitative analysis of the results obtained by Seq2Seg. The goal is to visualize the effect of each contribution, *i.e.* the impact of Level A and Level B contributions. In these visualization examples, the output of the object detector corresponds to the **predictions before the last NMS** was performed independently of the class of the predicted bounding box.

Table 4. Comparison to state-of-the-art approaches. Recognition is evaluated with Character Error Rate (CER) and Word Error Rate (WER) (lower values are better). Segmentation is evaluated with Intersection Over Union (IoU) and IoU$_{Pixel}$ (higher values are better). The *Average time* is the averaged number of seconds for a method to analyze a word. LM stand for "Language Model".

Method	LM	Recognition		Segmentation		Time
		CER (%)	WER (%)	IoU (%)	IoU$_{Pixel}$ (%)	Average Time (s)
Fusion competition [15]	Yes	**4.9**	**16.1**	**89.2**	90.5	5.07
Fusion competition (pruning) [15]	Yes	7.6	22.9	84.8	86.3	**0.72**
Seq2Seg (Our)	No	5.0	18.6	86.1	**88.9**	1.62

Figure 6 emphasizes the relevance of Level A contribution. The filtering process by the recognition model selects the correct number of letters by minimizing the overlap and maximizing the coverage rate. Moreover, the use of the segmentation lattice in Level B contribution produces a precise segmentation of the handwriting words especially in example 1 where the bounding boxes of the letters "i" and "t" overlap.

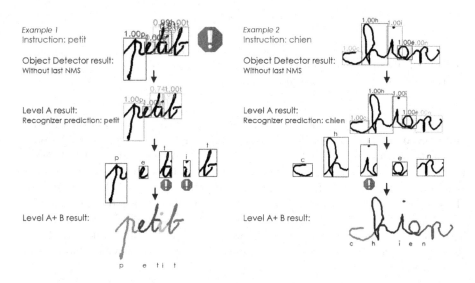

Fig. 6. Examples with an accurate recognition and a precise segmentation.

Figure 7 illustrates examples where the recognition model makes errors. In example 1, there is no matching between the prediction of the recognition model and the object detector. We can see that the Seq2Seq makes recognition errors and therefore its associated filtering would be wrong. In this case, the bounding boxes and labels predicted by the object detector are used and provide an accurate result in recognition and segmentation. Example 2 shows a case where the filtering by the recognition model leads to a segmentation error. In addition, we can note the omission of the drawing of the point of the "i" in example 2 which is quite common in a context of learning how to write.

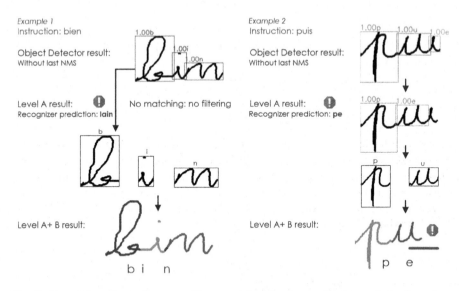

Fig. 7. Examples where the recognition model makes errors: in example 1, there is no matching between the recognition model and the object detector. In example 2 the filtering leads to an under-segmentation error.

Note that evaluating the quality of the handwriting segmentation with the currently used metrics is difficult. Indeed, it is not easy to define an absolute segmentation ground truth for some letters due to the ligature area between letters. Thus, a prediction can have an IoU lower than 100% with the ground truth while the associated segmentation is correct. Moreover, the ground truth class associated with a degraded letter can vary according to the annotator (confusion between the letter "e" and the letter "l", "a" and "o" ...). Taking into account the uncertainty in the predictions might be helpful to know when a (human) teacher should take over the automated system to provide a more useful advice to the children.

5 Conclusion

We presented Seq2Seg, an original combination strategy which uses a model dedicated to recognition as an oracle to filter out the segmentation predictions of an object detector and then refines the segmentation using an expert segmentation lattice. Seq2Seg produces the best of both worlds: the accurate recognition of a Seq2Seq and the precise segmentation provided by an R-CNN object detector. Seq2Seg is efficient enough to provide immediate feedback to children learning how to write and it outperforms the state of the art results on this task without the use of a language model. This last point makes Seq2Seg much more flexible to other learning contexts. Our future work will focus on evaluating and improving the quality of the feedback in school contexts. In particular, we plan to better leverage the uncertainty of the decisions (both for the Seq2Seq and the object detector), for example by allowing the system to reject hypotheses, to prevent giving erroneous feedback to the children.

References

1. Sayre, K.M.: Machine recognition of handwritten words: a project report. Pattern Recognit. **5**(3), 213–228 (1973)
2. Anquetil, E., Lorette, G.: Perceptual model of handwriting drawing application to the handwriting segmentation problem. In: 4th International Conference Document Analysis and Recognition (ICDAR 1997), 2-Volume Set, 18–20 August 1997, Ulm, Germany, Proceedings, p. 112. IEEE Computer Society (1997)
3. Anquetil, E., Lorette, G.: On-line handwriting character recognition system based on hierarchical qualitative fuzzy modelling. In: Progress in Handwriting Recognition, pp. 109–116 (1997)
4. Simonnet, D., Girard, N., Anquetil, É., Renault, M., Thomas, S.: Evaluation of children cursive handwritten words for e-Education. Pattern Recogn. Lett. **121**, 133–139 (2019)
5. Michael, J., Labahn, R., Grüning, T., Zöllner, J.: Evaluating sequence-to-sequence models for handwritten text recognition. In: 2019 International Conference on Document Analysis and Recognition, ICDAR 2019, Sydney, Australia, 20–25 September 2019, pp. 1286–1293. IEEE (2019)
6. Coquenet, D., Chatelain, C., Paquet, T.: End-to-end handwritten paragraph text recognition using a vertical attention network. In: IEEE Transactions on Pattern Analysis and Machine Intelligence (2022)
7. Kang, L., Riba, P., Rusiñol, M., Fornés, A., Villegas, M.: Pay attention to what you read: Non-recurrent handwritten text-line recognition. Pattern Recognit. **129**, 108766 (2022)
8. Barrere, K., Soullard, Y., Lemaitre, A., Coüasnon, B.: Transformers for Historical Handwritten Text Recognition. In: Doctoral Consortium - ICDAR 2021, Lausanne, Switzerland (2021)
9. Marti, U.-V., Bunke, H.: A full English sentence database for off-line handwriting recognition. In: Fifth International Conference on Document Analysis and Recognition, ICDAR 1999, 20–22 September 1999, Bangalore, India, pp. 705–708. IEEE Computer Society (1999)
10. Liwicki, M., Bunke, H.: IAM-OnDB - an on-line English sentence database acquired from handwritten text on a whiteboard. In: Eighth International Conference on Document Analysis and Recognition (ICDAR 2005), 29 August - 1 September 2005, Seoul, Korea, pages 956–961. IEEE Computer Society (2005)
11. Graves, A., Fernández, S., Gomez, F.J., Schmidhuber, J.: Connectionist temporal classification: labelling unsegmented sequence data with recurrent neural networks. In: Machine Learning, Proceedings of the Twenty-Third International Conference (ICML 2006), Pittsburgh, Pennsylvania, USA, 25–29 June 2006, vol. 148 of ACM International Conference Proceeding Series, pp. 369–376. ACM (2006)

12. Zeyer, A., Schlüter, R., Ney, H.: Why does CTC result in peaky behavior? CoRR, abs/2105.14849 (2021)
13. Liu, H., Jin, S., Zhang, C.: Connectionist temporal classification with maximum entropy regularization. In: Bengio, S., Wallach, H.M., Larochelle, H., Grauman, K., Cesa-Bianchi, N., Garnett, R., (edn.), Advances in Neural Information Processing Systems vol. 31, pp. 839–849 (2018)
14. Li, H., Wang, W.: Reinterpreting CTC training as iterative fitting. Pattern Recognit. **105**, 107392 (2020)
15. Krichen, O., Corbillé, S., Anquetil, E., et al.: Combination of explicit segmentation with Seq2Seq recognition for fine analysis of children handwriting. IJDAR 25, pp. 339–350 (2022). https://doi.org/10.1007/s10032-022-00409-4
16. Krichen, O., Corbillé, S., Anquetil, E., Girard, N., Nerdeux, P.: Online analysis of children handwritten words in dictation context. In: Barney Smith, E.H., Pal, U. (eds.) ICDAR 2021. LNCS, vol. 12916, pp. 125–140. Springer, Cham (2021). https://doi.org/10.1007/978-3-030-86198-8_10
17. He, K., Gkioxari, G., Dollár, P., Girshick, R.B.: Mask R-CNN. In: IEEE International Conference on Computer Vision, ICCV 2017, Venice, Italy, 22–29 October 2017, pp. 2980–2988. IEEE Computer Society (2017)
18. Ren, S., He, K., Girshick, R.B., Sun, J.: Faster R-CNN: towards real-time object detection with region proposal networks. In: Cortes, C., Lawrence, N.D., Lee, D.D., Sugiyama, M., Garnett, R., (edn.) Advances in Neural Information Processing Systems. vol. 28, pp. 91–99 (2015)
19. Pang, J., Chen, K., Shi, J., Feng, H., Ouyang, W., Lin, D.: Libra R-CNN: towards balanced learning for object detection. In: IEEE Conference on Computer Vision and Pattern Recognition, CVPR, pp. 821–830. Computer Vision Foundation/IEEE (2019)
20. Redmon, J., Divvala, S.K., Girshick, R.B., Farhadi, A.: You only look once: unified, real-time object detection. In: 2016 IEEE Conference on Computer Vision and Pattern Recognition, CVPR 2016, Las Vegas, NV, USA, 27–30 June 2016, pp. 779–788. IEEE Computer Society (2016)
21. Bochkovskiy, A., Wang, C.-Y., Liao, H.-Y.M.: YOLOv4: Optimal speed and accuracy of object detection. CoRR, abs/2004.10934 (2020)
22. Li, C., et al.: YOLOv6: A single-stage object detection framework for industrial applications. CoRR, abs/2209.02976 (2022)
23. Wang, C.-Y., Bochkovskiy, A., Liao, H.-Y.M.: YOLOv7: Trainable bag-of-freebies sets new state-of-the-art for real-time object detectors. CoRR, abs/2207.02696 (2022)
24. Lin, T.-Y., et al.: Microsoft COCO: common objects in context. In: Fleet, D., Pajdla, T., Schiele, B., Tuytelaars, T. (eds.) ECCV 2014. LNCS, vol. 8693, pp. 740–755. Springer, Cham (2014). https://doi.org/10.1007/978-3-319-10602-1_48
25. Mouchère, H., Bayoudh, S., Anquetil, E., Miclet, L.: Synthetic on-line handwriting generation by distortions and analogy. In: 13th Conference of the International Graphonomics Society (IGS2007), pp. 10–13, Melbourne, Australia, November 2007
26. Loshchilov, I., Hutter, F.: Decoupled weight decay regularization. In: 7th International Conference on Learning Representations, ICLR 2019, New Orleans, LA, USA, 6–9 May 2019. OpenReview.net (2019)
27. Wang, X., Song, J.-Y.: ICIoU: improved loss based on complete intersection over union for bounding box regression. IEEE Access **9**, 105686–105695 (2021)
28. Damerau, F.: A technique for computer detection and correction of spelling errors. Commun. ACM **7**(3), 171–176 (1964)

Fine-Tuning Vision Encoder–Decoder Transformers for Handwriting Text Recognition on Historical Documents

Daniel Parres[1]([✉]) [iD] and Roberto Paredes[1,2] [iD]

[1] PRHLT Research Center, Universitat Politècnica de València, Valencia, Spain
{dparres,rparedes}@prhlt.upv.es
[2] Valencian Graduate School and Research Network of Artificial Intelligence,
Camí de Vera s/n, 46022 Valencia, Spain

Abstract. Handwritten text recognition (HTR) has seen significant advancements in recent years, mainly due to the incorporation of deep learning techniques. One area of HTR that has garnered particular interest is the transcription of historical documents, as there is a vast amount of records available that have yet to be processed, potentially resulting in a loss of information due to deterioration.

Currently, the most widely used HTR approach is to train convolutional recurrent neural networks (CRNN) with connectionist temporal classification loss. Additionally, language models based on n-grams are often utilized in conjunction with CRNNs. While transformer models have revolutionized natural language processing, they have yet to be widely adopted in the context of HTR for historical documents.

In this paper, we propose a new approach for HTR on historical documents that involves fine-tuning pre-trained transformer models, specifically vision encoder–decoder models. This approach presents several challenges, including the limited availability of large amounts of training data for specific HTR tasks. We explore various strategies for initializing and training transformer models and present a model that outperforms existing state-of-the-art methods on three different datasets. Specifically, our proposed model achieves a word error rate of 6.9% on the ICFHR 2014 Bentham dataset, 14.5% on the ICFHR 2016 Ratspro-tokolle dataset, and 17.3% on the Saint Gall dataset.

Keywords: Transformers · Fine-tuning · Handwritten Text Recognition · Historical Documents

1 Introduction

Handwritten text recognition (HTR) is a rapidly advancing field within computer science, with a focus on document analysis and recognition. The goal of HTR is to transcribe the text in a document using machine learning techniques.

This research paper focuses on the application of HTR techniques on historical documents. In museums, archives, and libraries, a wide variety of ancient

G. A. Fink et al. (Eds.): ICDAR 2023, LNCS 14190, pp. 253–268, 2023.
https://doi.org/10.1007/978-3-031-41685-9_16

documents are stored, many of which are at risk of deterioration. This can lead to the loss of valuable information, including content and annotations or corrections that may be of historical or cultural significance. To preserve and make these documents more accessible for scientific and cultural research, it is crucial to develop technologies that can recognize and transcribe the handwritten text.

The current primary approach for HTR is based on using artificial neural network (ANN) algorithms [5, 20, 25]. These neural network models take images of a document as input and must be trained with their corresponding transcriptions. Typically, the page image is divided into individual lines of text, and the model learns to transcribe these lines and produce a sequence of words or tokens. However, historical documents present unique challenges, such as variations in writing styles, faint ink, stamps, skewed images, lines with considerable slope variation, and slanted scripts. These issues make it difficult for existing models to achieve accurate recognition, thus leading to a more challenging problem

2   Related Work

In recent years, the most successful approaches for HTR have come from using ANNs. While hidden markov models (HMM) have been used in the past [6, 13, 31], they have the drawback of only considering the current observation and not the context in which it occurs. In contrast, recurrent neural networks (RNN) such as long short term memory (LSTM) cells [12] can take into account the previous context internally and can memorize both long and short dependencies.

Connectionist temporal classification (CTC) [10] loss is a crucial component in current state-of-the-art HTR techniques. CTC-based architectures are the standard ANN models used for transcribing text. Early approaches utilized linear recurrent layers to process the input image and generate the output. However, recent state-of-the-art methods involve using convolutional neural networks (CNN) and RNNs, known as convolutional recurrent neural networks (CRNN) [5, 20, 25]. Additionally, language models based on character n-grams are applied to the ANN output to improve the results further [5, 20, 25].

The gated convolutional recurrent neural network (GCRNN) architecture is proposed by Bluche and Messina [5], building on the success of CRNNs in HTR. Using GCRNN topology and a large amount of external data in different languages, the authors achieved the best state-of-the-art results on the IAM [19] and Rimes [2] datasets. These results suggest a new methodology based on starting with a well-trained pre-existing model and fine-tuning it to a specific task.

Recently, there has been an increasing interest in applying transformer models [33] to HTR [3, 15, 24, 34]. These proposals typically use a transformer decoder to generate the character sequence, and a CNN-based encoder to extract image features. These extracted features are fed into a set of operators similar to those found in a transformer block. However, these proposals still rely on a CNN backbone and do not use a purely transformer-based design for end-to-end training. Despite the interest, these transformer-based ANNs for HTR perform on pair or slightly worse than traditional CRNN-based approaches [3, 15, 24, 34].

3   Our Approach

This paper proposes to use the vision encoder–decoder (VED) architecture, which employs pure transformer models, for HTR of historical documents. The VED architecture comprises two main parts: an optical encoder model that processes visual information and a generative text decoder model that produces the transcription based on the visual information provided by the encoder. The encoder aims to understand the image, and the decoder generates text transcription accordingly.

3.1   Encoder. Vision Transformer

In the field of computer vision (CV), a wide variety of problems involve image processing to solve a task, such as classification and segmentation. Currently, the state-of-the-art ANNs for image processing are CNNs. A CNN takes as input an image and uses learnable filters that slide along the image to extract relevant features and produce feature maps. Some of the most widely used CNN architectures in various tasks are ResNet [11], VGG [23], Inception [26], and EfficientNet [30].

Recently, transformers have also begun to be applied to CV problems and have shown to be able to achieve similar, or even superior results than CNNs [8,18,32]. One of the advantages of using transformers in vision problems is that they are less computationally expensive during training than CNNs. These architectures are called vision transformers, and the best-known proposal is ViT [8]. Inspired by the original transformer [33], ViT treats images as sequences by dividing the image into patches, linear projecting each patch, and feeding it into the model. The image patches or embeddings are essentially treated as tokens in natural language processing (NLP).

As vision transformers are a recent neural architecture, this paper proposes to analyze the performance of using pre-trained ViT as the optical model in our VED architecture, which the aim to improve the results and performance of HTR on historical documents.

3.2   Decoder. NLP Transformer

The transformer architecture has dramatically advanced the NLP field achieving state-of-the-art results by being trained on massive datasets. Given the success of transformers in various language tasks, it is worth exploring their potential use in HTR on historical documents.

Deep learning has increasingly relied on models trained on large datasets in recent years. However, training neural architectures on large datasets can be costly, resource-intensive, and not accessible to everyone. To overcome this issue, models trained on one task can be adapted for other or similar problems. This is because models can transfer their knowledge. Therefore, using pre-trained models on specific tasks can perform better than training models with random weight initialization.

An example of a pre-trained model in NLP is BERT [7], a transformer-based language representation model. BERT is trained on a large amount of text, considering the context, and can be fine-tuned for different NLP tasks. Similarly, RoBERTa [17] replicates BERT, but with optimized training strategies that result in a more robust model. It has been used for various NLP tasks such as [14,35]. Due to its flexibility and good results, this study proposes to use pre-trained RoBERTa as the decoder in our VED architecture for HTR on historical documents.

3.3   Vision Encoder–Decoder

The VED architecture proposed in this paper is a transformer-based model, in which the encoder is a pre-trained vision transformer (ViT), and the decoder is a pre-trained RoBERTa. Unlike other state-of-the-art HTR proposals that use transformers to estimate character-by-character transcriptions [3,15,24,34], our proposed model estimates transcriptions at the wordpiece level. Additionally, our proposed model does not use the CTC loss for transcription computation and does not rely on an external language model. This allows for end-to-end training without additional post-processing in the inference phase.

VED models typically consist of two stages. The first step involves the processing of the image by the encoder. The second stage implies the decoder, generating the transcription auto-regressively, considering the previous text input using a mask-attention mechanism and the encoder information using a cross-attention mechanism. In our VED, the decoder is a RoBERTa, where the cross-attention blocks have been added and initialized randomly.

Pre-trained models can achieve better results than untrained or randomly initialized models. However, fine-tuning the models for the specific task is crucial for success. Fine-tuning is a process that involves adjusting a pre-trained model for a different task. Nevertheless, without proper fine-tuning, the model can lose its learned knowledge and lead to poor results compared to a model initialized with random weights. Therefore, this paper aims to study the fine-tuning process of the pre-trained transformer model and demonstrate that it is not necessary to use large amounts of data to obtain similar or better results than state-of-the-art models.

4   Experiments and Results

This section presents the experiments conducted to perform an in-depth analysis of the VED model. Each experiment followed the standard practice of using a validation set and grid search to explore optimal parameterization. Grid search is a widely recognized technique for hyperparameter optimization in neural networks, allowing the exploration of parameter values to obtain the best model performance. Once the model has converged, the test set is evaluated, and the results are presented in the corresponding tables or figures.

In this study, we begin by introducing the historical text datasets used in our experiments. We then compare the performance of our VED model with state-of-the-art models on each dataset to demonstrate its superior performance. Next, we conduct a detailed fine-tuning analysis to achieve competitive results. Specifically, we investigate the optimal combination of learning rate and optimizer and analyze the importance of each model component, such as training the encoder, decoder, and cross-attention individually. Finally, we demonstrate that good fine-tuning enables the model to transcribe text in different languages without requiring a large amount of data.

4.1 Datasets

This study focuses on evaluating the performance of transformer models in the task of historical HTR on various databases. The first database used is one of the most widely extended datasets in this field, the Bentham Papers, presented at ICFHR 2014 [27]. This dataset comprises 433 pages of text written in English. We propose to investigate the model's adaptability in the second database by utilizing the Ratsprotokolle dataset presented at ICFHR 2016 [28], which comprises 450 pages of text in German. Additionally, we include the Saint Gall dataset [9] in our analysis. This dataset is a collection of Latin texts characterized by long sequences of lengthy words. All three datasets are divided into training, validation, and test sets, as outlined in Table 1, which presents the number of lines and partitions of each dataset used in this work.

Table 1. Number of lines per partition of each dataset.

Dataset Name	Train	Validation	Test
ICFHR 2014 Bentham [27]	9, 198	1, 415	860
ICFHR 2016 Ratsprotokolle [28]	8, 367	1, 043	1, 140
Saint Gall [9]	468	235	707

For the experiments, the lines of the datasets presented in Table 1 are segmented with corresponding transcripts. The images of the text lines and their transcriptions (or ground truth) are required to train an HTR system. During training, the optical model or encoder is fed with the images, while the decoder inputs are the transcriptions.

4.2 Model Initialization Analysis

This section presents an analysis of the best weight initialization and fine-tuning strategies for our VED model, which is composed of a ViT encoder and a RoBERTa decoder. Four experimental setups are evaluated, with all experiments being conducted on the ICFHR 2014 Bentham dataset. The learning rate is adjusted during training, decreasing linearly when results no longer improve after multiple epochs. Data augmentation techniques such as cutout, gaussian blur, erosion, dilation, underline, and rotation of the input images are applied to the training set. In contrast, no data augmentation is applied to the transcriptions. The loss function used throughout the experiments is cross-entropy.

Table 2. Analysis of the initialization of weights and fine-tuning strategies of the VED model in the ICFHR 2014 dataset.

Models	WER (%)	CER (%)
VED + fine-tunning	93.5	73.6
VED–frozen + fine-tunning	87.5	68.5
TrOCR	31.1	16.4
TrOCR + fine-tunning	6.9	2.7

In the first experiment, the VED model is initialized with the pre-trained weights of ViT (Imagenet) and RoBERTa, with the cross-attention layers being initialized with random values. The same pre-trained weights are used in the second experiment, but fine-tuning is performed in two steps. In the first step, the entire model is frozen except for the cross-attention layers. Once the cross-attention layers are trained, the whole model is unfrozen, and a second fine-tuning is performed. This second experiment is referred to as VED–frozen.

For the third experiment, the VED model is initialized with the weights of the optical character recognition (OCR) transformer model called TrOCR [16]. The TrOCR model is composed of a ViT encoder and a RoBERTa decoder. The critical point of this model is that it has undergone extensive pre-training on a diverse set of English text recognition tasks. The objective of this experiment is to evaluate the performance of the original TrOCR model on the ICFHR 2014 Bentham dataset without any fine-tuning.

In the final experiment, the VED model is initialized with the TrOCR weights and fine-tuned using the ICFHR 2014 Bentham dataset.

Table 2 summarizes the WER and character error rate (CER) of all four experiments on the ICFHR 2014 Bentham test dataset. The results indicate that fine-tuning the pre-trained VED (Imagenet and RoBERTa) on the ICFHR 2014 Bentham dataset yields the worst performance. On the other hand, fine-tuning the VED–frozen model slightly improves the previous result. Significant improvement is obtained using the original TrOCR model. Finally, using the TrOCR-initialized model and performing a fine-tuning on the ICFHR 2014 Bentham dataset provides the best performance.

Therefore, using and fine-tuning the TrOCR model weights allows us to transfer and adapt OCR knowledge to HTR on historical documents.

4.3   Benchmarking with the State of the Art

In this section, we compare the performance of the VED model initialized with TrOCR weights and fine-tuned to other state-of-the-art approaches on three different datasets: ICFHR 2014 Bentham, ICFHR 2016 Ratsprotokolle, and Saint Gall.

Table 3. WER and CER perfomance (%) of different state-of-the-art methods on ICFHR 2014 Bentham test.

Models	WER (%)	CER (%)
Discriminative HMMs + word 2-gram [31]	17.2	6.7
Bluche [5] + char 9-gram [25]	16.8	6.7
CRNN + regex/lexicon LM [27]	14.6	5.0
CRNN + word 2-gram [4]	14.1	5.0
CRNN (Laia) [20, 29]	12.7	6.2
LIMSI [27]	11.0	3.9
HTR-Flor + character 9-gram [25]	9.8	4.0
CRNN (Laia) [20] + character 7-gram [29]	9.7	5.0
A2IA [27]	8.6	2.9
TrOCR [16]	31.1	16.4
Ours	6.9	2.7

Table 3 presents the results for the ICFHR 2014 Bentham dataset. The table is divided into two parts, the first one presents the state-of-the-art results for ICFHR 2014 Bentham, and the second one contains the original TrOCR model and our fine-tuned version.

Among the state-of-the-art models, A2IA has the lowest error rates, with 8.6% WER and 2.9% CER. A2IA is an MDLSTM model that performs extensive pre-training with a large amount of external data. Additionally, it employs a hybrid word/character-gram language model and CTC loss.

The results show that using the original TrOCR model without fine-tuning performs worse than the state of the art, with a WER of 31.1% and a CER of 16.4%. On the other hand, fine-tuning the TrOCR model on the ICFHR 2014 Bentham dataset provides state-of-the-art results, outperforming A2IA with a WER of 6.9% and a CER of 2.7%.

The ICFHR 2016 Ratsprotokolle dataset is a collection of German documents, making it more interesting than many other HTR databases on historical manuscripts. Most of them are English or Latin texts. The first part of Table 4 shows the best state-of-the-art proposals for ICFHR 2016 Ratsprotokolle. Furthermore, the second part shows the original TrOCR model and our fine-tuned version.

As discussed above, A2IA is an MDLSTM model that uses word/character-gram language models. In this case, the A2IA model has been subjected to an

Table 4. WER and CER perfomance (%) of different state-of-the-art methods on ICFHR 2016 Ratsprotokolle test.

Models	WER (%)	CER (%)
ParisTech [28]	46.6	18.5
LITIS [28]	26.1	7.3
BYU [28]	21.1	5.4
A2IA [28]	21.0	5.1
CRNN + char 10-gram [28]	20.9	4.8
RWTH [28]	20.9	4.8
CRNN (Laia) [20, 29]	19.0	4.8
CRNN (Laia) [20] + character 8-gram [29]	17.5	4.5
TrOCR [16]	121.7	97.0
Ours	14.5	3.8

extensive pre-train of letters written in modern German. Despite this pre-train in German, the A2IA model does not obtain the best results. Instead, the CRNN model Laia, together with a character 8-gram, obtains the best error rates in the first part of the table (17.5% WER and 4.5% CER).

The error rates of the original TrOCR and the fine-tuned version are presented in the second part of Table 4. The original TrOCR obtains a poor performance for both WER and CER. We observed that the transcriptions generated by the model are longer than the ground truth. These transcriptions seem to be hallucinations of the model since, in most cases, the model proposes meaningless transcriptions. These poor results may be because the TrOCR model has worked exclusively with English text and has never seen anything in German.

In contrast, our fine-tuned version achieves the best state-of-the-art error rates with 14.5% WER and 3.8% CER. These results are striking because this model is initialized with the TrOCR weights, which produces poor transcriptions. However, fine-tuning can adapt the model capabilities to a task in another language producing the best results.

Table 5. WER and CER perfomance (%) of different state-of-the-art methods on Saint Gall test.

Models	WER (%)	CER (%)
Shonenkov [22]	26.2	3.7
Bluche [5] + char 11-gram [25]	23.7	6.0
CRNN (Laia) [20] + character 11-gram [25]	23.4	6.0
Abdallah [1]	23.0	7.3
HTR-Flor + character 11-gram [25]	21.1	5.2
Sai Suryateja + character 11-gram [21]	18.6	3.9
TrOCR [16]	103.9	43.5
Ours	17.3	2.5

Finally, like the previous two tables, Table 5 is divided into two parts. The word and character errors in the Saint Gall dataset are presented in Table 5. The proposal that achieves the best results in the first part of the table is from Sai Suryateja, with 18.6% WER and 3.9% CER. As for the results of the transformers, the original TrOCR again has very high error rates. However, fine-tuning the weights, our proposal fits the Latin model very well and outperforms all the state-of-the-art results with 17.3% WER and 2.5% CER. These results are remarkable for two factors: the first is that Saint Gall is a Latin database, and the second is that the sentences to be transcribed are considerably longer than usual.

As demonstrated, our proposed model can obtain the most competitive state-of-the-art results in English, German and Latin. It has also been shown that the original TrOCR model obtains high error rates; this is why a good fine-tuning strategy is mandatory.

4.4   TrOCR Fine-Tuning Analysis

The VED model outperforms the other state-of-the-art models in HTR on historical documents thanks to the original weights provided by TrOCR and the fine-tuning performed. Due to the importance of fine-tuning, the following experiments focus on analyzing the most critical hyper-parameters, such as learning rate, optimizer, dataset size and freezing strategies.

Learning Rate and Optimizer. When fine-tuning a pre-trained model, there are various parameters to configure, including the optimization algorithm and learning rate. These parameters play a crucial role in determining the performance of the fine-tuned model. Popular optimization algorithms for fine-tuning transformer models trained for NLP and CV include stochastic gradient descent (SGD), Adam, and AdamW. Therefore, it is of particular interest to investigate the model's behavior when using these three optimization algorithms.

Table 6. Fine-tuning analysis of transformers using as metric WER and CER in ICFHR 2014 Bentham test.

Optimizer	Learning rate	WER (%)	CER (%)
SGD	$5e^{-5}$	13.0	5.3
	$5e^{-6}$	20.7	9.5
	$5e^{-7}$	35.7	18.8
Adam	$5e^{-5}$	42.3	29.7
	$5e^{-6}$	7.8	3.2
	$5e^{-7}$	7.3	2.7
AdamW	$5e^{-5}$	44.0	28.3
	$5e^{-6}$	6.9	2.7
	$5e^{-7}$	7.3	2.8

Our study began with a pre-trained transformer and aimed to transfer its knowledge to historical HTR. The learning rate, in particular, is a critical factor that must be carefully chosen. Since we want to make the most of the pre-trained model, the learning rate values must be small. To this end, we analyzed different learning rate values to estimate the optimal value for knowledge transfer in the HTR task. Specifically, we propose to use three different learning rates: $5e^{-5}$, $5e^{-6}$, and $5e^{-7}$.

All of these experiments were conducted using the ICFHR 2014 Bentham database. Based on the results shown in Table 6, we found that SGD performed better with higher learning rates than Adam or AdamW. Adam, when using the learning rate that achieved the best results using SGD, resulted in the worst metrics. The same was true for AdamW. Additionally, we observed that for a learning rate of $5e^{-7}$, Adam and AdamW produced similar WER and CER scores. However, the best fine-tuning for the transformer model was achieved using AdamW with a learning rate of $5e^{-6}$.

Freezing Strategies. It is of interest to investigate the impact of different components of the VED architecture on fine-tuning. Specifically, identifying opportunities to optimize the training process and reduce the number of parameters that need to be adjusted for a given task is a crucial area to explore. To this end, we conducted three different experiments. All three experiments used frozen VED models and aimed to train only a single part of the transformer.

In the first experiment, we trained only the cross-attention layers. The encoder parameters were adjusted in the second experiment, and only the decoder was trained in the third experiment. These three experiments allowed us to study the critical components of the VED model. The fine-tuning performed in the three experiments consisted of training for 50 epochs using AdamW as the optimization algorithm, a learning rate of $5e^{-6}$, and a linear learning rate scheduler.

Table 7. Freezing study of ICFHR 2014 Bentham test dataset.

Dataset Name	Fine-tuning	WER (%)	CER (%)
ICFHR 2014 Bentham	Fine-tuning only cross-attention	13.5	6.2
	Fine-tuning only Encoder	7.1	2.5
	Fine-tuning only Decoder	14.5	6.8
	Fine-tuning whole model	6.9	2.7

We evaluated the performance of our VED model on the ICFHR 2014 Bentham, ICFHR 2016 Ratsprotokolle, and Saint Gall datasets and presented the results in Tables 7 to 9. These tables are divided into two parts. The first part shows the results of the three freezing experiments, and the second part presents the results when the entire transformer is trained.

Table 7 shows that training only the cross-attention layers or the decoder for the ICFHR 2014 Bentham dataset did not result in competitive results. However, freezing the entire VED except for the encoder resulted in an excellent performance, surpassing the CER of the VED model where the entire architecture was trained. This highlights the importance of the optical model of the VED for handwriting text recognition on historical document images. Furthermore, training only the encoder reduced the training time by more than half.

Table 8. Freezing study of ICFHR 2016 Ratsprotokolle test dataset.

Dataset Name	Fine-tuning	WER (%)	CER (%)
ICFHR 2016 Ratsprotokolle	Fine-tuning only cross-attention	44.9	18.8
	Fine-tuning only Encoder	24.8	6.6
	Fine-tuning only Decoder	44.6	21.7
	Fine-tuning whole model	14.5	3.8

Table 9. Freezing study of Saint Gall test dataset.

Dataset Name	Fine-tuning	WER (%)	CER (%)
Saint Gall	Fine-tuning only cross-attention	42.9	9.9
	Fine-tuning only Encoder	23.3	3.2
	Fine-tuning only Decoder	36.6	8.9
	Fine-tuning whole model	17.3	2.5

Table 8 presents the results of the experiments conducted on the ICFHR 2016 Ratsprotokolle dataset. The results of the experiments are similar in cases where only the cross-attention layers or the decoder are trained, as these tend to produce the highest error rates.

When training only the optical model, specifically the vision transformer, the VED model achieves competitive results. However, since this dataset is in German, it is not sufficient to adapt the optical model alone to overcome the performance of state-of-the-art models. It is necessary to train the entire network to obtain the best performance.

Table 9 presents the results obtained from the experiments conducted on the Saint Gall dataset. Similar to the results presented in Tables 7 and 8, training only the vision transformer allows us to achieve competitive results with the state of the art and reduces the training time by more than half. However, as the text in this dataset is in Latin, it is necessary to fine-tune the decoder to achieve results that surpass the state of the art.

The results of the three case studies demonstrate that the VED model's vision transformer component is the most critical component. Additionally, very competitive results can be obtained by focusing on training the encoder alone. However, when applying the VED model to tasks involving languages other

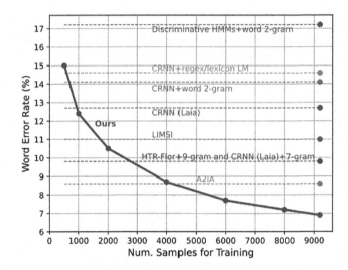

Fig. 1. Performance of the VED model versus the rest of the state-of-the-art models in the ICFHR 2014 Bentham test dataset.

than English, it is necessary to fine-tune the entire network to achieve optimal performance. This knowledge can significantly reduce training times by more than half.

Dataset Size. This section aims to examine the amount of data necessary for the VED model to produce results that compare favorably with state-of-the-art models. We used varying amounts of randomly selected samples from the training dataset to train the model.

We conducted seven experiments for the ICFHR 2014 Bentham database, each using a different number of samples: 500, 1000, 2000, 4000, 6000, 8000, and all available samples. The results of these experiments are illustrated in Fig. 1, along with their corresponding WERs.

Our VED model, trained on 4000 random samples, achieved a WER of 8.8%, which is comparable to the best state-of-the-art model, A2IA, which achieved a WER of 8.6% using all available ICFHR 2014 data, including its synthetic dataset. These results suggest that our VED model does not require the entire dataset to achieve competitive performance and that it can be trained with a relatively small amount of data.

Figure 2 presents the results of an experiment conducted on the ICFHR 2016 Ratsprotokolle database. The experiment aims to assess the performance of our model under different training conditions. To this end, we perform six experiments, each using a different subset of the data: 500, 1000, 2000, 4000, 6000, and all available samples.

We found that our model can achieve competitive results with state-of-the-art models when using only half of the total data. However, when we increased the

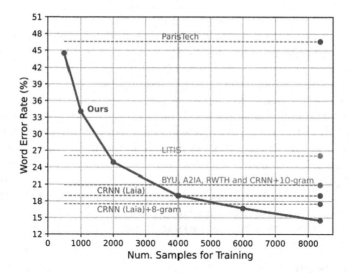

Fig. 2. Performance of the VED model versus the rest of the state-of-the-art models in the ICFHR 2016 Ratsprotokolle test dataset.

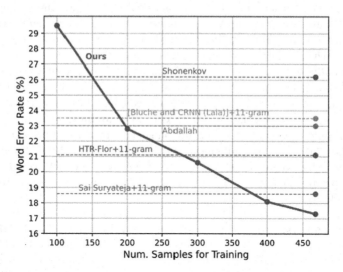

Fig. 3. Performance of the VED model versus the rest of the state-of-the-art models in the Saint Gall test dataset.

amount of data used to two-thirds of the total dataset, our model outperformed all other state-of-the-art models.

It is worth noting that the difference between VED and the other state-of-the-art models is less pronounced in Fig. 2 when compared to the experiment illustrated in Fig. 1. This may be because TrOCR weights are specialized for working with English text, and thus, the model might have an easier time working with this type of text.

The Saint Gall dataset presents a unique challenge, comprising only 468 training examples. To evaluate the efficacy of our model, we conduct experiments using a range of training data, specifically 100, 200, 300, 400, and all 468 lines. The results of these experiments are presented in Fig. 3.

We observe that when using only 300 training lines, our transformer model surpasses the performance of other state-of-the-art models, as evidenced by its lower WER. The only exception is the proposal by Sai Suryateja, which outperform our model in this particular scenario. However, when we increase the training data to 400 lines, our model achieved the most competitive results among the models evaluated.

The proposed VED model can outperform other state-of-the-art approaches, even when fine-tuned using a smaller dataset, demonstrating its flexibility and generalization capabilities. This suggests that the VED model can adapt to new languages and historical document tasks with limited training data.

5   Conclusion

The use of pre-trained models is a prevalent approach in deep learning. This practice is particularly crucial for HTR on historical documents due to the scarcity of large databases for training ANNs. The results of this work demonstrate that using the TrOCR weights initialization and an appropriate fine-tuning can surpass state-of-the-art results on three different datasets.

Additionally, we have also conducted an analysis of the fine-tuning process and found that using AdamW with a learning rate value of $5e^{-6}$ provides the best performance for adapting the model to recognize historical texts. Furthermore, our research shows that the proposed VED architecture can be adapted to different languages, and that focusing on specific modules during training can lead to competitive results while reducing the training time by half, particularly for English text recognition. Our analysis of the train dataset size also showed that appropriate fine-tuning enables the model to achieve competitive results with a smaller dataset.

Acknowledgements. Work partially supported by the Universitat Politècnica de València under the PAID-01-22 programme, by grant PID2020-116813RB-I00 funded by MCIN/AEI/ 10.13039/501100011033, by the support of valgrAI - Valencian Graduate School and Research Network of Artificial Intelligence and the Generalitat Valenciana, and co-funded by the European Union.

References

1. Abdallah, A., Hamada, M., Nurseitov, D.: Attention-based fully gated CNN-BGRU for Russian handwritten text. J. Imaging **6**, 141 (2020)
2. Augustin, E., Carré, M., Grosicki, E., Brodin, J.M., Geoffrois, E., Preteux, F.: RIMES evaluation campaign for handwritten mail processing. In: Proceedings of the International Workshop on Frontiers in Handwriting Recognition, pp. 231–235 (2006)

3. Barrere, K., Soullard, Y., Lemaitre, A., Coüasnon, B.: A light transformer-based architecture for handwritten text recognition. In: Proceedings of the Document Analysis Systems, pp. 275–290 (2022)

4. Bluche, T.: Deep neural networks for large vocabulary handwritten text recognition, Ph.D. thesis, Université Paris-Sud (2015)

5. Bluche, T., Messina, R.: Gated convolutional recurrent neural networks for multilingual handwriting recognition. In: Proceedings of the 14th IAPR International Conference on Document Analysis and Recognition, pp. 646–651 (2017)

6. Bunke, H., Roth, M., Schukat-Talamazzini, E.: Off-line cursive handwriting recognition using hidden markov models. Pattern Recogn. **28**, 1399–1413 (1995)

7. Devlin, J., Chang, M.W., Lee, K., Toutanova, K.: BERT: pre-training of deep bidirectional transformers for language understanding. arXiv preprint arXiv:1810.04805 (2018)

8. Dosovitskiy, A., et al.: An image is worth 16×16 words: transformers for image recognition at scale. arXiv preprint arXiv:2010.11929 (2020)

9. Fischer, A., Indermühle, E., Bunke, H., Viehhauser, G., Stolz, M.: Ground truth creation for handwriting recognition in historical documents. In: Proceedings of the 9th IAPR International Workshop on Document Analysis Systems, pp. 3–10 (2010)

10. Graves, A., Fernández, S., Gomez, F., Schmidhuber, J.: Connectionist temporal classification: labelling unsegmented sequence data with recurrent neural networks. In: Proceedings of the 23rd International Conference on Machine Learning, pp. 369–376 (2006)

11. He, K., Zhang, X., Ren, S., Sun, J.: Deep residual learning for image recognition. In: Proceedings of the IEEE Conference on Computer Vision and Pattern Recognition, pp. 770–778 (2016)

12. Hochreiter, S., Schmidhuber, J.: Long short-term memory. Neural Comput. **9**, 1735–1780 (1997)

13. Hu, J., Gek Lim, S., Brown, M.K.: Writer independent on-line handwriting recognition using an HMM approach. Pattern Recogn. **33**, 133–147 (2000)

14. Kamath, A., Singh, M., LeCun, Y., Synnaeve, G., Misra, I., Carion, N.: MDETR - modulated detection for end-to-end multi-modal understanding. In: Proceedings of the IEEE/CVF International Conference on Computer Vision, pp. 1780–1790 (2021)

15. Kang, L., Riba, P., Rusiñol, M., Fornés, A., Villegas, M.: Pay attention to what you read: non-recurrent handwritten text-line recognition. Pattern Recogn. **129**, 108766 (2022)

16. Li, M., et al.: TrOCR: transformer-based optical character recognition with pretrained models. arXiv preprint arXiv:2109.10282 (2021)

17. Liu, Y., et al.: RoBERTa: a robustly optimized BERT pretraining approach. arXiv preprint arXiv:1907.11692 (2019)

18. Liu, Z., et al.: Swin transformer: hierarchical vision transformer using shifted windows. In: Proceedings of the IEEE/CVF International Conference on Computer Vision, pp. 10012–10022 (2021)

19. Marti, U.V., Bunke, H.: The IAM-database: an English sentence database for offline handwriting recognition. Int. J. Doc. Anal. Recogn. **5**, 39–46 (2002)

20. Puigcerver, J.: Are multidimensional recurrent layers really necessary for handwritten text recognition? In: Proceedings of the 14th IAPR International Conference on Document Analysis and Recognition, pp. 67–72 (2017)

21. Sai Suryateja, S., Veerraju, P., Vijay Kumar Naidu, P., Ravi Kumar, C.V.: Improvement in efficiency of the state-of-the-art handwritten text recognition models. Turkish J. Comput. Math. Educ. **12**, 7549–7556 (2021)
22. Shonenkov, A., Karachev, D., Novopoltsev, M., Potanin, M., Dimitrov, D.: Stack-Mix and Blot augmentations for handwritten text recognition. arXiv preprint arXiv:2108.11667 (2021)
23. Simonyan, K., Zisserman, A.: Very deep convolutional networks for large-scale image recognition. arXiv preprint arXiv:1409.1556 (2014)
24. Singh, S.S., Karayev, S.: Full page handwriting recognition via image to sequence extraction. In: Proceedings of the Document Analysis and Recognition - International Conference on Document Analysis and Recognition, pp. 55–69 (2021)
25. de Sousa Neto, A.F., Bezerra, B.L.D., Toselli, A.H., Lima, E.B.: HTR-Flor: a deep learning system for offline handwritten text recognition. In: Proceedings of the 33rd Brazilian Symposium on Computer Graphics and Image Processing Conference on Graphics, Patterns and Images, pp. 54–61 (2020)
26. Szegedy, C., Vanhoucke, V., Ioffe, S., Shlens, J., Wojna, Z.: Rethinking the Inception architecture for computer vision. In: Proceedings of the IEEE Conference on Computer Vision and Pattern Recognition, pp. 2818–2826 (2016)
27. Sánchez, J.A., Romero, V., Toselli, A.H., Vidal, E.: ICFHR2014 competition on handwritten text recognition on Transcriptorium datasets (HTRtS). In: Proceedings of the 14th International Conference on Frontiers in Handwriting Recognition, pp. 785–790 (2014)
28. Sánchez, J.A., Romero, V., Toselli, A.H., Vidal, E.: ICFHR2016 competition on handwritten text recognition on the READ dataset. In: Proceedings of the 15th International Conference on Frontiers in Handwriting Recognition, pp. 630–635 (2016)
29. Sánchez, J.A., Romero, V., Toselli, A.H., Villegas, M., Vidal, E.: A set of benchmarks for handwritten text recognition on historical documents. Pattern Recogn. **94**, 122–134 (2019)
30. Tan, M., Le, Q.: EfficientNet: rethinking model scaling for convolutional neural networks. In: Proceedings of the 36th International Conference on Machine Learning, pp. 6105–6114 (2019)
31. Toselli, A.H., Vidal, E.: Handwritten text recognition results on the Bentham collection with improved classical N-Gram-HMM methods. In: Proceedings of the 3rd International Workshop on Historical Document Imaging and Processing, pp. 15–22 (2015)
32. Touvron, H., Cord, M., Douze, M., Massa, F., Sablayrolles, A., Jegou, H.: Training data-efficient image transformers & distillation through attention. In: Proceedings of the 38th International Conference on Machine Learning, pp. 10347–10357 (2021)
33. Vaswani, A., et al.: Attention is all you need. In: Proceedings of the Advances in Neural Information Processing Systems, pp. 5998–6008 (2017)
34. Wick, C., Zöllner, J., Grüning, T.: Transformer for handwritten text recognition using bidirectional post-decoding. In: Proceedings of the Document Analysis and Recognition - International Conference on Document Analysis and Recognition, pp. 112–126 (2021)
35. Zaheer, M., et al.: Big Bird: transformers for longer sequences. In: Proceedings of the Advances in Neural Information Processing Systems, pp. 17283–17297 (2020)

Fine-Tuning is a Surprisingly Effective Domain Adaptation Baseline in Handwriting Recognition

Jan Kohút[(⊠)] and Michal Hradiš

Faculty of Information Technology, Brno University of Technology,
Brno, Czech Republic
{ikohut,ihradis}@fit.vutbr.cz

Abstract. In many machine learning tasks, a large general dataset and a small specialized dataset are available. In such situations, various domain adaptation methods can be used to adapt a general model to the target dataset. We show that in the case of neural networks trained for handwriting recognition using CTC, simple fine-tuning with data augmentation works surprisingly well in such scenarios and that it is resistant to overfitting even for very small target domain datasets. We evaluated the behavior of fine-tuning with respect to augmentation, training data size, and quality of the pre-trained network, both in writer-dependent and writer-independent settings. On a large real-world dataset, fine-tuning on new writers provided an average relative CER improvement of 25% for 16 text lines and 50% for 256 text lines.

Keywords: Handwritten text recognition · OCR · Data augmentation · Fine-tuning

1 Introduction

In handwriting recognition, an OCR trained on a large and general dataset is often used to transcribe new writers. These writer-independent models provide good accuracy; however, when the writing style of the new writer differs from the general dataset, the transcription accuracy degrades and some form of domain adaptation may become necessary. In fact, we believe that some form of domain adaptation should be performed whenever a larger collection of consistent texts is to be transcribed. Although unsupervised strategies may be used, a couple of text lines from the target collection can be manually transcribed with minimal effort while often providing superior accuracy improvement.

In this paper, we explore domain adaption of large convolutional-recurrent CTC neural networks [5,10,29,32] from a large general dataset of mostly modern handwriting to specific documents written in various languages and scripts. Specifically, we fine-tune a general model to a small number of annotated text lines from a target document with practical strategies for early stopping. We

© The Author(s), under exclusive license to Springer Nature Switzerland AG 2023
G. A. Fink et al. (Eds.): ICDAR 2023, LNCS 14190, pp. 269–286, 2023.
https://doi.org/10.1007/978-3-031-41685-9_17

show that this simple approach is a surprisingly effective domain adaptation baseline, especially with suitable data augmentation, even for an extremely low amount of annotated target data. The proposed approach is stable, simple to implement, and provides consistent improvements in a wide range of situations. In fact, the fine-tuning approach is used in our text recognition web application PERO OCR[1] with great user feedback.

The specific contributions of this paper are as follows: (1) study of CTC network domain adaptation by fine-tuning on small datasets (1–256 text lines); (2) evaluation of possible variation of the improvement for multiple target documents written in different scripts and styles; (3) hyperparameter selection strategies suitable for fine-tuning in realistic scenarios; (4) convergence and overfitting analysis on small target datasets; (5) proposal of effective data augmentations and their evaluation; (6) strong evidence that fine-tuning is effective also in writer-dependent scenario (fine-tuning to documents or writer from the training set); (7) new dataset of 19 manuscripts suitable for domain adaptation experiments in various European languages and scripts with at least 512 hand-transcribed lines each.

2   Related Work

Modern handwritten text recognition approaches are either based on Connectionist Temporal Classification (CTC) [12] or are full seq2seq models with an autoregressive decoder. CTC models [5,10,29,32] are usually based on a stack of convolutional layers, followed by LSTM blocks [15]. Older seq2seq architectures [6,27] use encoders with similar architectures and decoders composed of LSTM blocks which are usually enhanced by various attention mechanisms. Lately, the recurrent layers were replaced by Transformers [36] blocks where information in a sequence is distributed purely by self-attention mechanism. Text recognition Transformers [3,9,16,22,37] similarly to other models use convolutional layers in the encoder. Based on the available literature, the mentioned architectures provide comparable transcription accuracy [9,16,27], while some works indicate that seq2seq model may prove to be superior as larger datasets become available [37].

Similar to our approach, several works [2,30,31,34] explored domain adaptation of CTC-based models by fine-tuning. However, the experiments did not explore the limits of such an approach (e.g. fine-tuning to less than a dozen lines), did not explore possible strategies for choosing hyperparameters, and did not explore the tendency of overfitting in these scenarios. Also, some of the findings and observed behaviors are not consistent (e.g. effect of data augmentation).

Aradillas et al. [2] experimented with domain adaptation from IAM [24] dataset to Washington [21] and Parzival [11] datasets, and between different partitions of the READ dataset [35]. Their conclusions are that it is better to fine-tune the network than selected network layers and that geometric augmentation [38] of the target domain degrades final accuracy. This is contrary to

---

[1] https://pero-ocr.fit.vutbr.cz.

our findings, where our data augmentation combined with fine-tuning brought substantial increases in accuracy. Soullard et al. [34] also experimented with fine-tuning on the READ dataset. Similar to us, they used cross-validation to estimate the optimal number of fine-tuning iterations. They used random rotation and scaling as data augmentation for both source and target domains. They also experimented with writers-specific language models which further improved results. Reul et al. [30] tested domain adaptation using fine-tuning on German medieval manuscripts in Gothic and Bastarda scripts. They utilized data augmentation in the form of several binarization strategies, both for source model training and fine-tuning. They used an ensemble of models combined with a voting strategy optimized with cross-validation. However, the stopping criterion of the fine-tuning was controlled by the testing datasets error. They observed that the closer the source model data was to the target data, the better the results after fine-tuning.

Bhunia et al. [4] approached domain adaptation to new writers as a meta-learning task, where the goal is to train a general model that can be effectively adapted to new domains with a single update and few words. They found that a single-shot adaptation of such a general model is superior to fine-tuning a model trained in a standard fashion. However, the experiments were restricted to word-level IAM and RIMES [13] datasets and 16 adaptation words images. Instead of training a general model, Kohut et al. [20] proposed a model with dedicated writer-dependent parameters which can handle multiple writers simultaneously. While adapting to a new writer, optimizing a new set of writer-dependent parameters brought worse performance than fine-tuning all parameters.

As speech and handwritten text recognition are closely related, we also present a short overview of domain adaptation from this field. Hank Liao [23] explored how a simple neural acoustic model may be adapted to speakers by fine-tuning the input layer, the output layer, or the entire network. Adapting the input layer was better than adapting the output layer, adapting all layers was even better. In order to overcome overfitting, some strategies [26,40] regularize the fine-tuning process by minimizing the divergence between the feature distributions of the original network and of the fine-tuned one, where the features might be taken from any layer. These approaches require evaluation of the original network while fine-tuning the new one. Dong Yu et al. [40] minimized the senone distributions divergence by adding the Kullback-Leibler term to the loss function, which is equivalent to constructing the fine-tuning ground truth as linear interpolating of the fine-tuned and original model senone distribution. Meng et al. [25,26] forced the distribution of hidden features to be close with an adversarial approach, which is equivalent to minimizing the Jensen-Shannon divergence.

In scenarios where no annotated data for the target domain are available, unsupervised approaches in the form of consistency regularization [1] and pseudo-labeling [18,28,39] may be used. In scenarios where both annotated and unannotated data are available, supervised and unsupervised approaches may be combined to get the best out of both worlds. For example, fine-tuning together with pseudo-labeling [7].

Fig. 1. Black, samples from the large general source CzechHWR dataset. ID with Color, representative words of 19 target writers. (Color figure online)

3   CzechHWR Dataset

We collected a large dataset of mainly 19th and 20th century Czech handwritten documents which, in our opinion, is a realistic example of a general dataset for training writer-independent models. The CzechHWR dataset was created from three main sources: documents processed by users of our text recognition web application PERO OCR, a collection of Czech letters transcribed by linguists [14], and Czech chronicles transcribed specifically for handwriting recognition. From the OCR application, we collected 295k text lines manually corrected by the users (after reviewing one or two pages from each user). The documents are mostly written in Czech modern cursive script, although a marginal part is written in German Kurrent and in several medieval scripts. The original sources are mainly military diaries, chronicles, letters, and notes. The Czech letters [14] consists of 2000 letters (87k text line annotations) from 20th century, mostly handwritten in Czech modern cursive with a limited amount of typeset ones. We manually annotated approximately 2 pages of 277 distinct Czech chronicles, resulting in 553 pages with 24k text lines.

The final CzechHWR dataset contains 406k annotated text lines and our estimate of distinct writers is 4.5k. The level of penmanship and readability differs, ranging from scribbles to calligraphy, although the tendency is towards fairly readable texts, see the left side of Fig. 1. The training (TRN) and testing (TST) subsets contain 379k and 5k lines. Due to the fact that writers with a small number of total lines are not sufficiently represented in TST, we created TST_W, which contains lines of all writers that have at least 20 lines in TRN. Table 1 shows the distribution of writers in the CzechHWR dataset according to the number of lines per writer with the respective amounts of lines for each subset.

We chose additional 19 writers from our PERO OCR web application as the small target datasets for fine-tuning[2], each writer is represented by at least 512 lines, and each line is at least 30 characters long. For each writer, an image of a

---

[2] https://pero.fit.vutbr.cz/handwriting_adaptation_dataset.

Table 1. The distribution of writers (NW) in the CzechHWR dataset according to the number of lines per writer (NWL) with the respective amounts of lines for each subset.

NWL	1–19	20–49	50–99	100–199	200–499	500–999	1000–	ALL
TRN	13k	79k	82k	43k	24k	16k	122k	379k
TST	169	1k	1.1k	566	287	198	1.7k	5k
TST$_W$	0	4.5k	6.2k	3.2k	2k	1.1k	5.4k	22.4k
NW	~1.1k	~2.3k	~1.2k	~322	~79	~21	~54	~5.1k

representative word is shown in Fig. 1, the colors match the colors in fine-tuning experiments graphs (Fig. 7 and Fig. 9), and the IDs match the IDs in Table 3. The scripts of these target writers range from some which are very similar to the majority of CzechHWR to some which are very different.

Our neural network architecture is similar to the state-of-the-art architectures for text recognition [5,10,29,32] trained with CTC loss [12]. It consists of a convolutional stage (CNN), inspired by the standard VGG arhitectures [33], and a parallel bidirectional LSTM [15] recurrent stage (RNN), which processes the input at multiple scales.

We trained the network with Adam [17] optimizer for 500k iterations up until convergence. We used polynomial warmup of a third order to gradually increase the learning rate from 0 to 3×10^{-4} in the first 10k iterations. At iterations 200k and 400k, we used the warmup again, but the learning rate maximums were 0.7×10^{-4} and 0.175×10^{-4}. The batch size was set to 32 and we used the B1C1G1M1 augmentation (see Sect. 4). The system reached CER of 0.51%, 2.17%, 2.26% on TRN, TST, and TST$_W$ subsets respectively, and the CER on augmented TRN subset was 2.4%. The distribution of test CER on the small target datasets, had a mean of 5.17%, a standard deviation of 4.82%, a minimum of 0.62%, and a maximum of 14.46%.

Architecture Details. The architecture is equivalent to our baseline TS-Net architecture [19], a more detailed description together with a detailed diagram can be found in the referenced work. CNN is a sequence of 4 convolutional blocks, where each block has 2 convolutional layers with numbers of output channels set to 64, 128, 256, and 512, respectively. All convolutional blocks except the last one are followed by a max pooling layer. The CNN subsamples an input text line image by a factor of 4 in width. The RNN consists of three parallel LSTM branches and one final LSTM layer. The branches process scaled variants of the input with two LSTM layers, the scaling factors are 1, 0.5, and 0.25. The outputs are upsampled back to the original dimension and their summation is processed by the final LSTM layer. Each LSTM layer is bidirectional and has a hidden feature size of 256 for both directions. The output of RNN is processed by a 1D convolutional layer with a kernel size of 3.

Table 2. Our augmentations as combinations of four basic ones: NoiseBlurGamma (B), Color (C), Geometry (G), and Masking (M). The number of dots specifies the level of augmentation intensity (1, 2, 3).

	NONE	B1	B1C1	B1G1	B1C1G1	B1C1G1M1	B2C1G1M1	B2C2G1M1	B2C1G2M1	B2C1G3M1	B2C2G2M1	B2C2G3M1	B2C2G2M2	B2C2G3M2
BlurNoiseGamma		•	•	•	•	•	••	••	••	••	••	••	••	••
Color			•		•	•	•	••	•	•	••	••	••	••
Geometry				•	•	•	•	•	••	•••	••	•••	••	•••
Masking						•	•	•	•	•	•	•	••	••

4   Data Augmentations

We chose various augmentations strategies to enlarge the amount of data artificially and to regularize the fine-tuning process. We used combinations of four basic augmentations: NoiseBlurGamma, Color, Geometry, and Masking. NoiseBlurGamma applies random motion blur, gauss noise, and gamma correction. Color randomly changes brightness, contrast, saturation, and hue changes. Geometry randomly adjusts text slant, horizontal scale, and vertical scale. Masking stands for random noise patch masking. The height of a noise patch is the same as the height of text line images, the width is chosen randomly up to the width of approximately two letters, and multiple masking patches can be applied to a single text line image. The intuition behind noise masking is to strengthen the language modeling capability of the system.

Table 2 shows the final augmentations in columns as combinations of the basic ones. If a basic augmentation is a part of the final one, the probability of applying it on the input is 0.2 for the NoiseBlurGamma, 0.333 for the Color, 0.66 for the Geometry, and 0.5 for the Masking, therefore all the augmentations allow the network to see the original text line images. The number of dots specifies the level of augmentation intensity, the higher the number, the greater the range of randomness in the respective image operations. There are two levels (1, 2) for NoiseBlurGamma (B), Color (C), and Masking (M) augmentations, and three levels (1, 2, 3) for Geometry (G) augmentation. We refer to the final augmentations with abbreviations e.g. augmentation B2C1G3M1 is a combination of NoiseBlurGamma level 2, Color level 1, Geometry level 3, and Masking level 1.

Figures 2, 3, and 4 show augmented versions of the top left text line image with NoiseBlurGamma, Color, and Geometry, respectively. For each level of augmentation intensity, there is a separate section of lines and only extreme samples are shown.

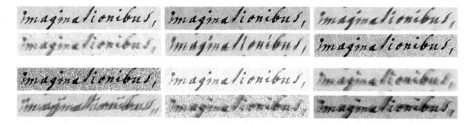

Fig. 2. Augmented versions of top left text line image with NoiseBlurGamma augmentation. Intensity 1 is shown in the top section and intensity 2 in the bottom one. Only the extreme samples of the distributions are shown.

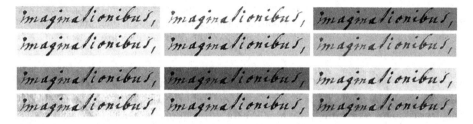

Fig. 3. Augmented versions of top left text line image with Color augmentation. Intensity 1 is shown in the top section and intensity 2 in the bottom one. Only the extreme samples of the distributions are shown. (Color figure online)

5 Writer-Independent Scenario

In writer-independent scenario experiments, we fine-tuned the source baseline model trained on the CzechHWR dataset to the 19 target writers. Experiments were based on writer fine-tuning runs. A writer fine-tuning run consisted of drawing 512 random lines of the respective target writer and splitting them into 256 testing and 256 adaptation ones. The adaptation lines were furthermore divided into 9 line clusters: 1, 2, 4, 8, 16, 32, 64, 128, and 256, where the numbers referred to the number of adaptation lines in them, and a smaller cluster was always a subset of all the larger ones. The numbers of fine-tuning iterations were 200, 200, 400, 800, 1000, 1500, 2000, 2500, and 3000 for 1, 2, 4, 8, 16, 32, 64, 128, and 256 adaptation lines, respectively. We run 10 fine-tuning runs for each writer resulting in total $19 \times 9 \times 10$ baseline model fine-tunings. Additionally, we run $19 \times 5 \times 10$ 4-fold cross-validations for the line clusters 16, 32, 64, 128, and 256, as cross-validation on less than 16 lines is not reliable.

We estimated the optimal number of fine-tuning iterations, with different estimation strategies (ET). The baseline estimation strategies were Last Iteration (L) and Oraculum (O). Last Iteration (L) returned the last/maximum iteration. Oraculum (O) returned the iteration of minimal CER on testing lines. As there are no testing lines in practice, we experimented with estimation strategies based on 4-fold cross-validation computed on adaptation lines. Minimum

Fig. 4. Augmented versions of top left text line image with Geometry augmentation. Intensity 1 is shown in the top section, intensity 2 in the middle one, and intensity 3 in the bottom one. Only the extreme samples of the distributions are shown.

Iteration Average Across Chunks (A) smoothed each of 4 cross-validation test loss curves with window size 4, averaged the smoothed loss curves, and returned the iteration of the minimum loss. Mean Minimum Iteration Per Chunk (M) smoothed each of 4 cross-validation test loss curves with window size 4, took the iteration of the minimum loss per each smoothed loss curve, and returned the mean of these iterations. Max Minimum Iteration Per Chunk (X) estimated the optimal fine-tuning iteration in the same way as M, but at the end, instead of mean, returned the maximum. Note, as we tested every 20 iterations, the window size of 4 spanned across 80 iterations.

We also experimented with a scenario, where there are multiple target writers with testing lines available and we want to assume a static number of fine-tuning iterations for a new target writer for which we do not have any testing lines. The optimal static iteration was estimated on the writers with testing lines as the iteration of the minimum value of the writers' fine-tuning test curve. The writers' fine-tuning test curve was calculated as an average of writer fine-tuning curves, which were normalized by their minimums. Each writer's fine-tuning curve was calculated as an average of 10 fine-tuning test loss curves (10 fine-tuning runs), which were smoothed with a window size of 4. We refer to this estimation strategy as Static Iteration (S). To compare it to others, we evaluate it in a 1 to N-1 manner, where N is the number of all target writers.

As the estimation based on loss often underestimated the number of optimal iterations (see Fig. 9), we also experimented with simple modifications of X and S, denoted as X_R and S_R, which multiplied the estimated iterations by a positive factor R of 1.5 and 3. Estimations of these strategies were limited by the actual number of fine-tuning iterations.

Choosing Augmentation for Fine-Tuning. Figure 5 compares the performance of models fine-tuned with different augmentations (see Sect. 4) to the performance of the baseline model, which served as the starting point for the

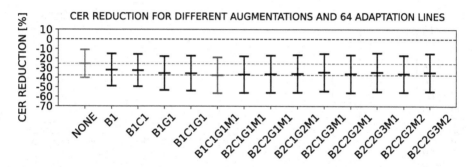

Fig. 5. The performance of models fine-tuned with different augmentations expressed as a relative reduction of the baseline model test CER. The means and the standard deviations represent the target writer distribution.

fine-tuning. The comparison is expressed as a relative reduction of the baseline model test CER and it is given by:

$$\frac{F - B}{B},\tag{1}$$

where B is the test CER of the baseline model and F is the test CER of its fine-tuned variant. Due to the high number of augmentations, we run fine-tuning runs just with the line cluster 64. $X_{1.5}$ was used as the estimation strategy for choosing the fine-tuning iteration of minimal CER. For each augmentation, the mean and the standard deviation of test CER reductions on all 19 target writers are shown. *In all experiments*, test CER reduction on a writer is the mean of test CER reductions across all 10 fine-tuning runs. Fine-tuned models consistently outperformed the baseline model on most of the target writers and they did not worsen the accuracy on any. NoiseBlurGamma, Geometry, and Masking augmentations improved the performance significantly, whereas Color augmentation had almost no effect. The higher levels of augmentation intensity (2, 3), did not bring any essential variations in performance. Even though the Geometry augmentations affected the writing style significantly (see Fig. 4), they consistently brought better performance for the fine-tuning. On average, for line cluster 64, fine-tuning without any augmentation (NONE) reduced the baseline CER by 25%, while combinations of all the basic augmentations reduced the CER by an additional 10%. Fine-tuning with data augmentations brought larger standard deviations across target writers. Furthermore, we only experiment with augmentations NONE and B1C1G1M1.

Pre-trained Quality of the Baseline Model. Figure 6 compares fine-tuning of baseline models trained for different amounts of iterations on the CzechHWR dataset. The performance is expressed as a relative reduction of the fully-trained baseline model test CER (500k). The boxplots represent the target writer distribution. As with the previous experiment, we run the fine-tuning runs only for line

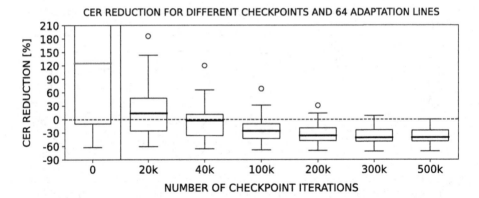

Fig. 6. Fine-tuning of baseline models trained for different amounts of iterations on the CzechHWR dataset. The performance is expressed as a relative reduction of the fully-trained baseline model (500k) test CER. The boxplots represent the target writer distribution. See the text for a description of the model fine-tuned from scratch (0).

cluster 64 and estimated the optimal fine-tuning iterations with $X_{1.5}$. The more well-trained the baseline model, the greater and more stable the performance across the target writers.

The architecture fine-tuned from scratch is almost identical to ours (described in Sect. 3), where the only essential difference is that the convolutional layers are initialized from VGG [33] architecture trained on ImageNet [8]. The fine-tuning was done on line cluster 256, for 10k iterations, and we estimated the number of optimal fine-tuning iterations with the Oraculum strategy (no cross-validation involved). In comparison to the well-trained baseline model, the performance was far worse for most of the writers, although there were exceptions among writers whose writing styles were not sufficiently represented in the CzechHWR dataset. Even though that VGG was trained on four times more lines and the Oraculum estimating strategy was used, it is surprising, that for writers with German Kurrent and Ghotic script, it evened out the fine-tuned well-trained baseline model.

Fine-Tuning Runs. Figure 7 shows relative test CER reductions of the baseline model test CER on the target writers for complete fine-tuning runs, with and without augmentation. The estimation strategy for choosing the fine-tuning iteration with minimal test CER was S_3 for line clusters 1, 2, 4, 8 and $X_{1.5}$ for line clusters 16, 32, 64, 128, and 256. Each writer is represented by a different color (see Fig. 1 for images of representative words). Static iteration setup improved the performance even for 1 adaptation line, however, it overfitted two writers for all cluster lines. Cross-validation setups improved the performance on all cluster lines, except for the same two writers in the case of line clusters 16 and 32.

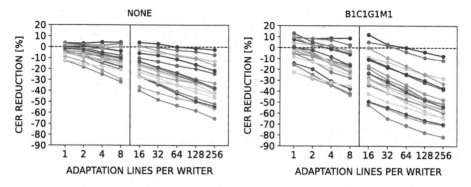

Fig. 7. Relative reductions of the baseline model test CER on the target writers for complete fine-tuning runs, with and without augmentation. The estimation strategy for choosing the fine-tuning iteration with minimal test CER was S_3 for 1–8 and $X_{1.5}$ for 16–256 adaptation lines.

Table 3. Test CER (in %) of the baseline model (0) and test CER after fine-tuning with B1C1G1M1 augmentation on 16 and 64 adaptation lines on all target writer datasets (the writer ID in the header corresponds to the ID in Fig. 1).

	0	1	2	3	4	5	6	7	8	9	10	11	12	13	14	15	16	17	18
0	6.5	2.3	2.5	1.0	12.7	4.1	1.8	10.2	1.4	1.3	11.0	9.4	1.2	2.0	2.1	14.5	0.6	0.6	13.2
16	3.3	2.6	1.3	1.0	8.3	3.7	1.6	4.9	1.2	1.0	7.4	6.8	1.0	1.4	1.2	12.9	0.4	0.6	10.1
64	2.4	2.3	1.0	1.0	6.6	3.1	1.5	2.9	0.9	0.8	5.5	5.2	0.8	1.2	1.0	11.0	0.3	0.5	7.8

Generally, the more adaptation lines, the greater the performance. Fine-tuning with B1C1G1M1 augmentation consistently outperformed fine-tuning without any augmentation, although there is a higher risk of worsening the performance when fine-tuning with smaller amounts of lines. The distribution across the target writers is Gaussian-like, while the augmentation shifts the mean, and stretches the standard deviation. The largest CER reductions (up to 82%) were achieved for distinct yet to some extent source-like writing scripts such as Kurrent or Czech block letters. The average CER reductions (up to 60%) were achieved for vastly different scripts such as Ghotic, and for harder-to-read source-like scripts. For easy-to-read source-like scripts, smaller CER reductions (up to 10%) were achieved for larger amounts of adaptation lines, whereas over-fitting led to worse performance (up to 15%) for smaller amounts of lines.

Table 3 shows the test CER of the baseline model (0) and the test CER after fine-tuning with B1C1G1M1 augmentation on 16 and 32 adaptation lines on all target writer datasets (the writer ID in the header corresponds to the ID in Fig. 1).

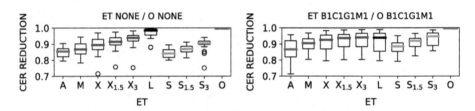

Fig. 8. Compares different estimation strategies (ET) for choosing the fine-tuning iterations with minimal test CER. The performance is expressed as a normalized relative reduction of the baseline test CER.

Figure 8 compares different estimation strategies for choosing the optimal number of fine-tuning iterations. An estimation strategy is shown as a distribution across the respective normalized writers' CER reductions, where the normalization is done across the estimate strategy dimension with the Oracle strategy, and the line cluster dimension is subsequently aggregated by mean. We calculated these statistics only on line clusters 16, 32, 64, 128, and 256, and we omitted three writers, as the normalization was not possible because some fine-tuned models worsen the performance of the baseline. The best estimation strategy for fine-tuning without adaptation is L. Generally, the estimation strategy which provides more fine-tuning iterations is better, we give our explanation of this phenomenon while discussing the fine-tuning curves. For fine-tuning with augmentation $X_{1.5}$ and X_3 brought the largest CER reductions among cross-validation approaches, and S_3 among the static iteration approaches.

For S_3, the ratios between the estimated number of fine-tuning iterations and the number of adaptation lines were 180, 90, 60, 38, 33, 30, 24, 15, and, 9, for 1, 2, 4, 8, 16, 32, 64, 128, and 256 lines, respectively. The difference between the ratios of different writers was negligible. This suggests that there is a fixed relation between the number of optimal fine-tuning iterations and the number of adaptation lines, which on average outperforms the cross-validation approaches. The target writer distribution for the L strategy is more skewed towards the poorer CER reductions, note that L can be seen as another variant of S.

Fine-Tuning Curves. To get a deeper insight into the fine-tuning process with B1C1G1M1 augmentation, we show aggregations of fine-tuning curves in Fig. 9. Line cluster fine-tuning curves for each line cluster in the left column graphs were computed with the Static Iteration (S) estimation strategy on all 19 target writers, for the CER graphs, the calculation is based on the test CER fine-tuning curves. The graphs in the right column show writer fine-tuning curves for line cluster 16 before aggregation, note that the colors match the colors in Fig. 1 and Fig. 7.

On average, for all line clusters, the fine-tuning curves had a U-like shape and the minimum test CER was always achieved later than the respective minimum test loss. For line cluster 16, the amount of fine-tuning iterations to achieve the optimal CER reduction varied among different writers, and some of them (darker

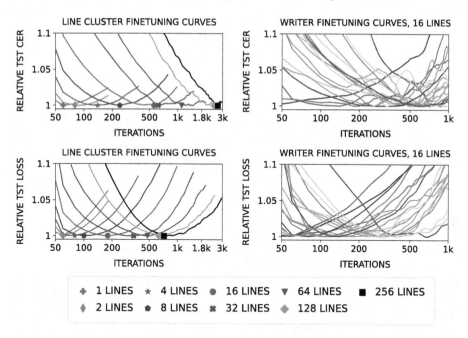

Fig. 9. Fine-tuning curves for B1C1G1M1 augmentation.

brown and blue) suffered from overtraining, this can also be seen in Fig. 7. The CER fine-tuning curves were smooth and had a negative slope up until the loss curves started to grow more dramatically, from this point they were prone to high noise. This phenomenon is more drastic for cross-validation, especially for a lower amount of adaptation lines. Therefore, the estimation of the minimal test CER fine-tuning iteration based on loss fine-tuning curves should be derived from iterations of a slightly uncertain region behind the minimum. Returning to the left graphs, we can see that on average the optimal level of uncertainty is higher for higher amounts of adaptation lines, which is the motivation behind X_R and S_R estimation strategies.

By inspecting the fine-tuning curves for the fine-tuning without augmentation, we found out that the baseline model quickly overfitted the adaptation lines. The loss on test lines got to the minimum around the first 100 iterations and started to increase afterward. Surprisingly, at this point, the CER saturated or even kept getting slightly better for the remaining iterations. This phenomenon might have been caused by the fact that after the model overfitted the adaptation lines the training loss was minimal and the subsequent iterations produced only slightly less confident models which turned out to be more accurate. This explains why using the L estimation strategy brought the best CER reductions.

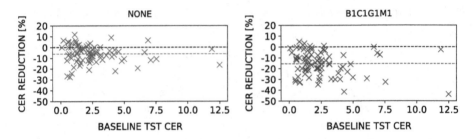

Fig. 10. Relative test CER reductions for fine-tuning in the writer-dependent scenario on 78 writers from the CzechHWR dataset together with the baseline test CER.

5.1 Writer-Dependent Scenario

This section describes fine-tuning of the baseline model on writers from the source CzechHWR dataset. To cover different numbers of training lines, we chose one random writer per each group of writers with the same number of lines, which resulted in 315 writers. The baseline model was fine-tuned for 1000, 2000, 3000, and 6000 iterations for writers with the number of lines more than or equal to 1, 100, 500, and 1000. To eliminate noise bias from the result statistics, we estimated a function that took the number of writer training lines as the input and output the number of fine-tuning iterations. The estimation was done as a polynomial fitting on a dataset of $(N, I)_W$ tuples, where W was the fine-tuned writer, N was the number of its training lines, and I was the fine-tuning iteration with minimal test CER. Polynomial fitting with additional parameters in the form of train and test loss/CER did not bring any improvements.

Figure 10 shows the relative CER reductions for 78 writers together with the baseline test CER. We do not show results for writers with less than 500 training lines, due to the insufficient number of testing lines in the CzechHWR dataset (see Table 1). The colored dashed lines are the means of the writers' CER reductions. Fine-tuning without augmentation was prone to overfitting but still brought a 6% CER reduction on average. Fine-tuning with B1C1G1M1 augmentation almost eliminated overfitting and brought a 15% CER reduction on average. These results show that our baseline model was not able to handle a vast number of writing styles present in the CzechHWR dataset, even though it was well-trained and for the last 100k iterations with a small learning rate did not bring any further improvements. We believe that fine-tuning in this writer-dependent scenario allows the model to adapt to otherwise ambiguous aspects of the text – that it is not just due to a low modeling capacity of the model with respect to the size and variability of the general dataset. An ensemble of writer-dedicated models, where each of these models would be a fine-tuned variant of the shared baseline model, seems to be a reasonable baseline for handwritten text recognition in the writer-dependent scenario.

6   Conclusion

Our experiments show that fine-tuning is a very efficient domain adaptation method for handwritten text recognition. In the writer-independent scenario, it improved the recognition accuracy of the baseline model by 20% to 45% relatively, for 16–256 adaptation lines, when choosing the number of fine-tuning iterations by cross-validation. We further showed that it is possible to estimate a fixed ratio between the number of fine-tuning iterations and the number of adaptation text lines, which outperformed the cross-validation technique. This indicates that in live handwriting recognition applications, this mapping can be estimated for a specific general model on a small number of exemplar documents and that fine-tunning for new documents can be performed with a predefined number of iterations conditioned only on the amount of available target data without risking overfitting or accuracy degradation. This fine-tuning with this fixed stopping criterion works even for a very small number of text lines. In our experiments, the improvements for 2–8 text lines were 10% to 20% on average, and even a single adaption text line without augmentations improved transcription accuracy by 5% on average. Fine-tuning was surprisingly resistant to overfitting even for an extremely low number of text lines and the region of an optimal number of fine-tuning iterations proved to be wide and easy to localize. Data augmentation proved to be an important component of the fine-tuning process with a combination of geometry, blur, and noise masking providing 1.5× larger improvement over fine-tuning without any augmentation.

Surprisingly, the fine-tuning was effective also on documents from the original training set (in the writer-dependent scenario) where the observed improvement reached 15%.

The experimental result reported in this paper has strong practical implications for handwriting recognition applications. The conclusion is that this type of fine-tuning should be always used and that it is safe to do so. We have already implemented this strategy in our text recognition web application PERO OCR[3], where users can repeatedly transcribe a document, where each transcription first fine-tunes the selected model to already corrected lines in the document.

We have performed preliminary experiments with Transformer-based sequence-to-sequence models. They tend to overfit the adaptation text lines while the noise masking augmentation makes the overfitting even worse. We presume that such behavior is due to the autoregressive decoder which learns the text of the adaptation lines. We are looking at several methods how to mitigate this behavior including self-training with dedicated language models, constraining the change of the model, and others.

Acknowledgment. This work has been supported by the Ministry of Culture Czech Republic in NAKI III project semANT - Semantic Document Exploration (DH23P03OVV060).

---

[3] https://pero-ocr.fit.vutbr.cz.

References

1. Aberdam, A., Ganz, R., Mazor, S., Litman, R.: Multimodal semi-supervised learning for text recognition. arXiv preprint arXiv:2205.03873 (2022)
2. Aradillas, J.C., Murillo-Fuentes, J.J., Olmos, P.M.: Boosting offline handwritten text recognition in historical documents with few labeled lines. IEEE Access **9**, 76674–76688 (2021)
3. Barrere, K., Soullard, Y., Lemaitre, A., Coüasnon, B.: A light transformer-based architecture for handwritten text recognition. In: Uchida, S., Barney, E., Eglin, V. (eds.) DAS 2022. LNCS, vol. 13237, pp. 275–290. Springer, Cham (2022). https://doi.org/10.1007/978-3-031-06555-2_19
4. Bhunia, A.K., Ghose, S., Kumar, A., Chowdhury, P.N., Sain, A., Song, Y.Z.: MetaHTR: towards writer-adaptive handwritten text recognition. In: Proceedings of the IEEE/CVF Conference on Computer Vision and Pattern Recognition, pp. 15830–15839 (2021)
5. Bluche, T., Messina, R.: Gated convolutional recurrent neural networks for multilingual handwriting recognition. In: 2017 14th IAPR International Conference on Document Analysis and Recognition (ICDAR), vol. 01, pp. 646–651 (2017). https://doi.org/10.1109/ICDAR.2017.111
6. Chowdhury, A., Vig, L.: An efficient end-to-end neural model for handwritten text recognition. arXiv preprint arXiv:1807.07965 (2018)
7. Das, D., Jawahar, C.V.: Adapting OCR with limited supervision. In: Bai, X., Karatzas, D., Lopresti, D. (eds.) DAS 2020. LNCS, vol. 12116, pp. 30–44. Springer, Cham (2020). https://doi.org/10.1007/978-3-030-57058-3_3
8. Deng, J., Dong, W., Socher, R., Li, L.J., Li, K., Fei-Fei, L.: Imagenet: a large-scale hierarchical image database. In: 2009 IEEE Conference on Computer Vision and Pattern Recognition, pp. 248–255. IEEE (2009)
9. Diaz, D.H., Qin, S., Ingle, R.R., Fujii, Y., Bissacco, A.: Rethinking text line recognition models. CoRR abs/2104.07787 (2021). https://arxiv.org/abs/2104.07787
10. Dutta, K., Krishnan, P., Mathew, M., Jawahar, C.V.: Improving CNN-RNN hybrid networks for handwriting recognition. In: 2018 16th International Conference on Frontiers in Handwriting Recognition (ICFHR), pp. 80–85 (2018). https://doi.org/10.1109/ICFHR-2018.2018.00023
11. Fischer, A., et al.: Automatic transcription of handwritten medieval documents. In: 2009 15th International Conference on Virtual Systems and Multimedia, pp. 137–142. IEEE (2009)
12. Graves, A., Fernández, S., Gomez, F., Schmidhuber, J.: Connectionist temporal classification: labelling unsegmented sequence data with recurrent neural networks. In: Proceedings of the 23rd International Conference on Machine Learning, pp. 369–376 (2006)
13. Grosicki, E., Abed, H.E.: ICDAR 2009 handwriting recognition competition. In: 2009 10th International Conference on Document Analysis and Recognition, pp. 1398–1402 (2009). https://doi.org/10.1109/ICDAR.2009.184
14. Hladká, Z.: 111 let českého dopisu v korpusovém zpracování (2013)
15. Hochreiter, S., Schmidhuber, J.: Long short-term memory. Neural Comput. **9**(8), 1735–1780 (1997)
16. Kang, L., Riba, P., Rusiñol, M., Fornés, A., Villegas, M.: Pay attention to what you read: non-recurrent handwritten text-line recognition. Pattern Recogn. **129**, 108766 (2022)

17. Kingma, D.P., Ba, J.: Adam: a method for stochastic optimization. In: Bengio, Y., LeCun, Y. (eds.) ICLR 2015, San Diego, CA, USA, 7–9 May 2015, Conference Track Proceedings (2015)

18. Kišš, M., Beneš, K., Hradiš, M.: AT-ST: self-training adaptation strategy for OCR in domains with limited transcriptions. In: Lladós, J., Lopresti, D., Uchida, S. (eds.) ICDAR 2021. LNCS, vol. 12824, pp. 463–477. Springer, Cham (2021). https://doi.org/10.1007/978-3-030-86337-1_31

19. Kohút, J., Hradiš, M.: TS-net: OCR trained to switch between text transcription styles. In: Lladós, J., Lopresti, D., Uchida, S. (eds.) ICDAR 2021. LNCS, vol. 12824, pp. 478–493. Springer, Cham (2021). https://doi.org/10.1007/978-3-030-86337-1_32

20. Kohút, J., Hradiš, M., Kišš, M.: Towards writing style adaptation in handwriting recognition (2023)

21. Lavrenko, V., Rath, T.M., Manmatha, R.: Holistic word recognition for handwritten historical documents. In: First International Workshop on Document Image Analysis for Libraries. Proceedings, pp. 278–287. IEEE (2004)

22. Li, M., et al.: TROCR: transformer-based optical character recognition with pre-trained models. In: AAAI 2023, February 2023. https://www.microsoft.com/en-us/research/publication/trocr-transformer-based-optical-character-recognition-with-pre-trained-models/

23. Liao, H.: Speaker adaptation of context dependent deep neural networks. In: 2013 IEEE International Conference on Acoustics, Speech and Signal Processing, pp. 7947–7951. IEEE (2013)

24. Marti, U.V., Bunke, H.: The Iam-database: an English sentence database for offline handwriting recognition. Int. J. Doc. Anal. Recogn. 5, 39–46 (2002)

25. Meng, Z., Gaur, Y., Li, J., Gong, Y.: Speaker adaptation for attention-based end-to-end speech recognition. CoRR abs/1911.03762 (2019), http://arxiv.org/abs/1911.03762

26. Meng, Z., Li, J., Gong, Y.: Adversarial speaker adaptation. CoRR abs/1904.12407 (2019). http://arxiv.org/abs/1904.12407

27. Michael, J., Labahn, R., Grüning, T., Zöllner, J.: Evaluating sequence-to-sequence models for handwritten text recognition. In: 2019 International Conference on Document Analysis and Recognition (ICDAR), pp. 1286–1293. IEEE (2019)

28. Nagai, A.: Recognizing Japanese historical cursive with pseudo-labeling-aided CRNN as an application of semi-supervised learning to sequence labeling. In: 2020 17th International Conference on Frontiers in Handwriting Recognition (ICFHR), pp. 97–102. IEEE (2020)

29. Puigcerver, J.: Are multidimensional recurrent layers really necessary for handwritten text recognition? In: 2017 14th IAPR International Conference on Document Analysis and Recognition (ICDAR), vol. 01, pp. 67–72 (2017). https://doi.org/10.1109/ICDAR.2017.20

30. Reul, C., Tomasek, S., Langhanki, F., Springmann, U.: Open source handwritten text recognition on medieval manuscripts using mixed models and document-specific finetuning. In: Uchida, S., Barney, E., Eglin, V. (eds.) DAS 2022. LNCS, vol. 13237, pp. 414–428. Springer, Cham (2022). https://doi.org/10.1007/978-3-031-06555-2_28

31. Reul, C., Wick, C., Nöth, M., Büttner, A., Wehner, M., Springmann, U.: Mixed model OCR training on historical Latin script for out-of-the-box recognition and finetuning. In: The 6th International Workshop on Historical Document Imaging and Processing, pp. 7–12 (2021)

32. Shi, B., Bai, X., Yao, C.: An end-to-end trainable neural network for image-based sequence recognition and its application to scene text recognition. CoRR abs/1507.05717 (2015). http://arxiv.org/abs/1507.05717

33. Simonyan, K., Zisserman, A.: Very deep convolutional networks for large-scale image recognition. arXiv preprint arXiv:1409.1556 (2014)

34. Soullard, Y., Swaileh, W., Tranouez, P., Paquet, T., Chatelain, C.: Improving text recognition using optical and language model writer adaptation. In: 2019 International Conference on Document Analysis and Recognition (ICDAR), pp. 1175–1180. IEEE (2019)

35. Strauß, T., Leifert, G., Labahn, R., Hodel, T., Mühlberger, G.: ICFHR 2018 competition on automated text recognition on a read dataset. In: 2018 16th International Conference on Frontiers in Handwriting Recognition (ICFHR), pp. 477–482. IEEE (2018)

36. Vaswani, A., et al.: Attention is all you need. In: Advances in Neural Information Processing Systems, vol. 30 (2017)

37. Wick, C., Zöllner, J., Grüning, T.: Transformer for handwritten text recognition using bidirectional post-decoding. In: Lladós, J., Lopresti, D., Uchida, S. (eds.) ICDAR 2021. LNCS, vol. 12823, pp. 112–126. Springer, Cham (2021). https://doi.org/10.1007/978-3-030-86334-0_8

38. Wigington, C., Stewart, S., Davis, B., Barrett, B., Price, B., Cohen, S.: Data augmentation for recognition of handwritten words and lines using a CNN-LSTM network. In: 2017 14th IAPR International Conference on Document Analysis and Recognition (ICDAR), vol. 1, pp. 639–645. IEEE (2017)

39. Wolf, F., Fink, G.A.: Self-training of handwritten word recognition for synthetic-to-real adaptation. In: 2022 26th International Conference on Pattern Recognition (ICPR), pp. 3885–3892. IEEE (2022)

40. Yu, D., Yao, K., Su, H., Li, G., Seide, F.: Kl-divergence regularized deep neural network adaptation for improved large vocabulary speech recognition. In: 2013 IEEE International Conference on Acoustics, Speech and Signal Processing, pp. 7893–7897. IEEE (2013)

Incremental Teacher Model with Mixed Augmentations and Scheduled Pseudo-label Loss for Handwritten Text Recognition

Masayuki Honda[1] , Hung Tuan Nguyen[1]([⊠]) , Cuong Tuan Nguyen[1] ,
Cong Kha Nguyen[2] , Ryosuke Odate[2] , Takashi Kanemaru[2] ,
and Masaki Nakagawa[1]

[1] Tokyo University of Agriculture and Technology, 2-24-16 Naka-Cho, Koganei, Tokyo, Japan
`s183611u@st.go.tuat.ac.jp`, `{fx7297,fx4102}@go.tuat.ac.jp`,
`nakagawa@cc.tuat.ac.jp`
[2] Hitachi Ltd., Tokyo, Japan
`{cong_kha.nguyen.zz,ryosuke.odate.qs,`
`takashi.kanemaru.kf}@hitachi.com`

Abstract. We propose a training framework for deep neural network-based handwritten text recognizers using both labeled and unlabeled data. The proposed framework is a semi-supervised learning (SSL) framework based on Mixed Augmentations and Scheduled Pseudo-Label loss. Mixed Augmentations provide weakly and strongly transformed variants from each original sample so that the pseudo-label loss is computed between these two variants. The Scheduled Pseudo-Label loss is used to gradually include the pseudo-label loss into the optimizer to avoid the negative effect of incorrect pseudo labels. First, a student model is pre-trained by labeled samples and used to initiate a teacher model. Subsequently, the teacher model predicts a pseudo label from every weakly transformed variant. On the other hand, the student model is trained using the Scheduled Pseudo-Label loss. Next, the teacher model is incrementally updated using the student model. Finally, it is used to evaluate. We term the framework Incremental Teacher Model. The proposed framework was applied to four architectures of distinct handwriting recognizers. For almost every architecture, the recognizer trained by our method outperforms those trained by well-known SSL methods, namely Mean Teacher, Pseudo-Labeling, and FixMatch, evaluated using different ratios of labeled training samples on the IAM handwriting database.

Keywords: Semi-Supervised Learning · Mixed augmentations · Scheduled Pseudo-Label loss · Training framework · Handwriting recognition

1 Introduction

Deep neural networks (DNNs) have been extensively studied in the past few decades and employed in multiple pattern recognition tasks owing to their high performance when large labeled datasets are available [1–3]. For handwriting recognition, DNNs

G. A. Fink et al. (Eds.): ICDAR 2023, LNCS 14190, pp. 287–301, 2023.
https://doi.org/10.1007/978-3-031-41685-9_18

have achieved increasing recognition accuracy [4–6] on many benchmark databases [7–11] of Latin, Arabic, Chinese, Indic, and Japanese scripts. These models require more labeled samples for training when the number of parameters is high [12, 13]. On the other hand, they do not take advantage of unlabeled samples. Unlabeled samples are easier to collect in large quantities and at a lower cost than labeled samples. For example, the two new databases of handwritten answers, namely SCUT-EPT [14] and NCUEE-HJA [15], have 40,000 labeled sentences and more than 190,000 unlabeled sentences, respectively. Only a few studies have utilized unlabeled samples for handwritten text recognition [16, 17]. Thus, we aim to create a generalized learning framework for any handwriting recognizer that satisfies two criteria (i) Trainable with as less labeled data as possible; (ii) Utilizable for unlabeled and labeled data.

Thus far, semi-supervised learning (SSL) methods have been established and developed to address the use of unlabeled data. Since the early deep learning era, Pseudo-Labeling has been proposed and extended for image classification tasks [18]. In the Pseudo-Labeling method, a pre-trained model is initialized using a small, labeled subset and is then used to predict the pseudo labels of a large unlabeled subset. Next, the unlabeled subset with the corresponding pseudo labels is used to re-train the model. Generally, Pseudo-Labeling is similar to the teacher-student training framework, where the initialized supervised pre-trained model is a teacher model while the training model is a student model. The teacher model provides pseudo labels for training a student model with unlabeled input samples. Thus, the handwriting recognizer is optimized on both the labeled and unlabeled samples using features from the unlabeled samples.

In fact, the Pseudo-Labeling method depends on the quality of the pseudo labels, as erroneous predictions often appear early in the training process [19]. Handwritten text recognition (HTR) is considered a sequential labeling task requiring a sequence of character predictions. It is difficult to employ Pseudo-Labeling for training HTR because misrecognized labels might lead to incorrect predictions in the rest of the sequence. Hence, we propose a framework, termed the Incremental Teacher Model, to gradually extend the effect of pseudo labels during the training process. The teacher model is incrementally updated after each epoch by its student model.

We have not focused on developing a novel handwriting recognizer in this work. Instead, we employ the proposed framework to train existing handwriting recognition architectures: Convolutional Recurrent Neural Network (CRNN) with connectionist temporal classification (CTC) [20], Attention-based Encoder-Decoder (AED) [21], and Self-Attention-based CRNN with CTC [22]. These handwriting recognition architectures utilize unlabeled data using the proposed SSL framework.

The rest of this paper is organized as follows: Sect. 2 reviews related studies on SSL methods. Section 3 presents our proposed framework with Mixed Augmentations and Scheduled Pseudo-Label loss. Section 4 presents the experiments and results of the proposed framework applied to different HTR architectures. In Sect. 5, we draw conclusions.

2 Related Works

Although DNNs have been continuously improved for higher performance, they strongly depend on large-scale labeled datasets for training. In fact, it is difficult to efficiently adapt them to new tasks, such as recognizing unseen or seen characters written in a new writing style. During the last few years, meta-learning has been widely studied to make DNNs to learn new patterns with a few training samples [23]. It is a wide field of machine learning that includes few-shot learning, one-shot learning, and domain adaptation [17, 24, 25]. Among them, the domain adaptation (DA) methods, particularly methods following the SSL approach, are promising to generalize a handwriting recognizer using both labeled and unlabeled data. Specifically, we focus on the inner-domain handwriting recognition task where training and testing sets have the same categories.

Two main approaches are studied based on these assumptions: consistency regularization and entropy minimization. Consistency regularization is mainly based on data augmentation and weight noise by dropout, as small changes should not significantly affect the prediction made by the network. The consistency loss measures the distance between the network predictions, with and without augmentations for input samples. Some well-known methods in this approach are the Π-Model [26], Temporal Ensembling [26], Mean Teacher [27], and Virtual Adversarial Training (VAT) [28].

The Π-Model employs stochastic augmentation to provide minor changes in each input sample. It also applies dropout to make noise on the weights of a given DNN model. The distance between the predictions of the original sample (without either augmentation or dropout) and its variant (with both augmentation and dropout) is then minimized. While the Π-Model requires two executions of the network for every sample, Temporal Ensembling keeps and updates the ensembled prediction of every sample during the training process; thus, its computation cost is lower than that of the Π-Model. Mean Teacher focuses on updating the ensembled model instead of tracing the ensembled patterns so that it helps converge faster than Temporal Ensembling. On the other hand, VAT approximates how augmentations to be employed on each input sample affect the output class distribution most significantly.

Entropy minimization prevents the decision boundary from lying near the low-confidence prediction region in the feature space. A simple loss term is commonly used to minimize the entropy for unlabeled data with all the classes. Two well-known methods based on entropy minimization are Pseudo-Labeling [18] and Label Propagation [29]. Pseudo-Labeling trains a student model based on a teacher model's predictions or pseudo labels, in which the teacher model is pre-trained using supervised learning. On the other hand, Label Propagation is to diffuse from labeled samples to unlabeled ones according to the propagation weights computed from pairwise similarity scores.

Recent studies have combined consistency regularization and entropy minimization, such as MixMatch [30] and FixMatch [31]. These methods apply multiple augmentations on a single unlabeled sample and force the model to predict these augmented input data similarly. By combining numerous augmentations, the trained model extracts invariant features to improve the overall performance even using a small number of labeled samples.

3   Methodology

By extending the Pseudo-Labeling method, we propose an SSL framework integrated with mixed augmentations and multiple losses, as shown in Fig. 1. First, an initial handwriting recognizer as a student model is prepared using labeled data by supervised learning. Second, mixed augmentations are applied to generate a weakly transformed variant and a strongly transformed variant from each original sample. Third, the teacher model produces a pseudo label from the weakly augmented variant and then computes a pseudo-label loss on the strongly augmented variant. For the prediction from the teacher model, the special tokens of padding or blank [PAD], start of sequence [SOS], and unknown [UNK] should not exist. These tokens are eliminated from the predictions to maintain the quality of the pseudo labels. Fourth, the student model is trained by minimizing both the supervised and pseudo-label losses with a flexible ratio. The ratio depends on the rate between labeled and unlabeled samples in a single training minibatch and the number of trained epochs. Note that the pseudo-label loss is gradually used to update the handwriting recognizer to avoid the negative effect of incorrect pseudo labels, termed the Scheduled Pseudo-Label loss. Finally, the teacher model was incrementally updated using the student model and used for evaluation.

Although the Mean Teacher and Pseudo-Labeling methods are the basis of this study, they follow different training schemes. Thus, we modified their training schemes similar to our model to achieve a fair comparison with the proposed framework in this study.

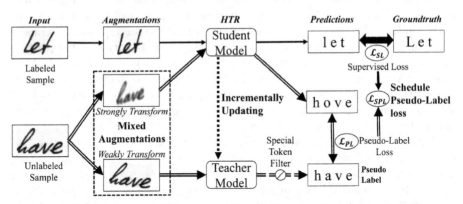

Fig. 1. Workflow of our proposed Incremental Teacher Model with Mixed Augmentations and Scheduled Pseudo-Label loss. The single-line arrows illustrate supervised learning using labeled samples, whereas the double-line arrows represent SSL with unlabeled samples.

3.1   Incremental Teacher Model

Updating of the models that generate pseudo labels is handled differently depending on the research and application. In [18], the teacher model is commonly pre-trained and fixed; therefore, the predicted pseudo labels are stable for training the student model. This approach is good in the case where the teacher model is sufficiently trained on labeled data. In practice, however, many labeled samples are not always available. On the other hand, methods that compute consistency regularization, such as Mean Teacher, can simultaneously train the student model and the teacher model that generates the pseudo labels in the training process. However, it might update the teacher model with a worse student model in the early stage of the training process. Thus, we propose to update the teacher model with the student model whenever the validation accuracy is improved at the end of each training epoch. The teacher model is updated by copying the weighted parameters from the student model. Finally, the teacher model was used for evaluation. To the best of our knowledge, this is the first work applying incremental updates of the teacher model for handwriting recognition using pseudo labels.

A well-initialized pre-trained model is essential to prepare a good teacher model to enhance the performance of the student model later. Because RotNet has been demonstrated to be effective for general images with complex background [32], we expected that it would be suitable for HTR with simple background. Moreover, the handwritten word image ratio was in range of general image ratio. Therefore, we employed RotNet, a self-supervised learning method for predicting the rotation of images, as a pretext task. This initialization method provides more general network weights to achieve a higher accuracy using supervised learning or SSL in the later training process.

3.2 Mixed Augmentations

In recent years, augmentation has played an important role in avoiding overfitting during the DNN training process [33] since it provides a large number of variants from a small number of samples. With more variants, a well-trained DNN model with augmentation tends to perform better extraction and focus on the invariant features. Since augmentation does not require newly collected data, it is commonly employed as an efficient method to improve the DNN performance. On the other hand, sequence-to-sequence contrastive learning (SeqCLR) has been proposed to employ stochastic image augmentation to generate two different variants from a single input sample [16]. Subsequently, the mapping between two extracted feature sequences is computed and considered the contrastive loss for optimization. In addition, augmentations are employed to generate multiple variants of a single sample for training based on prediction consistency [30].

In this study, we used multiple augmentation methods to generate two variants from a sample, which was named as "Mixed Augmentations". One variant used smaller deformations to obtain a pseudo label, while the other had larger deformations. Note that the stochastic image augmentation in SeqCLR randomly generates two variants of an original sample using a single transforming pipeline repeatedly. Owing to the asymmetry of the proposed framework, two generated variants in our method are normally generated by two different transforming pipelines (weak and strong).

Augmentations used in general image recognition, such as FixMatch [31], are composed of geometric transforms for weak and multiple mixed transformations for strong transforms. For handwriting recognition, however, geometric transforms are limited to maintain the readability of the augmented handwritten images. Thus, we use four augmentations, namely rotation, crop, perspective, and Gaussian blur, which are commonly employed in handwriting recognition studies, as shown in Table 1. These settings are based on comparative experiments and applied consistently in experiments with many HTR architectures and in different labeled ratio scenarios.

Table 1. Details of Mixed Augmentations.

Augmenta-tion	Description	Main parameter	Weak transformation	Strong transformation
Rotation	Randomly rotates the input text image between 0 and a parameter value	Rotation degree (deg)	15	15
Crop	Crops and enlarges a random area of the image by a specified percentage. Note that the aspect ratio is maintained	Crop percentage (%)	-	80
Perspective	Generates a perspective image with randomly transformed vertex positions according to the specified distortion ratio	Distortion percentage (%)	-	30
Gaussian Blur	Blurs an image by applying a Gaussian filter. The blur strength is specified by standard deviation	Sigma	-	2

3.3 Scheduled Pseudo-Label Loss

For training samples X with corresponding labels Y, the supervised loss is based on the negative log-likelihood as follows:

$$\mathcal{L}_{SL} = \sum_{(X,Y)} -\log p(Y|X) \tag{1}$$

The pseudo-label loss for the unlabeled training samples X is defined as follows:

$$\mathcal{L}_{PL} = \sum_{(X)} -\log p\left(\overline{Y}|\overline{\overline{X}}\right) \; with \; \overline{Y} = \text{teacher}(\overline{X}) \tag{2}$$

Here, $\overline{X}$ and $\overline{\overline{X}}$ are the weakly and strongly transformed variants from X, respectively. The pseudo labels $\overline{Y}$ are predicted by the teacher model on $\overline{X}$. Thus, the pseudo-label loss is based on the conditional probabilities of the pseudo-label $\overline{Y}$ for the strongly transformed variants $\overline{\overline{X}}$.

We introduce scheduling of the loss calculations for the pseudo labels of the unlabeled samples. It is aimed to avoid the problem that the target model does not converge due to the generation of incorrect pseudo labels in the early stages of training. Label scheduling has been proposed besides Pseudo-Labeling, and several derivations have been considered in other related studies. In this study, we applied the Scheduled Pseudo-Label loss as follows:

$$\mathcal{L}_{SPL} = \frac{1}{n}\mathcal{L}_{SL} + \alpha(t)\frac{1}{n'}\mathcal{L}_{PL} \tag{3}$$

where n is the total number of labeled samples, n' is the total number of unlabeled samples, t is the training epoch and $\alpha(t)$ is the scheduled weight for $\mathcal{L}_{PL}$ that depends on T_1, T_2, and A as shown below:

$$\alpha(t) = \begin{cases} 0 & t < T_1 \\ \frac{t-T_1}{T_2-T_1}A & T_1 \leq t < T_2 \\ A & T_2 \leq t \end{cases} \tag{4}$$

Thus, $\mathcal{L}_{PL}$ begins to affect $\mathcal{L}_{SPL}$ when the number of epochs crosses T_1 and monotonically increases until it reaches T_2; then, A is the highest weight of $\mathcal{L}_{PL}$. In this study, we applied T_1 of 50, T_2 of 250, and A of 1, so that $\mathcal{L}_{PL}$ is used from the midpoint of learning on the labeled data. Note that the current hyperparameters of the scheduled pseudo-label loss were experimentally chosen.

4 Experiments

4.1 IAM Handwriting Database and Scenarios for SSL

We used handwritten English word-level patterns of the IAM database for evaluation because they have been used as the benchmark for many HTR studies [7]. Although the SSL methods have been employed for many recognition tasks, they have not been widely applied in handwriting recognition as mentioned in the review section. For handwriting recognition, a sequence of characters is required for prediction instead of single characters. Thus, preliminary experiments at the word level are the most straightforward HTR task.

Table 2 shows four splitting scenarios derived from the RWTH Aachen University split[1] of the IAM handwriting database, where Words, Pages, and Writers denote the numbers of labeled and unlabeled samples in the training set, the number of samples

---

[1] https://www.openslr.org/56/.

in validation set, and that in the testing set, respectively. There is no writer duplication between the labeled and unlabeled samples. These scenarios are prepared to evaluate the SSL methods with our handwriting recognizers. These splitting scenarios satisfy the writer-independent requirement, which is commonly used to benchmark the handwritten English text recognizers.

Scenario 1 is the same as the supervised learning configuration without unlabeled samples. Scenarios 2, 3, and 4 are prepared to randomly select 50%, 10%, and 1% of the training set as the labeled training sets, respectively, while the rest is used as unlabeled training sets. Note that the labeled training set of Scenario 4 (1% labeled) does not include the eight character categories, which is over 10% of all character categories (8/79). Thus, Scenario 4 is the most challenging with unseen categories and writing styles.

Table 2. Details of SSL scenarios on IAM handwriting database.

Scenarios for SSL		IAM Subsets			
		Training set		Validation set	Testing set
		Labeled	Unlabeled		
Scenario 1 (100% labeled samples)	Words	55,081	0	8,895	25,920
	Pages	747	0	116	336
	Writers	283	0	56	161
Scenario 2 (50% labeled samples)	Words	27,727	27,354	Same as above	Same as above
	Pages	373	374		
	Writers	139	144		
Scenario 3 (10% labeled samples)	Words	5,364	49,717	Same as above	Same as above
	Pages	72	675		
	Writers	27	256		
Scenario 4 (1% labeled samples)	Words	551	54,530	Same as above	Same as above
	Pages	8	739		
	Writers	2	281		

4.2 Handwritten Text Recognition Architectures

As recognition models tested in the experiments, we used four architectures of handwriting recognizers. The first is a CRNN using ResNet as a feature extractor and Bidirectional Long Short-Term Memory (BLSTM) with CTC [20]. The second is another general encoder–decoder architecture, where an attention layer guides the decoder (AED) [21]. The third is a Deep Convolutional Recurrent Neural Network (DCRN) derived from AED with a simple Convolutional Neural Network (CNN) and a stacked BLSTM that provides a deeper sequential encoder [22]. The fourth is a CRNN using multiple Self-Attention layers for the sequential encoder (SelfAttn) [22]. These are listed in Table 3 with each major component.

Table 3. Main components of four HTR architectures.

Components	HTR Architectures			
	CRNN	AED	DCRN	SelfAttn
Feature Extractor (Local Encoder)	ResNet	ResNet	CNN	CNN
Sequential Encoder	BLSTM	BLSTM	Stacked BLSTM	BLSTM +SelfAttn
Sequential Decoder	CTC	LSTM +Attention	LSTM +Attention	CTC

4.3 Results of Different Recognition Architectures

To the best of our knowledge, no related research applied similar techniques to the HTR problem. The related studies were proposed for general image classification. For comparison, we experimented using Mean Teacher [27] and Pseudo-Labeling [18] because the proposed method is derived from them. Furthermore, we experimented using Fix-Match [31] as this is one of the most efficient SSL methods. Note that we modified these SSL methods to match with the training scheme used for our method.

Table 4 reports the results of four HTR architectures trained by different frameworks in each scenario. The baseline column shows the character accuracy rate (CAR) of the HTR architectures trained by only labeled samples, while the other columns show the CARs of trained HTR architectures using Mean Teacher, Pseudo-Labeling, FixMatch, and Incremental Teacher Model. For Pseudo-Labeling, we followed the default setting of scheduling parameters reported in [18]. Note that these reported results are on the IAM word-level testing set. The recognition rates shown here seem inferior to the state-of-the-art results [34] since these rates are obtained without word dictionaries and language models.

Overall, AED produced the best results in all scenarios with any training framework (bold), while CRNN typically produced the second-best results (underline). These results suggest that using a ResNet-based feature extractor seems to be better than the simple CNN. Moreover, the high complex sequential encoders of DCRN and SelfAttn did not achieve an accuracy as high as that of the simple sequential encoders of AED and CRNN. The performance of all the HTR architectures decreased significantly in Scenario 4 since the labeled training set did not cover the character set.

For the related SSL methods, Pseudo-Labeling outperformed Mean Teacher and FixMatch in almost all scenarios with all the HTR architectures. Note that in the case of the Mean Teacher and FixMatch methods, the performance of the HTR architecture is deteriorated in some cases, which is shown by ↓ in Table 4. Mean Teacher and FixMatch mainly rely on the loss calculated from the distribution comparison between pseudo labels and output, as the consistency cost is unsuitable for text line recognition. It is considered difficult to capture the consistency because the output before decoding is a time series of classification, which varies significantly depending on the augmentation

Table 4. Character accuracy rate (%) of HTR architectures trained by Supervised Learning, Mean Teacher, Pseudo-Labeling, FixMatch, and Incremental Teacher Model in four SSL scenarios.

Data Split	HTR Architecture	Supervised Learning (Baseline)	Mean Teacher [27]	Pseudo-Labeling [18]	Fix Match [31]	Incremental Teacher Model (Our)
Scenario 1 (100%)	CRNN	77.45				
	AED	**80.49**				
	DCRN	74.72				
	SelfAttn	74.19				
Scenario 2 (50%)	CRNN	72.29	74.36	76.05	73.68	76.22
	AED	75.86	76.14	77.84	76.00	**78.71**
	DCRN	71.30	71.19↓	74.54	72.25	75.18
	SelfAttn	68.73	68.70↓	70.83	72.00	71.22
Scenario 3 (10%)	CRNN	53.83	57.93	62.44	54.40	70.26
	AED	56.81	56.14↓	66.06	55.49↓	**72.57**
	DCRN	48.76	55.76	58.29	51.10	61.62
	SelfAttn	48.62	51.60	55.37	51.33	60.88
Scenario 4 (1%)	CRNN	20.42	21.10	22.21	19.52↓	24.99
	AED	23.55	22.80↓	25.10	22.43↓	**29.96**
	DCRN	21.33	20.85↓	22.11	21.82	24.13
	SelfAttn	20.89	21.33	22.78	21.98	21.70

with positional information. Therefore, a method that expands on the pseudo labels is effective, and additional study is required to introduce consistency costs.

For every architecture except SelfAttn, the recognizer trained by the Incremental Teacher Model outperforms the recognizers trained by the well-known SSL methods: Mean Teacher, Pseudo-Labeling, and FixMatch in every scenario using only 50%, 10%, or 1% labeled training samples on the IAM handwriting database, respectively. The SelfAttn architecture with a simple feature extractor and a complex sequential encoder does not perform well in Scenarios 2 and 4. Mixed Augmentations seem to be helpful for the feature extractor rather than the sequential encoder.

Figure 2 illustrates the changes in the recognition accuracy with the increase in the ratio of labeled data in the training set. The Incremental Teacher Model increases the accuracy of the AED architecture by at most 15.7 percentage points (p.p.) in Scenario 3. Despite using the 1% labeled samples for training, the accuracy of AED is increased by 6.4 p.p. Compared to Pseudo-Labeling, it improves the HTR accuracy by at least 0.9 and at most 6.5 p.p. in Scenarios 2 and 3, respectively. These results show that the proposed framework could leverage unlabeled data to improve the HTR efficiency. Moreover, they give a clue about the possibility of applying HTR in practice on an unlabeled dataset by labeling only a small portion of the dataset.

Table 5 lists six word-level samples from the IAM handwriting database with the predictions from four architectures trained by Incremental Teacher Model. For short words such as "of", "the" and "friend", CRNN and AED correctly predicted while DCRN and SelfAttn produced misrecognitions. For longer words, even CRNN and AED did not perform correctly. The predictions by AED differed from the ground truth by one to two

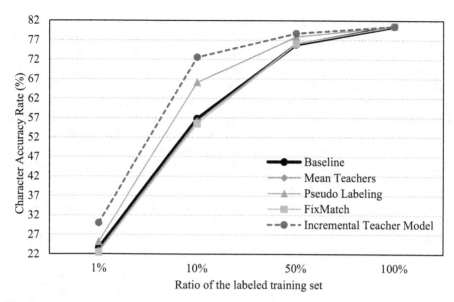

Fig. 2. Character accuracy rate (%) of AED trained by different methods in four SSL scenarios.

characters while those by CRNN had more differences. The predictions by DCRN were shorter than the ground truth which might suggest that the DCRN capability is limited in the length of its output sequences. The SelfAttn architecture performed well with its predictions being different from the ground truth by only one to two characters.

Table 5. IAM word-level samples with predictions from four architectures trained using Incremental Teacher Model in Scenario 3.

Samples						
Ground truth	of	the	friend	original	natural	respectability
CRNN	of	the	friend	oiginal	malural	eppectabet
AED	of	the	friend	original	natusal	expectability
DCRN	of	He	find	logial	what	repather
SelfAttn	of	he	friend	origimal	matual	nespectability

4.4   Results of Different Augmentation Configurations

Table 6 shows our search for weak/strong transformation settings, where we trained the AED architecture on Scenario 3 (10% of the training samples have been labeled). The most basic augmentation is rotation by at most 15 degrees (Rot15). Thus, we conducted a series of experiments with Rot15 as weak and strong transformations and inserted other augmentations into the strong transformation, such as Crop80 (randomly removed at most 20% of an image), Blur2 (randomly applied Gaussian blur with the highest value of sigma of 2), and Per30 (randomly and vertically distorted an image by at most 30%). By employing more augmentations on the strong transformation, the AED performance increases from *R1* to *R5*. Moreover, we tried to eliminate Rot15 from weak transformation; however, *R6* performs worse than *R5* at 2.3 p.p. Next, we modified the parameters used for augmentations from the settings of *R5* to make *R7*. The small changes in the parameters might reduce the final recognition accuracy. Moreover, we tested to include more augmentations in the weak transformation. As shown in the *R8* and *R9* rows, the recognition accuracy declines when more augmentations are applied.

Table 6. Ablation studies for different configurations of Mixed Augmentations in Scenario 3.

Weak transformation	Strong transformation	Character accuracy (%)	Result IDs
Rot15	Rot15	67.79	*R1*
Rot15	Rot15+Crop80	69.10	*R2*
Rot15	Rot15+Crop80+Blur2	69.88	*R3*
Rot15	Rot15+Crop80+Per30	70.15	*R4*
Rot15	**Rot15+Crop80+Per30+Blur2**	**72.57**	*R5*
–	Rot15+Crop80+Per30+Blur2	70.25	*R6*
Rot15	Rot30+Crop70+Per40+Blur2	70.70	*R7*
Rot15+Crop90	Rot15+Crop80+Per30+Blur2	68.55	*R8*
Rot15+Crop80+Per30+Blur2	Rot15+Crop80+Per30+Blur2	67.41	*R9*

Thus, we might assume that simple augmentations are suitable for weak transformations. Moreover, we still need to search for the optimal parameters of Mixed Augmentations.

4.5   Discussions

Based on the experiments, the AED model outperformed other models, which may be owing to its components of a ResNet-based feature extractor and an LSTM-based decoder with attention. These components are large and deep to extract useful features for recognition and correctly focus on character regions. Thus, they are commonly used to build handwriting recognizers. Because these experiments were on word-level patterns only, further experiments on sentence-level are required to verify the efficacy of the proposed framework. We believe that designing the consistency cost for long handwritten text is challenging. As it is impractical to investigate all types of augmentation in this study, we selected and applied the augmentations commonly used with better performance on HTR. However, we expect that other augmentations are also possible to be employed in the proposed framework.

5   Conclusions

We proposed Incremental Teacher Model and demonstrated its effectiveness. It produces a high recognition accuracy for handwritten text recognition even when only a part of the training set is labeled. It comprises Mixed Augmentations and Scheduled Pseudo-Label loss for handwritten text recognition. Instead of using a fixed pre-trained handwritten text recognition (HTR) model as a teacher model to generate pseudo labels, the proposed framework incrementally updates the teacher model using the latest recognizer. We applied the proposed framework to four DNN architectures for handwriting recognition and compared it with well-known semi-supervised learning methods: Mean Teacher, Pseudo-Labeling, and FixMatch. For almost every architecture, the recognizer trained by the Incremental Teacher Model outperforms the recognizers trained by other well-known SSL methods in every scenario when using only 50%, 10%, or 1% labeled training samples on the IAM handwriting database. However, we only confirmed the effectiveness of our framework for word-level English, so we plan to examine the framework for text-line-level English as well as for other languages in future works.

Acknowledgement. We thank anonymous reviewers for helpful comments on the manuscript. This work is partially supported by the joint research budget from Hitachi, Ltd. and Kakenhi (S) 18H05221.

References

1. Krizhevsky, A., Sutskever, I., Hinton, G.: ImageNet classification with deep convolutional neural networks. In: The 25th Neural Information Processing Systems, pp. 1106–1114 (2012). https://doi.org/10.1145/3065386
2. van den Oord, A., et al.: WaveNet: a generative model for raw audio. In: The 9th ISCA Speech Synthesis Workshop, p. 125 (2016). https://doi.org/10.1109/ICASSP.2009.4960364
3. Vinyals, O., Toshev, A., Bengio, S., Erhan, D.: Show and tell: lessons learned from the 2015 MSCOCO image captioning challenge. IEEE Trans. Pattern Anal. Mach. Intell. **39**, 652–663 (2017). https://doi.org/10.1109/TPAMI.2016.2587640
4. Graves, A., Schmidhuber, J.J.: Offline handwriting recognition with multidimensional recurrent neural networks. In: The 21st International Conference on Neural Information Processing Systems, pp. 545–552 (2008). https://doi.org/10.1007/978-1-4471-4072-6
5. Puigcerver, J.: Are multidimensional recurrent layers really necessary for handwritten text recognition? In: The 14th International Conference on Document Analysis and Recognition, pp. 67–72 (2017). https://doi.org/10.1109/ICDAR.2017.20
6. Bluche, T.: Joint line segmentation and transcription for end-to-end handwritten paragraph recognition. In: The 30th International Conference on Neural Information Processing Systems, pp. 838–846 (2016). https://doi.org/10.5555/3157096.3157190
7. Marti, U.V., Bunke, H.: The IAM-database: an English sentence database for offline handwriting recognition. Int. J. Doc. Anal. Recognit. **5**, 39–46 (2003). https://doi.org/10.1007/s10032020071
8. Shivram, A., Ramaiah, C., Setlur, S., Govindaraju, V.: IBM-UB-1: a dual mode unconstrained english handwriting dataset. In: The 12th International Conference on Document Analysis and Recognition, pp. 13–17 (2013). https://doi.org/10.1109/ICDAR.2013.12

9. Mahmoud, S.A., et al.: KHATT: an open Arabic offline handwritten text database. Pattern Recognit. **47**, 1096–1112 (2014). https://doi.org/10.1016/j.patcog.2013.08.009

10. Liu, C.L., Yin, F., Wang, D.H., Wang, Q.F.: CASIA online and offline Chinese handwriting databases. In: The 11th International Conference on Document Analysis and Recognition, pp. 37–41 (2011). https://doi.org/10.1109/ICDAR.2011.17

11. Kumar Bhunia, A., et al.: Handwriting trajectory recovery using end-to-end deep encoder-decoder network. In: The 24th International Conference on Pattern Recognition, pp. 3639–3644 (2018). https://doi.org/10.1109/ICPR.2018.8546093

12. He, K., Zhang, X., Ren, S., Sun, J.: Deep residual learning for image recognition. In: The 29th IEEE Conference on Computer Vision and Pattern Recognition, pp. 770–778 (2016). https://doi.org/10.1109/CVPR.2016.90

13. Huang, G., Liu, Z., Van Der Maaten, L., Weinberger, K.Q.: Densely connected convolutional networks. In: The 30th IEEE Conference on Computer Vision and Pattern Recognition, pp. 2261–2269 (2017). https://doi.org/10.1109/CVPR.2017.243

14. Zhu, Y., Xie, Z., Jin, L., Chen, X., Huang, Y., Zhang, M.: SCUT-EPT: new dataset and benchmark for offline Chinese text recognition in examination paper. IEEE Access. **7**, 370–382 (2019). https://doi.org/10.1109/ACCESS.2018.2885398

15. Nguyen, H.T., Nguyen, C.T., Oka, H., Ishioka, T., Nakagawa, M.: Handwriting recognition and automatic scoring for descriptive answers in Japanese language tests. In: Porwal, U., Fornés, A., Shafait, F. (eds.) ICFHR 2022. LNCS, vol. 13639, pp. 274–284. Springer, Cham (2022). https://doi.org/10.1007/978-3-031-21648-0_19

16. Aberdam, A., et al.: Sequence-to-sequence contrastive learning for text recognition. In: The 2021 IEEE/CVF Conference on Computer Vision and Pattern Recognition, pp. 15297–15307 (2021). https://doi.org/10.1109/CVPR46437.2021.01505

17. Kang, L., Rusiñol, M., Fornés, A., Riba, P., Villegas, M.: Unsupervised adaptation for synthetic-to-real handwritten word recognition. In: The IEEE/CVF Winter Conference on Applications of Computer Vision (2020). https://doi.org/10.1109/WACV45572.2020.9093392

18. Lee, D.-H.: Pseudo-label: the simple and efficient semi-supervised learning method for deep neural networks. In: ICML 2013 Workshop: Challenges in Representation Learning, pp. 1–6 (2013)

19. Rizve, M.N., Duarte, K., Rawat, Y.S., Shah, M.: In defense of pseudo-labeling: an uncertainty-aware pseudo-label selection framework for semi-supervised learning. In: The 9th International Conference on Learning Representations (2022). https://doi.org/10.48550/arXiv.2101.06329

20. Xie, Z., Sun, Z., Jin, L., Feng, Z., Zhang, S.: Fully convolutional recurrent network for handwritten Chinese text recognition. In: The 23rd International Conference on Pattern Recognition, pp. 4011–4016 (2016). https://doi.org/10.1109/ICPR.2016.7900261

21. Sueiras, J., Ruiz, V., Sanchez, A., Velez, J.F.: Offline continuous handwriting recognition using sequence to sequence neural networks. Neurocomputing **289**, 119–128 (2018). https://doi.org/10.1016/J.NEUCOM.2018.02.008

22. Ly, N.T., Ngo, T.T., Nakagawa, M.: A self-attention based model for offline handwritten text recognition. In: Wallraven, C., Liu, Q., Nagahara, H. (eds.) ACPR 2022. LNCS, vol. 13189, pp. 356–369. Springer, Cham (2022). https://doi.org/10.1007/978-3-031-02444-3_27

23. Munkhdalai, T., Yu, H.: Meta networks. In: The 34th International Conference on Machine Learning, pp. 2554–2563 (2017). https://doi.org/10.48550/arXiv.1703.00837

24. Souibgui, M.A., Fornés, A., Kessentini, Y., Megyesi, B.: Few shots are all you need: a progressive learning approach for low resource handwritten text recognition. Pattern Recogn. Lett. **160**, 43–49 (2022). https://doi.org/10.1016/J.PATREC.2022.06.003

25. Chakrapani Gv, A., Chanda, S., Pal, U., Doermann, D.: One-shot learning-based handwritten word recognition. In: Palaiahnakote, S., Sanniti di Baja, G., Wang, L., Yan, W.Q. (eds.) ACPR 2019. LNCS, vol. 12047, pp. 210–223. Springer, Cham (2020). https://doi.org/10.1007/978-3-030-41299-9_17

26. Laine, S., Aila, T.: Temporal ensembling for semi-supervised learning. In: The 5th International Conference on Learning Representations (2016). https://doi.org/10.48550/arXiv.1610.02242

27. Tarvainen, A., Valpola, H.: Mean teachers are better role models: Weight-averaged consistency targets improve semi-supervised deep learning results. In: The 31st International Conference on Neural Information Processing Systems, pp. 1195–1204 (2017). https://doi.org/10.48550/arxiv.1703.01780

28. Miyato, T., Maeda, S.I., Koyama, M., Ishii, S.: Virtual adversarial training: a regularization method for supervised and semi-supervised learning. IEEE Trans. Pattern Anal. Mach. Intell. **41**, 1979–1993 (2017). https://doi.org/10.48550/arxiv.1704.03976

29. Iscen, A., Tolias, G., Avrithis, Y., Chum, O.: Label propagation for deep semi-supervised learning. In: 2019 IEEE/CVF Conference on Computer Vision and Pattern Recognition, pp. 5065–5074 (2019). https://doi.org/10.1109/CVPR.2019.00521

30. Berthelot, D., Carlini, N., Goodfellow, I., Oliver, A., Papernot, N., Raffel, C.: MixMatch: a holistic approach to semi-supervised learning. In: The 33rd International Conference on Neural Information Processing Systems, pp. 5049–5059 (2019). https://doi.org/10.48550/arXiv.1905.02249

31. Sohn, K., et al.: FixMatch: simplifying semi-supervised learning with consistency and confidence. In: The 34th International Conference on Neural Information Processing Systems, pp. 596–608 (2020). https://doi.org/10.5555/3495724.3495775

32. Gidaris, S., Singh, P., Komodakis, N.: Unsupervised representation learning by predicting image rotations. In: The 6th International Conference on Learning Representations (2018). https://doi.org/10.48550/arXiv.1803.07728

33. Shorten, C., Khoshgoftaar, T.M.: A survey on image data augmentation for deep learning. J. Big Data **6**(1), 1–48 (2019). https://doi.org/10.1186/s40537-019-0197-0

34. Bhunia, A.K., Das, A., Bhunia, A.K., Kishore, P.S.R., Roy, P.P.: Handwriting recognition in low-resource scripts using adversarial learning. In: The IEEE Computer Society Conference on Computer Vision and Pattern Recognition, pp. 4762–4771 (2019). https://doi.org/10.1109/CVPR.2019.00490

AFFGANwriting: A Handwriting Image Generation Method Based on Multi-feature Fusion

Heng Wang[1,2,3], Yiming Wang[1,2,3], and Hongxi Wei[1,2,3]([✉])

[1] School of Computer Science, Inner Mongolia University, Hohhot 010021, China
[2] Provincial Key Laboratory of Mongolian Information Processing Technology,
Hohhot, China
[3] National and Local Joint Engineering Research Center of Mongolian Information
Processing Technology, Hohhot, China
cswhx@imu.edu.cn

Abstract. Recently, reliable quality images can be generated due to the development of adversarial generative networks. Nevertheless, computer-generated images are still not comparable to humans in terms of handwriting image generation. In this paper, a novel method (i.e. AFFGANwriting) based on multi-scale features fusion has been proposed for handwriting image generation. In AFFGANwriting, a style encoder based on VGG19 has been designed to extract features in different scales. In this way, a variety of global features (e.g. stroke thickness, inclination and so on) and local features (e.g. continuous strokes, personalized writing and so forth) can be obtained. After that, the global features and the local features can be fused together to generate much more realistic handwriting images by multiple feature fusion modules of AFFGANwriting. Experimental results demonstrate that the proposed method can be competent for the task of handwriting images generation and outperforms various baseline and state-of-the-art methods. The code is available at: https://github.com/wh807088026/AFFGanWriting.

Keywords: handwriting image generation · feature fusion · generative adversarial network · style encoder

1 Introduction

Handwriting image generation is an attractive field. Its aim is to generate realistic handwriting image according to a specified text and a certain style. The generated handwriting images can be used for data augmentation so as to improve the performance of handwritten text recognition. In the literature, the approaches of handwriting image generation can be divided into two categories: online generation and offline generation. With regard to the online handwriting generation approaches, the sequence-based models (e.g. RNN) were extensively used and trained by a dataset with stroke sequence information. For the offline handwriting generation approaches, generative adversarial networks (GANs) were

G. A. Fink et al. (Eds.): ICDAR 2023, LNCS 14190, pp. 302–312, 2023.
https://doi.org/10.1007/978-3-031-41685-9_19

generally utilized and trained by a dataset of word images and labels. Since a dataset of word images with labels are easier to be collected, offline handwriting generation gets more attention than online handwriting generation. Moreover, offline handwriting generation approaches are able to directly produce handwriting images.

In this paper, we concentrate on obtaining word images by means of offline handwriting generation. At present, the most representative word-level generation model is GANwriting [1]. However, the major drawback of GANwriting is suffering from simplicity of writing style. To solve the above problem of GANwriting, a novel handwriting image generation method (called AFFGANwriting) has been proposed in this paper. To be specific, a well-designed style encoder in AFFGANwriting was used for obtaining multi-scale features including global features and local features. Furthermore, multiple fusion modules in AFFGANwriting were utilized to combine the global features with the local features. It results in generating much more realistic handwriting images. The main contributions of our proposed method consist of the following two aspects:

(1) To avoid simplicity of writing style, multi-scale features can be obtained by a style encoder based on VGG19 in AFFGANwriting. These features can reflect the overall styles (i.e. global features) and partial styles (i.e. local features) of handwriting word images.
(2) Three kinds of feature fusion schemes, such as Concatenation Feature Fusion (CFF), Summation Feature Fusion (SFF) and Attentional Feature Fusion [2] (AFF), are compared with each other to improve the quality of the generated handwriting images.

2   Related Work

For the task of online handwriting generation, Graves et al. [3] firstly presented an approach based on Long Short-Term Memory (LSTM). The approach can predict the future position of pixels based on their previous location. After that, Aksan et al. [4] proposed a Variational RNN (VRNN) based method in this task. The inputs of VRNN are composed of text contents and writing styles. Subsequently, the VRNN was replaced with a Stochastic Temporal Convolutional Network by Aksan et al. in [5]. By this way, the problem of homogeneous writing style can be solved, which results in being superior to VRNN.

Compared with online handwriting generation, the task of offline handwriting generation only needs a dataset of handwritten images with labels for training, which is more convenient to collect such a dataset in practice. In recent years, GANs [6] have been successfully utilized to attain the aim of offline handwriting generation. Zhu et al. [7] designed a CycleGAN to generate Chinese handwriting images. Alonso et al. [8] also employed GANs to generate handwriting word images. But, it suffers from the problem of style collapse. Aiming at the above problem, a ScrabbleGAN was presented by Fogel et al. in [9] to synthesize handwriting words in random styles. In addition, Davis et al. [10] adopted GANs to generate text line images rather than word images.

Recently, Kang et al. [1] proposed a state-of-the-art model (called GAN-writing) for generating high-quality handwriting word images according to the provided styles. Whereafter, SmartPatch was put forward based on GANwriting in [11], in which a novel discriminator was introduced to improve the artifacts in GANwriting. Meanwhile, Gan et al. [13] proposed a new network structure called HiGAN, which can simultaneously learn multiple handwriting styles from different authors. And then, the different styles are randomly integrated together to generate handwritten word images. Bhunia et al. [12] proposed a transformer based network called Handwriting Transformers, in which transformers were used to solve the problem of styles entanglement. Kang et al. [14] proposed a method to produce long text-line samples with diverse handwriting styles, and used the new vFID as the evaluation standard. The work of Wang et al. [19] proposed the FDF module for learning local style variations and addressed the problem of artifacts by adding a focal frequency loss.

In the above-mentioned approaches, writing styles are only captured from whole word images. The corresponding styles are considered as global features of word images. However, some kind of local features of handwriting word images are ignored during the procedure of handwriting images generation. In contrast, several meaningful local features not only can be extracted by our proposed AFFGANwriting but also will be fused with the global features to generate more realistic handwriting word images.

3   Methodology

The work of this paper is a continuation of GANwriting. First, the GANwriting generator was added feature fusion modules, which enables the network to fuse local and global features. Then, the style encoder is improved so that the style encoder can extract features of different scales. Figure 1 depicts the improved network structure. The improved network is mainly composed of Style encoder, Content encoder, Generator and learning objectives.

3.1   Style Encoder

The style encoder S is to separate the author's handwriting style F_S. $N = 50$ word images $\mathcal{X}_i$ of an author $\mathcal{W}_i$ is randomly selected. The style encoder S mainly uses a VGG19 [15] network as the backbone network and encodes the word images $\mathcal{X}_i$ into the corresponding style-latent space. To obtain multi-scale features, the outputs of multiple intermediate layers (such as ReLU-5_1, ReLU-4_1 and ReLU-3_1) were fed into the generator simultaneously.

3.2   Content Encoder

The content encoder C can encode a one-hot encoded character vector of a given character into two vectors, F_C and f_C. The character-wise encoding F_C is combined with the handwriting style F_S output from the style encoder S

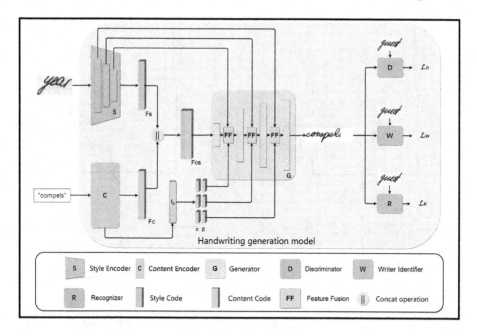

Fig. 1. The overview of the handwriting generation method

to create F_{CS}, which is sent to the generator. Global string encoding f_C is a one-dimensional vector that is divided into 3 pairs α and β injected into the generator.

3.3 Generator

This part mainly uses four bilinear interpolations as the upampling layer and finally passes through a tanh activation function layer. The model should assemble the semantic information of features at different levels, so the generator G must be able to accept features of different scales and fuse them. Furthermore, the desired content features cannot be lost in style feature fusion. Therefore, a feature fusion strategy has been designed to accomplish this aim.

3.4 Feature Fusion

Our approach considers both global style and local style. Let F_{CS} be the combination of the calligraphic style attributes and the textual content information character-wise. Let F_S be the calligraphy style feature map output by style encoder S. The style encoder S is used to extract multi-scale feature maps, extract the output F_S of ReLU-5_1, ReLU-4_1 and ReLU-3_1 layers and fuse with F_{CS}. AdaIN [16] is used as a normalization layer and combined with α and β of Global string encoding f_C segmentation to prevent the problem of losing the style of the content in multiple fusions. In addition, to take full advantage

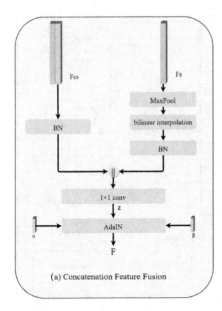 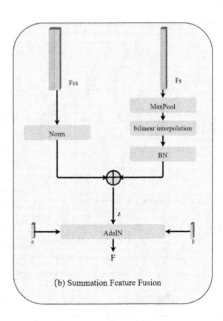

Fig. 2. (a) The structure of Concatenation Feature Fusion (CFF) module; (b) The structure of Summation Feature Fusion (SFF) module.

of the shallow and deep features, three feature fusion modules are aggregated in the generator. The three feature fusion modules are interspersed between the upsampling layers. The feature map size output by the style encoder S intermediate layer is aligned with the result size output by the upsampling layer. The results are entered sequentially. The feature F output by this module is the input F_{CS} of the next feature fusion module after the upsampling layer. This paper will introduce three different ways of feature fusion (as shown in Figs. 2 and 3).

3.4.1   Concatenation Feature Fusion (CFF)

Since the dimensions of F_{CS} and the middle layer's output F_S are different, to facilitate splicing, the output features of the middle layer first go through MaxPooling to align the dimensions of F_S and F_{CS}. Then perform Batch Normalization separately and finally splicing. The dimension of the spliced tensor is $2C \times H \times W$ and after C Point-wise Conv, the dimension is reduced to the input F_{CS} dimension $C \times H \times W$. Thus, the concatenation feature map $Z \in \mathbb{R}^{C \times H \times W}$ becomes:

$$Z = PWCov\left(Concat\left(BN\left(F_{CS}\right), BN\left(MP\left(F_S\right)\right)\right)\right), \tag{1}$$

PWCov is point-wise convolution, and BN denotes the batch normalization. MP is the max pooling operation, and Concat is the concatenation operation. After feature splicing and dimensionality reduction, content features are lost to a certain extent. GANwriting's improved AdaIN normalization is used to combine the fused style features with the α and β segmented by Global string encoding. Then, AdaIN is defined as:

$$AdaIN(Z, \alpha, \beta) = \alpha \left(\frac{Z - \mu(Z)}{\sigma(Z)} \right) \oplus \beta \tag{2}$$

$\oplus$ is broadcasting addition. μ is channel-wise mean and σ is standard deviations.

3.4.2 Summation Feature Fusion (SFF)

The summation feature fusion method is like the concatenation feature fusion method, except that the concatenation in CFF is turned into an addition operation with a weight ratio of 1:1. Of course, since the addition does not change the dimension of the feature map, there is no need to go through the Point-wise Conv layer after the operation. Thus, the summation feature map $Z \in \mathbb{R}^{C \times H \times W}$ becomes:

$$Z = BN\left(F_{CS}\right) \oplus BN\left(MP\left(F_S\right)\right), \tag{3}$$

The final representation of SFF is like CFF, and the output Z is finally fed into AdaIN to fuse with text features.

3.4.3 Attentional Feature Fusion (AFF)

When performing simple concatenation or summation feature fusion, we found that the features that need to be fused have significant semantic inconsistency, which significantly impacts the fusion weights of different layers on the quality of the generated images. To eliminate this affect, Attentional Feature Fusion (AFF) [2] is used to fuse features of different scales, which employs a Multi-scale Channel Attention Module (MS-CAM). Unlike the CFF fusion module, broadcasting addition operation can be used for aligning the dimensions of F_S and F_{CS}. And then, the result will be sent to MS-CAM. The MS-CAM is divided into two parts: Global channel Attention and Local channel Attention. The formula $L(X)$ for Local channel Attention is formally defined as:

$$L(X) = BN\left(PWConv_2\left(\delta\left(BN\left(PWConv_1(X)\right)\right)\right)\right) \tag{4}$$

where δ denotes the Rectified Linear Unit (ReLU), $PWConv_1$ and $PWConv_2$ are two different point-wise convolution (PWConv) layers. The difference between Global channel Attention and Local channel Attention is to perform a global average pooling operation on the input features first. Global Average Pooling (GAP). Therefore, the formula $G(X)$ of Global channel Attention is formally defined as:

$$G(X) = BN\left(PWConv_2\left(\delta\left(BN\left(PWConv_1(GPA(X))\right)\right)\right)\right) \tag{5}$$

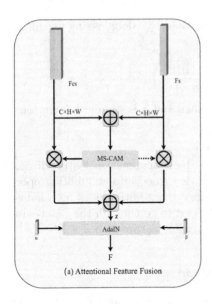

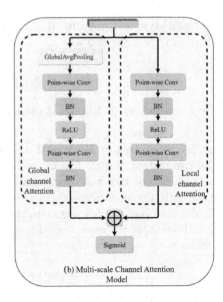

Fig. 3. (a) The structure of Attentional Feature Fusion (AFF) module; (b) The structure of Multi-scale Channel Attention Module (MS-CAM) in AFF.

The MS-CAM module adds the results of L(X) and G(X) and maps the results to $[0, 1]$ through the Sigmoid function. So the formula M(X) of MS-CAM is:

$$M(X) = sigmoid(L(X) \oplus G(X)) \tag{6}$$

$\oplus$ is broadcasting addition. Finally, the Attentional Feature Fusion feature map $Z \in \mathbb{R}^{C \times H \times W}$ is defined as:

$$Z(F_{CS}, F_S) = M(F_{CS} \oplus F_S) + (1 - M(F_{CS} \oplus F_S)) \otimes F_S \tag{7}$$

$\otimes$ is elementwise multiplication. The output of MS-CAM is the weight of feature fusion. To make a weighted average of F_S and F_{CS}, this value is subtracted from 1. Through training, let the network determine the respective weights. The final representation of AFF is like CFF and SFF, and the output Z is finally fed into AdaIN to fuse with text features.

3.5 Training and Loss Objectives

Our training is based on the original GANwriting paradigm. It contains three contents: (1) A discriminator model. It is used to distinguish the generated samples from actual samples; (2) A writer identifier model. It determines the generated image style attributes to guide the network to generate images with characteristic style attributes; (3) A Recognizer model. It was used to predict the correct text label for a given handwritten text image consisting of a VGG-19 network.

4   Experimental Results

4.1   Dataset and Experimental Settings

The IAM dataset [17] was used to evaluated our proposed method. The dataset was collected and released by the University of Bern's Computer Vision and Artificial Intelligence Research Group in 2002. The dataset consists of 9862 text lines and 62857 handwritten word images by 500 different authors. In our experiments, 44419 word images written by 340 authors are taken as the training set. The rest of 18436 word images are used as the testing set, which are all belong to out-of-vocabulary. In our experiments, the height and width of each word image is normalized to 64 pixels and 216 pixels, separately. Frechet Inception Distance (FID) [18] is used as an evaluation metric. Adam is selected as optimizer with an initial learning rate of 0.00001. The number of training epochs is set to 4000. The proposed method is implemented in PyTorch and trained on Tesla P40.

Table 1. The comparative results between the proposed method and baselines.

Method	FID
GANwriting [1]	51.16
HiGAN [13]	49.71
SmartPatch [11]	40.33
Handwriting Transformers [12]	42.05
Wang et al. [19]	33.96
SFFGANwriting (ours)	36.50
CFFGANwriting (ours)	33.69
AFFGANwriting (ours)	28.65

4.2   Baselines

GANwriting [1], SmartPatch [11], HiGAN [13], Handwriting Transformers [12] and Wang et al. [19] have been considered as our baselines. They are the most advanced and reproducible methods in the field of handwritten image generation. During the experiment, we found that different FID calculation methods have a great impact on the results. But many researchers ignored this problem and made unfair comparisons, such as HiGAN [13]. Only SmartPatch [11] describes the experimental device for FID calculation in the paper and code, which is consistent with the actual code, so we choose the experimental setup of SmartPatch as our FID calculation method. The specific FID calculation setting is based on 339 training set authors as style reference. Each author generates a fixed 114 words and performs FID calculation with the training set image.

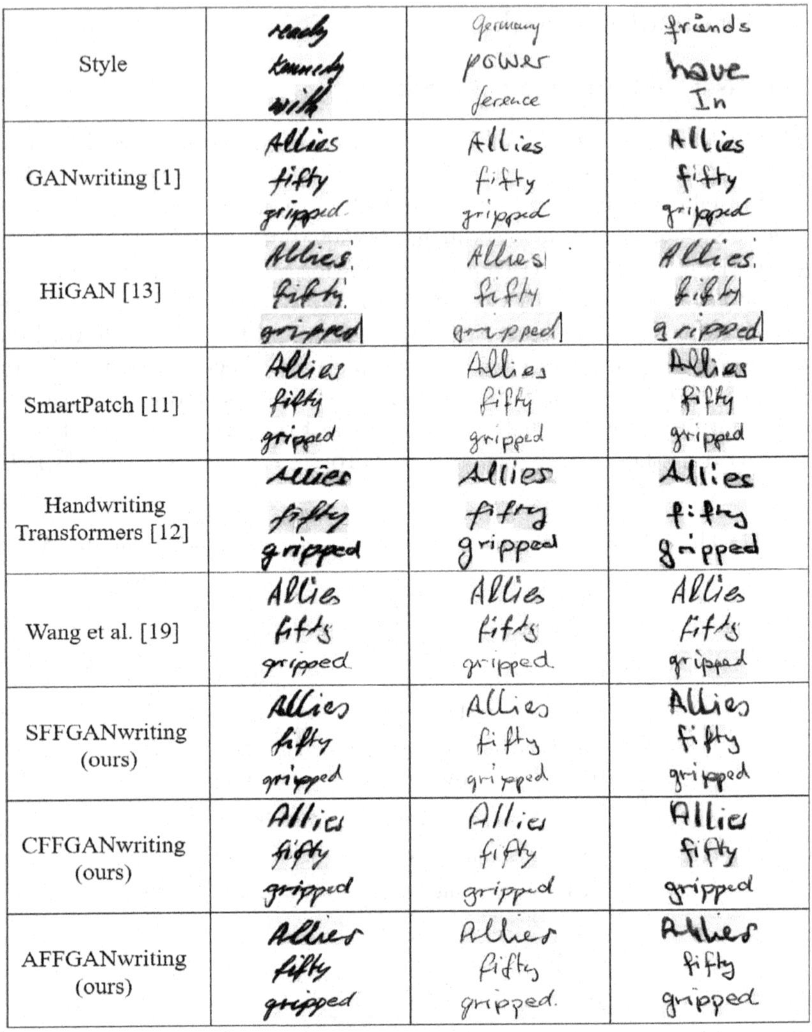

Style			
GANwriting [1]			
HiGAN [13]			
SmartPatch [11]			
Handwriting Transformers [12]			
Wang et al. [19]			
SFFGANwriting (ours)			
CFFGANwriting (ours)			
AFFGANwriting (ours)			

Fig. 4. Randomly chose to synthesize samples. The first column is the author's style, a total of three styles have been selected, and each style provides three reference images. Three words (i.e. Allies, fifty and gripped) are generated by different style. The second to eighth columns are the five baselines (i.e. GANwriting, HiGAN, SmartPatch, Handwriting Transformers and Wang et al.) and three proposed methods.

Experimental results are listed in Table 1. We can see that our proposed method is superior to the five baselines. Moreover, three kinds of feature fusion schemes have been compared with each other. The scheme of AFF outperforms SFF and CFF. Especially when testing HiGAN, the FID obtained by using the same weight according to the calculation method in his paper is 8.07, while using our method is 49.71. This also further proves that the FID calculation setting has a great impact on the results.

Finally, to further demonstrate the effectiveness of our method, we randomly sample some of the generated images and compare them (as shown in Fig. 4). A total of three different reference styles are selected, and each style gives three different authentic images. Using these style images, we generated three other words, Allies, fifty, and gripped, using seven methods for comparison.

According to the person's writing angle, the process of imitation should include the global writing style and the local writing style. The overall writing style should consist of the thickness, depth, inclination, roundness, etc., of the handwriting. Local writing styles should consist of continuous strokes, personalized writing of specific letters, etc. As shown in Fig. 4, in terms of global style, our method clearly shows the thickness and inclination of handwriting. Furthermore, in the performance of inclination, especially "Allies", our method significantly outperforms the other five methods. Moreover, in the representation of local writing style, our method also reflects the personalized writing of continuous strokes and specific letters, such as "es" in the second style. HiGAN can generate words of unspecified length, but it will randomly generate artifacts at the end of the word when generating reference styles.

5   Conclusion

In this paper, AFFGANwriting has been proposed to attain the aim of handwriting images generation. On one hand, multi-scale features can be obtained by a well-designed style encoder based on VGG19 to avoid simplicity of style. On the other hand, three kinds of fusion schemes (including SFF, CFF and AFF) have been adopted to fuse multi-scale features. Additionally, three kinds of fusion schemes have been compared with each other and the best one (i.e. AFF) has been determined. Through the above improvements, the proposed AFFGANwriting is the state-of-the-art approach, which results in generating much more realistic handwriting word images.

Acknowledgment. This study is supported by the Project for Science and Technology of Inner Mongolia Autonomous Region under Grant 2019GG281, the Natural Science Foundation of Inner Mongolia Autonomous Region under Grant 2019ZD14, and the Program for Young Talents of Science and Technology in Universities of Inner Mongolia Autonomous Region under Grant NJYT-20-A05.

References

1. Kang, L., Riba, P., Wang, Y., Rusiñol, M., Fornés, A., Villegas, M.: GANwriting: content-conditioned generation of styled handwritten word images. In: Vedaldi, A., Bischof, H., Brox, T., Frahm, J.-M. (eds.) ECCV 2020. LNCS, vol. 12368, pp. 273–289. Springer, Cham (2020). https://doi.org/10.1007/978-3-030-58592-1_17

2. Dai, Y., Gieseke, F., Oehmcke, S., et al.: Attentional feature fusion. In: Proceedings of the IEEE/CVF Winter Conference on Applications of Computer Vision, pp. 3560–3569 (2021)

3. Graves, A.: Generating sequences with recurrent neural networks. arXiv preprint arXiv:1308.0850 (2013)

4. Aksan, E., Pece, F., Hilliges, O.: DeepWriting: Making digital ink editable via deep generative modeling. In: Proceedings of the 2018 CHI Conference on Human Factors in Computing Systems, pp. 1–14 (2018)

5. Aksan, E., Hilliges, O.: STCN: stochastic temporal convolutional networks. arXiv preprint arXiv:1902.06568, 2019

6. Goodfellow, I., Pouget-Abadie, J., Mirza, M., et al.: Generative adversarial networks. Commun. ACM **63**, 139–144 (2020)

7. Zhu, J.Y., Park, T., Isola, P., et al.: Unpaired image-to-image translation using cycle-consistent adversarial networks. In: Proceedings of the IEEE International Conference on Computer Vision, pp. 2223–2232 (2017)

8. Alonso, E., Moysset, B., Messina, R.: Adversarial generation of handwritten text images conditioned on sequences. In: 2019 International Conference on Document Analysis and Recognition (ICDAR), pp. 481–486. IEEE (2019)

9. Fogel, S., Averbuch-Elor, H., Cohen, S., et al.: Scrabblegan: semi-supervised varying length handwritten text generation. In: Proceedings of the IEEE/CVF Conference on Computer Vision and Pattern Recognition, pp. 4324–4333 (2020)

10. Davis, B., Tensmeyer, C., Price, B., et al.: Text and style conditioned GAN for generation of offline handwriting lines. arXiv preprint arXiv:2009.00678 (2020)

11. Mattick, A., Mayr, M., Seuret, M., Maier, A., Christlein, V.: SmartPatch: improving handwritten word imitation with patch discriminators. In: Lladós, J., Lopresti, D., Uchida, S. (eds.) ICDAR 2021. LNCS, vol. 12821, pp. 268–283. Springer, Cham (2021). https://doi.org/10.1007/978-3-030-86549-8_18

12. Bhunia, A.K., Khan, S., Cholakkal, H., et al.: Handwriting transformers. In: Proceedings of the IEEE/CVF International Conference on Computer Vision, pp. 1086–1094 (2021)

13. Gan, J., Wang, W.: HiGAN: handwriting imitation conditioned on arbitrary-length texts and disentangled styles. In: Proceedings of the AAAI Conference on Artificial Intelligence, vol. 35, pp. 7484–7492 (2021)

14. Kang, L., Riba, P., Rusinol, M., et al.: Content and style aware generation of textline images for handwriting recognition. IEEE Trans. Pattern Anal. Mach. Intell. **44**(12), 8846–8860 (2021)

15. Simonyan, K., Zisserman, A.: Very deep convolutional networks for large-scale image recognition. arXiv preprint arXiv:1409.1556 (2014)

16. Huang, X., Belongie, S.: Arbitrary style transfer in real-time with adaptive instance normalization. In: Proceedings of the IEEE International Conference on Computer Vision, pp. 1501–1510 (2017)

17. Marti, U.V., Bunke, H.: The IAM-database: an English sentence database for offline handwriting recognition. Int. J. Doc. Anal. Recogn. **5**, 39–46 (2002)

18. Heusel, M., Ramsauer, H., Unterthiner, T., et al.: GANs trained by a two time-scale update rule converge to a local nash equilibrium. Adv. Neural Inf. Process. Syst. **30** (2017)

19. Wang, Y., Wang, H., Sun, S., et al.: An approach based on transformer and deformable convolution for realistic handwriting samples generation. In: 2022 26th International Conference on Pattern Recognition (ICPR), pp. 1457–1463. IEEE (2022)

SeamFormer: High Precision Text Line Segmentation for Handwritten Documents

Niharika Vadlamudi[(✉)] [ID], Rahul Krishna [ID], and Ravi Kiran Sarvadevabhatla [ID]

Centre for Visual Information Technology, International Institute of Information Technology, Hyderabad 500032, India
niharika.vadlamudi@research.iiit.ac.in, ravi.kiran@iiit.ac.in

Abstract. Historical manuscripts often contain dense unstructured text lines. The large diversity in sizes, scripts and appearance makes precise text line segmentation extremely challenging. Existing line segmentation approaches often associate diacritic elements incorrectly to text lines and also address above mentioned challenges inadequately. To tackle these issues, we introduce SeamFormer, a novel approach for high precision text line segmentation in handwritten manuscripts. In the first stage of our approach, a multi-task Transformer deep network outputs coarse line identifiers which we term 'scribbles' and the binarized manuscript image. In the second stage, a scribble-conditioned seam generation procedure utilizes outputs from first stage and feature maps derived from manuscript image to generate tight-fitting line segmentation polygons. In the process, we incorporate a novel diacritic feature map which enables improved diacritic and text line associations. Via experiments and evaluations on new and existing challenging palm leaf manuscript datasets, we show that SeamFormer outperforms competing approaches and generates precise text line segmentations.

Keywords: Text Line Segmentation · Historical Manuscripts

1 Introduction

Identifying text lines in ancient handwritten documents is an important problem in document image understanding [8,14,15,17,26,53]. Since historical documents usually contain text written in a highly unstructured manner with dense and non-standard layouts, the problem is challenging. The challenge aspect is particularly apparent for palm leaf manuscripts of South-East Asia and the Indian subcontinent. Western manuscripts predominantly use processed animal-skin (vellum) as their base material. Though these are not immune to ravages of time, palm leaf manuscripts are relatively more fragile. Also, palm leaves are thin, delicate and prone to damage. Moreover, the already faintly written text may fade over time and become indistinguishable from digitization noise. Document analysis tasks on palm leaf manuscripts involve characteristic challenges such as degradation,

© The Author(s), under exclusive license to Springer Nature Switzerland AG 2023
G. A. Fink et al. (Eds.): ICDAR 2023, LNCS 14190, pp. 313–331, 2023.
https://doi.org/10.1007/978-3-031-41685-9_20

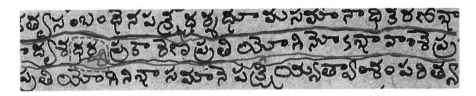

Fig. 1. An example to illustrate the importance of precise line segmentation in palm leaf manuscripts. The ground truth upper and lower portions of the enclosing line annotation are shown in red. The prediction is shown in blue. The green shaded portions indicate crucial text fragments omitted by prediction causing the semantic interpretation of text to change. For e.g., pink dashed region encloses a word ధర్మము from the Indic language Telugu which means '**moral duty**'. The incorrect boundary prediction causes the resulting line to contain a word ధర with a drastically different meaning. ధర means '**price**'. (Color figure online)

low contrast, variable inter-character and inter-line spacing and morphological distortions in character shapes [25,38,44]. The large diversity in spatial dimensions, languages, scripts, writing styles and presence of non-textual elements further compound the challenge for text line segmentation.

The output of text line segmentation is often processed by a subsequent Optical Character Recognition (OCR) module. Obtaining high precision segmentation maps of text lines which could be used as masks compactly enclosing the reference text is extremely crucial. Using such masks within the OCR pipeline reduces semantic noise from adjoining line fragments generally present in the text-line's bounding box and potentially increases OCR performance. Indic and South-East Asian manuscript texts are characterized by orthographic text fragments such as diacritics. These components typically exist at varying distances from the parent text line. Due to the semantics associated with such components, omission or incorrect association of diacritics to text lines during segmentation can result in a dramatically modified linguistic interpretation of the text (see Fig. 1). Therefore, it is essential to develop segmentation approaches for palm leaf manuscripts which are highly precise. The performance of existing line segmentation approaches fall short in this aspect.

To tackle the challenge, we propose SeamFormer, a robust text line segmentation framework for palm leaf manuscripts. SeamFormer is configured as a two stage pipeline (Sect. 3). In the first stage, the manuscript image is processed by a multi-task Transformer deep network to obtain the binarized image and coarse identifiers for each text line which we term 'scribbles' (Sect. 3.1). In the second stage (Sect. 3.2), the extracted scribbles, binarized image and custom-designed feature maps are fed to a scribble-conditioned seam generation algorithm which generates the desired tight fitting polygons enclosing the individual text lines. Via experiments and evaluations on new and existing palm leaf manuscript datasets, we show that SeamFormer generates significantly superior line segmentations compared to other competing approaches (Sect. 5).

The source code, pretrained models and associated material are available at https://ihdia.iiit.ac.in/seamformer.

2   Related Work

Many approaches have been proposed for text line segmentation in other (i.e. non palm leaf) historical documents. To encourage research, many historical document datasets with line segmentation annotations have been introduced and utilized in competitions at premier document analysis venues - refer to the comprehensive survey paper by Nikolaidou et al. [34] for details.

Early approaches favoured the use of classical digital image processing techniques followed by post processing. Alaei et al. [1] employ a painting technique for foregrounding smearing to tackle unconstrained handwritten text line segmentation for diverse languages. Grouping techniques utilizing nearest neighbor [35], learning algorithms [39], and heuristic rules [28] have also been employed for text line segmentation. Projection profiles are another popular top-down approach to isolate text lines [10,19,31,37,54]. However, profile-based approaches cannot cope with highly curved lines and uneven layouts. Adaptive Local Connectivity Map (ALCM) [45,46] is another technique for localizing and extracting text lines directly from gray-scale images. Generally, these approaches employ handcrafted processing elements with hyperparameters which do not generalize well across multiple datasets. The methods tend to require dataset specific techniques for isolating text line elements (e.g. strokes, diacritics) and often fail to disentangle touching components across consecutive text lines – a common occurrence in handwritten documents.

In recent years, a number of deep learning based approaches have been employed as well [6,7,9,27,29,30,36,40]. Most of these approaches use a variant of the popular U-Net [41] architecture. These methods have the appeal of being optimized end-to-end and work well on Western historical manuscripts. However, the approaches require drastic downsampling of input image which eliminates crucial inter-line information. Coupled with the boundary smoothing that occurs when the network predictions are upsampled, this leads to imprecise and unsatisfactory line segment boundary predictions for other types of historical manuscripts such as ours (i.e. palm leaf).

Relatively few works have tackled line segmentation for palm leaf manuscripts. In their survey paper, Kesiman et al. [25] consider palm-leaf manuscripts from South-East Asia and evaluate numerous line segmentation approaches developed for other (non-Asian) historical documents. Chamchong and Fung propose an adaptive partial projection (APP) technique [13], an improvement over their earlier partial projection approach [12] for line extraction in Thai manuscripts. Valy et al. [51] propose an approach which also employs connected components and projection profiles to determine medial positions of text lines followed by a path finding approach to mark the text line boundaries in Khmer manuscripts. Kesiman et al. [22] employ a similar approach for Balinese manuscripts. Apart from the assumption of a component-based script, these approaches inherit the shortcomings of projection-based works mentioned previously.

Works which employ deep neural networks for palm leaf manuscript text line segmentation are even fewer. Jindal and Ghosh [21] use a Faster-RCNN model to obtain bounding boxes for a collection of Indic palm leaf manuscripts. However, this approach cannot tackle the curvature of lines which is present in almost all manuscripts. Prusty et al. [38] and Sharan et al. [44] propose approaches which modify the Mask-RCNN [20] framework for segmenting various semantic regions including text lines in Indic manuscripts. Despite their relatively better performance and ability to tackle line curvature, these approaches produce overly smoothed line boundaries and even tend to have false negatives (i.e. missed lines) on some occasions.

Seam generation, an approach involving optimization over image-derived energy maps [5], is a popular approach for text line segmentation. Saabni and El-Sana introduce a seam generation algorithm based on an energy map calculated using Signed Distance Transform (SDT) for Arabic manuscripts [42]. However, the approach involves repeated energy map computations for each line and significant dataset-specific post-processing to tackle overlapping components and diacritics. Asi et al. [4] improve upon the aforementioned approach by replacing SDT with a geodesic distance transform energy map. This method fails to tackle elongated letters and widely separated diacritics. Nikolaos et al. [3] use a medial line obtained using a projection profile approach to guide seam generation for line segmentation in multiple historical datasets. However, the method requires dataset specific parameter tuning for various pipeline stages. Alberti et al. [2] first employ a deep network to obtain a binarized version of the image. Seam generation is applied on the binary image to obtain coarse region boundaries, followed by a graph-based connected component procedure to obtain the polygonal line boundaries. The approach is not suitable for highly skewed and unevenly curving text found in palm leaf manuscripts. For enhancing the seam generation process, Nguyen et al. [32] introduce an additional global cost function for better detection of ascenders, descenders and diacritics. The approach is not suitable for skewed or curved text and requires heavy dataset-specific parameter tuning for the cost functions.

In existing approaches [2–4,27,42], seam generation is generally used to separate text lines rather than *segment* them. As a result, extraneous isolated character fragments and noisy background elements present beyond the line's text content are often included as part of the line. In contrast, our approach generates polygons which compactly enclose the text lines. As a novel element, we introduce a custom energy map in our polygon generation stage which emphasizes proper association of diacritics to the parent text line. Another marked departure from existing methods is the absence of final post-processing. This enables our approach to generalize across multiple palm leaf manuscript datasets containing documents with varying scripts and text line densities.

3   Approach

Overview: Given the input palm leaf manuscript image, our objective is to generate tight-fitting polygons enclosing each of the text lines. Our processing pipeline has two stages – 'scribble generation' (Sect. 3.1) and 'text line polygon generation' (Sect. 3.2) – see Fig. 2. In the first stage, the manuscript image

Fig. 2. An outline of our SeamFormer pipeline for manuscript line segmentation (Sect. 3).

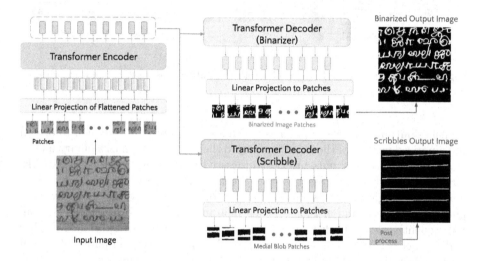

Fig. 3. Stage I: Scribble Generation Module - see Sect. 3.1.

is processed by a deep network which generates coarse binary medial blobs for each individual text line and a binarized version of the image. The medial blobs are further processed to extract coarse spatial identifiers for each line termed as 'scribbles'. In the second stage, scribbles from first stage and custom-designed feature maps derived from binarized image are fed to a seam generation algorithm which generates the desired tight-fitting polygons enclosing the individual text lines.

3.1 Stage I: Scribble Generation

We set up a multi-task variant of Vision Transformer (ViT) deep network architecture [16] to obtain two outputs - the binarized version of the input manuscript image and the medial blob masks for each text line (see Fig. 3). In a conventional ViT architecture, position-encoded patches of input image are processed within a Transformer [52] framework employing multi-head attention to obtain output patches. We extend the conventional setup to have two decoder branches. These

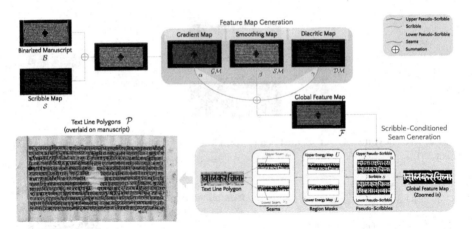

Fig. 4. Stage II: Text Line Polygon Generation Module - see Sect. 3.2 and Algorithm 1 for details.

branches output two sets of patches which are separately reassembled to obtain the binarized version of the input image and the medial blob masks binary image.

The blob mask outputs are post-processed to extract thin medial axis-like structures which cut across the line. We broadly classify our post-processing into local and global stages. In local post processing, we iteratively apply morphological dilation and erosion on each blob mask and perform skeletonisation. Subsequently, we apply skeleton pruning techniques to remove spurious branches and extract a clean medial fragment for each blob within the patch. We term these fragments as 'scribbles'. For the global post processing, we merge these patches to obtain a scribble map with the input image's dimensions. Given the fragments of scribbles, we group them based on distance thresholding technique as a function of its horizontal level.

The scribble, by nature of its construction, provides crucial information regarding local curvature of the text line. As we shall see, accurate determination of local curvature plays a key role for the next stage of processing and ultimately, for accurate text line segmentation.

3.2   Stage II: Text Line Polygon Generation

This stage involves two sub-stages – Feature Map Generation and Scribble-conditioned Seam Generation (see Fig. 4). For each scribble, we first generate a corresponding pair of *pseudo-scribbles* which are used at later stages of the pipeline (Sect. 3.2.1). Next, the scribbles are overlaid on binarized input image and the resulting scribble-overlaid image is used to create custom feature maps (Sect. 3.2.2). These feature maps are used as input to a seam generation procedure to generate the desired high-precision polygons enclosing the text lines (Sect. 3.2.3).

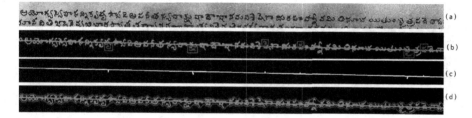

Fig. 5. Diacritic Map (Sect. 3.2.2) - (a) A text line from a palm leaf manuscript, (b) the reference text line is shown with the scribble overlaid. Pixels in green denote the text line connected by the scribble and pixels in red inside pink bounding boxes denote the corresponding diacritics of the parent text line (c) Diacritic Feature Map - note the tiny strokes extending out of the scribble to connect the diacritics with the main text line (d) final red seams enclosing the text line as a result of using the Diacritic map during seam generation - note that the aforementioned diacritics have been brought inside the enclosing seams.

3.2.1 Pseudo-Scribble Generation

As the first step, we sort the scribbles by the y-coordinate of the left-most point to obtain the sequence of scribbles S in a top-to-bottom order. Let $s_i \in S$ be a scribble. Let μ_{s_i} be the average of all y-coordinates of the scribble s_i's pixels. Let $\mu_{s_{i+1}}$ be a similar average for the neighboring scribble. The vertical offset between the scribble pair can be defined as $d(s_i, s_{i+1}) = |\mu_{s_i} - \mu_{s_{i+1}}|$. Let $\overline{d(S)}$ denote the average across all such vertical offsets within the set of scribbles. Define $\theta = \overline{d(S)} + \delta$ where δ is a fixed offset. For each scribble s, the upper pseudo-scribble (u) and lower pseudo-scribble (l), are obtained by vertically translating s by $+\theta$ and $-\theta$ pixels respectively – see the block 'Pseudo scribbles' which is part of 'Scribble-Conditioned Seam-Generation' (shaded blue) in Fig. 4.

3.2.2 Feature Map Generation

Gradient Map (GM): This feature map is obtained as the gradient magnitude map of the scribble-overlaid image. Using this map creates a high energy barrier between edges of characters in the text line and the background area immediately surrounding them. Employing this map in the subsequent seam generation stage enables seams to align closely with text letter boundaries, resulting in tight-fitting polygons around the text lines (ref. $\mathcal{GM}$ in Fig. 4).

Smoothing Map (SM): This feature map is obtained by applying a blur kernel on the scribble-overlaid image. Using this map increases the energy at horizontal inter-character text gaps and ensures that seams do not cut through the text (ref. $\mathcal{SM}$ in Fig. 4).

Diacritic Map (DM): This novel feature map specifically tackles the problem of diacritics not being enclosed within the polygons of corresponding parent text

Fig. 6. (a) A fragment from the top portion of a manuscript (b) Seams generated with Gradient and Smoothing Map, but without using scribble – the upper line boundary is missing (c) Seams when scribble is also added – upper line boundary is obtained, but diacritics are missed (d) Seams when Diacritic Map is also included – line boundaries properly enclose text and associated diacritic components.

lines - see Fig. 5. We first isolate the region around each text line with the help of upper and lower pseudo-scribbles as the demarcations. We overlay the corresponding scribble on the parent text-line and perform connected components analysis. This operation divides the components into three major groups: components connected to parent-line, disconnected diacritics and background noisy elements. We discard noise based on an area threshold. For each diacritic component, we connect its centroid and parent scribble via a perpendicular line. In effect, this line creates an energy barrier which forces the boundary generated during seam generation to move around the diacritic instead of separating the diacritic and its parent text line (ref. $\mathcal{DM}$ in Fig. 4). The utility of Diacritic Map is illustrated in Fig. 5. The neighborhood of a text line often contains text fragments from adjacent lines due to the uneven handwritten line orientation and dense handwriting. Our construction of the Diacritic Map actively prevents the neighboring text fragments from being picked up along with the diacritics.

The weighted combination of the above feature maps forms the final global feature map, i.e. $\mathcal{F} = \alpha\,\mathcal{GM} + \beta\,\mathcal{SM} + \gamma\,\mathcal{DM}$. Figure 6 illustrates the importance of using scribbles and the proposed combination of energy maps. It is important to note that unlike some of the existing seam-based approaches [42], we generate the feature map only once for the input image.

3.2.3   Scribble-Conditioned Seam Generation

For each scribble s, the paired end-points of the scribble and its corresponding upper pseudo-scribble u are connected to obtain an enclosed upper region U - see the block 'Region Masks' which is part of 'Scribble-Conditioned Seam-Generation' (shaded blue) in Fig. 4. The region's mask is applied to global feature map $\mathcal{F}$ and cropped to obtain the upper region feature map $\mathcal{F}_U$ for the scribble. To constrain the seams to lie within the masked portion, feature map values outside the mask are set to a fixed 'high energy' value. The upper region feature map is used during seam generation [5].

Algorithm 1 . Scribble-Conditioned Text Line Polygon Generation (Sect. 3.2.3)

1: ▷ **Input** binaryImage B and set of scribbles $\mathcal{S}$ from Stage I (Sect. 3.1)
2: ▷ **Output** Set of text line polygons $\mathcal{P}$
3: $\theta \leftarrow$ COMPUTEGAP($\mathcal{S}$)   ▷ Obtain interline gap using inter-scribble gap statistics
4: ▷ Feature Map Generation
5: $\mathcal{GM} \leftarrow$ GENERATEGRADIENTMAP($B, \mathcal{S}$)
6: $\mathcal{SM} \leftarrow$ GENERATESMOOTHINGMAP($B, \mathcal{S}$)
7: $\mathcal{DM} \leftarrow$ GENERATEDIACRITICMAP($B, \mathcal{S}$)
8: $\mathcal{F} \leftarrow$ GENERATEGLOBALFEATUREMAP($\mathcal{GM}, \mathcal{SM}, \mathcal{DM}$)
9: ▷ Scribble-conditioned Seam Generation
10: **for** s in $\mathcal{S}$ **do**   ▷ For each scribble
11:     $u, l \leftarrow$ GENERATEPSEUDOSCRIBBLES(s, θ)
12:     ▷ Generate upper seam
13:     $U \leftarrow$ GETREGION(s, u)
14:     $\mathcal{F}_U \leftarrow$ GETCROPPEDFEATUREMAP($U, \mathcal{F}$)
15:     $S_U \leftarrow$ GENERATESEAMS($\mathcal{F}_U$)
16:     ▷ Generate lower seam
17:     $L \leftarrow$ GETREGION(s, l)
18:     $\mathcal{F}_L \leftarrow$ GETCROPPEDFEATUREMAP($L, \mathcal{F}$)
19:     $S_L \leftarrow$ GENERATESEAMS($\mathcal{F}_L$)
20:     ▷ Generate the final text line polygon
21:     $P \leftarrow$ GENERATELINEPOLYGON(S_U, S_L)
22:     $\mathcal{P} \leftarrow \mathcal{P} \cup \{P\}$
23: **end for**
24: **return** $\mathcal{P}$

For a $M \times N$ image, a horizontal seam R is a connected sequence of pixels and can be defined as $R = (x_i, y_i); i = 1, 2, \ldots r, 1 \leqslant x_i \leqslant N, 1 \leqslant y_i \leqslant M$ where $x_1 = 1, x_r = N$ and $|x_i - x_{i-1}| \leqslant 1, i = 2, 3, \ldots r$. The 'energy cost' of the seam is defined as $U(R) = \sum_{i=1}^{r} \mathcal{F}_U(x_i, y_i)$. The seam with the minimum cost is defined as $S_U = \arg \min_{R} U(R)$ and is found using dynamic programming. In this context, feature map $\mathcal{F}_U$ has been constructed such that the minimum energy seam corresponds to tight upper boundary of the associated text line. Additionally, to enhance the tight-fit characteristic of the seam, we induce a bias in choosing the lowest energy path. During the seam propagation step, we greedily pick the lowest x or y coordinate value among potential energy paths. This choice results in energy seams circumscribing the character components tightly. A similar procedure as above is repeated with the lower pseudo-scribble l to obtain a tight lower boundary seam S_L for the text line. These seams (S_U, S_L) are connected at their paired endpoints to obtain the final high precision polygon P enclosing the text line.

It is important to note that the scribble generated in Stage-I determines the sub-image region in which seam generation operates. Confining seam generation by using scribble-based masks helps produce compact enclosing boundaries (see Fig. 6). This is unlike other seam-based methods which generate seams that go beyond actual extent of the text line. Algorithm 1 outlines the procedure for scribble-conditioned text line polygon generation.

4   Experiments

4.1   Datasets

We have tested the models on a selection of palm leaf manuscript datasets - Indiscapes2 [44], the datasets provided for the Challenge B (Text Line Segmentation) of the ICFHR 2018 Competition On Document Image Analysis Tasks for Southeast Asian Palm Leaf Manuscripts [24] containing manuscripts from Balinese, Khmer and Sundanese languages. In addition, a new manuscript collection called KgathaM has also been introduced.

Indiscapes2 [38]: This is the largest dataset for Indic palm leaf manuscripts and consists of manuscripts sourced from four distinct sources. Indiscapes2 comprises of 1275 documents with a large diversity in scripts, language,semantic regions, document dimensions, number of lines and text line density. It has 748 manuscript leaves for training and 258 leaves for the test split. The average manuscript dimension is 750×1900.

KgathaM: We introduce this new collection of palm leaf manuscript written in a classical component-based script of the Indic language Malayalam. The manuscript contains verses from a poem. A unique aspect is that the poem is written on manuscript leaves continuously and end to end, without spaces between words. It has a total of 392 pages with $8 - 12$ lines in each document. We have considered 313 pages for train split and 79 pages in the test split. The manuscript leaves are quite dense with an average of 9-10 text lines and contain extremely small character components. The average size of the manuscript page is 400×2800.

Balinese [23]: This consists of Balinese manuscripts. It has been extracted from the AMADI LontarSet [23], with 393 pages of palm leaf manuscripts from 23 different collections. In general, the documents have 4 text lines, most of them double-columned with occasional illustrations. One common characteristic of this manuscript is the variety of diacritics. The Challenge provides a total of 96 pages with 47 pages in the train split and 49 pages in the test split. In general the pages have 4 text lines. The average size of the manuscript page is 500×5000.

Khmer [50]: This set consists of Khmer (Cambodian) manuscripts. It has been extracted from the SleukRith Set [50], with 657 pages of Khmer palm leaf manuscript randomly selected from different sources. In general, the pages have 5 text lines. The Challenge provides a total of 250 pages with 50 pages in the train split and 200 pages in the test split. The average size of the manuscript page is 500×5500.

Sundanese [48]: This set consists of Sundanese manuscripts. It has been extracted from the Sunda Set [48], with 66 pages of Sundanese Lontar randomly selected from 27 collections. The Challenge provides a total of 61 pages with 31 pages in the train split and 30 pages in the test split. On average, the pages consist of 4 text lines. The mean size of the manuscript page is 350×3000.

Table 1. Comparison of SeamFormer with existing approaches on benchmark datasets (Sect. 5).

	Indiscapes2[38]	KGathaM	Bali[23]	Sunda[48]	Khmer[50]
	IoU ↑				
MMRCNN [38]	0.55	0.34	0.23	0.28	0.28
Palmira [44]	0.76	0.69	0.42	0.68	0.45
Doc-UFCN [9]	0.16	0.12	0.08	0.23	0.10
dhSegment [36]	0.34	0.12	0.03	0.12	0.08
LCG [2]	0.37	0.20	0.12	0.12	0.18
DocExtractor [30]	0.10	0.17	0.01	0.02	0.04
SeamFormer	**0.78**	**0.84**	**0.66**	**0.77**	**0.69**
	HD ↓				
MMRCNN [38]	447.58	855.76	2106.30	1147.30	1760.48
Palmira [44]	73.32	57.84	1699.58	130.34	1190.95
Doc-UFCN [9]	339.30	238.87	1873.00	630.79	2552.26
dhSegment [36]	295.58	216.79	2232.90	394.16	1560.45
LCG [2]	207.76	346.93	797.51	367.60	496.31
DocExtractor [30]	806.17	1423.26	3552.19	1865.25	3987.37
SeamFormer	**21.91**	**16.05**	**48.86**	**32.18**	**48.37**
	AvgHD ↓				
MMRCNN [38]	57.13	132.50	302.59	145.07	270.47
Palmira [44]	7.29	2.74	224.79	6.50	203.59
Doc-UFCN [9]	70.04	49.16	319.06	98.55	481.19
dhSegment [36]	60.33	43.60	415.24	66.77	319.57
LCG [2]	16.82	29.72	95.18	39.65	44.50
DocExtractor [30]	149.29	219.68	778.60	331.00	898.16
SeamFormer	**0.65**	**0.25**	**2.53**	**1.01**	**2.39**
	HD95 ↓				
MMRCNN [38]	355.74	702.45	1766.12	918.85	1449.68
Palmira [44]	42.47	21.49	1393.15	49.06	1019.12
Doc-UFCN [9]	304.73	214.35	1628.43	520.38	2271.27
dhSegment [36]	262.83	192.33	1967.72	329.84	1380.09
LCG [2]	99.77	197.94	390.21	231.38	191.88
DocExtractor [30]	595.61	1084.01	3255.44	1656.47	3654.05
SeamFormer	**4.59**	**1.96**	**19.49**	**7.77**	**18.83**

4.2   Implementation Details

Stage-I: For the ViT network, we use 256×256 overlapping manuscript patches with appropriate padding. Resampling is used to overcome the imbalance between text and empty (non-text) patches. For training the binarizer branch for South-East Asian datasets, we use the binary dataset from Challenge A of the ICFHR 2018 contest [24]. For other datasets, we use Sauvola-Niblack binarisation [33,43] as the ground truth. We initialize the binarization branch with pre-trained weights [47]. The learning rate is initialized to 0.05 and is decayed by Pytorch's learning scheduler, ExponentialLR with $\gamma = 0.8$. For training both of these branches we leverage the L2 loss. We adopt a training procedure where every individual branch is trained separately, while the other branch's weights are frozen. The optimizer used is stochastic gradient descent with $\gamma = 0.1$ and momentum of 0.9. We perform data-parallel optimization distributed across 2 GeForce RTX 2080 Ti GPUs for 40 epochs, with a fixed batch size of 4. We use random rotation augmentation $\alpha \in (-30, 30)$ to improve performance for non-axis oriented manuscripts. To tackle varied manuscript background textures and noise, we apply Gaussian Noise, AdvancedBlur, RandomColor, RandomFog, RandomBrightness and HueSaturations augmentations [11]. For post-processing, we apply erosion filters - a horizontal rectangular kernel 1×11 thrice, followed by a 1×1 dilation to separate any overlapping medial blobs. These blobs undergo a skeletonization procedure followed by pruning to remove any spurious branches with a minimum area threshold of 100 pixels. The post-processing is robust and does not need to be changed across datasets or approaches.

Fig. 7. A challenging manuscript from Indiscapes2 [44]. The figure shows insets of regions with ground-truth (blue) and predictions from SeamFormer (green) and Palmira [44] (red). The document level performance scores are shown in bottom right.

Stage-II: The offset for pseudo-scribble generation δ is set to 5. In the feature map generation pipeline, we use the standard 3×3 Sobel kernel for Gradient

Map. We apply a Gaussian blur kernel of 15×11 for high spatial coverage within the image to compute the Smoothing Map. The weights for various feature maps are empirically set to $\alpha = 0.4$ ($\mathcal{GM}$), $\beta = 0.6$ ($\mathcal{SM}$) and $\gamma = 1.0$ ($\mathcal{DM}$). The global feature map is normalised to $[0, 1]$ before the seam generation process.

5 Results

For quantitative evaluation of text line segmentation, we compare SeamFormer against various state-of-the-art approaches developed for palm-leaf and other types of manuscripts. The approaches were fine-tuned for each dataset. As performance measure, we use IoU. In their work, Trivedi et al. [49] show that Hausdorff Distance (HD) and its variants - HD_{95} and Average HD reflect the prediction performance for polygon boundary predictions better than area-based IoU metric. Therefore, we report these measures as well. Note that smaller the HD-based scores, better the text line polygon prediction.

Table 2. Ablation experiments using Indiscapes2. Proposed refers to design choices in SeamFormer.

Row-id	Stage I	Stage II	IoU ↑	HD ↓	HD_{95} ↓	Avg. HD ↓
1	Text Baseline	Proposed	0.63	62.59	8.29	35.86
2	ARU-Net [18]	Proposed	0.69	103.57	9.37	51.51
3	Proposed	$\mathcal{GM}$	0.76	23.83	0.78	5.15
4	Proposed	$\mathcal{GM}, \mathcal{SM}$	0.77	22.40	0.71	4.71
5	**Proposed**	**Proposed**	**0.78**	**21.91**	**0.65**	**4.59**

The overall quantitative results can be seen in Table 1. SeamFormer clearly outperforms the competing strong baseline approaches across all the datasets and across the performance measures. This shows the generalizability provided by our approach. Our consistently small HD scores are due to the high precision polygons generated by our custom scribble-conditioned seam generation pipeline. For some existing approaches, HD-based scores are one or two orders of magnitude higher due to low line accuracy. Most of these approaches resize the input image to a fixed size for optimal training of the neural network. However, due to the extremely large aspect ratio ($\approx 10 : 1$) and range in sizes for palm leaf manuscripts, the resizing causes text line polygon aliasing, causing poor performance. These factors are not an issue for SeamFormer since resizing is not a part of the pipeline. The second-best network Palmira [44] is competitive in terms of IoU for Indiscapes2 [44]. However, the performance gap is substantial for other datasets and other performance measures as well.

5.1   Ablation Study

We perform an ablation analysis with Indiscapes2 dataset to determine the contribution of various design choices within Stage-I (scribble generation) and Stage-II (seam generation). Instead of a medial scribble through the text, we tried the popular text underline (baseline) as an alternative. The bottom of the text line polygon is used as the baseline. However, this led to sub par performance since the baseline is not guaranteed to touch the text and does not prevent seams from cutting in between text components of the lines (row 1 of Table 2). In another experiment, we re-trained the popular ARU-Net [18] as an alternative to our ViT architecture for obtaining scribbles. ARU-Net produces disconnected scribbles which results in poor performance (row 2). Keeping Stage-I fixed, we also conducted experiments to determine the impact of each feature map (rows 3-4). We observe that the full set of feature maps (last row) provides the best performance – also see Fig. 6.

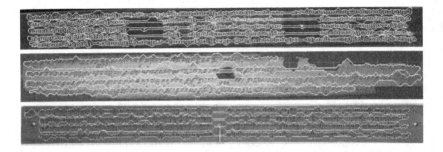

Fig. 8. SeamFormer predictions on South-East Asian Manuscripts [24] – Khmer (top), Sundanese (middle) and Balinese (top).

Fig. 9. SeamFormer predictions on Indiscapes2 [44] manuscripts.

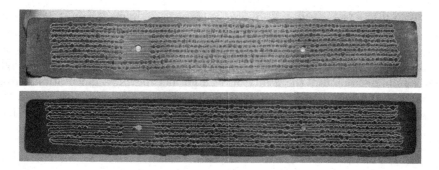

Fig. 10. SeamFormer predictions on manuscripts from the newly introduced KGathaM collection.

5.2   Qualitative Results

A visual comparison of performance between ground-truth and predictions by SeamFormer and the second-best model Palmira [44] can be seen in Fig. 7. The effect of resizing can be seen in Palmira's incorrect and coarse predictions. Despite the challenging nature of the manuscript (e.g. document tilt, dense and unevenly spaced text lines), SeamFormer predictions are significantly more accurate. This trend can also be seen in sample manuscripts from other datasets - see Figs. 8,9,10.

6   Conclusion

We introduce SeamFormer, a novel approach for high precision text line segmentation in handwritten documents. Instead of a monolithic framework, we tackle the challenge of text line segmentation using a divide-and-conquer two stage approach. The first stage generates medial line 'scribbles' which provide crucial information about the curvature of the text line and a binarized version of the input image. In the second stage, these scribbles and custom-designed feature maps derived from the binarized image are fed to a seam generation algorithm which generates the desired tight-fitting line polygons.

Our approach is a resizing-free method. As a result, text line gaps are not distorted or aliased, leading to significantly better results. Our novel inclusion of Diacritic Map in the second stage ensures complete and correct inclusion of diacritics within the predicted polygon. Also, pseudo-scribbles are a key innovation in our approach. The pseudo-scribbles serve as energy barriers during seam generation and ensure the seams do not cross the text line's spatial extents. The pseudo-scribbles also prevent the seams from deviating too much from the reference line unlike some existing approaches. The efficacy of our approach is evident from its comparatively superior performance across challenging datasets and metrics.

An additional advantage of our approach is that it enables interactive human-in-the-loop refinement. For instance, scribbles could be manually added for any missed lines followed by second stage processing. Another advantage is that unlike some existing approaches, no post-processing on the polygons is required. Our results demonstrate the utility of SeamFormer for line segmentation across multiple challenging datasets. Overall, SeamFormer is an attractive option for generating precise text line polygons in handwritten manuscript collections. The source code, pretrained models and associated material are available at https://ihdia.iiit.ac.in/seamformer.

Acknowledgement. This work is supported by Ministry of Electronics and Information Technology (MeiTY), Government of India. We also wish to thank BV Khadiravana for assistance related to KGathaM dataset.

References

1. Alaei, A., Pal, U., Nagabhushan, P.: A new scheme for unconstrained handwritten text-line segmentation. Pattern Recogn. **44**(4), 917–928 (2011)
2. Alberti, M., Vögtlin, L., Pondenkandath, V., Seuret, M., Ingold, R., Liwicki, M.: Labeling, cutting, grouping: an efficient text line segmentation method for medieval manuscripts. In: 2019 International Conference on Document Analysis and Recognition (ICDAR), pp. 1200–1206. IEEE (2019)
3. Arvanitopoulos, N., Süsstrunk, S.: Seam carving for text line extraction on color and grayscale historical manuscripts. In: 2014 14th International Conference on Frontiers in Handwriting Recognition, pp. 726–731. IEEE (2014)
4. Asi, A., Saabni, R., El-Sana, J.: Text line segmentation for gray scale historical document images. In: Proceedings of the 2011 workshop on historical document imaging and processing, pp. 120–126 (2011)
5. Avidan, S., Shamir, A.: Seam carving for content-aware image resizing. In: ACM SIGGRAPH 2007 papers, pp. 10-es (2007)
6. Barakat, B., Droby, A., Kassis, M., El-Sana, J.: Text line segmentation for challenging handwritten document images using fully convolutional network. In: 2018 16th International Conference on Frontiers in Handwriting Recognition (ICFHR), pp. 374–379. IEEE (2018)
7. Barakat, B.K., et al.: Unsupervised deep learning for text line segmentation. In: 2020 25th International Conference on Pattern Recognition (ICPR), pp. 2304–2311. IEEE (2021)
8. Barakat, B.K., El-Sana, J., Rabaev, I.: The pinkas dataset. In: 2019 International Conference on Document Analysis and Recognition (ICDAR), pp. 732–737. IEEE (2019)
9. Boillet, M., Kermorvant, C., Paquet, T.: Multiple document datasets pre-training improves text line detection with deep neural networks. In: 2020 25th International Conference on Pattern Recognition (ICPR), pp. 2134–2141. IEEE (2021)
10. Bruzzone, E., Coffetti, M.C.: An algorithm for extracting cursive text lines. In: Proceedings of the Fifth International Conference on Document Analysis and Recognition. ICDAR 1999 (Cat. No. PR00318), pp. 749–752. IEEE (1999)

11. Buslaev, A., Iglovikov, V.I., Khvedchenya, E., Parinov, A., Druzhinin, M., Kalinin, A.A.: Albumentations: fast and flexible image augmentations. Information **11**(2), 125 (2020). https://doi.org/10.3390/info11020125

12. Chamchong, R., Fung, C.C.: Character segmentation from ancient palm leaf manuscripts in Thailand. In: Proceedings of the 2011 Workshop on Historical Document Imaging and Processing, pp. 140–145 (2011)

13. Chamchong, R., Fung, C.C.: Text line extraction using adaptive partial projection for palm leaf manuscripts from Thailand. In: 2012 International Conference on Frontiers in Handwriting Recognition, pp. 588–593. IEEE (2012)

14. Clausner, C., Antonacopoulos, A., Derrick, T., Pletschacher, S.: ICDAR 2019 competition on recognition of early Indian printed documents-REID2019. In: 2019 International Conference on Document Analysis and Recognition (ICDAR), pp. 1527–1532. IEEE (2019)

15. Dolfing, H.J., Bellegarda, J., Chorowski, J., Marxer, R., Laurent, A.: The "scribblelens" dutch historical handwriting corpus. In: 2020 17th International Conference on Frontiers in Handwriting Recognition (ICFHR), pp. 67–72. IEEE (2020)

16. Dosovitskiy, A., et al.: An image is worth 16x16 words: transformers for image recognition at scale. ICLR (2021)

17. Grüning, T., Labahn, R., Diem, M., Kleber, F., Fiel, S.: Read-bad: a new dataset and evaluation scheme for baseline detection in archival documents. In: 2018 13th IAPR International Workshop on Document Analysis Systems (DAS), pp. 351–356. IEEE (2018)

18. Grüning, T., Leifert, G., Strauß, T., Michael, J., Labahn, R.: A two-stage method for text line detection in historical documents. Int. J. Doc. Anal. Recogn. (IJDAR) **22**(3), 285–302 (2019). https://doi.org/10.1007/s10032-019-00332-1

19. He, J., Downton, A.C.: User-assisted archive document image analysis for digital library construction. In: Seventh International Conference on Document Analysis and Recognition, 2003. Proceedings, pp. 498–502. IEEE (2003)

20. He, K., Gkioxari, G., Dollár, P., Girshick, R.: Mask R-CNN. In: ICCV (2017)

21. Jindal, A., Ghosh, R.: Text line segmentation in Indian ancient handwritten documents using faster R-CNN. Multimedia Tools Appl. **82**, 1–20 (2022)

22. Kesiman, M.W.A., Burie, J.C., Ogier, J.M.: A new scheme for text line and character segmentation from gray scale images of palm leaf manuscript. In: 2016 15th International Conference on Frontiers in Handwriting Recognition (ICFHR), pp. 325–330. IEEE (2016)

23. Kesiman, M.W.A., Burie, J.C., Wibawantara, G.N.M.A., Sunarya, I.M.G., Ogier, J.M.: Amadi_lontarset: the first handwritten balinese palm leaf manuscripts dataset. In: 2016 15th International Conference on Frontiers in Handwriting Recognition (ICFHR), pp. 168–173. IEEE (2016)

24. Kesiman, M.W.A., et al.: ICFHR 2018 competition on document image analysis tasks for southeast asian palm leaf manuscripts. In: 2018 16th International Conference on Frontiers in Handwriting Recognition (ICFHR), pp. 483–488 (2018). https://doi.org/10.1109/ICFHR-2018.2018.00090

25. Kesiman, M.W.A., et al.: Benchmarking of document image analysis tasks for palm leaf manuscripts from Southeast Asia. J. Imaging **4**(2), 43 (2018)

26. Kleber, F., Fiel, S., Diem, M., Sablatnig, R.: CVL-database: an off-line database for writer retrieval, writer identification and word spotting. In: 2013 12th International Conference on Document Analysis and Recognition, pp. 560–564. IEEE (2013)

27. Li, D., Wu, Y., Zhou, Y.: Linecounter: learning handwritten text line segmentation by counting. In: 2021 IEEE International Conference on Image Processing (ICIP), pp. 929–933. IEEE (2021)

28. Likforman-Sulem, L., Faure, C.: Extracting text lines in handwritten documents by perceptual grouping. Adv. handwriting drawing multi. approach, 117–135 (1994)
29. Mechi, O., Mehri, M., Ingold, R., Amara, N.E.B.: Text line segmentation in historical document images using an adaptive U-Net architecture. In: 2019 International Conference on Document Analysis and Recognition (ICDAR), pp. 369–374. IEEE (2019)
30. Monnier, T., Aubry, M.: docExtractor: an off-the-shelf historical document element extraction. In: ICFHR (2020)
31. Nagy, G., Seth, S.C., Stoddard, S.D.: Document analysis with an expert system. In: Pattern recognition in practice II, pp. 149–155 (1985)
32. Nguyen, T.N., Burie, J.C., Le, T.L., Schweyer, A.V.: An effective method for text line segmentation in historical document images. In: 2022 26th International Conference on Pattern Recognition (ICPR), pp. 1593–1599. IEEE (2022)
33. Niblack, W.: An introduction to digital image processing. Strandberg Publishing Company (1985)
34. Nikolaidou, K., Seuret, M., Mokayed, H., Liwicki, M.: A survey of historical document image datasets. Int. J. Doc. Anal. Recognit. **25**(4), 305–338 (2022). https:// doi.org/10.1007/s10032-022-00405-8
35. O'Gorman, L.: The document spectrum for page layout analysis. IEEE Trans. Pattern Anal. Mach. Intell. **15**(11), 1162–1173 (1993)
36. Oliveira, S.A., Seguin, B., Kaplan, F.: dhSegment: a generic deep-learning approach for document segmentation. In: 2018 16th International Conference on Frontiers in Handwriting Recognition (ICFHR), pp. 7–12. IEEE (2018)
37. Pavildas, T.: Page segmentation by white streams. In: Proceeding of the 1st International Conference Document Analysis and Recognition, pp. 945–953 (1991)
38. Prusty, A., Aitha, S., Trivedi, A., Sarvadevabhatla, R.K.: Indiscapes: instance segmentation networks for layout parsing of historical indic manuscripts. In: ICDAR, pp. 999–1006 (2019)
39. Pu, Y., Shi, Z.: A natural learning algorithm based on hough transform for text lines extraction in handwritten documents. Adv. Handwriting Recogn. **34**, 141–150 (1999). World Scientific
40. Renton, G., Soullard, Y., Chatelain, C., Adam, S., Kermorvant, C., Paquet, T.: Fully convolutional network with dilated convolutions for handwritten text line segmentation. Int. J. Doc. Anal. Recogn. (IJDAR) **21**(3), 177–186 (2018). https:// doi.org/10.1007/s10032-018-0304-3
41. Ronneberger, Olaf, Fischer, Philipp, Brox, Thomas: U-Net: convolutional networks for biomedical image segmentation. In: Navab, Nassir, Hornegger, Joachim, Wells, William M.., Frangi, Alejandro F.. (eds.) MICCAI 2015. LNCS, vol. 9351, pp. 234–241. Springer, Cham (2015). https://doi.org/10.1007/978-3-319-24574-4_28
42. Saabni, R., El-Sana, J.: Language-independent text lines extraction using seam carving. In: 2011 International Conference on Document Analysis and Recognition, pp. 563–568. IEEE (2011)
43. Sauvola, J., Pietikäinen, M.: Adaptive document image binarization. Pattern Recogn. **33**(2), 225–236 (2000)
44. Sharan, S.. P.., Aitha, Sowmya, Kumar, Amandeep, Trivedi, Abhishek, Augustine, Aaron, Sarvadevabhatla, Ravi Kiran: Palmira: a deep deformable network for instance segmentation of dense and uneven layouts in handwritten manuscripts. In: Lladós, Josep, Lopresti, Daniel, Uchida, Seiichi (eds.) ICDAR 2021. LNCS, vol. 12822, pp. 477–491. Springer, Cham (2021). https://doi.org/10.1007/978-3-030-86331-9_31

45. Shi, Z., Setlur, S., Govindaraju, V.: Text extraction from gray scale historical document images using adaptive local connectivity map. In: Eighth International Conference on Document Analysis and Recognition (ICDAR 2005), pp. 794–798. IEEE (2005)

46. Shi, Z., Setlur, S., Govindaraju, V.: A steerable directional local profile technique for extraction of handwritten arabic text lines. In: 2009 10th International Conference on Document Analysis and Recognition, pp. 176–180. IEEE (2009)

47. Souibgui, M.A., et al.: DocEntr: an end-to-end document image enhancement transformer. In: 2022 26th International Conference on Pattern Recognition (ICPR), pp. 1699–1705 (2022)

48. Suryani, M., Paulus, E., Hadi, S., Darsa, U.A., Burie, J.C.: The handwritten sundanese palm leaf manuscript dataset from 15th century. In: 2017 14th IAPR International Conference on Document Analysis and Recognition (ICDAR), vol. 1, pp. 796–800. IEEE (2017)

49. Trivedi, A., Sarvadevabhatla, R.K.: BoundaryNet: an attentive deep network with fast marching distance maps for semi-automatic layout annotation. In: Lladós, J., Lopresti, D., Uchida, S. (eds.) ICDAR 2021. LNCS, vol. 12821, pp. 3–18. Springer, Cham (2021). https://doi.org/10.1007/978-3-030-86549-8_1

50. Valy, D., Verleysen, M., Chhun, S., Burie, J.C.: A new khmer palm leaf manuscript dataset for document analysis and recognition: SleukRith set. In: Proceedings of the 4th International Workshop on Historical Document Imaging and Processing, pp. 1–6 (2017)

51. Valy, D., Verleysen, M., Sok, K.: Line segmentation for grayscale text images of khmer palm leaf manuscripts. In: 2017 Seventh International Conference on Image Processing Theory, Tools and Applications (IPTA), pp. 1–6. IEEE (2017)

52. Vaswani, A., et al.: Attention is all you need. Adv. Neural Inf. Process. Syst. **30** (2017)

53. Yalniz, I.Z., Manmatha, R.: A fast alignment scheme for automatic OCR evaluation of books. In: 2011 International Conference on Document Analysis and Recognition, pp. 754–758. IEEE (2011)

54. Zahour, A., Taconet, B., Mercy, P., Ramdane, S.: Arabic hand-written text-line extraction. In: Proceedings of Sixth International Conference on Document Analysis and Recognition, pp. 281–285. IEEE (2001)

SegCTC: Offline Handwritten Chinese Text Recognition via Better Fusion Between Explicit and Implicit Segmentation

Jiarong Huang[1], Dezhi Peng[1], Hongliang Li[1], Hao Ni[3], and Lianwen Jin[1,2(✉)]

[1] South China University of Technology, Guangzhou, China
[2] SCUT-Zhuhai Institute of Modern Industrial Innovation, Zhuhai, China
eelwjin@scut.edu.cn
[3] University College London, London, UK

Abstract. Handwritten Chinese text recognition (HCTR) is still a challenging and unsolved problem. The existing recognition methods are mainly categorized into two: explicit vs implicit segmentation-based methods. Explicit segmentation recognition methods use explicit character location information to train the recognizers. However, the widely used weakly supervised training strategy based on pseudo-label makes it difficult to get effective supervised training for difficult character samples. In contrast, the implicit segmentation recognition method use all transcript annotations for supervised training, but it is prone to misalignment problem due to the lack of explicit supervised information of character positions. To take advantage of the complementary nature of explicit and implicit segmentation approaches, we propose a new method, SegCTC, which better integrates these two approaches into a unified to be a more powerful recognizer. Specifically, we designed a hybrid Segmentation-based and Segmentation-free Feature Fusion Module (S^2FFM) to better fuse the features of both explicit and implicit segmentation-based recognition branches. Moreover, a co-transcription strategy is also proposed to better combine the predictions from different branches. Experiments on four widely used benchmarks including CASIA-HWDB, ICDAR2013, SCUT-HCCDoc and MTHv2 show that our method achieves state-of-the-art performance for the HCTR task under different scenarios.

Keywords: Handwritten Chinese text recognition · Branch feature fusion · Co-transcription

1 Introduction

Handwritten Chinese text recognition (HCTR) is still regarded as a challenging and unsolved problem mainly owing to numerous character categories, diverse writing styles, and frequent character touching or overlapping problem. The current mainstream recognition methods are mainly categorized into two: explicit segmentation-based and implicit segmentation-based recognition.

© The Author(s), under exclusive license to Springer Nature Switzerland AG 2023
G. A. Fink et al. (Eds.): ICDAR 2023, LNCS 14190, pp. 332–349, 2023.
https://doi.org/10.1007/978-3-031-41685-9_21

Implicit segmentation-based recognition methods based on hidden Markov model (HMM), connectionist temporal classification (CTC) [3], and attention mechanisms have achieved great success in text recognition of both scene and document scenarios. Compared with explicit segmentation-based recognition methods using over-segmentation strategy [21–23], implicit segmentation-based methods require only transcript annotations for training, and do not require the costly annotations of character bounding boxes. However, Chinese characters have more complex two-dimensional structures, and some components of Chinese characters can be used as separate characters, in which case the implicit segmentation-based recognition methods are prone to misalignment.

Unlike Latin script, Chinese characters are more independent from each other, so it is more intuitive to segment a text line into separate characters before recognizing them. The explicit segmentation methods are more in line with this practice. In addition, to overcome the reliance on character bounding box annotations of explicit segmentation-based recognition methods, weakly supervised training strategies based on pseudo-label have been proposed in recent years [1,14,31]. The existing weakly supervised training strategies based on pseudo-label use synthetic data with character bounding box annotations for pre-training. In the training phase, the real data samples without character bounding box annotations are first inferred, and then the "reliable" results filtered by certain rules are used as the pseudo-label to supervise the training. For characters that are difficult to classify, the prediction results have a high probability of being "unreliable", and therefore no corresponding pseudo-labels can be generated for further training. This makes it difficult to further improve the recognition ability for difficult samples.

To verify whether the above conjectures on the advantages and disadvantages of these two categories of recognition methods are correct, we trained two recognition models based on explicit segmentation [14] and implicit segmentation (CTC-based) respectively using the same backbone and data. The percentages of the three types of prediction errors (insert, delete, and substitution) using these two models inferring on ICDAR2013-Offline handwritten Chinese text line dataset are shown in Tabel 1. For misalignment errors such as "insert" and "delete", implicit segmentation-based model partitioning occurs more frequently, while explicit segmentation-based model is more prone to misclassification error "substitution". This validates our aforementioned assumption: for the implicit segmentation-based model, misalignment is the bigger problem; while the explicit segmentation-based model trained with pseudo-label is prone to misclassification errors owing to the incomplete generation of pseudo-label.

To exploit the complementary nature of the two categories of recognition methods, we design SegCTC, a new HCTR model integrating the advantages and complement the disadvantages of them. We propose hybrid Segmentation-based and Segmentation-free Feature Fusion Module (S^2FFM) to fuse explicit segmentation-based and implicit segmentation-based branch. S^2FFM enables two different types of supervision to be back-propagated to each others. In the inference phase, both branches of explicit and implicit segmentation can benefit from the features extracted from the other branch. Moreover, we introduced a

Table 1. The percentages of three types of prediction errors tested with explicit segmentation (Expl.) method and implicit segmentation (Impl.) method on ICDAR2013-Offline. Three types of errors are insert (I), delete (D), and substitution (S).

Expl. Method			Impl. Method		
I	D	S	I	D	S
4.04%	14.85%	81.11%	3.22%	19.42%	77.36%

co-transcription strategy (Co-T) to fully combine the prediction from the two different branches for more accurate results. The experiments are conducted on four widely used benchmarks including CASIA-HWDB, ICDAR 2013, SCUT-HCCDoc, and MTHv2. Our method achieves state-of-the-art performance on these datasets.

To summarize, the main contributions of this paper are as follows:

- We propose a new HCTR model, SegCTC, which combines the advantages of recognition branches based on explicit segmentation and implicit segmentation to be a more powerful recognizer.
- We design hybrid Segmentation-based and Segmentation-free Feature Fusion Module (S^2FFM) to fuse two different recognition branches, which is more fully fused two different branches than directly connected two branches in parallel behind the backbone.
- We introduce co-transcription strategy (Co-T) for a more accurate prediction text.
- Extensive experiments show that SegCTC achieves state-of-the-art performance on multiple offline Chinese handwritten text line benchmarks.

2   Related Works

The goal of handwritten Chinese text recognition (HCTR) is to transcribe Chinese handwritten text line images into the corresponding text. The current mainstream text recognition methods are roughly divided into two categories: explicit segmentation methods and implicit segmentation methods.

2.1   Explicit Segmentation Methods

Most of the previous explicit segmentation methods [21–23] were based on an over-segmentation strategy by first over-segmenting the text line images and then searching for the best segmentation-recognition path based on information such as classification results, language model, and geometric background. The explicit segmentation methods based on over-segmentation strategy are easily affected by touching or overlapping characters. In addition to the methods of using over-segmentation strategies, Peng et al. [15] propose a three-branch architecture for end-to-end handwritten Chinese text segmentation and recognition.

The previous methods described above usually requires character bounding box annotation, which is more time-consuming. To solve this problem, Wang et al. [25] proposed a over-segmentation-based model with weakly supervised training which minimizes the marginal log-likelihood on a string-level annotations. The method of Peng et al. [14] treats characters that match as "equal" in the calculation of the edit distance between the predicted text and the annotated text as "reliable" predictions for pseudo-label generation. The prerequisite for the pseudo-label to be generated is the correct character classification. For characters that are difficult to classify, their pseudo-labels cannot be generated, resulting in their inability to be further trained.

2.2 Implicit Segmentation Methods

Compared with the explicit segmentation recognition methods which require character bounding box annotations, the implicit segmentation recognition methods only require transcript annotations. There are three main categories of implicit segmentation recognition methods: hidden Markov model (HMM) based, connectionist temporal classification (CTC) [3] based, and attention mechanism based. The HMM-based methods [2,18,26,27] use a sliding-window manner for feature extraction, and cascading HMMs for modeling. Another category of implicit segmentation recognition is based on CTC. The methods of this category usually first extracts text line image into frames using CNN, then models the contextual relationship using RNN, and finally aligns using CTC. Messina et al. [12] combined multi-dimensional LSTM (MDLSTM) and CTC to solve the problem of HCTR. Wu et al. [28] introduced separable MDLSTM to reduce the computation consumption of RNN structure. Liu et al. [7] built a fast HCTR model with only convolutional layers and CTC loss, along with a Transformer-based language model with context beam search strategy applicable to CTC methods. The attention-based approaches [10,17,24] that are widely used for scene text recognition can also be used for HCTR. Xiu et al. [32] use a multi-level multi-modal fusion network to improve the attention mechanism-based decoder.

2.3 Combination of Explicit and Implicit Segmentation Methods

Since the recognition methods of explicit and implicit segmentation have different advantages, some scholars have tried to combine them together to improve the recognition performance. Zhu et al. [36] combine CTC-based and over-segmentation strategy-based recognition results using convolutional combination strategy. However, this method only uses the text predicted by the recognizers rather than probabilistic information, and it requires additional training of the combination network. To solve the problem that the over-segmentation methods is difficult to recognize overlapping and touching characters, Tanaka et al. [19] add a CTC recognition network to assist character segmentation. But this method does not feed back the segmentation results to the CTC recognition network to improve its recognition performance.

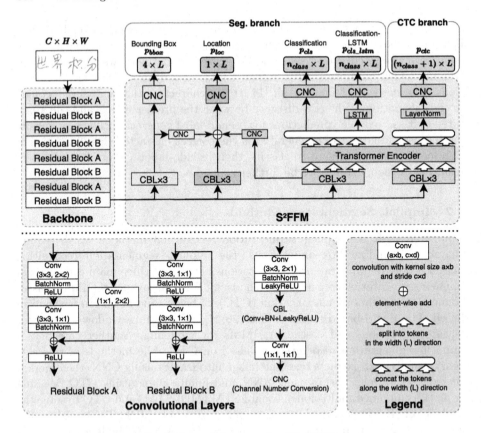

Fig. 1. Overall framework of SegCTC.

3   Proposed Methodology

3.1   Overall Framework

The overall structure of SegCTC is illustrated in Fig. 1. Our model consists of three parts: a backbone of CNN for extracting features, a hybrid Segmentation-based and Segmentation-free Feature Fusion Module (S²FFM) for integrating different types of features, and an output layer for outputting predicting results.

Following ResNet-18 [4], our backbone is stacked with eight residual blocks, as shown in Fig. 1. Given an input image $I \in \mathbb{R}^{C \times H \times W}$ (C, H, and W are the number of channels, height, and width of the image, respectively), the backbone will downsample the width and height by a factor of 16 to obtain feature maps $F \in \mathbb{R}^{512 \times \frac{H}{16} \times \frac{W}{16}}$.

The feature maps $F \in \mathbb{R}^{512 \times \frac{H}{16} \times \frac{W}{16}}$ extracted by the backbone will be fed into S²FFM to integrate different types of features (Subsect. 3.2). S²FFM will output four one-dimensional feature maps whose heights have been downsampled to 1 as

$$p_{loc} \in \mathbb{R}^{1 \times L}, p_{bbox} \in \mathbb{R}^{4 \times L}, p_{cls} \in \mathbb{R}^{nclass \times L}, p_{ctc} \in \mathbb{R}^{(nclass+1) \times L} \quad (1)$$

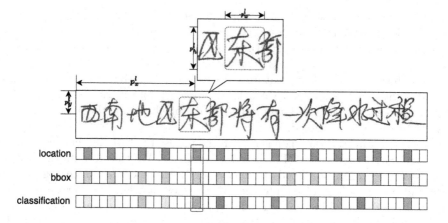

Fig. 2. The output of SegCTC's Seg. branch and the representation of the character bounding box.

where p_{loc}, p_{bbox}, and p_{cls} denote character location, character bounding box, and character classification outputs of explicit segmentation recognition branch (hereinafter called Seg. branch) following [14]. A schematic of these three outputs is shown in Fig. 2. The p_{ctc} is the output of implicit segmentation recognition branch (hereinafter called CTC branch). The L is the width of the four 1D feature maps, which can be considered to divide each feature map equally into L frames along the width direction. The $nclass$ denotes to the number of character categories.

It is same as general object detection for Seg. branch to obtain character bounding boxes. The frames with confidence less than the confidence threshold t_{conf} will first be discarded first. Then, non-maximum suppression (NMS) [13] with the threshold intersection over union (IoU) t_{IoU} will be used to remove duplicate bounding boxes. The remaining bounding boxes are sorted from left to right according to the corresponding p_x to obtain the final predicted text and the bounding box for each character. In our method, t_{conf} is set to 0.55, and t_{IoU} is set to 0.15.

3.2 Hybrid Segmentation-Based and Segmentation-Free Feature Fusion Module (SFFM)

As discussed in Sect. 1, explicit segmentation recognition methods can use character localization information to cope with the alignment failure problem, but suffers from the character classification error problem, while the implicit segmentation recognition method solves the character classification error problem by the full amount of transcript supervision, but suffers from the alignment failure problem. Owing to the complementary nature of these two recognition methods, we introduce two different branches in our model: Seg. branch and CTC branch, and as well propose hybrid Segmentation-based and Segmentation-free

Feature Fusion Module (S^2FFM). This module enables the back-propagation of supervision from CTC branch to Seg. branch as a complement to weakly supervision based on pseudo-label. Meanwhile, character location supervision from Seg. branch can be back-propagation to CTC branch.

As shown in Fig. 1, the feature map extracted by the backbone F is first divided into four heads, and then downsampled 8-fold in the height direction using three Conv + BN + LeakyReLU (CBL) blocks, while the width direction is not further downsampled.

A Transformer encoder [20] is introduced to fuse the two different recognition heads of Seg. branch and CTC branch after CBLs. We concatenate the feature maps of the recognition heads from Seg. branch and CTC branch together in the width direction and feed into the Transformer encoder. In other words, each frame in the width direction of the feature maps in the two heads is considered as a separate token. After the Transformer encoder output, we combine the tokens in the original order to revert to two recognition heads. The two revived recognition heads continue to complete the classification prediction through a channel number conversion block (CNC). The Transformer encoder can fully fuse each token through self-attention mechanism, allowing each head to be fully aware of information from tokens in the same and different heads when outputting. Both of the different types of losses can also be backward to each recognition head for complementary supervision.

To further the semantic capture capability of the model, we refer to the approach in [14] and add another output path via LSTM [5] in the character classification head of the Seg. branch. This output path with LSTM will only be kept during training and will be discarded during inference. In the Seg. branch, the location head will combine the features from the bounding box head and the character classification head for better character localization.

3.3 Weakly Supervision Strategy

The pseudo-label-based weakly supervised training strategy that we use for Seg. branch follows [14]. The whole training process is divided into two stages: pre-training phase and training phase.

In the pre-training phase, we need synthetic data with character bounding box annotations for training, so that the model has the basic character segmentation capability. The character bounding box and classification annotations will be directly involved in the calculation of the loss as

$$l_{loc} = -0.5[\frac{1}{|P_{loc}|} \sum_{l \in P_{loc}} \log(p_{loc}^l) + \frac{1}{|N_{loc}|} \sum_{l \in N_{loc}} \log(1 - p_{loc}^l)] \qquad (2)$$

$$l_{bbox} = \frac{1}{|P_{loc}|} \sum_{l \in P_{loc}} (g_{bbox}^l - p_{bbox}^l)^2 \qquad (3)$$

$$l_{cls} = -\frac{1}{|P_{loc}|} \sum_{l \in P_{loc}} \log(p_{cls}^{l,g_{cls}^l}) \qquad (4)$$

where P_{loc} indicates the frame set where the character centers is located, and N_{loc} is the complement of P_{loc}. Symbol p_{loc}^l, p_{bbox}^l, p_{cls}^l stand for character location, character bounding boxes, character classification outputs in frame l, and their corresponding ground truth are g_{bbox}^l, g_{cls}^l. Meanwhile, CTC loss will be used to supervise CTC branches as

$$l_{ctc} = \text{CTCLoss}(p_{ctc}, g_{ctc}) \tag{5}$$

where p_{ctc} and g_{ctc} stand for the output of CTC branch and the ground truth of text content. The total loss is given by

$$l_{total} = l_{loc} + l_{bbox} + l_{cls} + l_{ctc} \tag{6}$$

In the training phase, the dataset will comprise synthetic data with character bounding box annotations and real data with only transcript annotations. For samples with character bounding box annotation, the supervision is consistent with the supervision in pre-training phase. When encountering a sample with only text content annotations, we first infer on the sample to obtain the prediction results of character confidence, bounding box, and classification. Then, the edit distance between the ground truth text and the predicted text is calculated. Characters matched as "equal" are considered as "reliable" predictions, and their predicted bounding boxes are stored in the cache as pseudo-labels. When the pseudo-label of a character is not stored in the cache, the result of this inference is directly copied to the cache; otherwise, the result of this inference is weighted with the original result in the cache as

$$u_{bbox}^{i,j} = \lambda^{i,j} c_{bbox}^{i,j} + (1 - \lambda^{i,j}) p_{bbox}^{i,j} \tag{7}$$

$$u_{conf}^{i,j} = \lambda^{i,j} c_{conf}^{i,j} + (1 - \lambda^{i,j}) p_{conf}^{i,j} \tag{8}$$

where $u_{bbox}^{i,j}$, $c_{bbox}^{i,j}$, and $p_{bbox}^{i,j}$ indicate the bounding box to be updated to the cache, the original in the cache, and the bounding box obtained by this inference, respectively. $u_{conf}^{i,j}$, $c_{conf}^{i,j}$, and $p_{conf}^{i,j}$ are similar symbols for the confidence. i and j means the j-th character in the i-th sample. The weight $\lambda^{i,j}$ can be calculated as

$$\lambda^{i,j} = \frac{e^{10c_{score}^{i,j}}}{e^{10c_{score}^{i,j}} + e^{10p_{score}^{i,j}}} \tag{9}$$

3.4   Co-Transcription (Co-T)

During the transcription phase, we combine the output of the Seg. branch and the CTC branch for more accurate results. To keep the probability distributions of the character classification output from the two branches similar, we normalize them by

$$y_{cls} = \frac{p_{cls} - \mu_{cls}}{\sigma_{cls}}, y_{ctc} = \frac{p_{ctc} - \mu_{ctc}}{\sigma_{ctc}} \tag{10}$$

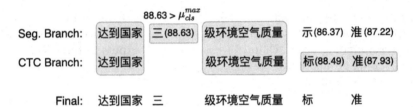

Fig. 3. Schematic diagram of co-transcription. The number in the brackets after a character indicates the normalized probability of the character.

where μ_{cls}, σ_{cls}, and μ_{ctc}, σ_{ctc} stand for the mean and variance of the probabilities of the character classification output from the two branch, which can be calculated as

$$\mu_{cls} = \frac{1}{|P_{loc}|} \sum_{l \in P_{loc}} p_{cls}^l, \sigma_{cls} = \sqrt{\frac{1}{|P_{loc}|} \sum_{l \in P_{loc}} (p_{cls}^l - \mu_{cls})^2} \qquad (11)$$

$$\mu_{ctc} = \frac{1}{|P_{nob}|} \sum_{l \in P_{nob}} p_{ctc}^l, \sigma_{ctc} = \sqrt{\frac{1}{|P_{nob}|} \sum_{l \in P_{nob}} (p_{ctc}^l - \mu_{ctc})^2} \qquad (12)$$

where P_{loc} indicates the frame set where the character centers is located in Seg. branch, and P_{nob} indicates the frame set of non-blank in CTC branch.

The lengths of the texts predicted by the two branches may be different, so the co-transcription algorithm is required to combine two different lengths of predicted texts. As shown in Fig. 3, we first calculate the edit distance between the two predicted texts, and include all the characters that match as "equal" ("达到国家" and "级环境空气质量" in both two branches as shown in Fig. 3) in the final predicted text. Among the characters that cannot be matched, if they are consecutive in the original predicted text, we merge them into a "block" ("三", "示准" in Seg. branch and "标准" in CTC branch). The subsequent processing will be performed in units of text blocks. For text blocks in the two branches, we one-to-one match them based on position ("示准" in Seg. branch is matched to "标准" in CTC branch). Some blocks may not match to the corresponding block in the other branch ("三" in Seg. branch). We calculate the average normalized probability for each text block (the average of the normalized probability of each character in a text block). Among a set of one-to-one matched text blocks, we select the text block with a higher average normalized probability and add it to the corresponding position of the final predicted text. For the text blocks that cannot be one-to-one matched, if their average normalized probability is greater than μ_{cls}^{max} in Seg. branch or μ_{ctc}^{max} in CTC branch, they are also included in the corresponding position of the final predicted text. μ_{cls}^{max} and μ_{ctc}^{max} can be calculated as

$$\mu_{cls}^{max} = \text{mean}(y_{cls}^{max}), \mu_{ctc}^{max} = \text{mean}(y_{ctc}^{max}) \qquad (13)$$

where y_{cls}^{max} and y_{ctc}^{max} denotes the maximum value of each frame in y_{cls} and y_{ctc}, respectively.

4   Experiments

4.1   Datasets

- **CASIA-HWDB** [8] is a widely used offline handwritten Chinese text line database written by 1,020 writers. This database includes CASIA-HWDB 1.0-1.2 containing 3,895,135 isolated characters and CASIA-HWDB 2.0-2.2 containing 52,230 text lines. Notably, the character samples of CASIA-HWDB 1.0-1.2 are not cropped from the text lines of CASIA-HWDB 2.0-2.2.
- **ICDAR2013-Offline** [33] is a competition dataset containing 3,432 offline handwritten Chinese text lines written by 60 writers.
- **SCUT-HCCDoc** [35] contains 12,253 offline handwritten Chinese document images captured by cameras with 116,629 text lines.
- **MTHv2** [11] is a Chinese historical document database comprising Tripitaka Koreana in Han (TKH) and the Multiple Tripitaka in Han (MTH). It contains 105,579 text lines with character bounding box annotations.

4.2   Evaluation Metrics

We adopted two commonly used evaluation metrics in HCTR called the accurate rate (AR) and correct rate (CR), which can be calculated as

$$AR = \frac{N_t - D_e - S_e - I_e}{N_t}, CR = \frac{N_t - D_e - S_e}{N_t} \tag{14}$$

where N_t represents the total number of characters in annotations, and D_e, S_e and I_e denote the total number of deletion, substitution and insertion errors, respectively.

4.3   Implementation Details

The input text line images are resized to 128 in height while maintaining the aspect ratio of the raw, and the RGB values are normalized to $[0, 1]$. For the synthetic images, we add Gaussian noise with mean 0 and variance 0.01 as data augmentation, while no data augmentation strategy is employed for the real data.

The batch size is set to 32. In the pre-traning phase, we train 37,500 iterations only on synthetic data. Adadelta [34] is adopted to optimize non-Transformer-encoder part of the model, with 0.33 as the initial learning rate and dropping to 0.1 after 10,000 iterations. In the training phase, both synthetic and real data are used. Non-Transformer-encoder part of the model is optimized by stochastic gradient descent (SGD) for 300,000 iterations, with an initial learning rate of 0.02. The learning rate is multiplied by 0.08 at 75,000, 150,000 and 225,000 iterations. The Transformer encoder part is optimized using AdamW [9] with the learning rate kept at 1e-5.

(a) For CASIA-HWDB

(b) For SCUT-HCCDoc

(c) For MTHv2

Fig. 4. Samples of synthetic data for CASIA-HWDB, SCUT-HCCDoc, and MTHv2.

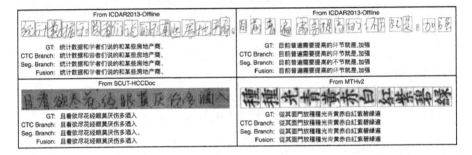

Fig. 5. SegCTC visualizations tested on ICDAR2013-Offline (without language model), SCUT-HCCDoc, and MTHv2 datasets.

4.4   Language Model

Recently, language models have become essential for improving the performance of HCTR recognizers. We adopt a Transformer-based language model which follows [7] to improve the performance of SegCTC. The Transformer-based language model was trained using a corpus from the same source as [14]. Because the language model [7] requires an input of CTC-style predictions, we only input the CTC branch output into the language model.

4.5   Experiments on ICDAR2013-Offline Dataset

Data Synthesis. The isolated character samples are collected from CASIA-HWDB1.0-1.2. During synthesizing, we simply paste the isolated character samples on a white background and record the bounding boxes of the characters (demo is shown in Fig. 4a). The content of text lines is obtained by random character sampling from character set or corpus described in Subsect. 4.4. In the pre-training phase, the corpus was used at a rate of 0.5, while in the training phase the corpus was used exclusively. The data is synthesized online during the training.

Data Preparation. In the pre-training phase, we use only synthetic data for training. In the training phase, both CASIA-HWDB 2.0-2.2 dataset and synthetic data are used for training, where the ratio of synthetic data is 0.5. The model is evaluated using the ICDAR2013-Offline dataset. We correct the angle of text line images of CASIA-HWDB 2.0–2.2 to make the texts horizontal referring to [15].

Table 2. Comparison with previous methods on ICDAR2013-Offline. "LM" denotes language model. **Bold** indicates state-of-the-art, while <u>underline</u> indicates the second best.

Method	Without LM		With LM	
	AR	CR	AR	CR
Messina et al. [12]	83.50	–	89.40	–
Du et al. [2]	83.89	–	93.50	–
Wang et al. [23]	88.79	90.67	94.02	95.53
Wu et al. [28]	86.64	87.43	90.38	–
Wang et al. [27]	89.66	–	96.47	–
Xiu et al. [32]	88.74	–	96.35	–
Peng et al. [15]	89.61	90.52	94.88	95.51
Xie et al. [30]	91.25	91.68	96.22	96.70
Wang et al. [25]	87.00	89.12	95.11	95.73
Zhu et al. [36]	90.86	–	94.00	–
Xie et al. [29]	91.55	92.13	96.72	96.99
Wang et al. [26]	91.58	–	96.83	–
Tanaka et al. [19]	91.00	–	94.63	–
Huang et al. [6]	91.82	92.13	–	–
Liu et al. [7]	93.62	–	97.51	–
Peng et al. [14]	94.50	94.76	**97.70**	**97.91**
Ours	**95.10**	**95.29**	<u>97.67</u>	<u>97.82</u>

Results and Analysis. A comparison of SegCTC with the previous method on ICDAR2013-Offline is shown in Tabel 2. SegCTC achieves state-of-the-art performance without a language model, and slightly lower than Peng et al.'s [14] method with a language model. SegCTC can also predict the bounding boxes of characters if output with Seg. branch even without the character bounding box annotations in the real data. The recognition and segmentation results of some samples are shown in Fig. 5.

4.6   Experiments on SCUT-HCCDoc Dataset

Data Preparation and Synthesis. We only use synthetic data in the pre-training phase, while in the training phase, SCUT-HCCDoc data and synthetic data is used at a ratio of 3:7 for training. The character samples for the synthetic data were taken from 101 font files, and it is also simple to paste the character samples on a white background to form a text line image. A sample of the synthetic data is shown in Fig. 4b.

Results and Analysis. Since the performance of the CTC branch is much better than the performance of the Seg. branch, the performance of the prediction output using the Co-T strategy is not as good as the performance of the CTC branch in this experiment. We provide the results of the output using the Co-T strategy and the output using the CTC branch in Tabel 3 and compare them with the previous methods. SegCTC achieves state-of-the-art performance in both AR metrics and CR metrics. The recognition and segmentation results of some samples are shown in Fig. 5.

Table 3. Comparison with previous methods on SCUT-HCCDoc. **Bold** indicates state-of-the-art, while <u>underline</u> indicates the second best.

Method	AR	CR
CTC-based*	87.46	88.83
Attention-based*	83.30	84.81
Wang et al. [24]**	83.53	85.41
Liu et al. [7]	89.06	90.12
Peng et al. [14]	90.71	92.01
Ours (Co-T output)	<u>91.49</u>	<u>92.93</u>
Ours (CTC branch output)	**92.08**	**93.40**

* Re-implemented by the author of [35] and the results were updated on their website at https://github.com/HCIILAB/ SCUT-HCCDoc_Dataset_Release. ** Re-implemented by the author of [14].

4.7   Experiments on MTHv2 Dataset

Data Preparation and Synthesis. Only synthetic data will be used in the pre-training phase. MTHv2 data and synthetic data are used in a ratio of 1:1 in training phase. Because the MTHv2 dataset has a character bounding box annotation, we use this annotation to cut out the character samples from the images. The cut-out character samples are with background, so we use the character samples from the same image to ensure the background similarity when stitching a text line image. A sample of the synthetic data is shown in Fig. 4c.

Results and Analysis. The MTHv2 data are character bounding box annotated. To verify the effectiveness of weakly supervised training, we removed the
character bounding box annotations in the training phase. A comparison of
SegCTC with the previous methods on MTHv2 is shown in Tabel 4. Our method
is only slightly lower than Peng et al.'s [14] method and performs better than
other methods. Recognition and segmentation results of some samples are shown
in Fig. 5.

Table 4. Comparison with previous methods on MTHv2. **Bold** indicates state-of-the-
art, while underline indicates the second best.

Method	AR	CR
Ma et al. [11]	95.52	96.07
Shi et al. [16]	96.94	97.15
Huang et al. [6]	97.42	97.62
Peng et al. [14]*	**97.89**	**97.94**
Ours	97.78	97.85

* Re-implemented by us.

4.8 Ablation Studies

In this section, we verify whether the recognition method of fusing explicit segmentation and implicit segmentation recognition can effectively improve the performance of the recognizer, and the effect of our proposed S^2FFM and Co-T on
the recognition performance enhancement through ablation experiments. The
ablation results on ICDAR2013-Offline are shown in Tabel 5. Experiments show
that the performance of the recognizer can be improved by directly connect-
ing two branches (without using Transformer encoder to fuse two recognition
branches). The addition of both S^2FFM and Co-T strategy can lead to a more
significant improvement in the performance of the recognizer.

Table 5. The ablation studies on ICDAR2013-Offline.

Method	CTC Branch		Seg. Branch	
	AR	CR	AR	CR
CTC	94.24	94.42	–	–
Seg	–	–	94.43	94.66
Direct Connection	94.49	94.65	94.43	94.66
+ S^2FFM	94.96	95.15	94.88	95.13
+ Co-T	**95.10**	**95.29**	**95.10**	**95.29**

To further verify that SegCTC that fuses the Seg. branch and CTC branch works better than using one branch alone, we further experimented on ICDAR2013-Offline, SCUT-HCCDoc and MTHv2 datasets. We use the same backbone and the same training strategy, the only difference is the retention or non-retention of the two branches. The experimental results are shown in Tabel 6. SegCTC outperformed the recognition model using only one of the branches on ICDAR2013-Offline and SCUT-HCCDoc datasets, and the performance is slightly lower than the recognizer with only Seg. branch in MTHv2 dataset.

Table 6. Comparison with recognizer with CTC branch only and with Seg. branch only. **Bold** indicates state-of-the-art, while <u>underline</u> indicates the second best.

Method	ICDAR2013-Offline		SCUT-HCCDoc		MTHv2	
	AR	CR	AR	CR	AR	CR
CTC	94.24	94.42	89.88	91.25	97.60	97.66
Seg	<u>94.43</u>	<u>94.66</u>	<u>90.52</u>	<u>91.86</u>	**97.89**	**97.94**
Ours*	**95.10**	**95.29**	**91.49**	**92.93**	<u>97.78</u>	<u>97.85</u>

* Output with Co-T strategy.

4.9   Limitation

According to the current research, SegCTC cannot guarantee that the output of every character comes from the Seg. branch when using the Co-T strategy, and therefore cannot guarantee that every character has the output of the bounding box. Additionally, if there is a large performance gap between the Seg. and CTC branches, the Co-T strategy may not outperform either branch. Therefore, the decision to use the Co-T strategy should be based on practical needs.

5   Conclusion

In this paper, we explore the strengths and weaknesses of explicit and implicit segmentation recognition models for the HCTR problem. Based on the complementary nature of explicit and implicit segmentation recognition methods, we propose a novel recognition model SegCTC to more fully fuse the two different recognition methods. Our proposed S^{2}FFM uses self-attention mechanism-based Transformer encoder to fuse the recognition headers from two branches more effectively. To more accurate predictions, a co-transcription strategy which combines the prediction from the two different branches is proposed. Experiments on ICDAR2013-Offline, SCUT-HCCDoc and MTHv2 datasets illustrate that SegCTC can achieve state-of-the-art performance in the HCTR task.

Acknowledgement. This research is supported in part by NSFC (Grant No.: 61936 003), Zhuhai Industry Core and Key Technology Research Project (no. 2220004002350), and Science and Technology Foundation of Guangzhou Huangpu Development District (No. 2020GH17) and GD-NSF (No.2021A1515011870).

References

1. Baek, Y., Lee, B., Han, D., Yun, S., Lee, H.: Character region awareness for text detection. In: Proceedings of the IEEE/CVF Conference on Computer Vision and Pattern Recognition (CVPR), pp. 9365–9374 (2019)
2. Du, J., Wang, Z.R., Zhai, J.F., Hu, J.S.: Deep neural network based hidden Markov model for offline handwritten Chinese text recognition. In: 2016 23rd International Conference on Pattern Recognition (ICPR), pp. 3428–3433. IEEE (2016)
3. Graves, A., Fernández, S., Gomez, F., Schmidhuber, J.: Connectionist temporal classification: labelling unsegmented sequence data with recurrent neural networks. In: Proceedings of the 23rd International Conference on Machine Learning (ICML), pp. 369–376 (2006)
4. He, K., Zhang, X., Ren, S., Sun, J.: Deep residual learning for image recognition. In: 2016 IEEE Conference on Computer Vision and Pattern Recognition (CVPR), pp. 770–778 (2016)
5. Hochreiter, S., Schmidhuber, J.: Long short-term memory. Neural Comput. **9**(8), 1735–1780 (1997)
6. Huang, Y., Jin, L., Peng, D.: Zero-shot Chinese text recognition via matching class embedding. In: Lladós, J., Lopresti, D., Uchida, S. (eds.) ICDAR 2021. LNCS, vol. 12823, pp. 127–141. Springer, Cham (2021). https://doi.org/10.1007/978-3-030-86334-0_9
7. Liu, B., Sun, W., Kang, W., Xu, X.: Searching from the prediction of visual and language model for handwritten Chinese text recognition. In: Lladós, J., Lopresti, D., Uchida, S. (eds.) ICDAR 2021. LNCS, vol. 12823, pp. 274–288. Springer, Cham (2021). https://doi.org/10.1007/978-3-030-86334-0_18
8. Liu, C.L., Yin, F., Wang, D.H., Wang, Q.F.: CASIA online and offline Chinese handwriting databases. In: 2011 International Conference on Document Analysis and Recognition (ICDAR), pp. 37–41. IEEE (2011)
9. Loshchilov, I., Hutter, F.: Decoupled weight decay regularization. arXiv preprint arXiv:1711.05101 (2017)
10. Luo, C., Jin, L., Sun, Z.: Moran: a multi-object rectified attention network for scene text recognition. Pattern Recognit. **90**, 109–118 (2019)
11. Ma, W., Zhang, H., Jin, L., Wu, S., Wang, J., Wang, Y.: Joint layout analysis, character detection and recognition for historical document digitization. In: 2020 17th International Conference on Frontiers in Handwriting Recognition (ICFHR), pp. 31–36. IEEE (2020)
12. Messina, R., Louradour, J.: Segmentation-free handwritten Chinese text recognition with LSTM-RNN. In: 2015 13th International Conference on Document Analysis and Recognition (ICDAR), pp. 171–175. IEEE (2015)
13. Neubeck, A., Van Gool, L.: Efficient non-maximum suppression. In: 18th International Conference on Pattern Recognition (ICPR). vol. 3, pp. 850–855. IEEE (2006)
14. Peng, D., Jin, L., Ma, W., Xie, C., Zhang, H., Zhu, S., Li, J.: Recognition of handwritten chinese text by segmentation: A segment-annotation-free approach. IEEE Trans, Multimedia (2022)

15. Peng, D., Jin, L., Wu, Y., Wang, Z., Cai, M.: A fast and accurate fully convolutional network for end-to-end handwritten Chinese text segmentation and recognition. In: 2019 International Conference on Document Analysis and Recognition (ICDAR), pp. 25–30. IEEE (2019)

16. Shi, B., Bai, X., Yao, C.: An end-to-end trainable neural network for image-based sequence recognition and its application to scene text recognition. IEEE Trans. Pattern Anal. Mach. Intell. **39**(11), 2298–2304 (2016)

17. Shi, B., Yang, M., Wang, X., Lyu, P., Yao, C., Bai, X.: ASTER: an attentional scene text recognizer with flexible rectification. IEEE Trans. Pattern Anal. Mach. Intell. **41**(9), 2035–2048 (2018)

18. Su, T.H., Zhang, T.W., Guan, D.J., Huang, H.J.: Off-line recognition of realistic Chinese handwriting using segmentation-free strategy. Pattern Recognit. **42**(1), 167–182 (2009)

19. Tanaka, R., Osada, K., Furuhata, A.: Text-conditioned character segmentation for CTC-based text recognition. In: Lladós, J., Lopresti, D., Uchida, S. (eds.) ICDAR 2021. LNCS, vol. 12823, pp. 142–156. Springer, Cham (2021). https://doi.org/10.1007/978-3-030-86334-0_10

20. Vaswani, A., et al.: Attention is all you need. In: Advances in Neural Information Processing Systems (NIPS). vol. 30 (2017)

21. Wang, D.H., Liu, C.L., Zhou, X.D.: An approach for real-time recognition of online Chinese handwritten sentences. Pattern Recognit. **45**(10), 3661–3675 (2012)

22. Wang, Q.F., Yin, F., Liu, C.L.: Handwritten Chinese text recognition by integrating multiple contexts. IEEE Trans. Pattern Anal. Mach. Intell. **34**(8), 1469–1481 (2011)

23. Wang, S., Chen, L., Xu, L., Fan, W., Sun, J., Naoi, S.: Deep knowledge training and heterogeneous CNN for handwritten Chinese text recognition. In: 2016 15th International Conference on Frontiers in Handwriting Recognition (ICFHR), pp. 84–89. IEEE (2016)

24. Wang, T., et al.: Decoupled attention network for text recognition. In: Proceedings of the AAAI Conference on Artificial Intelligence (AAAI). vol. 34, pp. 12216–12224 (2020)

25. Wang, Z.X., Wang, Q.F., Yin, F., Liu, C.L.: Weakly supervised learning for over-segmentation based handwritten Chinese text recognition. In: 2020 17th International Conference on Frontiers in Handwriting Recognition (ICFHR), pp. 157–162. IEEE (2020)

26. Wang, Z.R., Du, J., Wang, J.M.: Writer-aware CNN for parsimonious HMM-based offline handwritten Chinese text recognition. Pattern Recognit. **100**, 107102 (2020)

27. Wang, Z.-R., Du, J., Wang, W.-C., Zhai, J.-F., Hu, J.-S.: A comprehensive study of hybrid neural network hidden Markov model for offline handwritten Chinese text recognition. Int. J. Doc. Anal. Recogn. (IJDAR) **21**(4), 241–251 (2018). https://doi.org/10.1007/s10032-018-0307-0

28. Wu, Y.C., Yin, F., Chen, Z., Liu, C.L.: Handwritten Chinese text recognition using separable multi-dimensional recurrent neural network. In: 2017 14th IAPR International Conference on Document Analysis and Recognition (ICDAR). vol. 1, pp. 79–84. IEEE (2017)

29. Xie, C., Lai, S., Liao, Q., Jin, L.: High performance offline handwritten Chinese text recognition with a new data preprocessing and augmentation pipeline. In: Bai, X., Karatzas, D., Lopresti, D. (eds.) DAS 2020. LNCS, vol. 12116, pp. 45–59. Springer, Cham (2020). https://doi.org/10.1007/978-3-030-57058-3_4

30. Xie, Z., Huang, Y., Zhu, Y., Jin, L., Liu, Y., Xie, L.: Aggregation cross-entropy for sequence recognition. In: Proceedings of the IEEE/CVF Conference on Computer Vision and Pattern Recognition (CVPR), pp. 6538–6547 (2019)
31. Xing, L., Tian, Z., Huang, W., Scott, M.R.: Convolutional character networks. In: Proceedings of the IEEE/CVF International Conference on Computer Vision (ICCV), pp. 9126–9136 (2019)
32. Xiu, Y., Wang, Q., Zhan, H., Lan, M., Lu, Y.: A handwritten Chinese text recognizer applying multi-level multimodal fusion network. In: 2019 International Conference on Document Analysis and Recognition (ICDAR), pp. 1464–1469. IEEE (2019)
33. Yin, F., Wang, Q.F., Zhang, X.Y., Liu, C.L.: ICDAR 2013 Chinese handwriting recognition competition. In: 2013 12th International Conference on Document Analysis and Recognition (ICDAR), pp. 1464–1470. IEEE (2013)
34. Zeiler, M.D.: Adadelta: an adaptive learning rate method. arXiv preprint arXiv:1212.5701 (2012)
35. Zhang, H., Liang, L., Jin, L.: SCUT-HCCDoc: a new benchmark dataset of handwritten Chinese text in unconstrained camera-captured documents. Pattern Recognit. **108**, 107559 (2020)
36. Zhu, Z.Y., Yin, F., Wang, D.H.: Attention combination of sequence models for handwritten Chinese text recognition. In: 2020 17th International Conference on Frontiers in Handwriting Recognition (ICFHR), pp. 288–294. IEEE (2020)

Adversarial Attacks on Convolutional Siamese Signature Verification Networks

Maham Jahangir[1]([✉]), Muhammad Imran Malik[1], and Faisal Shafait[1,2]([✉])

[1] School of Electrical Engineering and Computer Science (SEECS), National University of Sciences & Technology (NUST), Islamabad, Pakistan
{mjahangir.phdcs17seecs,faisal.shafait}@seecs.edu.pk
[2] Deep Learning Laboratory, National Center of Artificial Intelligence (NCAI), Islamabad, Pakistan

Abstract. A handwritten signature serves as an important biometric modality to identify individuals. The state-of-the-art methods for signature verification employ deep learning networks to perform the classification task. However, deep neural networks can be fooled by adversarial attacks that introduce small imperceptible perturbations to the input images. In this paper, we explore the vulnerability of signature verification systems against adversarial attacks. The state-of-the-art attacks developed by the machine learning community to fool image classifiers are unsuitable for attacking document classifiers as they are applied to the background of signature images making them quite perceptible. To overcome this challenge, we design an attack based on dictionary learning with the goal to perturb the foreground (strokes) of the signature image. The proposed method is evaluated in terms of attack success rate and imperceptibility. The experimental results on the benchmark CEDAR dataset using Siamese Deep Signet Model highlight the efficacy of the proposed approach as compared to other methods by achieving 95% and 98% attack success rates with our proposed approach.

Keywords: Adversarial Attack · Sparse Encoding · Dictionary Learning · Signature Verification

1 Introduction

Biometric Systems are widely used to recognize individuals in legal, financial, and administrative matters [7,15]. Handwritten signatures serve as one such biometric which are required especially during financial transactions to identify and verify an individual. The Signature verification systems can be offline (static) and online (dynamic). The offline systems identify individuals from a signature image (spatial information) containing handwriting strokes whereas the online system's recognition is based on the signature generation process (considering spatial and temporal information). Offline systems are used widely due to low cost and convenience. Moreover, there are scenarios where offline signature verification is inevitable for example during cheque transactions. The traditional

G. A. Fink et al. (Eds.): ICDAR 2023, LNCS 14190, pp. 350–365, 2023.
https://doi.org/10.1007/978-3-031-41685-9_22

systems relied on handcrafted features for signature verification but lately, most of the research efforts on offline signature verification systems are based on deep neural networks. These systems work under two approaches a) writer independent and b) writer dependent. The writer-independent approach is generally considered more practical as the systems based on writer dependent approach need to be updated every time a new writer is registered [4]. This research article also considers writer-independent offline signature verification scenarios to evaluate the robustness of signature verification systems.

We have used SigNet: Convolutional Siamese Network [3] in this study. The available data is divided into train and test sets with a couple of image pairs such as (genuine, genuine) and (genuine, forged) labeled as, similar and dissimilar classes. Siamese networks can efficiently model such problems. Siamese networks are based on twin convolutional networks which accept two images that can either be similar or dissimilar. Since Deep Neural Networks (DNNs) are employed here for signature verification, unfortunately, DNNs are vulnerable to adversarial examples [19]. These examples are generated by imposing carefully crafted perturbations to clean input images. This research area gained quick popularity since its advent [19] and a lot of attacks have been proposed to exploit the vulnerabilities of deep neural network-based systems. However, attacking signature images is a relatively different and challenging task when compared to other fields. The vulnerability of signature verification systems against adversarial attacks has not been explored thoroughly and only a handful of research is available on the topic. In this article, we present the first attempt to particularly attack Siamese network-based signature verification system.

It should be noted that attacking verification systems is very different from attacking classification systems and presents challenges not present in classification systems. First, when a new user gets registered a new unseen class and unseen examples are introduced to the system. Second, for signature verification systems the background and foreground are clearly separated and a verification system clearly uses the foreground information (strokes) to extract features and then classify the image as genuine or forged. The state-of-the-art attacks impose perturbations on the background making them perceptible and since background information is not used by the system, therefore, the attack success rate is greatly reduced. Further, in the model used in this article, the images are inverted during pre-processing making it even harder to attack. The third problem is that most of the state-of-the-art methods specifically gradient-based methods applied to signature verification systems are white box in nature (they require full information on the training set, the model used, and parameters learned in order to attack a system). These systems are well protected by organizations and such information is unknown to attackers. So traditional white-box attack methods are not practical.

In view of the above-mentioned problems, this research article proposes a black-box attack method to attack signature verification systems using ideas from sparse representation. Our recent work explored the idea of dictionary learning to craft sparse adversarial attacks for image classification [8]. Formally, we used the idea of sparse representation to craft adversarial images using feature maps of an image. In this research, we have extended the idea and developed a novel approach to learn a dictionary on feature descriptor (foreground extraction) and improved sparse representation to create adversarial attacks with the goal to perturb the foreground (strokes) of the signature image. The sparse representation includes dictionary learning and sparse coding stages to generate perturbations that can be induced in the original images making them adversarial. Dictionary learning is a transformation process that transforms an image to its linear combination of basic elements called atoms. Sparse Coding is a method for learning a sparse representation of the input using dictionary learning [13]. In this paper, a novel feature descriptor approach is used to learn the dictionary and improve sparse representation quality. In this regard, we used the Grab cut algorithm [18] to extract the foreground of the signature images and then learn the dictionary. This is an attempt to learn only important and relevant information. The proposed technique is evaluated on the benchmark publicly available CEDAR Signature Dataset and is also compared with the state-of-the-art methods.

The main contributions and findings include:

1. The proposed model generates adversarial perturbations to fool signature verification systems with minimum ℓ_2-norm and maximum attack success rates of 95% and 98% respectively.
2. We introduce improved sparse representation quality by learning a dictionary on a feature descriptor (foreground extraction) rather than original unprocessed images.
3. We attacked a convolution-based Siamese network for a handwriting signature verification system not attacked before.
4. Our experiments show that attacking strokes of signature is important as attacks on the background won't produce desirable results.

The structure of the paper is as follows. Section 2 describes the related works. Section 3 details the problem, threat model, and methodology of the proposed approach. Section 4 defines the experimental protocol. Section 5 presents experimental results and analysis. Section 6 concludes the paper.

2   Related Work

Adversarial examples are manipulated input images with perturbations that fool the classifiers. The concept of adversarial attacks was introduced by Szegedy et al. [19] in 2013. Since then a lot of attacks have been proposed by the machine learning community to evaluate the robustness of deep networks. Among the pioneers is Fast Gradient Sign Method (FGSM) [5]. This is a gradient-based method that maximizes the loss of the classifier to craft adversarial examples. Later iterative methods like Deep Fool [17], Basic Iterative method (BIM) [9], and Carlini and Wagner (C&W) [2] were also introduced. Universal adversarial attacks create a single adversarial perturbation that fools the classifier with high probability and generalizes well across different neural networks [16].

Projected Gradient Descent [12] is a well-optimization method essentially similar in behavior to iterative FGSM with the difference that it initializes the input sample to a random point in the ball of interest. On the other hand, Boundary Attack [1] is one of the decision-based attacks which follows the decision boundary between adversarial and non-adversarial examples using a simple rejection sampling algorithm.

In the context of adversarial attacks against signature verification systems Hafemann [6] explored the vulnerability of these systems against adversarial attacks. They attacked the system using existing adversarial attacks, like FGSM and C&W and presented two types of threats to these systems hence two types of attacks. Type: I, where an adversary manipulates a genuine signature to be misclassified by the system (False Rejection). Type: II where a forged signature is manipulated to be classified as genuine by the systems (False Acceptance). The authors point out that Type: I attacks are easy to generate as compared to Type: II. These perturbations were introduced on the background of the images making them quite perceptible and requiring perfect knowledge of the system under attack which is not practical. In another research, Li et al. [10] proposed a gradient-free black-box attack against signature verification systems by restricting the area of perturbations to the region of strokes. Their attack method is not applicable to binary images as the perturbation intensity of each pixel is not continuously adjustable. Therefore, selecting optimal pixels for perturbations will not be possible.

To the best of the authors' knowledge, these two research articles explored the vulnerability of signature verification systems against adversarial attacks. This area still needs a lot of exploration and presents great room for improvement. None of the above-mentioned researchers tested their proposed methods on Siamese Networks. Attacking Siamese networks is much more challenging than other classification systems. It is evident from the results section that state-of-the-art attack methods couldn't attack these networks efficiently. The attack success rates of the state-of-the-art are quite low when compared with literature where they showed good performance while they attacked other signature verification systems. Siamese Networks are widely used and acquired state-of-the-art performance on signature verification systems. That is why they have gained fast-growing popularity in signature verification systems. These systems serve the rightful purpose of comparing the images and then identifying them as genuine or forged based on their similarity or dissimilarity. Therefore, in this research, we explored the vulnerability of the Convolutional Siamese Networks against adversarial attacks. We designed a black-box attack (information on the training set, the model used by a verifier, and parameters learned are not required) based on the sparse representation of foreground features of images. The experimental results prove the efficacy of the proposed method.

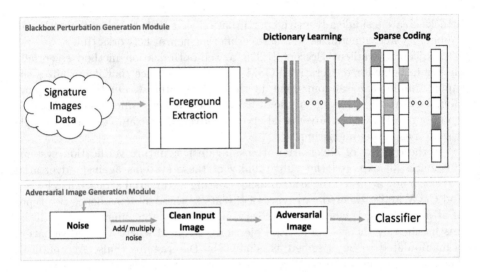

Fig. 1. General Framework of the Proposed Attack Model

3   Methodology

This section explains our methodology in detail as illustrated in Fig. 1. The first step in the proposed approach is to extract the foreground from the signature images. These images are fed to a dictionary learning algorithm to learn sparse representation. The sparse representation is then used as a perturbation to manipulate the original input image to fool the classifier. Below we discuss the Siamese Network under attack, followed by the problem statement, foreground extraction, sparse representation, and adversarial image generation.

3.1   Siamese Network

In this research, we evaluated the robustness of the Siamese network named: Signet [3] against adversarial attacks. The Siamese networks are very popular among signature verification systems and to the best of our knowledge are not yet studied for robustness against adversarial attacks. One of the reasons behind their popularity is their ability to learn from minimum data. They need only a few images to make better predictions and data is not abundant in various problems including signature verification [14]. The Siamese networks are based on twin CNN architectures with shared weights joined at the output by a loss function. The goal is to find similarities between the two images. They learn a feature space when similar observations are placed in proximity and are used to evaluate whether a given signature is genuine or forged. This is achieved by exposing the network to both similar and dissimilar pairs and the network maximizes the Euclidean distance between dissimilar pairs whereas minimizes the distance between similar pairs. The popular loss function used by Siamese networks is contrastive loss and is defined as follows:

$$L(a, b, y) = \alpha(1 - y)D_w^2 + \beta y \max(0, m - D_w)^2 \quad \text{where} \quad a, b \in X \quad (1)$$

a and b are input samples that belong to the set X. They can be genuine signatures or forged entries in the system. y is a binary indicator that indicates whether the given two signatures belong to the same class or not. α and β are two constants whereas, m indicates the margin i.e. 1 in this case. $D_w = \| f(a; w_1) - f(b; w_2) \|_2$. It is the Euclidean distance computed in feature space, f is a function that maps a signature image to its real vector space through CNN whereas, w_1 and w_2 are learned weights of that particular layer of the network. The training of Siamese networks involves pairwise learning so the classifier won't output probabilities of the prediction but the distance from each class. We have reported this distance in our experiments of the proposed approach as well as for the state-of-the-art methods. The threshold of 0.5 is selected to determine if the output of the Siamese network is the same or not.

3.2 Problem

A typical Siamese-based offline signature verification model under attack is depicted in Fig. 2. The model takes signature images as input. These signature images can be genuine – by authentic users or can be forgeries – entered into the system by a skilled forger. The forgers generate signature images that resemble original images from the same user in an attempt to fool the system. Since the system is trained on skilled forgeries as well, Signature verification systems successfully recognize the forgeries. However, these systems are still vulnerable to two main threats. First, an original authentic signature image can be modified in a way that system rejects the original image that is **Type: I, False Rejection (FR)**. The second form of attack is the one in which the forged signature images are modified in a way that gets accepted by the system termed as **Type: II, False Acceptance (FA)**. Some previous researchers consider that the second type of adversarial attack is harder to generate [6,10] as compared to the first one. However, in the case of Siamese networks, our experiments show that Type: I attacks are harder to generate. In this paper, we considered both of these adversarial attacks for evaluation purposes. Adversarial examples are images similar to the true data distribution but fool the system. These images are generated by adding small perturbations to the original data. If we denote X as input space and a function $F(X)$ maps these input to a label Y then the adversarial examples X_{adv} that are visually similar to clean samples X_{org} but fools the classifier that is $F(X_{\text{adv}} \neq Y)$. In the case of the Siamese network

$$L(a, b_{\text{adv}}) \neq y \quad (2)$$

where,

$$b_{\text{adv}} = b + \epsilon p \quad \text{and} \quad d(b_{\text{adv}}, b) < \epsilon \quad (3)$$

where, ϵ is the magnitude of perturbation p added in the image. The distance d between original signature image b and adversarial image b_{adv} should be minimum.

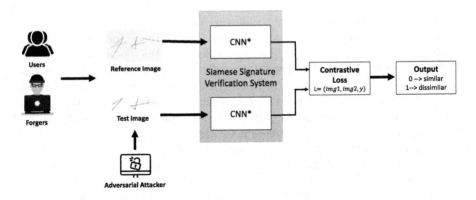

Fig. 2. Siamese Network-based Signature Verification System and Threat Model

3.3 Foreground Extraction

The first step in our proposed approach is to extract the foreground of the signature image. The foreground contains the signature strokes. Our goal is to learn a dictionary on these strokes as they are the only important and relevant information that we need from the signature image. Background doesn't hold any detail in signature verification systems. In order to learn specific features we intend to learn the dictionary on the foreground of the image rather than the full image. The background pixels are changed to 0-pixel value whereas, the foreground to 1. Let pixels covering the foreground be denoted as F_d and that of the background as B_d.

$$X' = F_d + B_d \quad \text{where,} \quad F_d = 1 \text{ and } B_d = 0 \tag{4}$$

For the above-mentioned purpose, we used the GrabCut algorithm [18] to extract the foreground of the image which can be used to learn the dictionary and its corresponding sparse representation. It is a graph cuts-based image segmentation method. It uses a Gaussian mixture model to separate the background and the target object.

3.4 Sparse Representation (Dictionary Learning and Sparse Coding)

The next step is to learn the sparse representation of the processed images from the last section. The foreground extraction serves as an important feature descriptor to improve the quality of learned representations. The goal is to improve the feature descriptor of the signature images by keeping specific and minimal information. Sparse coding is an encoding process where a sparse representation of input images is learned using a linear combination of basic elements. These elements are called atoms and they combine to form a dictionary. Let X' denote the foreground extracted images from the previous step. A transformation operator to learn sparse representation is applied to it and denoted as $T(X')$. The optimization function to learn dictionary and sparse representation proposed by Mairal et al. [13] and is given as

$$T(X') = D\alpha \tag{5}$$

$$\min_{D,\alpha} \frac{1}{2} \parallel x' - D\alpha \parallel_2^2 + \lambda \parallel \alpha \parallel_1 \quad s.t. \parallel D_k \parallel_2 = 1 \quad \forall \quad k \in [0, n] \tag{6}$$

where, x' is the pre-processed signature image and λ is a regularization parameter, α is the sparse representation, D is the dictionary learned, and n is the number of dictionary atoms. The algorithm explaining the steps of this section is listed in Algorithm: 1.

Algorithm 1: Adversarial Dictionary Learning

Input: $X' \rightarrow$ Set of pre-processed original signature images;
Result: D $\rightarrow$ Learned Dictionary , $T(X') \rightarrow$ Sparse representation
$D \rightarrow$ Initial Dictionary ;
$OMP \rightarrow$ Orthogonal Matching Pursuit() ;
$k \rightarrow$ Sparsity ;
$n \rightarrow$ no. of atoms ;
for $t = 1$ to iterations **do**
  $\quad T(X') \leftarrow OMP(D, X')$;
  $\quad$ **Dictionary Update Stage**;
  $\quad D = \min_{D,\alpha} \frac{1}{2} \parallel x' - D\alpha \parallel_2^2 + \lambda \parallel \alpha \parallel_1 \qquad s.t. \parallel D_k \parallel_2 = 1 \quad \forall \quad k \in [0, n]$;
Return D **Return** $T(X')$

3.5 Tuned Adversarial Signature Image Generation

This is the final stage where an adversarial image is generated. A dictionary of perturbations is learned and saved by the dictionary learning algorithm as discussed above. These perturbations have a different effect on the attack success rate. So in this step, the adversarial signature image is tuned for all the available perturbations. The perturbations that maximize the loss of the classifier and achieve the highest attack success rate are selected. The complete process of adversarial image generation involving all sections is defined step by step in Algorithm 2. The first step is to extract the foreground of signature images. For Type: I attack the forged samples of images are used to learn the dictionary whereas, genuine samples in the case of Type: II attacks. Next, we learn the dictionary and compute sparse representation. This sparse representation is basically our noise/perturbation to be used to manipulate the original image. As we discussed earlier, contrary to some findings in literature the Type: I attack was much more challenging than anticipated in the case of Siamese networks. With reference to Siamese networks, the additive noise model couldn't attack the genuine image to be declared as forged by the classifier. Therefore, inspired by recent work on multiplicative noises [11] we multiplied the noise perturbation with the original image to craft our adversarial example. The experimental results prove the effectiveness of multiplicative noise over additive. Detailed analysis of multiplicative and additive noises for the Type: I attack is discussed in Sect. 5.

Algorithm 2: Tuned Adversarial Signature Image Generation

Result: $X_{adv} \rightarrow$ Tuned Adversarial Image
Input: $X_{org} \rightarrow$ legitimate source input image;
$X_{forg} \rightarrow$ skilled forged signature input image;
$\epsilon \rightarrow$ magnitude of noise ;
$L \rightarrow$ classifier's loss;
if $attack = type : I$ **then**
$\quad | \quad X' = Grabcut(X_{forg})$;
else
$\quad \lfloor \quad X' = Grabcut(X_{org})$;
$T(X') = DictLearningAlgo(X')$;
$P = T(X')$;
for $i < size(X_{org})$ **do**
$\quad$ **if** $attack = type : I$ **then**
$\quad \quad | \quad \max_{L(X_{org},X_{adv},Y)} X_{adv\,i} = X_{org\,i} * \epsilon P_i$;
$\quad$ **else**
$\quad \quad \lfloor \quad \min_{L(X_{org},X_{adv},Y)} X_{adv\,i} = X_{org\,i} + \epsilon P_i$;
Return X_{adv}

---

4 Experimental Protocol

The experimental design and detail to evaluate the proposed methodology are discussed in this section.

4.1 Dataset

We conducted the experiments on the widely used benchmark signatures dataset, CEDAR signature Database[1]. We have used this dataset as it is quite well-known and used by almost all the articles we reviewed during this research.

---

[1] http://www.cedar.buffalo.edu/NIJ/data/signatures.rar.

Table 1. Attributes of CEDAR dataset used in the experiments to define training and test splits. Note that the splits were carefully done in a way that the users in dictionary learning, training Siamese network, and testing were mutually exclusive.

Attributes	Count
Number of users	55
Users in the training set	28
Users in the test set	12
Users to train the dictionary	15
Genuine signatures per user	24
Forgeries per user	24

Moreover, it contains signatures of 55 users from different ethnic and professional backgrounds. Each user signed 24 genuine signatures with a difference of 24 minutes in between. Forgers copied the signatures of 3 genuine users, 8 times each. Hence, each user has 24 genuine and 24 forged signatures. A total of $55 \times 24 = 1320$ genuine and 1320 forged signatures are available in this dataset. The total number is $1320 \times (2) = 2640$. These images are available in grayscale mode.

We divided the dataset into training and test sets as shown in Table: 1. The system is trained and tested using signatures from 40 users with a train test split of 70% : 30%. We also reserved some signature images which were not part of the training or testing of the model. This allows us to define a black-box attack scenario to evaluate our approach where the attacker has no access to the training or test data or the model used by the signature verification system. The remaining signatures from 15 users are used to simulate the environment where an attacker has a dataset of his own with some genuine signatures by users and the respective forgeries. These images are used to train the dictionary and learn sparse representations. These sparse representations are added as perturbations to the test set of the dataset to create adversarial examples.

4.2   Pre-processing and Performance of Signet-Siamese Network

The model is trained and tested as per the guidelines outlined in the paper [3]. The same pre-processing steps are employed. The publicly available implementation of the model architecture is used to carry out the training[2]. The images are resized to a fixed size (155×220) and then inverted to get a black background with pixel values: 0. Finally, all the images are normalized. The detail on the Siamese network has been provided in Sect. 3.1. We trained the network for 80 epochs. The training loss equal to 0.3 and accuracy of 85% are calculated respectively. The test loss and accuracy were 0.015 and 97% respectively.

---

[2] https://github.com/AtharvaKalsekar/SigNet/.

4.3   Metrics

The contrastive loss of the classifier, attack success rate, and mean and median ℓ_2-norm are calculated during experimentation. The attack success rate defines the number of genuine signatures that failed to pass through the system and the number of forged signatures that successfully passed through the system. The ℓ_2-norm is a standard method to compute the length of a vector in Euclidean space. We use it to find the similarity between two images. Here it is the squared distance between the adversarial and original clean image. A lower distance means that the two images appear the same and the noise in adversarial images is imperceptible. We have calculated the mean and median values of ℓ_2-norm.

4.4   State-of-the-art Adversarial Attacks

We compared our approach with state-of-the-art methods. The adversarial robustness toolbox[3] was used to conduct experiments for the state-of-the-art. We evaluated the proposed systems against Fast Gradient Sign Method (FGSM) [5], Basic Iterative Method (BIM [9], Projected Gradient Descent (PGD) [12], and Boundary Attack Method [1]. These are all baseline attack methods that achieved state-of-the-art attack success rates in traditional image classification systems. These systems are gradient-based evasion attacks that are white-box in nature (where the attacker has access to the training or test data or the model used by a signature verification system). Epsilon ϵ refers to the magnitude of noise introduced to the original clean image to create an adversarial image. Our proposed method relies on a very small magnitude of noise in order to attack the system. The other state-of-the-art methods don't attack the system at all if the ϵ is kept very low. Therefore, we cannot test the system for the same values of ϵ. We have used $\epsilon = 0.3$ for the state-of-the-art to conduct the experiments.

5   Results and Discussion

This section explains the results reported when the proposed approach is applied to the CEDAR signature dataset and compared with the state-of-the-art methods. Moreover, the effect of perturbations on strokes of signatures images is discussed with reference figures and examples.

5.1   Type: I Attack (False Rejection)

In this attack, perturbation is applied to genuine signatures images such that the system fails to verify the image as genuine. Contrary to the popular opinion in the literature where attacking genuine signatures(Type: I) is argued to be an easy task, we found the Type: I attack to be equally challenging as that of Type: II specifically in the case of Siamese networks. Since the model pre-processes the image where the background is black and the signature strokes are white. This

---

[3] https://adversarial-robustness-toolbox.readthedocs.io/en/latest/.

makes it hard to add noise to the strokes. The background noise fails to attack the system. This is evident from results tabulated in Table 2. Only the proposed method is able to attack successfully with a success rate of 95% and with the lowest ℓ_2-norm value of 0.09. The first row of Fig. 3 illustrates the example images generated through our proposed approach as well as the state-of-the-art. It can be clearly seen that almost all baseline methods attack the background of the image, therefore, their attack success rates are very low, and ℓ_2-norm is quite high.

Table 2. The magnitude of noise ϵ, Loss of Classifier (higher the value more successful the attack is), Attack Success Rate, Mean and median ℓ_2-norm (lower the value more imperceptible the attack is) values reported for Type: I attack for our proposed method and state-of-the-art.

Method	Epsilon(ϵ)	Loss	Attack Succ. (%)	Mean ℓ_2-norm	Median ℓ_2-norm
FGSM [5]	0.3	0.19	29	0.37	0.37
BIM [9]	0.3	0.05	8	0.13	0.12
PGD [12]	0.3	0.04	7	0.13	0.12
Boundary Attack [1]	–	0.01	2	0.42	0.43
Proposed	**0.002**	**1.50**	**95**	**0.09**	**0.09**

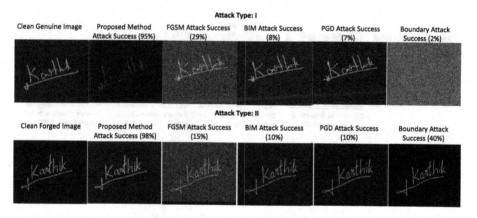

Fig. 3. Clean and Adversarial Image examples from the results of experiments reported in Table:2 and Table:3 for Type-I and Type-II attacks

5.2   Type: II Attack (False Acceptance)

In this attack, the perturbation is applied to forged signature images such that the system accepts them during the verification and classifies them as genuine which was previously declared forged by the system. The results for this type are tabulated in Table 3. The proposed approach successfully attacks the system with an attack success rate of 98% using a very low magnitude of noise $\epsilon = 0.0004$. The ℓ_2-norm is also the lowest among all baseline methods which is 0.07. The second row of Fig. 3 illustrates the example images of the proposed method and all other methods. Again the other methods fail to attack the system significantly as they attack the background of the image except for the Boundary Attack. Nevertheless, its attack success rate is still very low (attack success rate of 40%, and the ℓ_2-norm of 0.17) compared to the proposed method (attack success rate of 98%, and the ℓ_2-norm of 0.07).

Table 3. The magnitude of noise ϵ, Loss of Classifier (lower the value more successful the attack is), Attack Success Rate, Mean and median ℓ_2-norm (lower the value more imperceptible the attack is) values reported for Type: II attack for our proposed method and state-of-the-art.

Method	Epsilon(ϵ)	Loss	Attack Succ. (%)	Mean ℓ_2-norm	Median ℓ_2-norm
FGSM	0.3	0.88	15	0.36	0.36
BIM	0.3	1.27	10	0.12	0.12
PGD	0.3	1.27	10	0.12	0.12
Boundary Attack	–	0.92	40	0.17	0.17
Proposed Method	0.0004	0.01	98	0.07	0.07

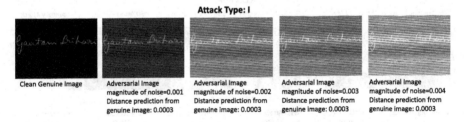

Attack Type: I

Clean Genuine Image

Adversarial Image magnitude of noise=0.001 Distance prediction from genuine image: 0.0003

Adversarial Image magnitude of noise=0.002 Distance prediction from genuine image: 0.0003

Adversarial Image magnitude of noise=0.003 Distance prediction from genuine image: 0.0003

Adversarial Image magnitude of noise=0.004 Distance prediction from genuine image: 0.0003

Fig. 4. Effect of magnitude of noise on the prediction of the model in case of Type: I attack

5.3 Effect of Magnitude of Noise on the Prediction of the Model on Genuine Signatures Images

As discussed above, Type: I attack, which were generally considered as an easy target [6,10], have been proven challenging while attacking the Siamese network. Figure 4 illustrates the effect of the increasing magnitude of noise on the genuine signatures. It can be seen that even increasing the magnitude of noise causes no effect on the prediction of the model. It still declares the image as genuine. This is because the strokes of the images remain intact and the model only used the information of strokes to learn features and classify them.

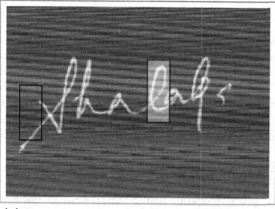

(a) Signature adversarial image with additive noise –minimum to no perturbations on strokes

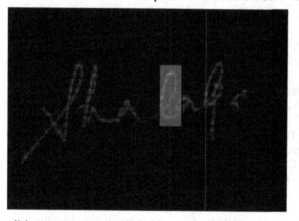

(b) Signature adversarial image with multiplicative noise – perturbations on strokes

Fig. 5. Additive and Multiplicative Noise Adversarial Example Images with same values of epsilon and their effect on strokes of the signature image

5.4 Effect of Multiplicative and Additive Noise on Genuine Signatures

We have used multiplicative noise in the case of the Type: I attack for the proposed method. Figure 5 illustrates how multiplicative noise attacks the strokes of the signature image while additive noise just disrupts the background. We have shown a zoomed version of the portion of the stroke to illustrate our point. This is why we chose multiplicative noise rather than popular additive noise to craft our adversarial examples.

6   Conclusion

In this research, we attacked a convolutional Siamese signature verification network using sparse representation and dictionary learning. A novel algorithm to learn a dictionary from an important feature descriptor that extracts foreground is proposed. The attack proposed is black-box in nature that doesn't require information about the signature verification model used, its weights, or training or test data. The experimental results show that our proposed method outperforms all the baseline methods and achieves attack success rates of 95% and 98% for Type: I and Type: II adversarial attacks, respectively.

In the future, we will test our proposed method with more datasets and evaluate its performance for transferability across other deep networks. We shall also evaluate our proposed approach against defense methods. The improvement of sparse representation quality in terms of improved feature descriptors should be studied too.

References

1. Brendel, W., Rauber, J., Bethge, M.: Decision-based adversarial attacks: reliable attacks against black-box machine learning models. arXiv preprint arXiv:1712.04248 (2017)
2. Carlini, N., Wagner, D.: Towards evaluating the robustness of neural networks. In: 2017 IEEE Symposium on Security and Privacy, pp. 39–57. IEEE (2017)
3. Dey, S., Dutta, A., Toledo, J.I., Ghosh, S.K., Lladós, J., Pal, U.: SigNet: convolutional Siamese network for writer independent offline signature verification. CoRR abs/1707.02131 (2017). http://arxiv.org/abs/1707.02131
4. Diaz, M., Ferrer, M.A., Impedovo, D., Malik, M.I., Pirlo, G., Plamondon, R.: A perspective analysis of handwritten signature technology. ACM Comput. Surv. 51(6), 1–39 (2019)
5. Goodfellow, I.J., Shlens, J., Szegedy, C.: Explaining and harnessing adversarial examples. arXiv preprint arXiv:1412.6572 (2014)
6. Hafemann, L.G., Sabourin, R., Oliveira, L.S.: Characterizing and evaluating adversarial examples for offline handwritten signature verification. IEEE Trans. Inf. Forensics Secur. 14(8), 2153–2166 (2019)
7. Hameed, M.M., Ahmad, R., Kiah, M.L.M., Murtaza, G.: Machine learning-based offline signature verification systems: a systematic review. Signal Process. Image Commun. 93, 116139 (2021)
8. Jahangir, M., Shafait, F.: Adversarial attack using sparse representation of feature maps. IEEE Access 10, 120724–120734 (2022)
9. Kurakin, A., Goodfellow, I.J., Bengio, S.: Adversarial examples in the physical world. In: Artificial Intelligence Safety and Security, pp. 99–112. Chapman and Hall/CRC (2018)
10. Li, H., Li, H., Zhang, H., Yuan, W.: Black-box attack against handwritten signature verification with region-restricted adversarial perturbations. Pattern Recogn. 111, 107689 (2021)
11. Lo, S.Y., Patel, V.M.: MultAV: multiplicative adversarial videos. In: 2021 17th IEEE International Conference on Advanced Video and Signal Based Surveillance, pp. 1–6. IEEE (2021)

12. Madry, A., Makelov, A., Schmidt, L., Tsipras, D., Vladu, A.: Towards deep learning models resistant to adversarial attacks. arXiv preprint arXiv:1706.06083 (2017)
13. Mairal, J., Bach, F., Ponce, J., Sapiro, G.: Online dictionary learning for sparse coding. In: Proceedings of the 26th Annual International Conference on Machine Learning, pp. 689–696 (2009)
14. Malik, J., Elhayek, A., Ahmed, S., Shafait, F., Malik, M.I., Stricker, D.: 3DAirSig: a framework for enabling in-air signatures using a multi-modal depth sensor. Sensors 18(11), 3872 (2018)
15. Malik, M.I., Liwicki, M., Dengel, A.: Part-based automatic system in comparison to human experts for forensic signature verification. In: 2013 12th International Conference on Document Analysis and Recognition, pp. 872–876. IEEE (2013)
16. Moosavi-Dezfooli, S.M., Fawzi, A., Fawzi, O., Frossard, P.: Universal adversarial perturbations. In: Proceedings of the IEEE Conference on Computer Vision and Pattern Recognition, pp. 1765–1773 (2017)
17. Moosavi-Dezfooli, S.M., Fawzi, A., Frossard, P.: DeepFool: a simple and accurate method to fool deep neural networks. In: Proceedings of the IEEE Conference on Computer Vision and Pattern Recognition, pp. 2574–2582 (2016)
18. Rother, C., Kolmogorov, V., Blake, A.: GrabCut interactive foreground extraction using iterated graph cuts. ACM Trans. Graph. 23(3), 309–314 (2004)
19. Szegedy, C., et al.: Intriguing properties of neural networks. arXiv preprint arXiv:1312.6199 (2013)

A System for Processing and Recognition of Greek Byzantine and Post-Byzantine Documents

Panagiotis Kaddas[1,2]([✉]), Konstantinos Palaiologos[1,3], Basilis Gatos[1], Vassilis Katsouros[4], and Katerina Christopoulou[1,5]

[1] Computational Intelligence Laboratory, Institute of Informatics and Telecommunications, NCSR "Demokritos", 15310 Athens, Greece
{pkaddas,k.palaiologos,bgat,achristopoulou}@iit.demokritos.gr

[2] Department of Informatics and Telecommunications, University of Athens, 15784 Athens, Greece

[3] Hellenic Institute, Royal Holloway, University of London, Egham Hill, Egham TW20 0EX, Surrey, UK

[4] Institute for Language and Speech Processing, Athena Research Center, Athens, Greece
vsk@athenarc.gr

[5] School of Environment, Geography and Applied Economics, Department of Economics & Sustainable Development, Harokopio University, 17676 Athens, Greece

Abstract. Processing and recognition of Greek Byzantine and Post-Byzantine (old Greek) Documents has been proven to be a tedious task in the domain of Historical Document Image Processing. Several unique characteristics of these documents (existence of character ligatures, abbreviations, lack of clear word division, existence of symbols or punctuations in an arbitrary position) impose significant difficulties for current processing and recognition tools. In this work, we introduce a system for processing and recognition of old Greek documents and give details about all the components that comprise it. These include an image pre-processing, a text line segmentation and a recognition module. In order to test the proposed system, we introduce and provide publicly a new dataset of old Greek Documents that includes text line images and the corresponding transcription. Using this dataset, we evaluate the embedded recognition engine of the proposed system which is the open-source Calamari-OCR engine employing a variety of configurations. The best result corresponded to a character error rate less than 1.5% which is acceptable and promising. Finally, we also achieved promising results when comparing the embedded OCR engine with other recognition methods already proposed for the recognition of old Greek Documents.

Keywords: Document Analysis System · Deep Neural Networks · Calamari-OCR · Greek Byzantine and Post-Byzantine Documents · Text Line Recognition

© The Author(s), under exclusive license to Springer Nature Switzerland AG 2023
G. A. Fink et al. (Eds.): ICDAR 2023, LNCS 14190, pp. 366–376, 2023.
https://doi.org/10.1007/978-3-031-41685-9_23

1 Introduction

Old Greek Documents are an important source of historical information for scholars related to our cultural heritage conservation. In this paper, we focus on processing and recognition of Greek Byzantine and Post-Byzantine (old Greek) documents dated from the 12th to the 16th century. As it can be observed in the sample of Fig. 1, old Greek documents of this period have some unique characteristics such as the existence of character ligatures (neighboring character maybe joined together), abbreviations, lack of clear word division, existence of symbols or punctuations in an arbitrary position, which impose significant difficulties for current optical character recognition tools.

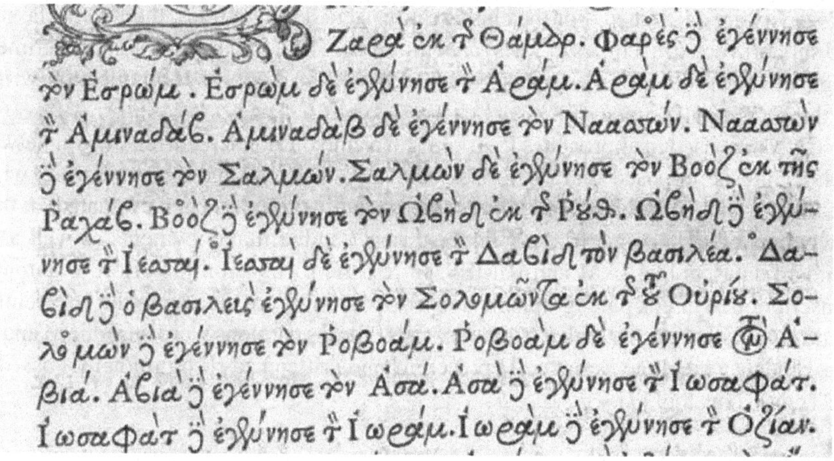

Fig. 1. A sample of Greek early printing document (*grecs du roi* typeface, Greek New Testament, published by Robert Estienne in 1550).

In this work, we introduce a system for processing and recognition of old Greek documents. We give details about all the components that comprise it focusing both on the automatic procedures for image pre-processing, text line segmentation and recognition, as well as on the semi-automatic procedures for correcting the text line segmentation and recognition result. The embedded recognition engine of the proposed system is the open-source, TensorFlow-based Calamari-OCR engine [1] that uses an advanced deep neural network. In order to test the recognition engine, we introduce and provide publicly a new dataset of old Greek Documents [2] that includes text line images and the corresponding transcription. By employing a variety of configurations on the recognition engine, we demonstrate that we can achieve very promising results using a small number of images for training. The best result obtained corresponds to a character error rate less than 1.5%. Finally, we also achieved promising results when comparing the embedded OCR engine with other recognition methods already proposed for the recognition of old Greek Documents.

The rest of the paper is organized as follows. In Sect. 2, the related work is presented, Sect. 3 introduces the proposed system, Sect. 4 demonstrates our experimental results and Sect. 5 presents the conclusion of this work.

2 Related Work

Processing and recognition of old Greek Documents has not attracted lot of attention in the literature. There are some approaches that follow more traditional image processing techniques based on feature extraction and some more recent techniques based on Convolutional Neural Networks (CNN).

In approaches [3, 4], the document image is first binarized, enhanced and skeletonized. Next, the open and closed cavities of the skeletonized characters are detected and a feature extraction step is applied in order to provide the input for the recognition process. Finally, the individual cavities are recognized on the basis of their features. At the feature extraction step, all segments that belong to a protrusion of an isolated character's cavity are calculated. For the classification step, decision trees, the K-NN classifier and support vector machines (SVMs) are employed. The corpus used for the experiments originates from the Sinaitic Codex Number Three, the Book of Job collection written by three different writers.

CNNs are used in approaches [5, 6]. In [5], a convolutional recurrent neural network architecture is proposed that comprises octave convolution and recurrent units which use effective gated mechanisms. The proposed architecture has been evaluated on three newly created collections from Greek historical handwritten documents as well as on standard datasets like IAM and RIMES. In [6], the focus is on the effort to automate transcription of Greek paleographic manuscripts dating from the 10th to the 16th century. To this end, two datasets with a parallel corpus of transcriptions were introduced and the experiments were done using an AI powered handwritten text recognition tool based on the Transkribus tool [7].

3 Proposed System

In this Section, we give details about all the components that comprise the proposed system for processing and recognition of old Greek documents. This includes the image pre-processing component that performs image binarization, as well as the text line detection and the text line recognition components applied both in an automatic and a semi-automatic way.

3.1 Image Pre-processing

Image pre-processing includes image binarization that refers to the conversion of the grayscale or color image to a binary image. Having the binary version of the image mainly helps our system to re-define the text line polygons that are automatically extracted (see Sect. 3.2) in order to exclude non-text areas or to fully include text areas that lie in the polygon limits.

The binarization method used in the proposed system is fully described in [8] and consists of five distinct steps: a preprocessing procedure using a low-pass Wiener filter, a rough estimation of foreground regions using Niblack's approach, a background surface calculation by interpolating neighboring background intensities, a thresholding by combining the calculated background surface with the original image and, finally, a postprocessing step that improves the quality of text regions and preserves stroke connectivity. An example of the binarization pre-processing step is demonstrated in Fig. 2.

3.2 Text Line Segmentation

Image pre-processing is followed by the step of text line segmentation, which is a necessary procedure in order to obtain precise and accurate recognition results. Automatic text line segmentation was carried out by the use of a variation of the well-known YOLOv5 [9] Deep Neural Network model (YOLOv5-OBB[1]). In order to automatically edit detection results acquired from YOLOv5-OBB and to efficiently apply OCR, the

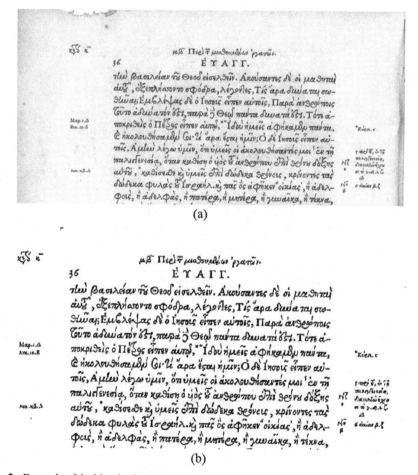

(a)

(b)

Fig. 2. Example of the binarization pre-processing method. (a) original image (b) resulting binary image.

---

[1] https://github.com/hukaixuan19970627/yolov5_obb

detected polygons are sorted using Density-based spatial clustering (DBSCAN) in order to preserve the correct reading order of the text lines.

At a first step, the automatic text line detection procedure results to a set of polygons that surround each text line of the document, as shown in Fig. 3. Then, the user can correct the segmentation results. The system provides the user with the following functions:

- To correct a polygon by moving the desired points in the right position.
- To add a new polygon before or after a polygon.
- To delete a polygon.
- To connect a polygon with another polygon.

Fig. 3. The result of the text line segmentation shown in the proposed system.

In Fig. 4(a) one can see the result of the text line segmentation of the system, while in Fig. 4(b) the corrected polygon by the user. In Fig. 5 we present another example of the functionality of the system where the user can connect two parts of the same text line and create a polygon for the whole line.

When the procedure of text line detection and correction is completed, the system creates text line images with the image parts inside each polygon. These images will be then used as input for the text line recognition module.

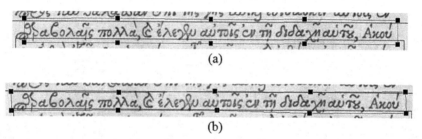

(a)

(b)

Fig. 4. An example of the text line detection (a) and the correction of the line polygon by the user (b).

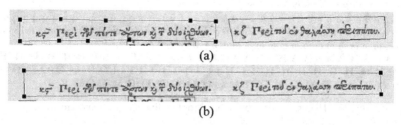

(a)

(b)

Fig. 5. An example of two polygons (a) that should be connected in order to form one text line (b).

3.3 Text Line Recognition

Having completed the text line segmentation, the user can move to the next step which is the text line recognition. This is first done automatically by the open-source, TensorFlow-based Calamari-OCR engine [1] that uses advanced deep neural network. As it is explained in Sect. 4, the best configuration for the recognition engine is selected after experimentation with a new database for old Greek documents.

At a next step, the user can correct the predicted OCR lines (see Fig. 6). A part of the image is presented with the text line of the prediction enclosed in a red box, thus placing the text line into context in relation to the surrounding text. What follows is a detailed image of the text line, which helps the user to focus on the correction. The predicted text appears in a box while a virtual keyboard appears below in order to help the user to make corrections. The user is able to navigate to the next or previous lines using the arrows.

The user checks the text thoroughly for errors and can perform corrections with the following ways. Using the cursor, he/she can select the erroneous character and either type the correct one, or he/she can choose to use the virtual keyboard below, containing a comprehensive list of Greek polytonic characters (see Fig. 7).

At a next phase, the corrected text together with the corresponding corrected text line polygon are used for network re-training.

Fig. 6. The user interface provided for text line recognition correction.

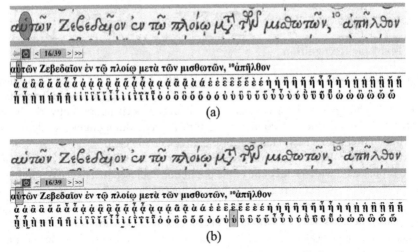

Fig. 7. Correction of recognition results: (a) the user spots the error, (b) the user corrects the error by selecting the correct character using the virtual keyboard.

4   Experimental Results

In order to test the proposed system, we introduce and provide publicly a new dataset of old Greek Documents [2] (see Fig. 1). This dataset consists of 57 pages containing the complete Gospel of Matthew, from the third edition of the Greek New Testament published in 1550 by Robert Estienne (1503–1559). Estienne, also known in his activities with the name Stephanus, was appointed "Royal Typographer" during the rule of King of France François I (1494–1547). As a typographer and scholar, he published a number of classical texts including Greek and Latin translations of the Bible. His first edition of the Greek New Testament was in 1546 and the text was based on that printed by Erasmus in Basel in 1516. For his most significant edition, known as *Editio Regia* or the "Royal Edition" printed in 1550, Estienne used fifteen additional Byzantine manuscripts and presented for the first time a textual apparatus listing the variant readings of the different manuscripts he examined. The edition printed in large folio size using the *grecs du roi* typeface. This typeface, which attempts to imitate the Greek handwriting of this period, includes a large number of ligatures and abbreviations. Produced by Claude Garamont on the basis of the Greek minuscule style of the calligrapher Angelos Vergikios (1505–1569) from Crete, who was active copying Greek manuscripts in Venice and France, it became the most widely used Greek typeset for European printers. The dataset consists of 2045 text lines, 1431 used for training, 204 for validation and 410 for test. The text lines were produced automatically by our system based on [9] and then corrected by a user as it is described in Sect. 3.2. Also, the corresponding transcription was corrected following the procedure described in Sect. 3.3.

In order to test the recognition accuracy of the proposed system, we evaluated the embedded recognition engine which is the open-source Calamari-OCR engine [1] using a wide variety of configurations. In Table 1, we present selected results that show high performance and are associated with the predefined network architecture and the application or not of data augmentation during training. The evaluation presented in Table 1 uses as metrics the Character Error Rate (CER) and Word Error Rate (WER). Concerning the network architecture, we present results for:

def: The default Calamari with one BiLSTM layer and.

htr +: An adaptation of the standard network structure of the Transkribus platform [7].

def is the default Calamari network follows a *CONV1- > MAXPOOL- > CONV2 > MAXPOOL- > BiLSTM* scheme, where: *CONV1* is Convolutional Layer with 40 filters, stride 1 and 3x3 receptive field, followed by *ReLU* activation. *CONV2* is similar to *CONV1* but with 60 filters. *MAXPOOL* is a $2x2^2$ max pooling layer and *BiLSTM* is a Bidirectional Long Short-term Memory layer with 200 hidden nodes. After the *BiLSTM* layer, Dropout is applied with a skip ratio of 0.5.

The *htr +* Calamari network follows a *CONV1- > CONV2- > MAXPOOL1- > CONV3 > MAXPOOL2- > BiLSTM1- > BiLSTM2- > BiLSTM3* scheme, where *CONV1* is a Convolutional Layer with 8 filters, stride 2x4 and 2x4 receptive field, followed by leaky *ReLU* activation. *CONV2* is a Convolutional Layer with 32 filters, stride 2x4 and 1x1 receptive field, followed by leaky *ReLU* activation. *MAXPOOL1* is a 2x4 max pooling layer with stride 2x4. *CONV3* is a Convolutional Layer with 64 filters, stride 1x1 and 3x3 receptive field, followed by leaky *ReLU* activation. *MAXPOOL2* is a 2x1 max pooling layer with stride 2x1. *BiLSTM1*, *BiLSTM2* and *BiLSTM3* are Bidirectional Long Shortterm Memory layers with 256 hidden nodes respectively. After each BiLSTM layer, Dropout is applied with a skip ratio of 0.5. For both architectures, CTC Loss is calculated as a scoring function.

---

[2] When notation AxB is used, A is for the horizontal axis of a layer (x-width) and B for the vertical axis (y-height)

Calamari also includes network architecture *deep3* which is not included in our results because of lower performance. Data augmentation includes padding, distortions, blobs, and multiscale noise.

Table 1. Experimental results (CER% / WER%) on the new dataset of old Greek Documents [2].

Network architecture:	*htr +*	*def*
Augmentation:		
NO	5.12 / 23.61	2.08 / 11.82
YES	3.66 / 18.06	**1.45 / 8.53**

As it can be observed in Table 1, the *def* architecture outperforms the *htr +* architecture while data augmentation improves the results significantly. The best results correspond to CER of 1.45% and WER of 8.53% which are acceptable and promising having in mind the small number of images included in the training set.

In order to compare the embedded OCR engine using best configuration (*def* architecture + augmentation in training) with other recognition methods applied on old Greek Documents, we also trained and tested on the datasets presented in [4]. Table 2 presents the results for these 4 datasets compared to the approach of [4] using best settings (Deslanting) as well as to the approaches of de Sousa Neto et al. [10] and Puigcerver [11]. As is can be observed in Table 2, the embedded OCR engine of the proposed system is comparable or outperforms existing approaches taking into account the CER performance.

Table 2. Comparative experimental results (CER% / WER%) on the datasets presented in [4].

Dataset	Tsochatzidis et al. [4]	de Sousa Neto et al. [10]	Puigcerver [11]	OCR Engine embedded in the proposed system
χφ53	**6.77 / 30.09**	7.85 / 34.63	10.45 / 30.20	8.04 / 35.64
χφ79	6.51 / 28.51	7.75 / 33.13	10.33 / 28.55	**5.14 / 28.25**
χφ114	7.71 / 34.30	8.03 / 36.72	10.19 / 34.58	**7.01 / 41.79**
Eparchos	**4.53 / 20.03**	4.95 / 21.91	5.18 / 22.21	32.48

5 Conclusions

In this work, we present a system for processing and recognition of Greek Byzantine and Post-Byzantine Documents. This includes modules for (i) image pre-processing for binarization, (ii) text line segmentation that is first done automatically by employing a variation of the well-known YOLOv5 [9] Deep Neural Network model (YOLOv5OBB) and then corrected manually using a user-friendly interface and (iii) text line recognition that is provided by an advanced deep neural network using the open-source Calamari engine [1]. The correction of the OCR is done in an efficient way and by using a virtual keyboard. Moreover, we introduce and provide publicly a new dataset of old Greek Documents [2] that includes text line images and the corresponding transcription. This dataset helps us (i) to find the best configuration of the recognition network and (ii) to assess the accuracy the embedded OCR engine. Therefore, it proved to be acceptable and promising for the case of old Greek Documents since it resulted to a character error rate less than 1.5% by using a relatively small set of images for training. Promising results were also achieved when comparing the embedded recognition engine with other recognition methods already proposed for the recognition of old Greek Documents. In all cases, the proposed system is comparable or outperforms existing approaches taking into account the CER performance. Future work includes testing of other deep neural network architectures for the task of text line segmentation and recognition and also the creation of a larger dataset that can be used for training.

Acknowledgments. This research has been partially co-financed by the European Union and Greek national funds through the Operational Program Competitiveness, Entrepreneurship and Innovation, under the call "RESEARCH-CREATE-INNOVATE", project Culdile (Cultural Dimensions of Deep Learning, project code: T1EΔK-03785) and the Operational Program Attica 2014–2020, under the call "RESEARCH AND INNOVATION PARTNERSHIPS IN THE REGION OF ATTICA", project reBook (Digital platform for re-publishing Historical Greek Books, project code: ATTP4–0331172).

References

1. Wick, C., Reul, C., Puppe, F.: Calamari - A High-Performance Tensorflow-based Deep Learning Package for Optical Character Recognition. Digit. Humanit. Q. **14**(1) (2020)
2. https://zenodo.org/record/7876098#.ZEvjNtJBxNh
3. Ntzios, K., Gatos, B., Pratikakis, I., Konidaris, T., Perantonis, S.J.: An old Greek handwritten OCR system based on an efficient segmentation-free approach. Int. J. Doc. Anal. Recogn. (IJDAR) **9**(2–4), 179–192 (2007). special issue on historical documents
4. Gatos, B., Ntzios, K., Pratikakis, I., Petridis, S., Konidaris, T., Perantonis, S.J.: An efficient segmentation-free approach to assist old Greek handwritten manuscript OCR. Pattern Anal. Appl. (PAA) **8**(4), 305–320 (2006)
5. Tsochatzidis, L., Symeonidis, S., Papazoglou, A., Pratikakis, I.: HTR for Greek historical handwritten documents. J Imaging **7**, 260 (2021)
6. Platanou, P., Pavlopoulos, J., Papaioannou, G.:. Handwritten paleographic greek text recognition: a century-based approach. In: Proceedings of the Thirteenth Language Resources and Evaluation Conference, pp. 6585–6589. European Language Resources Association, Marseille (2022)
7. https://readcoop.eu/transkribus/
8. Gatos, B., Pratikakis, I., Perantonis, S.J.: Adaptive degraded document image binarization. Pattern Recogn. **39**, 317–327 (2006)
9. https://github.com/ultralytics/yolov5

10. de Sousa Neto, A.F., Bezerra, B.L.D., Toselli, A.H., Lima, E.B.: HTR-Flor: a deep learning system for offline handwritten text recognition. In: Proceedings of the 33rd SIBGRAPI Conference on Graphics, Patterns and Images, pp. 54–61. Recife/Porto de Galinhas (2020)
11. Puigcerver, J.: Are multidimensional recurrent layers really necessary for handwritten text recognition? In: Proceedings of the 14th IAPR International Conference on Document Analysis and Recognition, pp. 67–72. Kyoto (2017)

Towards Writing Style Adaptation
in Handwriting Recognition

Jan Kohút[(✉)][iD], Michal Hradiš[iD], and Martin Kišš[iD]

Faculty of Information Technology, Brno University of Technology,
Brno, Czech Republic
{ikohut,ihradis,ikiss}@fit.vutbr.cz

Abstract. One of the challenges of handwriting recognition is to transcribe a large number of vastly different writing styles. State-of-the-art approaches do not explicitly use information about the writer's style, which may be limiting overall accuracy due to various ambiguities. We explore models with writer-dependent parameters which take the writer's identity as an additional input. The proposed models can be trained on datasets with partitions likely written by a single author (e.g. single letter, diary, or chronicle). We propose a Writer Style Block (WSB), an adaptive instance normalization layer conditioned on learned embeddings of the partitions. We experimented with various placements and settings of WSB and contrastively pre-trained embeddings. We show that our approach outperforms a baseline with no WSB in a writer-dependent scenario and that it is possible to estimate embeddings for new writers. However, domain adaptation using simple fine-tuning in a writer-independent setting provides superior accuracy at a similar computational cost. The proposed approach should be further investigated in terms of training stability and embedding regularization to overcome such a baseline.

Keywords: Handwritten text recognition · OCR · Domain adaptation · Domain dependent parameters · Finetuning · CTC

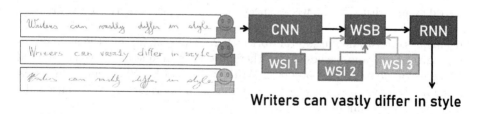

Fig. 1. Our proposed Writer Style Block (WSB) learns to utilize various writer styles based on writer-style identifiers (WSI).

G. A. Fink et al. (Eds.): ICDAR 2023, LNCS 14190, pp. 377–394, 2023.
https://doi.org/10.1007/978-3-031-41685-9_24

1   Introduction

Handwritten text of multiple writers can vastly differ in style, for example, the degree of slant, the way letters are joined (cursive or block letters), spacing between letters and words, similarity to printed text, the width of the stroke, etc. In fact, some characters may not be recognizable without the knowledge of the writer's identity. To achieve sufficient text recognition accuracy, the state-of-the-art neural networks [3,8,14,23,27,32,37] must adapt to a large number of writer styles. While transcribing, these architectures have to rely only on the image of a text line, which may not provide sufficient context. The improper adaptation may lead to wrong interpretations of ambiguities, which naturally arise among multiple writers.

Figure 1 illustrates WS-Net, which is a standard CTC-based [3,8,27,32] architecture enhanced by our proposed Writer Style Block (WSB). Apart from the text line image, WS-Net takes an additional input in the form of a Writer Style Identifier (WSI). WSI serves as an index into a WSB writer-style embedding table, where each writer is represented by a single embedding. WSB is an adaptive instance normalization [13] conditioned on writer-style embeddings. The adaptive instance normalization can modulate how the network processes information and which features become important and as such it provides WS-Net the ability to adapt to a vast amount of different writers.

The specific contributions of this paper are as follows: (1) Writer Style Block (WSB), an adaptive instance normalization conditioned by writer-style embeddings, which can enhance any standard text handwritten recognition network with the ability to explicitly utilize writing styles in the training dataset; (2) extensive evaluation of WSB for various embedding dimensions 16–256, both in standard and pre-trained mode; (3) evaluation of WSB in writer-dependent and writer-independent scenarios.

2   Related Work

In the state-of-the-art approaches to handwritten text recognition, the writer's identity information is usually not utilized. However, providing the recognition model with such information might be useful as the model can then better handle different writing styles, writer-specific patterns, etc. Closely related methods to the model adaptation, such as transfer learning, fine-tuning, and other similar techniques, are studied in several works [19,28,29,33,33,40], but the main disadvantage of these approaches is that they produce a unique model when adapted to a specific domain. Our proposed Writer Style Block (WSB) enhances a text recognition network to explicitly use writer-style information, so it can handle multiple writers simultaneously.

An approach similar to ours was proposed by Wang et al. [36]. It consists of two neural networks – writer-style extractor network and text recognition network, where the first network is trained to classify the writers' identities and the second one is trained to recognize the text content, each network accept text line image as input. Local writer-style representations, obtained as output features of the last recurrent layer of the writer-style extractor network, are aggregated by a mean pooling layer into a global writer-style representation. The writer's identity is utilized by aggregating the global writer-style representation and the local style representations by linear layers and connection operation into a single vector which is then aggregated into features of the text recognition network. When compared to our method, we do not extract the writer-style representation in a feed-forward manner, but instead, we learn a fixed number of representative embeddings on the training dataset, which allows the network to utilize all available writer-style information, not only the writer-style information of currently processed text line image.

Bhunia et al. [2] proposed writer-style adaptation as a meta-learning task. The goal is to train a general model which can be effectively adapted to a new writer, the adaptation should be fast and only a few labeled samples from the target domain should be needed. They used a Seq2Seq model expanded of special meta parameters in the form of a gamma layer and dedicated learning rates for each layer. During the adaptation process, the gamma layer should allow the model to focus more on problematic/unknown characters and ignore the already well-learned ones, while the dedicated learning rates should provide the model with the ability to prefer/ignore the adaptation of certain layers. The meta parameters together with the general model parameters are optimized with a meta-learning process consisting of two phases, inner and outer. The inner phase fine-tunes dedicated writer models (always initialized from the general one) on writer-specific data. The outer phase evaluates the dedicated writer models on respective writer-held-out support sets and updates the meta parameters together with the general model's parameters with all the dedicated writer models gradients. The inner and outer phases are repeated. Instead of training a general model which can be effectively adapted to new writers, resulting in a dedicated model for each, our goal is to train a single model with writer-dedicated parameters, where the adaptation to a new writer is done by optimizing a new set of writer-dedicated parameters.

More extensive model adaptation research can be found in the speech recognition area (ASR) [1]. Structure transform approaches represent the most relevant domain adaptation methods to our work, the general idea is to build an architecture with a small set of speaker-dependent (SD) parameters for each speaker while keeping most of the parameters shared – the speaker-independent (SI) parameters. While training, both the SD and the SI parameters are updated, but during the adaptation to a new writer, only the SD parameters are optimized. In previous works, the SD parameters include the input layer (linear input network, LIN [26]), the hidden layer (linear hidden network, LHN [10]), and the output layer (linear output network, LON [20]). Such adaptation has several drawbacks, mainly the large number of adapted parameters which results in a slow adaptation process and model overfitting if strong regularization is not used. Also, as the speaker information is typically discarded in the latter layers of the network [24], in more recent approaches the SD parameters are located more toward the beginning of the network. Many approaches aim to speed up the slow adaptation and suppress the overfitting problems mentioned above by reducing the number of adapted SD parameters. Parametrization of activation functions with SD parameters was proposed by Zhang et al. [39]. In other works, the SD parameters are represented by scales and/or offsets in various layers. Namely, in Learning Hidden Unit Contributions (LHUC) [34], every kernel is followed by an SD scale parameter. Another option is to use the scales and offsets in batch normalization layers as the SD parameters [22,35]. Approaches by Zhao et al. [42,43] and Samarakoon et al. [30] propose to factorize weight matrices of SD linear layers as most of the information is stored in diagonals. Utilization of such decomposed matrices results in fewer SD parameters.

Approaches proposed by Cui et al. [5] and Delcroix et al. [6] use an auxiliary SI network that generates SD parameters based on a small SD input (e.g. i-vector, learned speaker embedding, or similar features). In the first approach, the auxiliary network generates SD scales and offsets for hidden activations. The latter approach proposes to train a recognition network with several branches under the assumption that the different branches learn to specialize in different types of speakers. The auxiliary network is then used to generate weights for aggregation of the outputs produced by the individual branches. Similarly to Wang et al. [36], some of the existing methods [16,31,38] do not explicitly utilize the speaker's identity, instead, they extract global style features from the entire utterance, that is being currently recognized, and incorporate these features into the recognition process of its parts.

Our WS-Net architecture utilizes embeddings as writer-dependent (WD) parameters, while the rest of the parameters are shared among all writers (WI). The writer-dependent embeddings are part of the Writer Style Block (WSB), which is an adaptive instance normalization connected to the writer-independent part of the WS-Net. Because the adaptive instance normalization is conditioned on the writer-dependent embeddings, the information about a writer is utilized by the writer-independent part of the network. WS-Net is inspired by style transfer approaches and style-dependent Generative Adversarial Networks, which use an adaptive instance normalization (AdaIN) to broadcast information about the desired output style across a whole image [7,11,13,15]. Instead of using the AdaIN layers to broadcast the information about the style, Murase et al. [25] used it to broadcast the information about the content while optimizing an autoencoder for writer verification. In this way, the autoencoder can focus to extract only the writer-style information which is important for successful writer verification. In our previous work [18] we introduced TS-Net, where the only difference to WS-Net is the location of the AdaIN layer.

3   Writer Style Block

We propose Writer Style Block (WSB), which allows WS-Net, described in detail in Sect. 5, to learn dedicated writer parameters in the form of writer-style

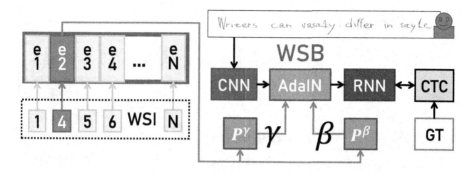

Fig. 2. Our proposed neural network (WS-Net) consists of a convolutional part (CNN), a recurrent part (LSTM), and Writer Style Block (WSB).

embeddings. Figure 2 shows WSB as part of the WS-Net. The WSB is an adaptive instance normalization layer AdaIN [13]:

$$\text{AdaIN}(\mathbf{X}_c, \gamma_c, \beta_c) = \gamma_c \left(\frac{\mathbf{X}_c - \mu(\mathbf{X}_c)}{\sigma(\mathbf{X}_c) + \epsilon} \right) + \beta_c, \qquad (1)$$

where $\mathbf{X}$, γ, β, μ, σ, ϵ stand for input, scales, offsets, mean, standard deviation and a small positive constant, c specifies the channel dimension. The adaptive scales γ and offsets β are given by two affine projections P^γ and P^β:

$$\gamma = P^\gamma(\mathbf{e}) = \mathbf{W}^\gamma \mathbf{e} + \mathbf{b}^\gamma, \beta = P^\beta(\mathbf{e}) = \mathbf{W}^\beta \mathbf{e} + \mathbf{b}^\beta, \qquad (2)$$

where $\mathbf{W}^\gamma$, $\mathbf{W}^\beta$, $\mathbf{b}^\gamma$, $\mathbf{b}^\beta$ are projection matrices and biases and $\mathbf{e}$ is a writer-style embedding specified by the corresponding Writer Style Identifier (WSI). Therefore, in our architecture, each writer has dedicated parameters in the form of a writer-style embedding $\mathbf{e}$, while all the other parameters are shared. While training, a writer-style embedding $\mathbf{e}$ is updated only on the respective writer training data, all the other parameters of WS-Net are updated on all writers.

Initialization. We initialize WSB similarly to the standard instance normalization. Each writer-style embedding $\mathbf{e}$ is initialized from the standard normal distribution $\mathcal{N}(0, 1)$. The projection matrices $\mathbf{W}^\gamma$, $\mathbf{W}^\beta$ are initialized from an uniform distribution $\mathcal{U}(-\sqrt{\text{ED}} \times \tau, \sqrt{\text{ED}} \times \tau)$, where ED stands for embedding dimension and τ is a small positive constant. Conditioning the initialization on the square root of the ED results in the same standard deviation of the scales γ and the offsets β across all ED. The magnitude of this standard deviation can be manipulated with τ. In conducted auxiliary experiments we found that a reasonable value of the standard deviation is 0.1 and it is achieved by setting τ to 0.174. The $\mathbf{b}^\gamma$ and $\mathbf{b}^\beta$ are set to ones and zeros, respectively. For all ED, this way of initialization ensures, that both the scales γ and the offsets β have

Fig. 3. Left, samples from the CzechHWR dataset. Right, representative words of 19 writers from Handwriting Adaptation Dataset.

the same small standard deviation, while the scales are centered around 1 and the offsets around 0. Instead of initializing the embeddings from the standard normal distribution, pre-trained embeddings can be used (see Sect. 6).

4   CzechHWR Dataset

Our dataset CzechHWR consists of triplets: a text line image, WSI, and the ground-truth transcription of the text line image. It was created from three main sources: our text recognition web application PERO OCR[1], a collection of Czech letters [12], and Czech chronicles. So far, users of the web application have uploaded and annotated documents containing about 295k handwritten text lines. Most of the documents were written in Czech modern cursive script, although, a marginal amount of documents in different scripts such as Gothic or German Kurrent is also present. They are mainly composed of military diaries, chronicles, letters, and notes. Based on document or page level manual inspection of the annotated documents, we assigned approximately 2.6k WSI. Czech letters is a collection of 2k letters, where the number of text lines is 87k. Most of them were written in Czech modern cursive, while a minimum amount was typeset. As it can be reasonably assumed that a handwritten letter has only one author, we assigned a distinct WSI to each letter. We manually annotated approximately 2 pages of 277 distinct Czech chronicles, resulting in 553 annotated pages with 24k text lines. As each page was chosen from a visually different place in the chronicle, we assigned a different WSI to each page.

The final CzechHWR dataset contains 406k annotated text line images with 5.1k WSI. The level of penmanship/readability differs, ranging from scribbles to calligraphy, although due to the origin of the data, the tendency is towards fairly readable texts (see the left side of Fig. 3). There are two issues resulting from the WSI assigning process. First, there is no assurance that multiple WSI does not identify the same writer because it is not possible for the annotator to keep track

---

[1] https://pero-ocr.fit.vutbr.cz.

Table 1. The number of lines and writers (WSI) for each cluster in the CzechHWR dataset. For a detailed description, see the text.

	1	20	50	100	200	500	1000	ALL
TRN	13k	79k	82k	43k	24k	16k	122k	379k
TST	169	1k	1.1k	566	287	198	1.7k	5k
TST$_W$	0	4.5k	6.2k	3.2k	2k	1.1k	5.4k	22.4k
WSI	1.1k	2.3k	1.2k	322	79	21	54	5.1k

of thousands of writing styles. Second, as we assigned the WSI on the document or page level, there is a possibility, that some text lines of the document/page belonged to distinct writers. The first issue should not present a problem for our system, as it will try to learn the same writer-style embedding e multiple times. The second issue should present only a slight regularization to the optimization process, as the vast majority of text lines were assigned correctly. The exact number of distinct writers is unknown, but a reasonable lower estimate is 4.5k. To avoid overcomplicated statements we often use "writer lines" to refer to data samples (triplets) that share the same WSI.

We divided the CzechHWR dataset for training and testing in the following way. We randomly draw 5k lines for testing (TST). The remaining lines were divided into seven clusters: 1, 22, 55, 110, 225, 550, and 1100, according to their writers (WSI). The cluster numbers specify the minimum number of lines for each writer to belong to that cluster. A writer belongs only to the cluster with the highest possible number, the resulting clusters are disjoint. Out of clusters 22, 55, 110, 225, 550, and 1100, for each writer (WSI), we respectively took 2, 5, 10, 25, 50, and 100 lines for testing clusters TST$_C$. The remaining lines formed the training clusters TRN$_C$. Out of convenience, we refer to both testing and training clusters with the minimum number of lines belonging to each writer in the training clusters: 1, 20, 50, 100, 200, 500, and 1000, e.g. TST$_{20}$ or TRN$_{500}$. For example, a CER measured on TST$_{50}$ set shows an error for writers that have at least 50 lines in the training dataset. The number of lines and WSI for each cluster is shown in Table 1, ALL stands for all clusters.

For training, we merged all training clusters into a single training dataset (TRN). TST allows us to inspect the average CER because the distribution of WSI is the same as in TRN. TST$_C$ and TRN$_C$ clusters allow us to measure how the number of data samples per WSI affects the WSB performance.

For experiments with writer-independent scenario we used our Handwriting Adaptation Dataset[2] (HAD) [19] which consists of 19 writers, right side of Fig. 3 shows sample words for each writer. In comparison with CzechHWR dataset, HAD contains both similar and vastly different scripts.

---

[2] https://pero.fit.vutbr.cz/handwriting_adaptation_dataset.

5   Writer Style Network

In this section, we describe the proposed WS-Net and its training procedure. We first introduce the baseline network architecture and later we specify the changes that lead to the proposed WS-Net.

The baseline architecture is similar to text recognition state-of-the-art architectures trained with CTC loss function [3,8,27,32]. It consists of a convolutional part (CNN) and a recurrent part (RNN). The CNN part is a sequence of 4 convolutional blocks, where each has 2 convolutional layers with numbers of output channels set to 64, 128, 256, and 512, respectively. All convolutional blocks except the last one are followed by a max pooling layer, while the input width subsampling factor of the CNN is 4. The RNN part processes three scaled versions of the WSB output with three branches, the scaling factors are 1, 0.5, and 0.25 and each branch has two LSTM layers. The outputs are upsampled back to the original resolution, summed, and processed with a final LSTM layer. Each LSTM layer is bidirectional and has a hidden feature size of 256 for each dimension. The output of the RNN block is processed by a final 1D convolutional layer. The baseline architecture has 5 instance normalization layers, the first four are in the convolutional part after each convolutional block (CNN), and the last is after the recurrent block (RNN). A more detailed description together with a detailed diagram can be found in our previous work [18].

WS-Net (see Fig. 2) is the baseline architecture enhanced with WSB, which replaces one or multiple standard instance normalization layers with adaptive instance normalization layers (AdaIN) conditioned on writer embeddings. Each AdaIN has its own projection matrices $\mathbf{W}^\gamma$, $\mathbf{W}^\beta$, and biases $\mathbf{b}^\gamma$, $\mathbf{b}^\beta$, but the writer-style embeddings $\mathbf{e}$ are shared. We experimented with two variants of WS-Net: Single AdaIN and All AdaIN. Single AdaIN is WS-Net, where the adaptive normalization layer (AdaIN) is placed after the convolutional block (CNN), and all the rest are standard instance normalization layers. All AdaIN is WS-Net, where every normalization layer is AdaIN.

Motivation behind AdaIN placements. The motivation behind the AdaIN layer placement in Single AdaIN architecture is based on auxiliary experiments with All AdaIN architecture trained in multiple embedding dimension (ED) settings. By fixing scales and offsets for various AdaIN layers of a trained All AdaIN system, we simulated all possible settings of AdaIN layers. We fixed the respective AdaIN by conditioning the scales and offsets on the mean writer embedding. Out of all placements, we noticed a significantly poorer performance for every setting where the AdaIN layer after the CNN block was fixed, which suggests that the All AdaIN system benefited most from this adaptive normalization layer. Furthermore, in speech recognition, A Mohamed et al. [24] found that information about speakers generally vanishes toward the end of the network, which suggests that placing the AdaIN layer near the end of the network might have no effect. On the other hand, placing the AdaIN near the begging of the network might have no effect too as the activation in the first layers usually bears only low-level information. Finding the best possible setting for AdaIN layers would mean training 31 different settings from scratch, for each embedding size ED. Additionally, we experimented with a setup where the AdaIN layer was behind the CNN and the RNN block, which turned up to be unstable and therefore we do not discuss the respective results in more detail.

Training. WS-Net is trained jointly with Adam optimizer [17], and the CTC loss function. The training data consists of triplets: a text line image, WSI, and the ground-truth transcription. Because the training samples are drawn randomly, embeddings that have more data samples assigned to them are updated more frequently. We mitigate this by normalizing embedding batch gradients by the frequency of their WSI in the batch. We trained Single AdaIN and All AdaIN, with the embeddings initialized from the standard normal distribution (normal embeddings), for 500k iterations up until convergence. We used polynomial warmup of a third order to gradually increase the learning rate from 0 to 3×10^{-4} in the first 10k iterations. At iterations 200k and 400k, we used the warmup again, but the learning rate maximums were 7×10^{-5} and 1.75×10^{-5}. The batch size was set to 32. The baseline was trained with the same strategy.

Additionally, we trained Single AdaIN and All AdaIN, with the pre-trained embeddings (described in Sect. 6), for 675k iterations in three consecutive steps. For each step, we trained the model up until convergence. First, we optimized all the parameters except the embeddings for 400k iterations. We used the warmup strategy in iterations 0, 200k, and 300k with the learning rate maximums being 3×10^{-4}, 1.5×10^{-4}, and 7×10^{-5}. Second, for the next 100k iterations, we optimized just the embeddings. We used the warmup strategy in iterations 400k and 450k and kept the learning rate the same. Third, for the final 175k iterations, we finetuned all the parameters. We used the warmup strategy at iterations 500k, 550k, 600k, and 650k with the learning rate maximums being 7×10^{-5}, 3×10^{-5}, and 1.5×10^{-5}. The model was significantly more accurate after each step.

We used data augmentation including color changes, noise, blur, and various geometric transformations to simulate different backgrounds, slants, spacing between characters, etc. Additionally, we mask the text line images with a random number of noise patches. The height of a noise patch is the same as the height of the text line image, the width is chosen randomly up to the height of the text line image forming at most a square patch. In this setting, the noise patch usually masks a maximum of two letters. The intuition behind masking is to strengthen the language modeling capability of the system. A more detailed description of the used augmentation together with examples can be found in our work [19], where the respective augmentation is B1C1G1M1.

6   Pre-training Writer-Style Embeddings

Instead of using embeddings initialized from the standard normal distribution, we also use pre-trained ones. We implemented a contrastive learning approach, where the encoder is a stack of four convolutional layers with output channel dimensions 32, 64, 128, and 512 respectively, followed by three multi-head attention blocks with 4 heads and 512 channels, average pooling over the width dimension, and L2 normalization. The encoder generates an embedding $\mathbf{q}$ for each text line image input. We used the normalized temperature-scaled cross entropy (NT-Xent) loss function [4]:

$$\mathcal{L}_{\mathbf{qp}} = -log \frac{\exp(\mathbf{q} \cdot \mathbf{p}/\tau)}{\sum_{j=0}^{N} \exp(\mathbf{q} \cdot \mathbf{n}_j/\tau) + \exp(\mathbf{q} \cdot \mathbf{p}/\tau)}, \tag{3}$$

where $\mathbf{q}$ forms a positive pair with the embedding $\mathbf{p}$, and negative pairs with embeddings $\mathbf{n}_j$. Embeddings of a positive pair are generated from image text lines belonging to the same writer, while embeddings of a negative pair are generated from image text lines belonging to distinct writers. The final NT-Xent loss is given by $\mathcal{L} = \mu(\mathcal{L}_{\mathbf{qp}}), (\mathbf{q}, \mathbf{p}) \in \mathbf{P}$, where $\mathbf{P}$ is a set of all positive pairs. As the final layer of the encoder is L2 normalization, the dot product will produce cosine similarity. The temperature parameter τ affects the strictness of the NT-Xent loss function. Values closer to 0 make it more strict, whereas higher values make it looser. Higher strictness force the encoder to produce closer cosine similarities, which means closer embeddings. We set τ to 0.15.

As the encoder can only provide an embedding $\mathbf{q}$ for a text line image, we extract the final writer-style embedding $\mathbf{e}$ by aggregating output embeddings for 32 distinct and random text line images belonging to the respective writer. The aggregation is done by choosing the embedding $\mathbf{q}_i$ which has the largest sum of above-average cosine similarities. The cosine similarity of two embeddings $\mathbf{q}_i$ and $\mathbf{q}_j$ is above average if it is larger than the average cosine similarity between all 32 embeddings. No data augmentation is used during the extracting process.

Training Writer-Style Encoder. The encoder was trained with AdamW [21] optimizer for 20k iterations, with a batch size of 180, and learning rate 2×10^{-4}. We use the same dataset as for the standard training, the augmentations are similar, but stronger, without any geometry transformation and patch noise masking. The NT-Xent loss is evaluated for all positive pairs in the batch. We checked the convergence by visual inspection of text line images that belonged to the k-nearest neighbors embeddings according to cosine similarity.

7   Writer-Dependent Scenario

In this section, we describe the experiments conducted with the WS-Net on the CzechHWR dataset. Specifically, we compare Single AdaIN and All AdaIN variants of WS-Net to the baseline network. We trained architectures in both normal and pre-trained embedding setups for embedding dimensions (ED): 16, 32, 64, 128, 256, and 512, and a separate writer-style encoder was trained for each ED. Figure 4 shows test character error (CER) for Single AdaIN, All AdaIN, and the baseline on the testing set TST. The left graph shows architectures initialized with normal embeddings. The Single AdaIN performed better for ED

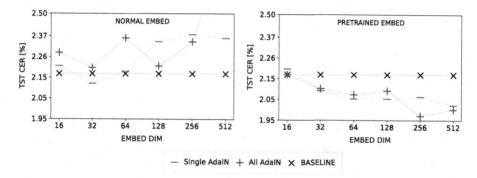

Fig. 4. Character error rate (CER) for Single AdaIN, All AdaIN, and the baseline on the test set. The graphs show CER for different embedding dimensions (ED) and for different initialization: randomly initialized (left) and pre-trained (right).

16, 32, 64, and 512, and brought the best result for ED 32, All AdaIN was inconsistent across ED. All settings, except Single AdaIN with ED 32, brought worse performance than the baseline. The right graph shows architectures initialized with pre-trained embeddings. All settings consistently outperformed the baseline, with the exception of ED 16, which brought similar CER. Both AdaIN settings brought similar results, except for the ED 256, where All AdaIN had the lowest CER of all settings and decreased the test CER of the baseline by 9.22% relatively. All AdaIN is more stable across different ED, while initialized with pre-trained embeddings. When initialized with pre-trained embeddings, Single AdaIN brought progressively better performance with increasing ED, whereas initialization with normal embeddings had the opposite tendency since the ED 32. In further experiments, we do not show results for All AdaIN, as it did not bring any significant improvement over Single AdaIN.

Figure 5 shows Single AdaIN CER measured on various testing and training clusters (see Table 1 and Sect. 4) for ED 16, 32, and 256. The top graphs show the CER for Single AdaIN initialized with normal embeddings. We do not show the results for ED 64, 128, and 512, as the CER for ED 64 had the same tendency as ED 32, whereas ED 128 and 512 had the same tendency as ED 256. For larger clusters 100, 200, 500, and 1000, the test CER was smaller or similar to the baseline, while for smaller clusters 20 and 50, it was worse or similar. Only ED 32 and 64, outperformed the baseline for the larger clusters and evened out the baseline for smaller ones. As there was no noticeable relative decrease in train CER for smaller clusters (20, 50), the respective writer-style embeddings probably overfitted in the wrong way and therefore brought poor generalization. Although we trained ED 256 up until convergence, the train CER suggests that it was not trained properly, as it should have overfit more than the smaller ED and the baseline (the same applied for ED 128 and 512).

The bottom graphs in Fig. 5 show the results for Single AdaIN initialized with pre-trained embeddings. For all clusters, all ED except 16 had consistently

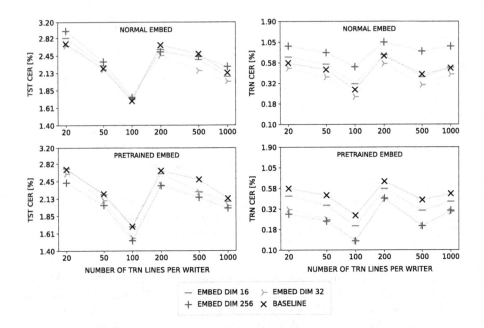

Fig. 5. Character error rate (CER) measured on various testing and training clusters for Single AdaIN with randomly initialized embeddings (top) and pre-trained embeddings (bottom).

smaller test CER than the baseline. The general tendency among all clusters, both for the testing and the training clusters, is that a higher ED has always lower CER than a smaller ED. Surprisingly, ED 16 brought significantly better performance for writers in cluster 500, this is probably due to the fact that the writer-style encoder was able to find the unique properties of their writing styles and encode them even to smaller embeddings. Although all ED were fairly overfitted, they were able to generalize well among all the clusters.

The pre-trained embedding initialization variants generally outperformed the normal ones, while the largest differences were on smaller clusters. If we compare the best ED out of each variant, ED 32 for normal embeddings to ED 256 for pre-trained embeddings, we can see that the latter was better only on the smaller clusters. Single AdaIN trained from scratch (initialized with normal embeddings) does not guarantee in any way that the learned embeddings would represent the respective writer styles and therefore it is prone to overfit on writer irrelevant details, especially for smaller amount of writer lines. On the other hand, Single AdaIN initialized with the pre-trained embeddings is forced to learn the proper utilization of writer styles in relation to the handwritten text recognition task, as the pre-trained embeddings are fixed for the first 400k training iterations. As the writer-style encoder was directly trained to encode the writing style, it learned to extract robust and representative embeddings even for writers from smaller clusters.

For pre-trained embeddings and normal embeddings with ED 128 and 256, the t-SNE projections showed semantically meaningful clusters, where writers of similar scripts were grouped together. For normal embeddings with ED 16, 32, and 64, the projections did not show visible clusters.

8   Writer-Independet Scenario

For new writers, our architecture cannot be used in a simple feed-forward manner. However, as all parameters except the writer-style embeddings are shared among the writers, we should be able to adapt to a new writer by finding a new representative embedding. We experimented with two approaches. The first selected the new embedding out of the existing ones according to CER on adaptation lines As there are more than 5k existing embeddings, we clustered the existing embedding space with the k-mean algorithm into 50 clusters and evaluated only one random embedding from each cluster. The second optimized a new embedding with 150 LBFGS iterations, the adaptation text line images were augmented in the same way as the training ones. To inspect the quality of selection and optimization for different numbers of adaptation lines, we define writer adaptation runs. A writer adaptation run consists of adapting the respective writer on 5 line clusters: 16, 32, 64, 128, and 256, where the numbers refer to the number of lines in them, a smaller cluster is always a subset of a larger one, and the lines of the largest cluster are drawn randomly from all available lines. We run 23 adaptation runs for 19 new writers of HAD dataset and 3 fully-trained Single AdaIN architecture setups: normal embeddings with ED 32, pre-trained embeddings with ED 32, and pre-trained embeddings with ED 256, resulting in final $23 \times 19 \times 3$ runs for both the selection and optimization approach. We always chose the best-performing embedding on the adaptation lines and inspected CER on test lines, there are 256 randomly drawn testing lines for each adaptation run.

The selection approach performed worse than the baseline even for higher amounts of adaptation lines. Generally, for all Single AdaIN setups, the selection was more accurate for higher numbers of adaptation lines. Surprisingly, the best-performing setup of Single AdaIN was normal embeddings ED 32, while for some writers and selections based on 256 adaptation lines, it performed similarly to the baseline. We do not show detailed results, as the selection approach did not outperform the baseline and we did not notice any interesting properties apart from the already described. For the optimization approach, we tried to optimize from the selected embeddings, but the mean of the existing ones turned out to be a better starting point.

Figure 6 shows the performance of our optimization approach expressed as relative test CER reductions of the baseline. A boxplot represents the distribution of the 19 writers' CER reductions, and a writer's CER reduction is the mean of the writer's CER reductions measured on 23 runs. More precisely, the CER reduction is given by:

$$\frac{A - B}{B}, \tag{4}$$

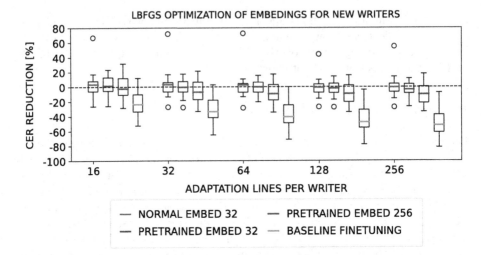

Fig. 6. The performance of our optimization approach expressed as relative test CER reductions of the baseline compared to baseline finetuning.

where A is the test CER of the adapted Single AdaIN, and B is the test CER of the baseline. Generally, for all Single AdaIN setups, the optimization was more accurate for higher numbers of adaptation lines. By further inspecting the results in relation to Fig. 3, we noticed that the performance varied across writers in relation to their scripts. CER reductions were largest for scripts that were not sufficiently represented in the CzechHWR dataset, such as German Kurrent or Ghotic, while the performance worsen for the CzechHWR-like scripts. On average, the pre-trained embeddings ED 256 setup provided the best performance, although it performed worse than the respective ED 32 setup for some writers. The normal embeddings ED 32 setup brought the lowest CER reductions, while it vastly worsen the performance for some writers.

Based on these results, WS-Net is able to learn writer-style space that can significantly boost the transcription accuracy of new writers, however, finding representative embeddings is not straightforward. We argue that this is to some extent caused by the sensitivity of WS-Net to precise writer-style embeddings. We evaluated the sensitivity of WS-Net by randomly shuffling WSI in testing datasets, the test CER increased 2 times for the normal setting and 4 times for the pre-trained setting. Therefore, simply selecting an existing embedding brought worse performance for all new writers, as WS-Net was too overfitted to existing writers and none of them had an extremely similar style to the new ones. Optimization was difficult and unstable, as both the mean and the selected embeddings did not provide a good starting point.

Note that for Single AdaIN pre-trained ED 256 setup we evaluated WS-Net with embeddings provided by the writer-style encoder in an unsupervised manner, both without and with further optimization, but the results were comparable to supervised selection and optimization from the mean. Furthermore, as the embeddings might have slightly changed in the last phase of the training, we tried to boost the performance of this approach by finetuning the writer-style encoder on the WS-Net writer-style embeddings using the L2 loss, but even this setup failed.

So far, the only reasonable solution for new writers is to find representative embeddings with optimization on larger numbers of adaptation lines, however for such scenarios, a simple finetuning of the baseline brought significantly better test CER reductions. Note that we estimated the optimal number of finetuning iterations with 4-fold cross-validation. An extensive analysis of the finetuning approach can be found in our work [19], where we showed that significant CER reductions can be obtained even with less than 16 adaptation lines. So far, WS-Net is only suitable for the writer-dependent scenario where it consistently outperformed the baseline, whereas it is not a suitable choice for writer-independent scenarios, especially if enough annotated data for new writers are available.

In our future work, we plan to redesign the WSB block, so it can be used in writer-independent scenarios without any adaptation lines. The idea is based on an attention mechanism, which would take the hidden features of the processed text line image as queries, and embeddings of the writer-style space as keys and values. In speech recognition, similar ideas were proposed by Zhao et al. [41] and Fan et al [9].

9 Conclusion

We showed that a standard CTC-based neural network enhanced by proposed Writer Style Block (WSB) can utilize a vast number of writing styles. While initializing WSB with embeddings pre-trained in an unsupervised contrastive manner, the enhanced architecture was able to consistently outperform the baseline version even for writers that were poorly represented in the training set. On the other hand, training the WSB writer-style embeddings from scratch, led to worse performance on average. Although we were unable to find an appropriate way to estimate representative embeddings for new writers, we confirmed their existence in the WSB writer-style embedding space by optimization on 256 adaptation lines. This suggests that the WSB provides superior performance even for new writers. In future work, we plan to learn the appropriate estimation in a supervised manner by extending the WSB with an attention mechanism, which would take the hidden features of a processed text line image as queries and the writer-style embeddings as keys and values.

Acknowledgment. This work has been supported by the Ministry of Culture Czech Republic in NAKI III project Machine learning for printed heritage digitisation (DH23P03OVV066).

References

1. Bell, P., Fainberg, J., Klejch, O., Li, J., Renals, S., Swietojanski, P.: Adaptation algorithms for speech recognition: an overview (2020)
2. Bhunia, A.K., Ghose, S., Kumar, A., Chowdhury, P.N., Sain, A., Song, Y.Z.: Metahtr: towards writer-adaptive handwritten text recognition. In: Proceedings of the IEEE/CVF Conference on Computer Vision and Pattern Recognition, pp. 15830–15839 (2021)
3. Bluche, T., Messina, R.: Gated convolutional recurrent neural networks for multilingual handwriting recognition. In: 2017 14th IAPR International Conference on Document Analysis and Recognition (ICDAR), vol. 01, pp. 646–651 (2017). https://doi.org/10.1109/ICDAR.2017.111
4. Chen, T., Kornblith, S., Norouzi, M., Hinton, G.: A simple framework for contrastive learning of visual representations. In: International Conference on Machine Learning, pp. 1597–1607. PMLR (2020)
5. Cui, X., Goel, V., Saon, G.: Embedding-based speaker adaptive training of deep neural networks. CoRR abs/1710.06937 (2017)
6. Delcroix, M., Kinoshita, K., Ogawa, A., Huemmer, C., Nakatani, T.: Context adaptive neural network based acoustic models for rapid adaptation. IEEE/ACM Trans. Audio Speech Lang. Process. **26**(5), 895–908 (2018)
7. Dumoulin, V., Shlens, J., Kudlur, M.: A learned representation for artistic style. CoRR abs/1610.07629 (2016)
8. Dutta, K., Krishnan, P., Mathew, M., Jawahar, C.V.: Improving CNN-RNN hybrid networks for handwriting recognition. In: 2018 16th International Conference on Frontiers in Handwriting Recognition (ICFHR), pp. 80–85 (2018). https://doi.org/10.1109/ICFHR-2018.2018.00023
9. Fan, Z., Li, J., Zhou, S., Xu, B.: Speaker-aware speech-transformer. In: 2019 IEEE Automatic Speech Recognition and Understanding Workshop (ASRU), pp. 222–229. IEEE (2019)
10. Gemello, R., Mana, F., Scanzio, S., Laface, P., De Mori, R.: Linear hidden transformations for adaptation of hybrid ANN/HMM models. Speech Commun. **49**(10–11), 827–835 (2007)
11. Ghiasi, G., Lee, H., Kudlur, M., Dumoulin, V., Shlens, J.: Exploring the structure of a real-time, arbitrary neural artistic stylization network. CoRR abs/1705.06830 (2017)
12. Hladká, Z.: 111 let českého dopisu v korpusovém zpracování (2013)
13. Huang, X., Belongie, S.J.: Arbitrary style transfer in real-time with adaptive instance normalization. CoRR abs/1703.06868 (2017)
14. Kang, L., Riba, P., Rusiñol, M., Fornés, A., Villegas, M.: Pay attention to what you read: non-recurrent handwritten text-line recognition. Pattern Recogn. **129**, 108766 (2022)
15. Karras, T., Laine, S., Aila, T.: A style-based generator architecture for generative adversarial networks. CoRR abs/1812.04948 (2018)
16. Kim, T., Song, I., Bengio, Y.: Dynamic layer normalization for adaptive neural acoustic modeling in speech recognition. CoRR abs/1707.06065 (2017)
17. Kingma, D.P., Ba, J.: Adam: a method for stochastic optimization. In: Bengio, Y., LeCun, Y. (eds.) ICLR 2015, San Diego, CA, USA, 7–9 May 2015, Conference Track Proceedings (2015)
18. Kohút, J., Hradiš, M.: Ts-net: Ocr trained to switch between text transcription styles. In: International Conference on Document Analysis and Recognition, pp. 478–493. Springer (2021)

19. Kohút, J., Hradiš, M.: Finetuning is a surprisingly effective domain adaptation baseline in handwriting recognition (2023)
20. Li, B., Sim, K.C.: Comparison of discriminative input and output transformations for speaker adaptation in the hybrid NN/HMM systems. In: Eleventh Annual Conference of the International Speech Communication Association (2010)
21. Loshchilov, I., Hutter, F.: Fixing weight decay regularization in adam. CoRR abs/1711.05101 (2017). http://arxiv.org/abs/1711.05101
22. Mana, F., Weninger, F., Gemello, R., Zhan, P.: Online batch normalization adaptation for automatic speech recognition. In: IEEE ASRU 2019, pp. 875–880. IEEE (2019)
23. Michael, J., Labahn, R., Grüning, T., Zöllner, J.: Evaluating sequence-to-sequence models for handwritten text recognition. In: 2019 International Conference on Document Analysis and Recognition (ICDAR), pp. 1286–1293. IEEE (2019)
24. Mohamed, A.R., Hinton, G., Penn, G.: Understanding how deep belief networks perform acoustic modelling. In: IEEE ICASSP 2012, pp. 4273–4276. IEEE (2012)
25. Murase, K., Nakatsuka, S., Hosoe, M., Kato, K.: Handwriting feature extraction method for writer verification independent of character type by using adabn and adain. In: International Workshop on Advanced Imaging Technology (IWAIT) 2020, vol. 11515, pp. 11–14. Spie (2020)
26. Neto, J., Almeida, L., Hochberg, M., Martins, C., Nunes, L., Renals, S., Robinson, T.: Speaker-adaptation for hybrid HMM-ANN continuous speech recognition system (1995)
27. Puigcerver, J.: Are multidimensional recurrent layers really necessary for handwritten text recognition? In: 2017 14th IAPR International Conference on Document Analysis and Recognition (ICDAR), vol. 01, pp. 67–72 (2017). https://doi.org/10.1109/ICDAR.2017.20
28. Reul, C., Tomasek, S., Langhanki, F., Springmann, U.: Open source handwritten text recognition on medieval manuscripts using mixed models and document-specific finetuning. In: DAS 2022, pp. 414–428. Springer, Cham (2022). https://doi.org/10.1007/978-3-031-06555-2_28
29. Reul, C., Wick, C., Nöth, M., Büttner, A., Wehner, M., Springmann, U.: Mixed model OCR training on historical Latin script for out-of-the-box recognition and finetuning. In: The 6th International Workshop on Historical Document Imaging and Processing, pp. 7–12 (2021)
30. Samarakoon, L., Sim, K.C.: Factorized hidden layer adaptation for deep neural network based acoustic modeling. IEEE/ACM Trans. Audio Speech Lang. Process. **24**(12), 2241–2250 (2016)
31. Sarı, L., Thomas, S., Hasegawa-Johnson, M., Picheny, M.: Speaker adaptation of neural networks with learning speaker aware offsets. Interspeech (2019)
32. Shi, B., Bai, X., Yao, C.: An end-to-end trainable neural network for image-based sequence recognition and its application to scene text recognition. CoRR abs/1507.05717 (2015). http://arxiv.org/abs/1507.05717
33. Soullard, Y., Swaileh, W., Tranouez, P., Paquet, T., Chatelain, C.: Improving text recognition using optical and language model writer adaptation. In: : 2019 International Conference on Document Analysis and Recognition (ICDAR), pp. 1175–1180 (2019)
34. Swietojanski, P., Li, J., Renals, S.: Learning hidden unit contributions for unsupervised acoustic model adaptation. CoRR abs/1601.02828 (2016)
35. Wang, Z.Q., Wang, D.: Unsupervised speaker adaptation of batch normalized acoustic models for robust ASR. In: IEEE ICASSP 2017, pp. 4890–4894. IEEE (2017)

36. Wang, Z.R., Du, J.: Fast writer adaptation with style extractor network for handwritten text recognition. Neural Networks **147**, 42–52 (2022). https://doi.org/10.1016/j.neunet.2021.12.002. https://www.sciencedirect.com/science/article/pii/S0893608021004755

37. Wick, C., Zöllner, J., Grüning, T.: Transformer for handwritten text recognition using bidirectional post-decoding. In: Lladós, J., Lopresti, D., Uchida, S. (eds.) ICDAR 2021. LNCS, vol. 12823, pp. 112–126. Springer, Cham (2021). https://doi.org/10.1007/978-3-030-86334-0_8

38. Xie, X., Liu, X., Lee, T., Wang, L.: Fast DNN acoustic model speaker adaptation by learning hidden unit contribution features. In: INTERSPEECH, pp. 759–763 (2019)

39. Zhang, C., Woodland, P.C.: Parameterised sigmoid and relu hidden activation functions for DNN acoustic modelling. In: Sixteenth Annual Conference of the International Speech Communication Association (2015)

40. Zhang, Y., Nie, S., Liu, W., Xu, X., Zhang, D., Shen, H.T.: Sequence-to-sequence domain adaptation network for robust text image recognition. In: Proceedings of the IEEE/CVF Conference on Computer Vision and Pattern Recognition, pp. 2740–2749 (2019)

41. Zhao, Y., Ni, C., Leung, C.C., Joty, S.R., Chng, E.S., Ma, B.: Speech transformer with speaker aware persistent memory. In: INTERSPEECH, pp. 1261–1265 (2020)

42. Zhao, Y., Li, J., Gong, Y.: Low-rank plus diagonal adaptation for deep neural networks. In: IEEE ICASSP 2016, pp. 5005–5009. IEEE (2016)

43. Zhao, Y., Li, J., Kumar, K., Gong, Y.: Extended low-rank plus diagonal adaptation for deep and recurrent neural networks. In: IEEE ICASSP 2017, pp. 5040–5044. IEEE (2017)

Historical Document Image Segmentation Combining Deep Learning and Gabor Features

Maroua Mehri[1,2]([✉]) [iD], Akrem Sellami[3] [iD], and Salvatore Tabbone[2,4] [iD]

[1] Université de Sousse, Ecole Nationale d'Ingénieurs de Sousse, LATIS-Laboratory of Advanced Technology and Intelligent Systems, 4023 Sousse, Tunisie
[2] Université de Lorraine, IDMC-Institut des sciences du Digital, Management & Cognition, Pôle Herbert Simon, 13 Rue Michel Ney, 54000 Nancy, France
`maroua.mehri@eniso.u-sousse.tn`, `salvatore.tabbone@univ-lorraine.fr`
[3] Université de Lille, CNRS, UMR 9189 CRIStAL, Campus scientifique, Bâtiment ESPRIT, Avenue Henri Poincaré, 59655 Villeneuve d'Ascq, France
`akrem.sellami@univ-lille.fr`
[4] Université de Lorraine, CNRS, LORIA, UMR 7503, Campus Scientifique, 615 Rue du Jardin-Botanique, 54506 Vandœuvre-lès-Nancy, France

Abstract. Due to the idiosyncrasies of historical document images (HDI), growing attention over the last decades is being paid for proposing robust HDI analysis solutions. Many research studies have shown that Gabor filters are among the low-level descriptors that best characterize texture information in HDI. On the other side, deep neural networks (DNN) have been successfully used for HDI segmentation. As a consequence, we propose in this paper a HDI segmentation method that is based on combining Gabor features and DNN. The segmentation method focuses on classifying each document image pixel to either graphic, text or background. The novelty of the proposed method lies mainly in feeding a DNN with a Gabor filtered image (obtained by applying specific multichannel Gabor filters) instead of an original image as input. The proposed method is decomposed into three steps: a) filtered image generation using Gabor filters, b) feature learning with stacked autoencoder, and c) image segmentation with 2D U-Net. In order to evaluate its performance, experiments are conducted using two different datasets. The results are reported and compared with those of a recent state-of-the-art method.

Keywords: Historical document image segmentation · Gabor filters · Deep neural networks · Stacked autoencoder · 2D U-Net architecture

1 Introduction

Over the past three decades, many open questions related to analyzing images of historical printed and handwritten manuscripts have been pointed out by archivists, historians, human and social science researchers. These questions are related to the non-availability of high-performance computer-aided systems that ease the tasks of text transcription, indexation and keyword searching in historical documents. The existing systems are hindered by many issues related to the unsatisfactory performances of the optical character recognition tools. These performances are mainly due to the particularities of the digitized documents, such as the different levels and types of degradation present in these documents, the scanning defects (e.g. curvature) or the superimposition of information layers (e.g. noise, back-to-front interference, handwritten notes at the margins, stamps). Besides, the large variability and complex page layouts (e.g. several columns with irregular sizes, irregular spacings) of these documents makes necessary to tackle the layout analysis issue (i.e. physical structure identification). The layout analysis task is usually posed as an image segmentation issue that focuses on splitting a document image into homogeneous regions, such as paragraph, text line, background, table, figure caption and decoration. Image segmentation has long been considered as an important prerequisite task for further document image analysis (DIA) steps, such as text recognition, script/font analysis, image recognition, document indexing and retrieval. Furthermore, to achieve high accuracy rate of text recognition, it is important to ensure that the analyzed document image has been correctly segmented by identifying the textual and graphical contents [44]. Nevertheless, historical document image segmentation (HDIS) also remains a challenging preliminary step in DIA systems due the aforementioned particularities [11]. As a consequence, to address the above questions researchers keep proposing more efficient HDIS solutions able to extract the textual content from these document images [41].

There are two main categories of approaches that are both the most efficient and the most widely used to address the HDIS issue. The first category is based on extracting and analyzing low-level features, and particularly the texture descriptors. The second category represents the deep neural networks (DNN) based approaches. The recent trend in solving the HDIS problem is to use a DNN that learns the representation directly from the image pixels rather than using low-level extracted features. It has been shown that DNN have increased the HDIS performances by their ability to learn comprehensive visual features [41]. To the best of our knowledge, using both texture features and DNN in the same architecture to solve the HDIS issue has not been yet tackled. Therefore, in this paper we propose a HDIS method using Gabor filter (GF) based DNN. The novelty of this work lies in using the images filtered by GF as input of a DNN instead of feeding it with original images (ad-hoc approach) to solve the HDIS issue. Thanks to the high feature representation capacity of the Gabor filtered images that is obtained by applying specific multi-channel GF, we have shown that the proposed method outperforms the ad-hoc one for the task of segmenting HDI at pixel level. The segmentation method focuses on classifying each document image pixel to either graphic, text or background. The proposed method is conceptualized by the three following processes: *a)* filtered image generation using GF, *b)* feature learning with stacked autoencoder (SAE), and *c)* image segmentation with 2D U-Net. We applied GF on HDI in order to capture texture features to be learned by a stacked autoencoder and subsequently used for pixel-level segmentation by a 2D U-Net architecture (with a reduced number of parameters). The proposed method has been evaluated on two different datasets: a large-scale synthetic dataset having $12k$ pixel-wise annotated historical document images (HDI) [37] and 64 book images of the HBA dataset [18]. The conducted experiments demonstrate that the proposed method provides interesting segmentation results and outperforms a recent state-of-art method.

The remainder of this paper is structured as follows. Section 2 reviews the main methods proposed in the literature for HDIS. Section 3 details the proposed method. Section 4 provides the conducted experiments and the achieved results. Finally, the conclusions and further work are given in Sect. 5.

2 Related Work

In this section, we firstly report a review of the main recent state-of-the-art methods that are based on the multi-channel GF and DNN, with a particular focus on those related to different document image analysis sub-fields and tasks.

The state-of-the-art methods addressing the HDIS methods can be categorized into three classes: classical, texture, and DNN based approaches.

First, the classical approaches can be also classified into three categories: projection based (e.g. XY-CUT), smearing-based (e.g. RLSA), connected component based (e.g. Delaunay triangulation), and hybrid methods. The classical approaches usually deal with printed documents that have predominantly textual content and simple layout, and consequently they are not well-suited to HDI [19].

Second, since the early 2000's the texture based approaches have been extensively used to tackle different sub-fields and tasks related DIA, such as script identification [7], font recognition [10], text line, word and character segmentation [5] or HDIS [9]. For the HDIS task, the texture features have been extracted and analyzed in order to generate a partition of the analyzed document image into homogeneous or similar regions. The texture based methods have been considered as a consistent choice for meeting the need to segment a document image having various degradation levels and different noise types [30]. Besides, they are known to be more robust to different document layouts [1]. There are five main categories of texture based approaches that are defined according to the properties or characteristics of the extracted texture features: statistical (e.g. auto-correlation [13], Tamura [3]), geometric (e.g. difference-of-Gaussian filter [2]), model (e.g. LBP [6]), spectral (e.g. GF [24]), extensively used in a wide array of applications related to the DIA fields, such as document binarization [32], ground truth generation [21], detection of main text area from side-notes [8], text line segmentation [9], etc. There are many state-of-the-art methods based on extracting and analyzing Gabor features that were proposed for the HDIS task. For instance, Asi et al. [8] proposed a learning-free approach to detect the main text area from side-notes in ancient manuscripts based on coarse-to-fine scheme. First, a coarse segmentation of the main text area was processed by using GF. Then, the segmentation outputs were refined by using the energy minimization and graph cut techniques.

Finally, the last category of the existing HDIS methods are based on DNN. Since the last two decades, many solutions proposed for different tasks related to DIA center almost around DNN [35,38,41,43]. For instance, Oliveira et al. [27] presented a generic framework, called dhSegment, that addresses multiple tasks simultaneously, such as page extraction, baseline extraction, layout analysis or multiple typologies of illustrations and photograph extraction. The convolutional neural network (CNN) was firstly used in the dhSegment framework due to the heavy lifting of predicting pixel-wise characteristics. Then, simple image processing techniques were carried out to extract the components of interest (e.g. boxes, polygons, lines, masks). The dhSegment model is composed of an encoder (called the contracting path) that follows the deep residual network ResNet-50 architecture [14], and a decoder (called the expansive path) that maps the low resolution encoder feature maps to full input resolution feature maps. Liebl and Burghardt [36] compared the pixel-wise segmentation performances of 11 different DNN backbone architectures (including dhSegment) and 9 different tiling and scaling configurations for separating text, tables or table column lines in historical newspapers. Chen et al. [17] proposed a pixel labeling method for handwritten HDIS based on using CNN. In another work, Droby et al. [34] set a siamese neural network for segmenting handwritten document images into regions. For the task of layout analysis at pixel level, Alberti et al. [16] showed the effectiveness of using the linear discriminant analysis (LDA) for initializing the weights of a CNN layer wise. CNN based method has also been used for the automatic processing of music score documents[29]. Furthermore, Wei et al. [22] proposed a layout analysis method of HDI using the sequential forward selection algorithm and the autoencoder technique as a DNN for feature selection and learning. For page segmentation into text zones a deep convolutional encoder-decoder network was proposed by Kaddas and Gatos [23] and a fully convolutional neural network (FCN) for segmenting HDI by Wick and Puppe [28]. In the same vein, Mechi et al. [42] proposed an adaptive 2D U-Net architecture for segmenting text lines in HDI. In addition, Deep convolutional and recurrent neural networks were used by Weinman et al. [33] for text detection and recognition in historical maps. Recently, Monnier and Aubry [37] proposed an encoder-decoder architecture (called ResUNet) that combines the descriptive power of ResNet [14] with the localization recovery capacity of U-Net [12] for extracting different historical document contents (e.g., text line, caption, title, image drawing, glyph, table, background).

It is worth noting that researchers working on HDIS stress mostly on using deep feature representation in DNN as a set of low-level feature extractor. On the other side, there are recently developed methods for tackling large variety of pattern recognition fields and computer vision tasks using both DNN and low-level features [45], and particularly those extracted from GF. These methods have been shown to yield state-of-the-art performance [25]. However, to the best of our knowledge, using both texture features and DNN in the same architecture to solve the HDIS issue has not been yet tackled. Luan et al. [26] showed that by incorporating GF into the convolution filter of CNN, a reduced complexity, a high feature representation capacity and an enhanced CNN robustness (invariant to image transformations, such as scale changes and rotations) were achieved for object recognition task. In addition, Gabor based DNN were proposed for automatic detection of mine-like objects in sonar imagery [39]. These networks have the particularities to embed steerable Gabor filtering modules within the cascaded layers to enhance the scale and orientation decomposition of images. For identifying fingerprint images, Alam et al. [40] proposed a trainable DNN using the fusion of both CNN and Gabor features, and showed that better accuracy was achieved through extracting and fusing features from two different feature extractors. For face recognition, Dumitrescu and Dumitrache [31] extracted first multi-level feature vectors using GF, and then trained the SAE using the extracted feature vectors. Yao et al. [15] proposed Gabor feature based CNN for object recognition in natural scene. First, they generated three gray-level Gabor feature maps from each input images at three directions (0, $\pi/4$ and $\pi/2$). Second, they combined these Gabor feature maps into a 3-channel image. Finally, the obtained 3-channel images were used to pretrain CNN.

3   Proposed Method

In this section, we present the proposed method which aims to segment HDI at pixel level. The proposed method is based on feeding a Gabor filtered image that is obtained by applying specific multi-channel GF, instead of an original image as input to DNN. The proposed method relies on three modules: *a)* filtered image generation with GF, *b)* feature learning with SAE, and *c)* image segmentation with 2D U-Net. Figure 1 reports the three modules of the proposed method.

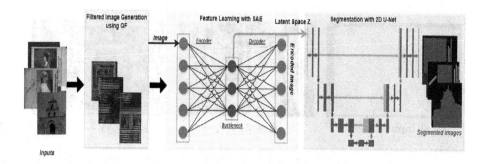

Fig. 1. Scheme of the proposed method used for HDIS.

3.1   Filtered Image Generation Using GF

Mehri et al. [19] showed that GF are the highest performing among nine texture based feature sets (Tamura, LBP, GLRLM, auto-correlation, GLCM, GF, and three wavelet based approaches) for segmenting different content types, without formulating a hypothesis concerning neither the document layout nor its content. Besides, the GF are characterized by their high discriminant capability for recognition based applications, as well as its translation, scale and rotation invariance properties. Thus, we use the multi-channel GF in this paper.

The Gabor filtered images are proceeded by tuning the analyzed image to many combinations of a specific frequency and a defined orientation in a narrow range which are referred to channels and interpreted as band-pass filters. A GF is considered as a linear selective band-pass filter that characterizes a specified channel formed by two parameters: spatial frequency f_g and orientation θ_g. The spatial frequency f determines the distance from the Gaussian centers to the origin, while the orientation θ_g specifies the angle from the horizontal axis (i.e. α-axis to the Gaussian centers). In this work, the space of GF is firstly set to $\sigma_g = \sigma_x = \sigma_y = 1$. Then, the magnitude responses of the Gabor functions are computed. If the specified GF matches a particular texture, the output magnitude will have a significant value. Otherwise, low response to the specified GF is explained by a mismatch of the dominant texture properties of the analyzed image to the set of the spatial frequency components of the fixed GF.

In our work, 24 GF are applied (4 different orientations $\theta_g=\{0, \pi/4, \pi/2, 3\pi/4\}$ and 6 different spatial frequencies $f_g=\{2\sqrt{2}, 4\sqrt{2}, 8\sqrt{2}, 16\sqrt{2}, 32\sqrt{2}, 64\sqrt{2}\}$). By convolving an image with 24 Gabor channels, 24 Gabor filtered images are generated (cf. Figs. 2(b) and 2(c)).

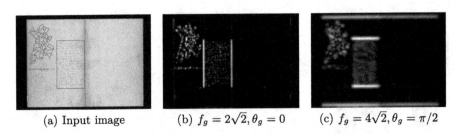

(a) Input image   (b) $f_g = 2\sqrt{2}, \theta_g = 0$ (c) $f_g = 4\sqrt{2}, \theta_g = \pi/2$

Fig. 2. Samples of the generated filtered images obtained by applying GF.

3.2   Feature Learning Using SAE

The 24 Gabor filtered images which are obtained from the first module by applying the 24 specified GF, are fed as input of the SAE model. The aim of using the SAE model in our work consists in encoding the Gabor filtered images with a relevant latent representation from which an accurate segmentation model could be guaranteed. The obtained latent representation has the advantage to have more selective and pertinent high-level features for the HDIS task. It has been widely argued that using the SAE as a learning representation model is a consistent choice for preserving the pertinent features in the intermediate layer through its encoding and decoding layers by minimizing the reconstruction error between the input (filtered images) and the output (reconstructed images). However, by using the other conventional representation learning techniques, such as the principal component analysis (PCA) and the independent component analysis (ICA), the obtained latent representations are not sufficiently accurate for segmentation, and the initial information could not be preserved. Hence, the SAE model is more suitable to determine a relevant latent representation from the input data by learning non-linear transformations with a non-linear activation function and multiple layers.

The SAE model is a multi-layer neural network that projects non-linearly the input data (i.e. Gabor filtered images) into a novel formed representation space. It contains two main functions: the encoder (g_θ) and decoder $(f_{\theta'})$ functions. From an input x, the encoder function $(g_\theta(x))$ is equal to $\Phi_g(W_x + b)$, where Φ_g is the activation function, W is the weight, and b is the bias. Usually, the encoder layer learns a latent representation (i.e. the bottleneck layer $z = g_\theta(x)$ from x). The decoder function $(f_{\theta'})$ aims at reconstructing the latent representation into the reconstructed input $x' = f_{\theta'}(z)$, where $f_{\theta'}(z) = \Phi_f(W'_z + b')$. Φ_g and Φ_f are the activation functions of the encoder and decoder layers, respectively. In our work, the SAE model includes one encoder noted Enc_r for the Gabor filtered image inputs. Besides, z_g denotes the corresponding representation output by Enc_r, where $z_g = Enc_r(x_g)$.

A latent representation (z) is computed from the encodings of the Gabor filtered images. Then, this latent representation is fed as input to a decoder (Dec_g) that aims at reconstructing the Gabor filtered images $(\hat{x}_g = Dec_r(g))$. The decoder is also a non-linear neural network, including one to three hidden layers. The learning criterion of the SAE model is the reconstruction error which is computed using the mean squared error (MSE). MSE is defined according to the following equation:

$$\mathcal{L}_{SAE}(\theta, \theta') = \frac{1}{2n} \sum_i^n \|x^{(i)} - f_{\theta'}(g_\theta(x^{(i)}))\|^2 \tag{1}$$

where n denotes the number of samples (i.e. pixels) [20].

3.3   Image Segmentation Using 2D U-Net

The latent space z obtained from the second module is fed as input of the last module of the proposed method. The last module is based on using the 2D U-Net architecture for pixel-wise image segmentation.

The 2D U-Net architecture is a variant of FCN that was previously proposed for medical image segmentation [12]. It learns segmentation in an end-to-end setting by assembling a set of convolutional and max-pooling layers in order to provide a feature map of an image and to decrease its dimensionality, and subsequently the network complexity (i.e. the number of its learned parameters). One of the main advantages of choosing the 2D U-Net architecture in our work for the segmentation stage is its computational efficiency.

In the used 2D U-Net architecture, the input layer contains 256×256 pixels with 6 channels (extracted features). The convolution is applied by means of the "Conv2D" operation using a 3×3 filter to the input. We also use the padding operation to fill the perimeter of the input with zero, and subsequently to compensate the size of the feature maps performed by the convolution. The "MaxPooling2D" operation is applied by selecting the maximum value from each region of the feature map in order to reduce the complexity and the dimension of the feature maps. We use the "Conv2DTranspose" operation to reconstruct the input, which is the opposite of that used for "Conv2D", whereby the feature map dimensionality is increased using a 3×3 filter. The "batch-normalization" is applied in the hidden layers of the network in order to normalize all pixels and weights. The dropout layer is used in the 2D U-Net network to solve the overfitting issues by randomly deactivating some neurons of each layer. The random deactivation is achieved in 50% of the units. Finally, the concatenation layer is used to connect the input feature maps.

4   Experiments and Results

In this section, we focus on evaluating the proposed method by presenting the experimental corpus, the implementation details, the hyperparameter settings, the computed performance evaluation metrics and the obtained results.

4.1   Experimental Corpus

In our experiments, two different datasets are used: SynDoc12k and HBA. HDI examples of the SynDoc12k and HBA datasets with their corresponding ground truths are illustrated in Figs. 3(a) and 3(b), respectively. The proposed method has been trained on the SynDoc12k dataset, and tested on both the SynDoc12k and HBA datasets.

The SynDoc12k dataset is collected using docExtractor[1] which is a fast and scalable synthetic document generation engine [37]. The docExtractor engine provides accurate pixel-wise annotations of the generated synthetic document images. The choice of the SynDoc12k dataset is firstly justified by the context of our work. Since the proposed method is based on building deep models, it is hence necessary to bring together large masses of precise ground truthed data for the training phase. To the best of our knowledge, the SynDoc12k dataset is among the more recent datasets that is a large-scale (composed of $12k$ annotated HDI) and has accurate pixel-wise annotations. Furthermore, the SynDoc12k dataset is characterized by strong heterogeneity (one-page and double pages), with differences in layout, typography, illustration style, complex layouts (e.g. irregular

---

[1] https://github.com/monniert/docExtractor.

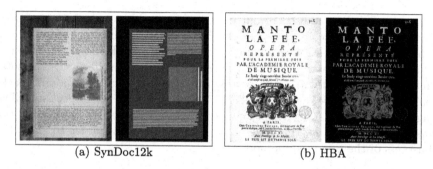

(a) SynDoc12k   (b) HBA

Fig. 3. Document image examples with ground truth (graphic and text).

spacing, varying text column widths, marginal notes). The resolutions of the SynDoc12k images range from 1192×1192 to 2384×2175 pixels.

The training and validation sets of the SynDoc12k dataset contain $10,000$ and $1,000$ images, respectively. The remaining $1,000$ images are used for the test phase. The ground truth annotation of the SynDoc12k dataset[2] contains more than 31 billion annotated pixels. Table 1 details the distribution of the annotation classes (background, graphic, and text) of the SynDoc12k dataset in terms of the number of annotated pixels. 78.33% of the total number of the annotated pixels represents the background, while 14.34% and 7.33% represent the graphic and textual contents, respectively. 83.39%, 8.15%, and 8.47% of the total number of the annotated pixels represent those of the train, validation and test subsets, respectively.

Table 1. Ground truth statistics of the SynDoc12k dataset in terms of the number of annotated pixels.

	Train	Validation	Test	Total
Background	20 387 408 136	1 993 995 994	2 060 892 339	**24 442 296 469** **(78.33%)**
Graphic	3 726 528 197	357 559 908	391 627 876	**4 475 715 981** **(14.34%)**
Text	1 906 266 235	190 676 138	189 930 417	**2 286 872 790** **(7.33%)**
Total	**26 020 202 568** **(83.39%)**	**2 542 232 040** **(8.15%)**	**2 642 450 632** **(8.47%)**	**31 204 885 240**

---

[2] The annotations of the SynDoc12k dataset are available at this url.

The second dataset used in our experiments is composed of 64 pixel-wise annotated real HDI collected from the HBA dataset. The HBA dataset was released at the ICDAR2017 competition on historical book analysis[3]. The HDI of the HBA dataset were collected from the French digital library[4]. The total numbers of the annotated pixels representing the graphical, textual and background contents are 4 789 604, 19 805 571, and 322 815 560, respectively. The images of the HBA dataset are only used in the test phase. We have tested the proposed method on unseen HDI in the training phase in order to show the robustness of the proposed method.

4.2 Implementation Details

The two used deep architectures in the proposed method, 2D U-Net and SAE, are implemented using the Darknet and Keras frameworks, respectively. They are trained and tested on Tesla $K80$ GPU with $12GB$ memory. We have trained for 80 epochs using the Adam optimizer with an initial learning rate of 10^{-3}, and a batch size of 64 training samples. Moreover, the SAE model is built using 3 hidden layers $[24, 12, 6, 12, 24]$ in which 6 dimensions of the latent space (enc) have been used for the segmentation phase. In order to ease the training task of the deep models, we have extracted a set of patches with size of 256×256 for each image. Hence, 335 129 images are used in our experiments.

4.3 Hyperparameter Settings

In this section, we have conducted a thorough experimental study on the performances of the used representation learning model (SAE) in order to select its best hyperparameter configuration for the training phase.

First, the SAE model has been trained using different pairs of activation functions for the hidden layers and output layer: *(relu, linear)*, *(relu, relu)*, *(linear, linear)*, *(tanh, linear)*, and *(tanh, relu)*.

Second, we have compared the performances of the different defined configurations of the used SAE model with the PCA and ICA methods. Figure 4 reports the MSE values *versus* the encoding dimensions learned on the training data. We deduce that the SAE model having the *(relu, relu)* configuration can easily reconstruct initial data with few features ($enc = 6$), where the MSE is equal to 0.07.

Third, we have tested different architectures of the used SAE model in order to select its best configuration (i.e. hyperparameter settings) for the training phase according to the reconstruction error (i.e. the most minimal value of MSE). In fact, we have set the number of extracted features (enc), and we have varied the number of hidden layers from 1 to 3 hidden dense layers using different pairs of activation functions. In Fig. 5, we report the average MSE after 5 cross-fold validation with an encoding dimension equals to 15. According to the obtained

---

[3] http://icdar2017hba.litislab.eu/.

[4] https://gallica.bnf.fr/ark:/12148/bpt6k840383d/f1.planchecontact.r.

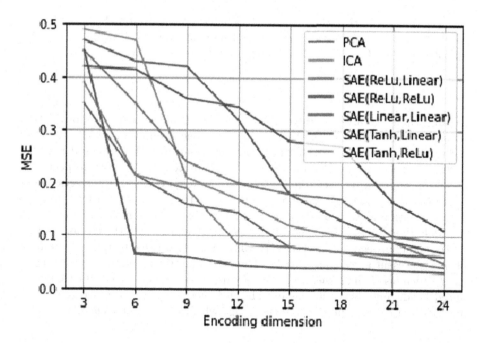

Fig. 4. MSE of reconstruction using different representation learning methods (PCA, ICA, and SAE with different pairs of activation functions).

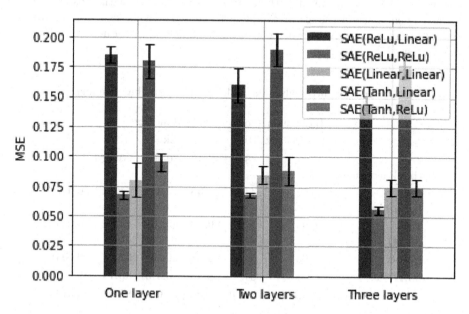

Fig. 5. MSE of reconstruction after 5 cross-fold validation using different numbers of layers.

results, we observe that the best MSE is achieved with the SAE model having the (*relu, relu*) configuration and using 3 hidden layers ($\mathbf{Z}^{(1)} = 20$, $\mathbf{Z}^{(2)} = 18$, and $\mathbf{Z}^{(3)} = 15$) which is equal to $0.054(\pm0.004)$.

4.4   Evaluation Metrics

In this paper, we are interested in the task of semantic segmentation in HDI images which consists in predicting the class (background, graphic or text) of each pixel in an image. To evaluate the performance of the proposed method, several performance evaluation metrics are computed in this work respectively, the mean intersection over union ($mIoU$), the weighted mean intersection over union ($wmIoU$), and the mean accuracy (mA). These scores are the main common evaluation metrics used for assessing the pixel-wise semantic segmentation methods. The intersection over union (IoU) metric measures the number of pixels common between the target and prediction masks divided by the total number of pixels present across both masks. The $mIoU$ metric is obtained by averaging the IoU values of all classes. The $mwIoU$ metric is computed to deal with imbalanced headcounts between classes. The mA metric denotes the average across all classes of the rate of pixels in the image which are correctly classified [37].

4.5   Results

Table 2 reports a comparison of the achieved performances with those obtained using a recent state-of-the-art method which is based on a classical deep approach (based on applying DNN on original images), and was introduced by Monnier and Aubry [37].

Table 2. Performance evaluation of the proposed and baseline methods.

	Baseline method		Proposed method	
	SynDoc12k	HBA	SynDoc12k	HBA
IoU (background)	98.20	88.07	**98.47** (±0.001)	**90.12** (±0.042)
IoU (graphic)	96.92	44.58	**97.36** (±0.012)	**51.22** (±0.031)
IoU (text)	86.95	30.99	**89.19** (±0.001)	**43.02** (±0.026)
$mIoU$	94.03	54.55	**96.51** (±0.001)	**62.19** (±0.003)
$wmIoU$	97.21	84.22	**98.40** (±0.013)	**89.71** (±0.002)
mA	96.38	81.00	**97.52** (±0.001)	**86.08** (±0.002)

First, we observe that the proposed method outperforms Monnier and Aubry [37]' solution with weighted mean intersection over union ($wmIoU$) and mean accuracy (mA) equal to 98.40% and 97.52% for the SynDoc12k dataset, respectively. Hence, we note overall performance gains of 1.19%($wmIoU$) and 1.14%(mA) for the SynDoc12k dataset. Similarly, better performances are obtained when using the proposed method on the HBA dataset compared to Monnier and Aubry [37]' solution (89.71% and 86.08% are noted for the $wmIoU$ and mA metrics respectively). Overall performance gains of 5.49%($wmIoU$) and 5.08%(mA) are obtained for the HBA dataset. Furthermore, we note that the proposed method (86.08%) also outperforms the approach proposed in [18] where a mA of 71.86% on the HBA dataset is reported. Therefore, we show that feeding DNN with Gabor filtered image instead of an original image as input strengthens the learning of representative information. Nevertheless, we observe significant drops in the IoU values of the graphic and text classes, when applying both the proposed and baseline methods on the HBA dataset compared to the Syn-Doc12k dataset. This can be justified by the particularities of the ground truth annotations of the HBA datasets (cf. Fig. 6).

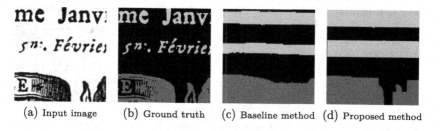

(a) Input image   (b) Ground truth   (c) Baseline method   (d) Proposed method

Fig. 6. Segmentation outputs of the proposed and baseline methods applied on a patch image of the HBA dataset.

The HBA ground truth is obtained by annotating each foreground pixel retrieved using a standard parameter-free binarization method, while the Syn-Doc12k ground truth has been outlined by using rectangular regions drawn around content regions. Besides, the two evaluated methods have been trained using the SynDoc12k ground truth annotations (bounding boxes), and have been tested on the HBA dataset that has pixel level annotations.

5   Conclusions and Further Work

In this paper, we propose a HDIS method using both GF and DNN. The proposed method starts by generating filtered images obtained by applying specific multi-channel GF. Then, a feature learning step using the SAE model is performed. Finally, an image segmentation task is applied using the 2D U-Net architecture to classify document image pixels. The obtained results on two benchmark datasets demonstrate that the proposed method outperforms a recent method based on applying DNN on original images. Besides, the experimental results demonstrate the robustness of the proposed method by proposing generic DNN trained on a wide variety of HDI and tested on unseen pages of the HBA dataset.

As future work, there are essentially four main streams. First, we will integrate an ablation study with a particular focus on the numerical complexity to show the effectiveness of each stage of the proposed method. This study is an ongoing work that will better clarify the strengths and weaknesses of the proposed method. Second, we will evaluate the proposed method on other public pixel-wise annotated datasets. Third, we will extend the segmentation classes to more specified ones (table, title, header, paragraph, separator, etc.). Finally, we will explore the multi-view representation learning techniques in order to propose multi-modal DNN that focus on combining different input modalities (low-level features, original, and filtered images).

Acknowledgments. This work has been funded under the "19PEJC-08-02" grant agreement number by the Tunisian Ministry of Higher Education and Scientific Research that is gratefully acknowledged.

The authors would like also to thank IDMC-Institut des sciences du Digital, Management & Cognition for supporting this research work.

References

1. Okun, O., Pietikäinen, M.: A survey of texture-based methods for document layout analysis. In: Series in Machine Perception and Artificial Intelligence: Texture Analysis in Machine Vision, pp. 165–177 (2000)
2. Nicolas, S., Kessentini, Y., Paquet, T., Heutte, L.: Handwritten document segmentation using hidden Markov random fields. In: International Conference on Document Analysis and Recognition, pp. 212–216 (2005)
3. Keysers, D., Shafait, F., Breuel, T.: Document image zone classification - a simple high-performance approach. In: International Conference on Computer Vision Theory and Applications, pp. 44–51 (2007)
4. Journet, N., Ramel, J., Mullot, R., Eglin, V.: Document image characterization using a multiresolution analysis of the texture: application to old documents. Int. J. Doc. Anal. Recogn. **11**(1), 9–18 (2008)
5. Nikolaou, N., Makridis, M., Gatos, B., Stamatopoulos, N., Papamarkos, N.: Segmentation of historical machine-printed documents using adaptive run-length smoothing and skeleton segmentation paths. Image Vis. Comput. **28**(4), 590–604 (2010)
6. Bhowmik, T., Kar, M.: Text localization in historical document images with local binary patterns and variance models. In: International Conference on Pattern Recognition and Machine Intelligence, pp. 501–508 (2013)
7. Ferrer, M., Morales, A., Pal, U.: LBP based line-wise script identification. In: International Conference on Document Analysis and Recognition, pp. 369–373 (2013)
8. Asi, A., Cohen, R., Kedem, K., El-Sana, J., Dinstein, I.: A coarse-to-fine approach for layout analysis of ancient manuscripts. In: International Conference on Frontiers in Handwriting Recognition, pp. 140–145 (2014)
9. Chen, K., Wei, H., Liwicki, M., Hennebert, J., Ingold, R.: Robust text line segmentation for historical manuscript images using color and texture. In: International Conference on Pattern Recognition, pp. 2978–2983 (2014)

10. Nicolaou, A., Slimane, F., Märgner, V., Liwicki, M.: Local binary patterns for Arabic optical font recognition. In: International Workshop on Document Analysis Systems, pp. 76–80 (2014)
11. Saabni, R., Asi, A., El-Sana, J.: Text line extraction for historical document images. Pattern Recogn. Lett. **35**, 23–33 (2014)
12. Ronneberger, O., Fischer, P., Brox, T.: U-net: convolutional networks for biomedical image segmentation. In: International Conference on Medical Image Computing and Computer-Assisted Intervention, pp. 234–241 (2015)
13. Grana, C., Serra, G., Manfredi, M., Coppi, D., Cucchiara, R.: Layout analysis and content enrichment of digitized books. Multimedia Tools Appl. **75**(7), 3879–3900 (2016)
14. He, K., Zhang, X., Ren, S., Sun, J.: Deep residual learning for image recognition. In: IEEE Conference on Computer Vision and Pattern Recognition, pp. 770–778 (2016)
15. Yao, H., Chuyi, L., Dan, H., Weiyu, Y.: Gabor feature based convolutional neural network for object recognition in natural scene. In: International Conference on Information Science and Control Engineering, pp. 386–390 (2016)
16. Alberti, M., Seuret, M., Pondenkandath, V., Ingold, R., Liwicki, M.: Historical document image segmentation with LDA-initialized deep neural networks. In: International Workshop on Historical Document Imaging and Processing, pp. 95–100 (2017)
17. Chen, K., Seuret, M., Hennebert, J., Ingold, R.: Convolutional neural networks for page segmentation of historical document images. In: International Conference on Document Analysis and Recognition, pp. 965–970 (2017)
18. Mehri, M., Héroux, P., Mullot, R., Moreux, J., Coüasnon, B., Barrett, B.: HBA 1.0: a pixel-based annotated dataset for historical book analysis. In: International Workshop on Historical Document Imaging and Processing, pp. 107–112 (2017)
19. Mehri, M., Héroux, P., Gomez-Krämer, P., Mullot, R.: Texture feature benchmarking and evaluation for historical document image analysis. Int. J. Doc. Anal. Recogn. **20**(1), 1–35 (2017)
20. Tang, X., Hao, K., Wei, H., Ding, Y.: Using line segments to train multi-stream stacked autoencoders for image classification. Pattern Recogn. Lett. **94**, 55–61 (2017)
21. Wei, H., Seuret, M., Liwicki, M., Ingold, R.: The use of Gabor features for semi-automatically generated polyon-based ground truth of historical document images. Digit. Scholarsh. Human. **32**, i134–i149 (2017)
22. Wei, H., Seuret, M., Liwicki, M., Ingold, R., Fu, P.: Selecting fine-tuned features for layout analysis of historical documents. In: International Conference on Document Analysis and Recognition, pp. 281–286 (2017)
23. Kaddas, P., Gatos, B.: A deep convolutional encoder-decoder network for page segmentation of historical handwritten documents into text zones. In: International Conference on Frontiers in Handwriting Recognition, pp. 259–264 (2018)
24. Kim, N., So, H.: Directional statistical Gabor features for texture classification. Pattern Recogn. Lett. **112**, 18–26 (2018)
25. Liu, C., Ding, W., Wang, X., Zhang, B.: Hybrid Gabor convolutional networks. Pattern Recogn. Lett. **116**, 164–169 (2018)
26. Luan, S., Chen, C., Zhang, B., Han, J., Liu, J.: Gabor convolutional networks. IEEE Trans. Image Process. **27**(9), 4357–4366 (2018)
27. Oliveira, S.A., Seguin, B., Kaplan, F.: dhSegment: a generic deep-learning approach for document segmentation. In: International Conference on Frontiers in Handwriting Recognition, pp. 7–12 (2018)

28. Wick, C., Puppe, F.: Fully convolutional neural networks for page segmentation of historical document images. In: International Workshop on Document Analysis Systems, pp. 287–292 (2018)

29. Zaragoza, J., Castellanos, F., Vigliensoni, G., Fujinaga, I.: Deep neural networks for document processing of music score images. Appl. Sci. **8**(5), 654 (2018)

30. Do, T., Terrades, O., Tabbone, S.: DSD: document sparse-based denoising algorithm. Pattern Anal. Appl. **22**(1), 177–186 (2019)

31. Dumitrescu, C., Dumitrache, I.: Combining deep learning technologies with multi-level Gabor features for facial recognition in biometric automated systems. Stud. Inform. Control **28**(2), 221–230 (2019)

32. Sehad, A., Chibani, Y., Hedjam, R., Cheriet, M.: Gabor filter-based texture for ancient degraded document image binarization. Pattern Anal. Appl. **22**(1), 1–22 (2019)

33. Weinman, J.J., Chen, Z., Gafford, B., Gifford, N., Lamsal, A., Staab, L.: Deep neural networks for text detection and recognition in historical maps. In: International Conference on Document Analysis and Recognition, pp. 902–909 (2019)

34. Droby, A., Barakat, B., Madi, B., Alaasam, R., El-Sana, J.: Unsupervised deep learning for handwritten page segmentation. In: International Conference on Frontiers in Handwriting Recognition, pp. 240–245 (2020)

35. Lombardi, F., Marinai, S.: Deep learning for historical document analysis and recognition - a survey. J. Imaging **6**(10), 110 (2020)

36. Liebl, B., Burghardt, M.: An evaluation of DNN architectures for page segmentation of historical newspapers. In: International Conference on Pattern Recognition, pp. 5153–5160 (2020)

37. Monnier, T., Aubry, M.: docExtractor: an off-the-shelf historical document element extraction. In: International Conference on Frontiers in Handwriting Recognition, pp. 91–96 (2020)

38. Saire, D., Tabbone, S.: Documents counterfeit detection through a deep learning approach. In: International Conference on Pattern Recognition, pp. 3915–3922 (2020)

39. Thanh Le, H., Phung, S.L., Chapple, P.B., Bouzerdoum, A., Ritz, C.H., Tran, L.C.: Deep Gabor neural network for automatic detection of mine-like objects in sonar imagery. IEEE Access **8**, 94126–94139 (2020)

40. Alam, N., Ahsan, M.M., Based, M.A., Haider, J., Kowalski, M.: An intelligent system for automatic fingerprint identification using feature fusion by Gabor filter and deep learning. Comput. Electr. Eng. **95**, 107387 (2021)

41. Aubry, M.: Deep learning for historical data analysis. In: Workshop on Structuring and Understanding of Multimedia heritAge Contents (2021)

42. Mechi, O., Mehri, M., Ingold, R., Amara, N.: A two-step framework for text line segmentation in historical Arabic and Latin document images. Int. J. Doc. Anal. Recogn. **24**(3), 197–218 (2021)

43. Sellami, A., Tabbone, S.: EDNets: deep feature learning for document image classification based on multi-view encoder-decoder neural networks. In: International Conference on Document Analysis and Recognition, pp. 318–332 (2021)

44. Markewich, L., et al.: Segmentation for document layout analysis: not dead yet. Int. J. Doc. Anal. Recogn. **25**(2), 67–77 (2022)

45. Sellami, A., Tabbone, S.: Deep neural networks-based relevant latent representation learning for hyperspectral image classification. Pattern Recogn. **121**, 108224 (2022)

Group, Contrast and Recognize: A Self-supervised Method for Chinese Character Recognition

Xinzhe Jiang[1], Jun Du[1(✉)], Pengfei Hu[1], Mobai Xue[1], Jiefeng Ma[1], Jiajia Wu[2], and Jianshu Zhang[2]

[1] University of Science and Technology of China, Hefei, China
{xzjiang,hudeyouxiang,xmb15,jfma}@mail.ustc.edu.cn, jundu@ustc.edu.cn
[2] iFLYTEK Research, Hefei, China
{jjwu,jszhang6}@iflytek.com

Abstract. Chinese character recognition has been a challenging problem in the field of computer vision, attracting significant research attention due to its widespread applications and technical complexity. However, previous methods rely heavily on manual annotations to guide model learning, without considering self-supervised representation learning. Motivated by the educational approach of teaching pupils to recognize Chinese characters through grouping and differentiation, we introduce a novel self-supervised method that employs clustering and contrastive learning to group similar characters and separate them. Our proposed objective consists of two components: intra-group and inter-group contrastive objectives. The intra-group objective distinguishes the target character from similar characters within the group, while the inter-group objective encourages the model to encode the discriminative semantic structure of each group. The experimental results demonstrate the advantages of our self-supervised representation over previous methods, as well as its superior performance on benchmark comparisons.

Keywords: Chinese character recognition · Self-supervised learning · Contrastive learning · Clustering

1 Introduction

Chinese characters play an irreplaceable role in the transmission of Chinese culture and in the interaction of the Chinese people. Over the years, great efforts have been made to study the problem of Chinese character recognition, and the ability to recognize Chinese characters has become the cornerstone of many commercial applications [20,23].

In the era of deep learning, there are three main categories of Chinese Character Recognition (CCR) methods: character-based ones [27,29,35], radical-based ones [2,24,26], and stroke-based ones [4,37]. These methods vary in terms of their modeling granularity, with character-based methods being the coarsest and stroke-based methods being the finest. Character-based methods treat CCR as a typical classification task, with the goal of finding the category to which each Chinese character image belongs. Radical-based methods, on the other hand, analyze characters by their internal components and present them as a sequence of radicals. Stroke-based methods decompose characters at the stroke level and determine the recognition result through a combination of similarity and edit distance matching. These methods require character, radical, and stroke level annotations as supervision, which require a large amount of labeled data for training. However, labeling these types of data is expensive and time-consuming, making it more cost-effective to find alternatives through self-supervised representation learning without human annotation.

For self-supervised Chinese character recognition, SAE [7] decomposes a character into individual stroke images generated from a predetermined writing sequence. However, this approach only takes printed character images as input and mainly focuses on reconstructing the stroke sequence, ignoring the writing style variations in the real world. In this paper, we present Group, Contrast and Recognize (GCR), a novel self-supervised method for Chinese character recognition that incorporates semantic knowledge from real-world unlabeled character images. The method combines clustering and contrastive learning, drawing inspiration from educational practices where presenting characters in radical-based groups can help beginning learners distinguish and memorize Chinese characters [31].

Typical contrastive learning uses an instance discrimination pre-text task to obtain useful representations by maximizing the agreement between positive pairs and disagreement between negative pairs. However, this approach has two drawbacks: (1) training with easily-distinguishable negative samples can lead to a shortcut solution [30], and (2) the semantic structure of negative samples is neglected [14]. To address these challenges, this paper introduces the hard negative sampling strategy and semantic structure through dynamic clustering during contrastive learning.

First, the method employs the k-means clustering to divide all negative samples in the dictionary queue into different clusters. Second, the proposed model considers both the intra-group comparison and the inter-group comparison as the optimization objective. The intra-group objective requires the model to distinguish the input from hard similar samples within close neighbor clusters, using two augmented views of the input as the positive pair and the input and samples within close neighbor clusters as the negative pairs. The inter-group objective incorporates the semantic structure of negative samples into representation learning by using the input and its closest centroid as the positive pair and the input and other centroids as the negative pairs.

The main contributions of our work are summarized as:

- We introduce a self-supervised method named Group, Contrast and Recognize (GCR) for Chinese character recognition, which leverages both clustering and discrimination to derive meaningful representation from unlabeled data.

– We propose two contrastive objectives, including the intra-group and inter-group objectives, which enhance the model's ability to learn more discriminative representations and better semantic structure.
– Extensive experiments on public benchmarks validate the advantages of GCR, which result in substantial improvements in accuracy compared to supervised baselines.

2   Related Works

2.1   Chinese Character Recognition

The Chinese character recognition problem has been researched for decades. Before the popularity of deep learning, early approaches [3,12,21] used morphology-based observations to obtain hand-crafted features for the CCR task. After that, the deep learning based methods can be categorized into three types: character-based, radical-based, and stroke-based ones. Character-based ones recognize the input image via classification. ATR-CNN [27] proposes relaxation convolution and alternate training to solve the slow convergence and over-fitting problems. DirectMap [35] combines the traditional normalization-cooperated direction-decomposed feature map with the deep convolutional neural network. [29] proposes the template-instance loss functions to alleviate the imbalance problem between easy and difficult character instances. Radical-based ones describe a Chinese character by its internal radicals and structures under the artificial rules. DenseRAN [26] designs an attention-based encoder-decoder model to recognize the radicals and structures of character. FewShotRAN [24] proposes the radical aggregation module to learn robust radical feature and the character analysis decoder to avoid the inflexible match decoding. HDE [2] integrates the tree-based decomposition of Chinese characters into model and learns the compatibility between the input image and the knowledge-based representation. Stroke-based ones adopt the smaller modeling unit and regard the character as a stroke sequence following the writing order. [4] proposes a stroke-based method which decomposes a character into a sequence of five stroke categories, which solves the character zero-shot and radical zero-shot problems. Besides that, it uses a matching-based strategy to acquire the final result in the test stage to overcome the one-to-many problem.

2.2   Self-supervised Contrastive Learning

Recently, self-supervised contrastive learning has achieved success on various vision tasks such as image classification and object detection. It intends to learn an embedding space with alignment and uniformity [25], where two augmentation views of the same instance attract each other while the sample embeddings from different instances are repelled. Specifically, the positive and negative pairs are indispensable for building the contrastive InfoNCE objective [19]. There exists a lot of methods varied with the augmentation and negative sampling strategies. SimCLR [5] generates the instance features within the mini-batch samples, exempt from the requirements of specialized architectures and memory bank. MoCo [8] adopts a momentum-updated encoder as one branch and maintains a dictionary queue of the past instance features. With the projection head and strong augmentation of SimCLR integrated into the vanilla MoCo, MoCo v2 [6] leads to better performance. [28] proposes a ring discrimination method to construct a conditional distribution for hard negative examples, proving the tradeoff between bias and variance. PCL [14] introduces prototypes as latent variables into contrastive learning by the ProtoNCE loss, which can capture high-level semantics. Nevertheless, self-supervised contrastive learning for CCR has rarely been researched.

2.3 Self-supervised Learning for Text Recognition

In order to leverage the potential of unlabeled data, many researchers have turned to self-supervised learning techniques for text recognition. One such method is SeqCLR [1], which is the first self-supervised representation learning approach for text recognition. By dividing the feature map into different instances and conducting sequence-to-sequence contrastive learning, SeqCLR can learn effective self-supervised representations. Another promising approach is PerSec [16], which utilizes dual context perceivers to contrast and learn latent representations from both low-level stroke and high-level semantic contextual spaces simultaneously through hierarchical contrastive learning. Inspired by the reading and writing behaviors of humans, [32] proposes DiG to enhance the performance of text recognition and other text-related tasks. By integrating contrastive learning and masked image modeling, DiG can effectively learn discrimination and generation, ultimately leading to the acquisition of useful representations. These methods are primarily focused on text line recognition rather than isolated Chinese character recognition.

3   Methodology

Our approach adheres to the standard two-stage workflow for self-supervised representation learning, consisting of pre-training and fine-tuning. The fine-tuning stage starts with initializing the encoder with pre-trained backbone weights. In Sect. 3.1, we present our observations and motivation. In Sect. 3.2, we introduce the architecture of the proposed Group, Contrast and Recognize (GCR) method. The intra- and inter-group contrastive objectives are explained in Sect. 3.3 and Sect. 3.4 respectively. Finally, the algorithmic implementation is outlined in Sect. 3.5.

3.1 Observation and Motivation

In exploring the potential of pre-training for CCR, we first aim to examine the distribution of pre-trained features in the latent space. For this purpose, we input a set of labeled character images into the MoCo pre-trained DenseNet encoder, extract their features without fine-tuning, and use k-means clustering to assign the sample features into different clusters based on Euclidean distance. The labels are used for demonstration purposes only and not for training supervision. As shown in Fig. 1, we randomly select some clusters and display their corresponding labels. Our observations are as follows: (1) similar characters with identical components tend to be clustered together, and (2) the major shared component of each cluster is different.

The first observation reveals the fact that it is difficult to distinguish the characters with similar appearances, and the second observation shows the inherent semantics of Chinese characters. As explored in [31], presenting characters with shared radicals in groups can enhance a learner's semantic understanding of Chinese characters. With this in mind, we hypothesize that discrimination among similar characters within a group and among different semantic groups can help the model learn more discriminative features. To achieve this, we leverage the combination of clustering and contrastive learning to mimic the grouping and distinguishing processes.

Clusters	Sample Labels within One Cluster	Major Shared Component
cluster-1	茌 苗 莲 葆 芭 芴 芮 菏 芄 芺 茯 喏 送	艹
cluster-2	湿 淘 澄 浇 渴 溪 浚 溟 谟 璜 躅 缧 遒	氵
cluster-3	捩 拉 拓 拘 掊 捉 相 稍 秋 斌 炉	扌
cluster-4	锹 锡 镇 铌 镔 锭 锯 锐 辗 跟 玻	钅
cluster-5	桩 枷 枳 枸 权 积 忙 恢 敉 菥	木

Fig. 1. The sample labels of some clusters.

3.2 Architecture

The architecture of our proposed GCR is depicted in Fig. 2. The input image x is augmented to create two views, x_a and x_b, which are then processed by the query encoder and the momentum key encoder, with the query instance q and key instance k obtained. The query encoder f_q consists of the backbone $F(\cdot)$ and the projection head $P(\cdot)$, and the momentum key encoder f_k consists of the $F_m(\cdot)$ and $P_m(\cdot)$. θ_q and θ_k are the parameters of the query encoder f_q and momentum key encoder f_k, respectively. Additionally, a dictionary queue Q is maintained, where the encoded momentum representations of the current batch are stored, and the oldest are removed. Finally, the acquired clusters and instance features in the dictionary queue are utilized to achieve both intra-group and inter-group contrastive learning.

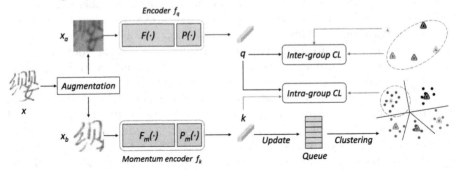

Fig. 2. Architecture of Group, Contrast, and Recognize (GCR). The green lines and red lines indicate the source of positive and negative samples in contrastive learning, respectively. CL is the abbreviation for contrastive learning. The triangles with various colors represent the cluster centroids. (Color figure online)

3.3   Intra-group Contrastive Learning

A long-standing issue in CCR is the tendency of similar characters to be easily confused. This is due to their encoded features often being close in the embedding space, making it difficult for predictors to correctly recognize them. This mirrors the common experience of beginning learners who frequently struggle to distinguish among similar characters. To address this, we sample hard negatives for contrastive learning in order to magnify the differences among similar characters. The key point is how to effectively sample the required hard negatives.

Based on the first observation in Sect. 3.1, we introduce the concept of intra-group contrastive learning, as depicted in Fig. 2. During model pre-training, we dynamically cluster the momentum representations of instances in the dictionary queue into M clusters. With these clusters, we divide the negatives into M subsets $\mathbf{C} = \{\mathbf{C}_1, \mathbf{C}_2, \ldots, \mathbf{C}_M\}$. The proposed intra-group contrastive loss is given by:

$$\mathcal{L}_{\text{intra}} = -\log \frac{\exp\left(q \cdot k/\tau\right)}{\exp\left(q \cdot k/\tau\right) + \sum_{\mathbf{C}_i \subset \mathbf{G}_q} \sum_{k_j \in \mathbf{C}_i} \exp\left(q \cdot k_j/\tau\right)} \tag{1}$$

where τ is the temperature, and $\mathbf{G}_q$ is the sampled group that includes the hard negatives. We choose the union of the top-D closest clusters as the hard negatives group $\mathbf{G}_q$. Notably, $\mathbf{G}_q$ excludes the closest cluster, since the closest cluster probably contains the identical instances of q, which are the false negatives and cause the model to discard semantic information [11].

In this section, we distinguish the query instance from the similar instances in the hard negatives group to ensure local uniformity in the embedding space. By sampling hard negatives, the model receives more discriminative information, leading to improved representation.

3.4   Inter-group Contrastive Learning

In this section, we aim to ensure the distinction among different semantic groups by enlarging the separation among all cluster centroids. To achieve this, we introduce an inter-group contrastive loss. This loss minimizes the distance between the query instance and its corresponding centroid in the embedding space, while pushing other centroids away.

In detail, we assign one centroid to each query instance and calculate the inter-group contrastive loss as follows:

$$\mathcal{L}_{\text{inter}} = -\log \frac{\exp\left(q \cdot c_q^{\text{s}}/\tau\right)}{\sum_{c_j^{\text{s}} \in \mathbf{C}^{\text{s}}} \exp\left(q \cdot c_j^{\text{s}}/\tau\right)} \tag{2}$$

where c_q^{s} is the closest centroid to the query q, c_j^{s} is the cluster centroid of $\mathbf{C}_j$ and $\mathbf{C}^{\text{s}}$ is the union set of c_j^{s}. Note that the centroid embedding is calculated as the average of the instance embeddings within the cluster. With this objective, we aim to encourage global uniformity in the embedding space by treating each cluster as a single group.

3.5   Network Training

The procedure for self-supervised pre-training of the GCR framework is outlined in Algorithm 1. Unlike DnC [22], our clustering process is integrated seamlessly into the contrastive learning process, instead of being separated into several steps. In addition to the intra- and inter-group objectives, we also incorporate the vanilla InfoNCE loss as formulated in Eq. 3, to ensure local smoothness and support the clustering bootstrapping, following the strategy of PCL [14].

$$\mathcal{L}_{\text{infonce}} = -\log \frac{\exp\left(q \cdot k/\tau\right)}{\exp\left(q \cdot k/\tau\right) + \sum_{k_j \in Q} \exp\left(q \cdot k_j/\tau\right)} \tag{3}$$

The final loss function is a combination of all these objectives, formulated as follows:

$$\mathcal{L}_{\text{total}} = \lambda_1 \mathcal{L}_{\text{infonce}} + \lambda_2 \mathcal{L}_{\text{intra}} + \lambda_3 \mathcal{L}_{\text{inter}} \tag{4}$$

where λ_1, λ_2, λ_3 are coefficients that control the contribution of each part to the total loss.

4   Experiments

In this section, we outline the experimental setup, including details on the datasets, baseline models, and implementation specifications. Subsequently, we perform extensive experiments on benchmark datasets to evaluate the GCR from both qualitative and quantitative perspectives.

Algorithm 1 Main algorithm of GCR

1: Input unlabeled image x, temperature τ, mini-batch size N, query encoder f_q, key encoder f_k, momentum coefficient m, number of desired clusters M, clustering interval r, total training steps s, loss coefficients λ_1, λ_2 and λ_3
2: Randomly initialize parameters θ_q and θ_k, $\theta_q = \theta_k$
3: Randomly initialize the queue Q of negative instances k_j
4: **for** $step \in s$ **do**
5:     **if** $step\%r == 0$ **then**
6:         $\mathbf{C}^s$, $\mathbf{C} \leftarrow$ K-means Clustering on Q for M clusters
7:     **end if**
8:     **for** $x \in$ mini-batch **do**
9:         $x_a = Aug_1(x)$
10:        $q = f_q(x_a)$
11:        $x_b = Aug_2(x)$
12:        $k = f_k(x_b)$
13:        $\mathcal{L}_{\text{intra}} = -\log \frac{\exp(q\cdot k/\tau)}{\exp(q\cdot k/\tau) + \sum_{\mathbf{C}_i \subset \mathbf{G}_q} \sum_{k_j \in \mathbf{C}_i} \exp(q\cdot k_j/\tau)}$
14:        $\mathcal{L}_{\text{inter}} = -\log \frac{\exp(q\cdot c_q^s/\tau)}{\sum_{c_j^s \in \mathbf{C}^s} \exp(q\cdot c_j^s/\tau)}$
15:        $\mathcal{L}_{\text{infonce}} = -\log \frac{\exp(q\cdot k/\tau)}{\exp(q\cdot k/\tau) + \sum_{k_j \in Q} \exp(q\cdot k_j/\tau)}$
16:        $\mathcal{L}_{\text{total}} = \lambda_1 \mathcal{L}_{\text{infonce}} + \lambda_2 \mathcal{L}_{\text{intra}} + \lambda_3 \mathcal{L}_{\text{inter}}$
17:     **end for**
18:     update f_q by back-propagation
19:     update f_k with momentum from f_q: $\theta_k \leftarrow m\theta_k + (1-m)\theta_q$
20:     enqueue the keys k to Q
21:     dequeue the oldest keys
22: **end for**

4.1   Datasets

We utilize a collection of 3 million scanned and camera images of handwritten Chinese characters for pre-training, which we name the SC3M dataset. For fine-tuning, we use the HWDB1.0-1.1 dataset [15], consisting of 2.73 million offline handwritten Chinese character images from 720 writers. To evaluate the performance of the GCR framework, we conduct experiments on the ICDAR2013 benchmark [33], which includes 224,419 offline handwritten Chinese characters from 60 writers with 3755 classes. We also evaluate the model's ability to recognize printed artistic characters using the Printed Artistic dataset [4], which

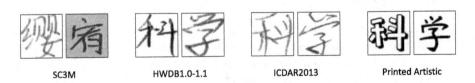

SC3M   HWDB1.0-1.1   ICDAR2013   Printed Artistic

Fig. 3. Some examples in the datasets.

contains 3755 characters in 105 printed artistic fonts. An illustration of some examples from these datasets can be seen in Fig. 3.

4.2   Baselines

In this work, we have constructed three baseline models for CCR, including character-level, radical-level, and stroke-level models. Our first model, called DenseClassifier, is a character-level model that combines a CNN encoder with a linear classifier. The second model, RAN, is a radical-level method that utilizes an encoder-decoder architecture with a coverage attention mechanism. Finally, our stroke-level model, SLD, is comprised of an image-to-feature encoder and a feature-to-stroke decoder that employs a matching-based strategy. Both the DenseClassifier and RAN models use a modified DenseNet [10] as their backbone for feature extraction, while SLD uses a modified ResNet [9]. The character accuracy is employed as the evaluation metric for the downstream CCR task.

4.3   Implementation Details

In the pre-training stage, we follow the configuration of the vanilla MoCo v2 [6] and apply random crop, random color jittering, random grayscale conversion, and random Gaussian blur. The optimization algorithm used is SGD with a momentum of 0.9, a weight decay of 0.0001, and a batch size of 3200. The temperature τ is set to 0.2, the queue size K to 65536, the number of clusters M to 1500, the number of hard negative clusters D to 5, the momentum coefficient m to 0.999, and the clustering interval steps r to 30. The coefficients for each loss λ_1, λ_2, λ_3 are set to 1, 0.5, and 0.5, respectively. The dynamic k-means clustering is implemented using the efficient faiss tool [13]. The initial learning rate is set to 0.03 and adjusted using a cosine scheduler. The experiments are run on 16 NVIDIA Tesla V100 (24GB RAM) GPUs.

In the fine-tuning stage, we use the plateau scheduler and Adadelta optimizer with an initial learning rate of 0.0001, a weight decay of 0.0001, and a batch size of 96 for the DenseClassifier and RAN. For the SLD model, the Adadelta optimizer is used with an initial learning rate of 1.0 and a weight decay of 0.0001. The input images for DenseClassifier and RAN are resized to 64 × 64, while the input for SLD is resized to 32 × 32. The experiments are conducted on 4 NVIDIA Tesla V100 (12GB RAM) GPUs.

4.4   Representation Quality of Self-supervised Pre-training

In order to assess the impact of self-supervised learning on representation quality in CCR, we use DenseClassifier as our baseline model and carry out experiments with different pre-text tasks for pre-training. The representation quality is evaluated by freezing the weights of the pre-trained encoder and training a randomly initialized linear layer on the entire HWDB1.0-1.1 dataset, followed by testing on ICDAR2013. The results, as shown in Table 1, demonstrate that incorporating prior knowledge from the pre-text tasks can improve the representation quality and overall performance of the model. Among the various approaches, our proposed GCR, which combines contrastive learning and clustering, achieves the best result and outperforms the MoCo method by 5.19%.

Table 1. Performance comparison in the frozen setting of different pre-training methods. 'None' means the encoder is randomly initialized and frozen, with the single linear classifier trained.

Pre-train Method	None	Jigsaw [18]	MoCo [8]	GCR
Accuracy	0.05%	74.91%	78.90%	**84.09%**

To evaluate the performance of self-supervised pre-training in low-resource scenarios, we conduct N-shot experiments where the training set includes N images per character. The pre-trained encoder is utilized for initialization and then fine-tuned. As shown in Table 2, the results indicate that self-supervised methods are capable of improving model performance when training data is limited. Our GCR method consistently enhances the supervised baseline performance, outperforming the Jigsaw and MoCo methods when N is set to 1, 3, 5, and 10. As N increases to 10, the performance gain of pre-training methods reaches a limit.

Table 2. Performance comparison in N-shot setting of different pre-training methods. 'None' means the encoder is randomly initialized and trained, i.e., supervised baseline.

Pre-train Method	1-shot	3-shot	5-shot	10-shot
None	0.10%	17.32%	67.54%	91.38%
Jigsaw [18]	7.81%	69.84%	88.03%	93.10%
MoCo [8]	8.52%	73.49%	89.11%	93.16%
GCR	**17.18%**	**79.61%**	**90.10%**	**93.38%**

To further demonstrate the discrimination power of GCR, we conduct an experiment using two sets of similar characters with 60 images per character, which are selected from ICDAR2013. We utilize the pre-trained encoder to extract the self-supervised features and average-pool them into vectors. These vectors are then embedded into a 2-D space using t-SNE visualization [17]. As shown in Fig. 4, each color represents a different character, with the shared component of the top and bottom rows being 'kou' and 'zou' respectively. Our results indicate that compared to MoCo and Jigsaw, GCR provides more discriminative features for similar characters, resulting in better cluster separation.

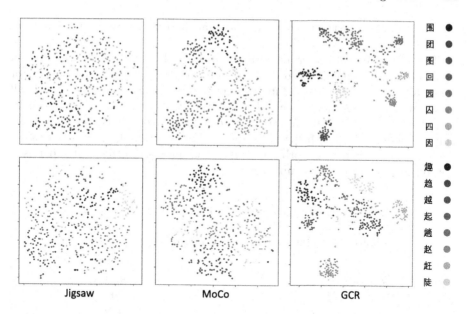

Fig. 4. T-SNE visualization of the self-supervised learned representation of two sets of similar characters. Left: Jigsaw; Middle: MoCo; Right: GCR (ours). Colors represent character classes.

4.5 Handwritten Benchmark Comparison in Zero-Shot Setting

Performance Comparison: We conduct experiments on handwritten characters in the zero-shot setting. We fine-tune the RAN and SLD baseline models. For the training set, we select the first m classes of 3755 characters from HWDB1.0-1.1, where m ranges in $\{500, 1000, 1500, 2000, 2755\}$. The test set consists of samples with labels from the last 1000 classes of the ICDAR2013 dataset. Note

Table 3. Performance comparison in the character zero-shot setting on the handwritten benchmark.

Handwritten	Pre-train	Character Zero-Shot Setting				
		500	1000	1500	2000	2755
DenseRAN [26]	None	1.70%	8.44%	14.71%	19.51%	30.68%
HDE [2]	None	4.90%	12.77%	19.25%	25.13%	33.49%
ACPM [37]	None	9.72%	18.50%	27.74%	34.00%	42.43%
RAN [34]	None	2.65%	10.10%	16.92%	21.56%	31.78%
	MoCo [8]	2.96%	10.14%	17.78%	21.86%	32.59%
	GCR	**3.99%**	**10.17%**	**18.67%**	**23.59%**	**33.29%**
SLD [4]	None	5.60%	13.85%	22.88%	25.73%	37.91%
	SAE [7]	5.91%	14.35%	24.32%	30.17%	40.22%
	MoCo [8]	5.70%	16.60%	24.62%	29.47%	38.90%
	GCR	**6.45%**	**21.03%**	**28.11%**	**33.00%**	**42.01%**

that our partition method is the same as that used in the SLD for a fair comparison.

The results of our experiments are summarized in Table 3. In the case of RAN, both MoCo and GCR are able to improve the baseline performance. For SLD, GCR makes a substantial improvement to the baseline accuracy across all partition settings, outperforming MoCo and SAE. The success of GCR can be attributed to its ability to capture more discriminative details and distinguish similar characters, which helps SLD perform better. Overall, GCR achieves the best results compared to both supervised baselines and other self-supervised methods, demonstrating its superiority.

Table 4. Ablation study on each part of pre-training objectives.

InfoNCE loss	Inter-group loss	Intra-group loss	Accuracy
			37.91%
✓			38.90%
✓	✓		40.60%
✓		✓	41.17%
✓	✓	✓	42.01%

Ablation Study: Since the total loss consists of three parts, it is necessary to investigate whether the proposed intra-group and inter-group contrastive loss can improve the capability of feature representation. To this end, we conduct experiments on the SLD baseline model under the zero-shot partition 2755.

The results, presented in Table 4, show that incorporating the vanilla InfoNCE loss into the SLD model leads to an improvement of 0.99%. By removing the inter-group loss, the fine-tuning accuracy decreases by 0.84% (from 42.01% to 41.17%). The removal of the intra-group loss results in a decrease of 1.41% in fine-tuning accuracy (from 42.01% to 40.60%). Our experiments reveal that the intra-group loss has a more significant impact than the inter-group loss, as the former is designed to distinguish similar characters, which is more crucial for unseen character recognition. Finally, when both the intra-group and inter-group losses are employed, the accuracy improves by 3.11% (from 38.90% to 42.01%), further confirming their advantages.

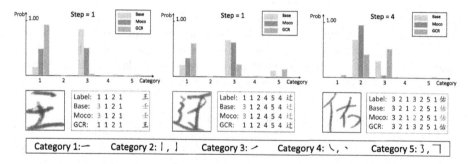

Fig. 5. Case Study. The bar charts depict the probabilities of predictions for each category of strokes, excluding the categories 'sos' and 'eos'. A single category comprises multiple instances of strokes. The utilized SLD model decomposes a character into a series of stroke categories. The red bold number signifies an incorrect recognition result, which is represented by the red area in the image. (Color figure online)

Qualitative Analysis: As seen in Fig. 5, we can qualitatively observe how the proposed GCR captures the detail information and correctly recognize the unseen characters, compared with the baseline SLD model and MoCo. The cases are from the zero-shot partition 2755 experiment. Taking the left 'wang' as an example, the confusing region is wrongly recognized as category 3 by the baseline and MoCo at the first decoding step, and the final result is 'ren' which is similar to the character 'wang' and has appeared in the training set. However, our GCR can correctly recognize it with high confidence, which suggests the capability of GCR to distinguish similar characters.

4.6 Printed Artistic Benchmark Comparison in Zero-Shot Setting

Besides the handwritten characters, we also conduct experiments with printed artistic characters in the zero-shot setting. The dataset is Printed Artistic and

Table 5. Performance comparison in the character zero-shot setting on the Printed Artistic benchmark.

Printed Artistic	Pre-train	Character Zero-Shot Setting				
		500	1000	1500	2000	2755
DenseRAN [26]	None	0.20%	2.26%	7.89%	10.86%	24.80%
HDE [2]	None	7.48%	21.13%	31.75%	40.43%	51.41%
RAN [34]	None	0.83%	19.13%	28.49%	43.57%	56.85%
	MoCo [8]	4.55%	22.19%	30.20%	45.80%	57.10%
	GCR	**7.12%**	**24.11%**	**31.24%**	**48.25%**	**59.35%**
SLD [4]	None	7.03%	26.22%	48.42%	54.86%	65.44%
	SAE [7]	8.25%	32.24%	50.72%	57.13%	68.88%
	MoCo [8]	10.81%	36.50%	53.85%	60.56%	69.22%
	GCR	**11.85%**	**41.14%**	**55.46%**	**63.04%**	**70.69%**

the partition manner is the same as that of SLD. The SLD fine-tuned from GCR outperforms not only the supervised baselines and self-supervised methods, but also other previous methods, as shown in Table 5. Compared with handwritten characters, printed artistic characters have more clear strokes and fixed writing styles relatively, which are easier to be correctly recognized.

Table 6. The results in seen character setting on ICDAR2013.

Method	Decomposition	Accuracy
HCCR-GoogLeNet [36]	Character	96.35%
DirectMap+ConvNet+Adaptation [35]	Character	97.37%
DenseRAN [26]	Radical	96.66%
FewShotRAN [24]	Radical	96.97%
HDE [2]	Radical	97.14%
template+instance [29]	Character	97.45%
SLD [4]	Stroke	96.74%
ACPM [37]	All	97.80%
RAN [34]	Radical	96.61%
RAN+MoCo	Radical	96.67%
RAN+GCR	Radical	96.79%
DenseClassifier	Character	97.23%
DenseClassifier+MoCo	Character	97.34%
DenseClassifier+GCR	Character	**97.51%**

4.7   Handwritten Benchmark Comparison in Seen Setting

The results of our experiments under the seen character setting are presented in Table 6. In line with previous studies, we use the ICDAR2013 dataset as the test set, where all labels have appeared in the training set HWDB1.0-1.1, without any zero-shot challenge. Our results indicate that the RAN model benefits from both MoCo and GCR, with accuracy improvements of 0.06% and 0.18% respectively over the baseline. Similarly, the DenseClassifier model shows improvements with MoCo and GCR, yielding accuracy improvements of 0.11% and 0.28% respectively. Notably, the DenseClassifier fine-tuned from GCR is only second to the state-of-the-art model ACPM which incorporates multi-level decomposition information.

5   Conclusion and Future Work

In this paper, we propose GCR, a novel self-supervised method for CCR. By combining clustering and contrastive learning, and optimizing the proposed inter-group and intra-group contrastive objectives, GCR significantly enhances the representation ability compared to the baseline model and other self-supervised methods. Consequently, our GCR achieves obvious performance improvements on the benchmark datasets ICDAR2013 and Printed Artistic. The key takeaway is that the hard similar negatives and semantic structure of the unlabeled data can be utilized to obtain useful self-supervised representations for the downstream CCR task. In the future, we will further evaluate the generalization capability of GCR for other languages, such as Korean.

References

1. Aberdam, A., et al.: Sequence-to-sequence contrastive learning for text recognition. In: Proceedings of the IEEE/CVF Conference on Computer Vision and Pattern Recognition, pp. 15297–15307 (2021)
2. Cao, Z., Lu, J., Cui, S., Zhang, C.: Zero-shot handwritten Chinese character recognition with hierarchical decomposition embedding. Pattern Recogn. **107**, 107488 (2020)
3. Chang, F.: Techniques for solving the large-scale classification problem in chinese handwriting recognition. In: Proceedings of the 2006 Conference on Arabic and Chinese Handwriting Recognition, pp. 161–169 (2006)
4. Chen, J., Li, B., Xue, X.: Zero-shot Chinese character recognition with stroke-level decomposition. In: Proceedings of the Thirtieth International Joint Conference on Artificial Intelligence, IJCAI-21, pp. 615–621. International Joint Conferences on Artificial Intelligence Organization (2021)
5. Chen, T., Kornblith, S., Norouzi, M., Hinton, G.: A simple framework for contrastive learning of visual representations. In: International Conference on Machine Learning, pp. 1597–1607. PMLR (2020)
6. Chen, X., Fan, H., Girshick, R., He, K.: Improved baselines with momentum contrastive learning. arXiv preprint arXiv:2003.04297 (2020)
7. Chen, Z., Yang, W., Li, X.: Stroke-based autoencoders: self-supervised learners for efficient zero-shot Chinese character recognition. Appl. Sci. **13**(3), 1750 (2023)
8. He, K., Fan, H., Wu, Y., Xie, S., Girshick, R.: Momentum contrast for unsupervised visual representation learning. In: Proceedings of the IEEE/CVF Conference on Computer Vision and Pattern Recognition, pp. 9726–9735 (2020)
9. He, K., Zhang, X., Ren, S., Sun, J.: Deep residual learning for image recognition. In: Proceedings of the IEEE Conference on Computer Vision and Pattern Recognition, pp. 770–778 (2016)
10. Huang, G., Liu, Z., Van Der Maaten, L., Weinberger, K.Q.: Densely connected convolutional networks. In: Proceedings of the IEEE Conference on Computer Vision and Pattern Recognition. pp. 2261–2269 (2017)
11. Huynh, T., Kornblith, S., Walter, M.R., Maire, M., Khademi, M.: Boosting contrastive self-supervised learning with false negative cancellation. In: Proceedings of the IEEE/CVF Winter Conference on Applications of Computer Vision, pp. 986–996 (2022)
12. Jin, L.W., Yin, J.X., Gao, X., Huang, J.C.: Study of several directional feature extraction methods with local elastic meshing technology for HCCR. In: Proceedings of the Sixth International Conference for Young Computer Scientist, pp. 232–236 (2001)
13. Johnson, J., Douze, M., Jégou, H.: Billion-scale similarity search with GPUs. IEEE Trans. Big Data **7**(3), 535–547 (2021)

14. Li, J., Zhou, P., Xiong, C., Hoi, S.: Prototypical contrastive learning of unsupervised representations. In: International Conference on Learning Representations (2021)
15. Liu, C.L., Yin, F., Wang, D.H., Wang, Q.F.: Online and offline handwritten Chinese character recognition: benchmarking on new databases. Pattern Recogn. **46**(1), 155–162 (2013)
16. Liu, H., et al.: Perceiving stroke-semantic context: hierarchical contrastive learning for robust scene text recognition. In: Proceedings of the AAAI Conference on Artificial Intelligence, vol. 36, pp. 1702–1710 (2022)
17. Van der Maaten, L., Hinton, G.: Visualizing data using t-SNE. J. Mach. Learn. Res. **9**, 2579–2605 (2008)
18. Noroozi, M., Favaro, P.: Unsupervised learning of visual representations by solving jigsaw puzzles. In: Leibe, B., Matas, J., Sebe, N., Welling, M. (eds.) ECCV 2016. LNCS, vol. 9910, pp. 69–84. Springer, Cham (2016). https://doi.org/10.1007/978-3-319-46466-4_5
19. Oord, A.v.d., Li, Y., Vinyals, O.: Representation learning with contrastive predictive coding. arXiv preprint arXiv:1807.03748 (2018)
20. Qian, R., Zhang, B., Yue, Y., Wang, Z., Coenen, F.: Robust Chinese traffic sign detection and recognition with deep convolutional neural network. In: 11th International Conference on Natural Computation (ICNC), pp. 791–796. IEEE (2015)
21. Su, Y.M., Wang, J.F.: A novel stroke extraction method for Chinese characters using gabor filters. Pattern Recogn. **36**(3), 635–647 (2003)
22. Tian, Y., Hénaff, O.J., Oord, A.v.d.: Divide and contrast: self-supervised learning from uncurated data. In: Proceedings of the IEEE/CVF International Conference on Computer Vision, pp. 10043–10054 (2021)
23. Wang, J., et al.: Towards robust visual information extraction in real world: new dataset and novel solution. In: Proceedings of the AAAI Conference on Artificial Intelligence, vol. 35, pp. 2738–2745 (2021)
24. Wang, T., Xie, Z., Li, Z., Jin, L., Chen, X.: Radical aggregation network for few-shot offline handwritten Chinese character recognition. Pattern Recogn. Lett. **125**, 821–827 (2019)
25. Wang, T., Isola, P.: Understanding contrastive representation learning through alignment and uniformity on the hypersphere. In: International Conference on Machine Learning, pp. 9929–9939. PMLR (2020)
26. Wang, W., Zhang, J., Du, J., Wang, Z.R., Zhu, Y.: DenseRAN for offline handwritten Chinese character recognition. In: 16th International Conference on Frontiers in Handwriting Recognition (ICFHR), pp. 104–109. IEEE (2018)
27. Wu, C., Fan, W., He, Y., Sun, J., Naoi, S.: Handwritten character recognition by alternately trained relaxation convolutional neural network. In: 14th International Conference on Frontiers in Handwriting Recognition, pp. 291–296. IEEE (2014)
28. Wu, M., Mosse, M., Zhuang, C., Yamins, D., Goodman, N.: Conditional negative sampling for contrastive learning of visual representations. In: International Conference on Learning Representations (2021)
29. Xiao, Y., Meng, D., Lu, C., Tang, C.K.: Template-instance loss for offline handwritten chinese character recognition. In: International Conference on Document Analysis and Recognition (ICDAR), pp. 315–322. IEEE (2019)
30. Xie, J., Zhan, X., Liu, Z., Ong, Y.S., Loy, C.C.: Delving into inter-image invariance for unsupervised visual representations. Int. J. Comput. Vision **130**(12), 2994–3013 (2022)

31. Xu, Y., Chang, L.Y., Perfetti, C.A.: The effect of radical-based grouping in character learning in Chinese as a foreign language. Mod. Lang. J. **98**(3), 773–793 (2014)
32. Yang, M., et al.: Reading and writing: discriminative and generative modeling for self-supervised text recognition. In: Proceedings of the 30th ACM International Conference on Multimedia, pp. 4214–4223 (2022)
33. Yin, F., Wang, Q.F., Zhang, X.Y., Liu, C.L.: ICDAR 2013 Chinese handwriting recognition competition. In: 12th International Conference on Document Analysis and Recognition, pp. 1464–1470. IEEE (2013)
34. Zhang, J., Du, J., Dai, L.: Radical analysis network for learning hierarchies of Chinese characters. Pattern Recogn. **103**, 107305 (2020)
35. Zhang, X.Y., Bengio, Y., Liu, C.L.: Online and offline handwritten Chinese character recognition: a comprehensive study and new benchmark. Pattern Recogn. **61**, 348–360 (2017)
36. Zhong, Z., Jin, L., Xie, Z.: High performance offline handwritten Chinese character recognition using GoogLeNet and directional feature maps. In: 13th International Conference on Document Analysis and Recognition (ICDAR), pp. 846–850. IEEE (2015)
37. Zu, X., Yu, H., Li, B., Xue, X.: Chinese character recognition with augmented character profile matching. In: Proceedings of the 30th ACM International Conference on Multimedia, pp. 6094–6102 (2022)

Content-Aware Urdu Handwriting Generation

Zeeshan Memon[1(✉)], Adnan Ul-Hasan[2], and Faisal Shafait[1,2(✉)]

[1] School of Electrical Engineering and Computer Science (SEECS), National University of Sciences and Technology (NUST), Islamabad, Pakistan
zeeshan.bese20seecs@seecs.edu.pk
[2] Deep Learning Laboratory, National Center of Artificial Intelligence (NCAI), Islamabad, Pakistan
{adnan.ulhassan,faisal.shafait}@seecs.edu.pk

Abstract. The performance of handwriting recognition systems has undergone significant improvement in recent years. However, the accuracy of these systems for multiple cursive scripts, including Arabic and Urdu, is still limited due to the lack of labeled training data. Handwriting generators are a potential solution to this problem. Previous research on Urdu handwriting generation has primarily focused on generating realistic ligatures using Generative Adversarial Networks (GANs) with common adversarial loss but has not addressed the issue of maintaining content and generated image entanglement. This paper aims to address this gap by proposing a content-controlled training approach for Urdu Handwriting Generation with pre-trained recognizer loss. Our generation model is trained on a diverse set of printed ligatures and then fine-tuned with transfer learning on handwritten images. In this paper, a new metric for evaluating the performance of handwriting generation systems is also suggested, which is specifically tailored to the context of handwriting generation tasks. To our knowledge, this is the first Urdu handwriting generation system that is capable of generating content-controlled images.

Keywords: Handwriting Generation · Recognition Loss · Generated Adversarial Network

1 Introduction

Handwriting is a fundamental aspect of human communication and has played an important role in the documentation of human history. From ancient civilizations to the present day, handwriting has been used to record personal and public events, preserving them for future generations. The use of digital devices has made it easier and more efficient to produce and share written materials. Moreover, more people now rely on typing rather than handwriting to create written documents [10]. Handwriting is still used today for record-keeping in certain fields such as medicine and therapy [12], where it is important to have legible, accurate records of patient information. Additionally, handwriting is also significantly used in education as teachers and students often take notes by hand and it is seen as a personal expression.

© The Author(s), under exclusive license to Springer Nature Switzerland AG 2023
G. A. Fink et al. (Eds.): ICDAR 2023, LNCS 14190, pp. 428–444, 2023.
https://doi.org/10.1007/978-3-031-41685-9_27

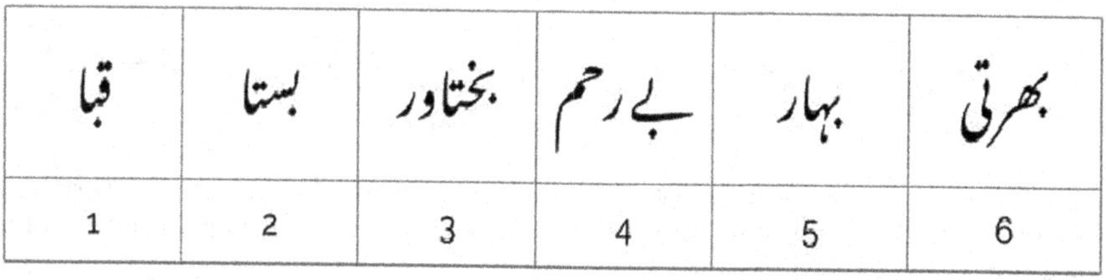

قبا	بستا	بخْتاور	بے رحم	بہار	بھرتی
1	2	3	4	5	6

Fig. 1. The figure illustrates the contextual variations of the Urdu character *'bay'* based on its position (initial, medial, or final) and its combination with other characters [2].

Optical Character Recognition systems (OCRs) have achieved great performance against printed text but still lacks behind in handwritten text due to limited data and diverse writing styles, specifically for Arabic scripts including Urdu.

Urdu, Farsi, Sindhi, Pashto, and Punjabi are all written in scripts that are derived from the Arabic script, which has unique characteristics due to its cursive nature [1]. The shape of characters within the script is dependent on their position within a given word as shown in Fig. 1. Additionally, the use of diacritics and dots serves to indicate grammar and pronunciation [1]. There also exists diverse variations when it comes to inter-word and intra-word spacing within the script with overlapping characters, which adds more to its complexity as highlighted in Fig. 2.

Furthermore, Arabic script is cursive in nature, where a single word consists of one or more ligatures. A ligature is a combination of two or more characters that are merged into a single more complex shape [11]. These factors, in conjunction with individual variations in handwriting style, contribute to the complexity of processing and analyzing text written in the Arabic script.

Given the challenges posed by complex and limited data, it has become increasingly evident that the current recognition systems are not meeting the desired standards of performance. Recently, there has been a growing interest in the field of handwriting generation as a potential solution for the data limitation challenge. Variational Autoencoders (VAEs) [19] and Generative Adversarial Networks (GANs) have emerged as popular research areas in this field. However, a major challenge in these studies, particularly for Arabic scripts, is the absence of controlled text generation [5]. Controlled text generation is a critical aspect of handwriting generation for OCR systems, as it is necessary for training OCR models and obtaining annotated data that accurately reflects the needs of the OCR systems [18].

From the current literature, two significant research gaps have been identified. Firstly, the lack of content-controlled generation for complex script handwriting is a major challenge [10]. This means that there is difficulty in generating images that accurately represent the intended content when dealing with scripts that have a high level of complexity. Secondly, even when content control is attempted, the concurrent training of the recognition system with the GANs presents another challenge. This is because the recognition system can validate substandard generated images as readable, which does not ensure that

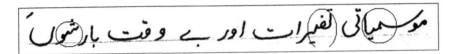

Fig. 2. Illustrating the Complexity of Urdu Handwriting: A Sample of Overlapping Characters in Urdu Script. The image shows the cursive and overlapping nature of the Urdu script.

the generator will converge towards producing realistic and readable handwriting images [20]. This can result in the generation of images that do not accurately represent the intended content.

In this paper, an alternative approach is proposed that presents a novel solution to the challenges of content-controlled and readable generation of handwriting images. By incorporating a pre-trained recognizer network into a generative end-to-end architecture, the generational power of GANs is leveraged while addressing the issue of model failure in producing readable text images. For any previously unseen unicode Urdu string given as input, our approach generates an image of the corresponding Urdu text in a handwriting style.

This paper is further divided into different sections. Section 2 summarizes the relevant work done previously. Section 3 provides a detailed methodology of the proposed approach. Section 4 presents experimental configurations and Sect. 5 discusses the results and compares them with other similar works. Section 6 finally concludes the paper with a summary and future directions.

2   Previous Work

Handwritten text generation techniques utilizing deep learning can be broadly classified into two categories: online and offline generation techniques. Online techniques typically utilize temporal data obtained from the sequential recording of real handwriting samples (in vector form) via the use of a digital stylus [4]. Alternatively, recent generative offline handwritten text generation techniques [5] focus on the direct generation of text through training on offline handwriting images.

Graves et al. [3] present the very first approach utilizing a Recurrent Neural Network (RNN) with Long-Short-Term Memory (LSTM) cells for predicting future stroke points based on previous pen positions and input text. Further, Ji et al. [6] extended the method presented in [3] by incorporating a GAN framework with a discriminator. The introduction of a disentanglement mechanism in DeepWriting [4] allows for greater control over the generation of style without affecting the content. Haines et al. [7] proposed a method for author-specific handwriting generation, which requires a significant amount of character-level annotation for each new sample.

Recent advancements in handwriting generation have aimed to improve control over both content and style. One such approach is presented by Alonso et al. [5], which utilizes a GAN architecture composed of a discriminator and generator, as well as two additional networks: a bidirectional LSTM and a CNN with LSTM layers at the end. This approach specifically focuses on the generation of fixed-length and width handwritten strings in French and Arabic. The generated images were incorporated into an existing dataset, resulting in improved accuracy of recognition systems.

In another study, Fogel et al. [8] presents a novel method for generating images from text, referred as ScrabbleGAN, which incorporates a correlation between the width of the generated image and the length of the input text. The results of ScrabbleGAN demonstrate its proficiency in generating high-quality images that are semantically consistent with the input text. Farooqui et al. [9] presents an approach for improving handwriting recognition of the Urdu language by generating additional data samples using different GAN variants. Seven different GAN architectures were implemented for the generation of handwritten Urdu ligatures, with each GAN trained to produce a specific class of ligatures or at most 10 classes for class-conditioned variants. The goal is to increase the amount of training data to improve the accuracy of word-spotting tasks. Sharif et al. [10] present a GAN-based approach for Urdu Handwriting Generation, which produces realistic Urdu ligatures. Three different GANs variants were evaluated, with WGANs showcasing the best performance. It basically utilizes the receptive power of a deep convolutional generator to generate complex overlapping ligatures, but it does not ensure content-controlled generation.

Similar to ScrabbleGAN [5], we also investigate the problem of content-controlled generation and propose an approach, utilizing a pre-trained recognizer network with frozen weights during training instead of training along with GANs, to ensure a stable mapping between input character embeddings and generated images. This approach is intended to overcome the limitations present in current handwriting generation techniques.

3   Proposed Methodology

In this study, we propose to use a GAN-based approach as shown in Fig. 3, where in addition to the discriminator, the generated image is also evaluated by a recognizer network. The purpose of the discriminator is to promote the realistic appearance of handwriting styles, while the recognizer network serves to ensure the generated image is readable and accurately represents the input text. Each component is discussed in the following subsections.

3.1   Fully Convolutional Generator

The fundamental principle guiding our proposed model is the realization that handwriting is a localized process, meaning that each letter is influenced only by the letters preceding and succeeding it. This is also supported by Graves et al. [3], who employed recurrent neural networks for the task at hand.

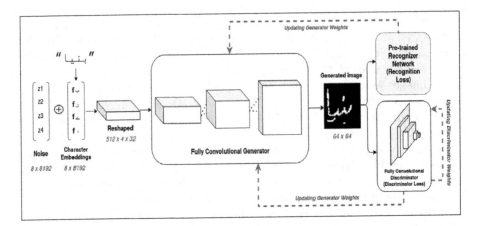

Fig. 3. Proposed approach for Ligature Image Generation. Embeddings for each character in a given ligature are combined with noise before being fed into the network. The image is generated through upsampling and convolutional layers and then passed through a convolutional network and recognizer network, resulting in discriminator loss and recognition loss, which guide the learning of the overall architecture (as shown by the dashed blue line). (Color figure online)

By analyzing Urdu ligature formation, we posit that the shape of characters in Urdu is heavily influenced by the surrounding characters within the ligature. This suggests that this characteristic can be effectively learned through the use of convolutional neural networks. The proposed generator utilizes character embeddings and maps them onto the generated image through a series of convolutional layers.

The generator can be conceptualized as one that generates individual character-wise patches, rather than generating complete words in their entirety. A combination of convolutional layers and upsampling layers in each layer of the generator is employed which increases the overlap between neighboring characters, thereby expanding the receptive field. This facilitates interactions among neighboring characters, resulting in a smoother transition within ligatures. For every character in a given ligature, a character embedding is combined with a noise vector in order to account for natural variations in handwriting. The resulting embeddings are then passed through a fully convolutional generator, where the region generated by each character filter is of the same dimension and the receptive fields of adjacent filters overlap to generate the ligature image.

3.2   Fully Convolutional Discriminator

In the traditional GAN architecture, the role of the discriminator is to accurately distinguish between original data samples and those generated by the generator. In the proposed model, a discriminator is also utilized to score images as either real or fake. The discriminator is a fully convolutional neural network, similar to the generator, but with an architecture opposite of the generator. Both actual handwritten samples and generated samples are provided as input to the discriminator, which then evaluates these images and produces an output. This output is subsequently used in the loss function to update the weights of both the generator and discriminator.

3.3   Recognizer Network

The recognizer network evaluates generated images on the basis of readability by comparing the recognized text from the recognizer network with the input label provided to the generator. For the recognizer network, we have used the state-of-the-art recognition system for Urdu Handwriting proposed by Riaz et al. [12], which combines the capabilities of a convolutional neural network (CNN) and a transformer (Conv-Transformer). The CNN component extracts visual features from the input image, which are then passed to a full transformer consisting of three encoder-decoder layers. The model employs a cross-entropy loss function to measure the difference between the predicted text and the text labels.

For our task, we trained the Conv-Transformer architecture as suggested by Riaz et al. [12] on NUST-UHWR Dataset [18] with the objective of achieving generalizability of recognizer network.

3.4   Optimization Functions

Three distinct learning objectives are discussed in this work. Specifically, discriminator loss, generator loss, and recognition loss are utilized. The utilization of these three objectives aims to enforce the content-controlled generation of images, thus enhancing their overall quality.

Discriminator Loss. We employ a discriminator model to estimate the probability of whether a given sample is from the training data (X) or from the artificially generated distribution. The optimization problem is formulated as a min-max problem, where the generative network (G) and the discriminator (D) are trained in competition with each other. Formally, it can be defined as

$$\min_{L_D} L_D = D((G(Z, L)), 0) - D(X, 0)) \tag{1}$$

where Z and L represent noise vector and text label. The G represents the generator network, which generates text images given Z and L.

Generator Loss. It is typically a function that measures how realistic generated images are. It can be formally defined as

$$\min_{L_G} L_G = -D((G(Z, L)), 0) \tag{2}$$

Recognition Loss. A pre-trained state-of-the-art Urdu recognition is utilized as a handwritten text recognizer network (R) that guides the generation of synthetic

Fig. 4. Text, its rendered and augmented Versions, where (**a**) displays the input text, (**b**) presents the rendered version using Pango with the 'Pak Nastaleeq'font and (**c**) showcases additional augmentations that mimic real handwriting variations.

word images with specific textual content. As the recognizer network is frozen during training, this loss optimizes the weights of the generator network only. This is a fundamental minimization problem, which can be defined as

$$\min_{L_R} L_R = R((G(Z, L)), L) \tag{3}$$

The overall architecture is trained using a combination of three proposed loss functions keeping the weights of the recognizer network freeze. The three losses represented by Eqs. (1), (2), and (3) are combined arithmetically to yield the overall loss.

$$\min_{L} L = L_D + L_G + L_R \tag{4}$$

The weights of the generator and discriminator are updated in an alternating fashion to ensure the stability of the overall learning process.

4   Experiment Configuration

Several experiments were conducted to evaluate the effectiveness of the proposed model for generating Urdu ligatures and to compare its results with those of existing baselines and state-of-the-art approaches. The specifications of the datasets used implementation details, and hyperparameter settings are thoroughly discussed below.

4.1   Datasets Used

In order to assess the performance of the proposed model, both rendered and real handwriting datasets were utilized. A brief overview of the database is provided below:

UCOM Database: In order to evaluate the proposed network, the UCOM database [13] is utilized. The dataset consists of 48 distinct lines of Urdu text, authored by 100 different individuals. The Urdu language comprises 36 unique alphabets, with standalone Urdu alphabets also being considered as ligatures of a single character. Following the methodology outlined in [9], the 317 unique ligatures are extracted from images of Urdu sentences through binarization, segmentation, and resizing to obtain ligatures of a fixed dimension. By utilizing data augmentation 32,000 samples were obtained.

Center of Language Engineering (CLE) Database: The CLE database [23], developed by the Center for Research in Urdu Language Processing (CRULP) [24], primarily consists of 18,000 frequent Urdu ligatures in Unicode format. These ligatures are organized based on the number of characters, ranging from 2 to 8 characters. We used ligatures consisting of up to 4 characters, yielding a total of 10,012 unique ligatures. These ligatures are rendered using Pango [14] as shown in Fig. 4.

4.2 Pre-processing of Dataset

We argue that better formation of ligatures can be learned through a large corpus of ligatures, and this knowledge can be transferred and refined to a specific handwriting dataset. In order to achieve this, augmentations that aim to make the rendered images as similar as possible to real handwriting are utilized. These augmentations include erosion, dilation, rotation, and shear transformation as shown in Fig. 4. A dataset of 30,036 images is generated by using these augmentations on 10,012 unique ligatures from the CLE database.

4.3 Implementation Details and Hyper Parameters

The architecture of the network is configured to generate fixed-size images of 64 × 64 pixels. The input ligature is padded up to a sequence length of eight characters and then passed through the embedding layer of *Generator (G)* to generate embeddings of shape 8×8192 for each sample. As illustrated in Fig. 3, for the generation of an n-character ligature, 'n'number of character embeddings are generated according to the characters. These embeddings are reshaped into $512 \times 4 \times 32$ and subsequently passed through convolutional layers, followed by an upsampling layer. Leaky ReLU (LReLU) and batch normalization [15] is applied between these layers, and a sigmoid activation function is utilized to produce the final output of size 64 × 64. Table 1 shows the detailed architecture of the generator with corresponding output shapes.

The *Discriminator (D)* network is essentially the inverse of the generator network, with the exception of the absence of the spatial embeddings layer. An image of 64 × 64 pixels is provided as input to the discriminator, which is then processed through a series of layers, including the convolutional layer, Leaky ReLU (LReLU) layer, batch normalization, and max pool layer. The final layer is a linear layer that outputs a single value, representing the score or probability of the image being real or fake. The detailed architecture of the discriminator, encompassing dimensions and activation functions at each layer, is presented in Table 1. A batch size of 32 was employed, where the input labels' sequence length was padded up to the maximum sequence length of 8. The Adam optimizer with a learning rate of $2e^{-4}$ was utilized for the training of our architecture. For every generator update, the discriminator is updated 7 times, and recognition loss is optimized on every 5th training step of the epoch.

Hyperparameter Tuning. The stability of GAN training is a well-known challenge in the field of generative modeling. The stability of GANs is influenced by several factors, including the learning rate and the number of times the discriminator is trained compared to the generator, as stated in the seminal work by [16]. In practice, there is no standard set of hyperparameters that works for all models and datasets.

In our study, we conducted an extensive exploration of the hyperparameters to stabilize the training of GANs. The learning rate was varied from $1e^{-4}$ to $1e^{-5}$ with intervals, while the number of times the discriminator was trained relative to the generator, represented by the parameter 'k', was varied from 2 to 10. Our results showed that a learning rate of $2e^{-5}$ and a value of k=7 were the optimal hyperparameters for our proposed approach.

4.4   Experiments Performed

We have executed three distinct variations of experiments, considering the dataset or approach employed. Our results have been benchmarked against the state-of-the-art in Urdu Handwriting Generation [10] and are thoroughly discussed in subsequent sections.

Performance on CLE Database. The proposed model was trained on the CLE database from scratch, utilizing 300 epochs with hyperparameters in accordance with the specifications described in Sect. 4.3. However, the discriminator was trained seven times that of the generator, as advised in [16]. Out of 30,036 samples, 28,512 were designated for training and the rest for testing. The effectiveness of the proposed model was evaluated using the Fréchet Inception Distance (FID), Geometric Score (GS), and Recognition Accuracy, as outlined in Table 2. The Recognition Accuracy was determined through the Character Accuracy Rate, which reflects the number of characters that can be recognized from the generated images through the use of a state-of-the-art Urdu handwriting recognition system.

Performance on UCOM Database. In a similar fashion to the training on the CLE database, the proposed model underwent 300 epochs of training with hyperparameters consistent with those specified for the CLE database, except for

Table 1. Configurations of custom Generator and Discriminator blocks. Each convolution layer has ReLU activation except the last one which has Sigmoid Activation.

Generator layer	Output shape	Discriminator layer	Output shape
Embedding Layer + Noise	8×8192	Input Image Vector	$1 \times 64 \times 64$
Embedding Layer + Noise	$512 \times 4 \times 32$	Convolution	$32 \times 64 \times 64$
(Reshaped)		Convolution	$32 \times 64 \times 64$
Convolution	$256 \times 8 \times 32$	Convolution	$32 \times 64 \times 64$
Batch Normalization	$256 \times 8 \times 32$	Convolution	$64 \times 32 \times 32$
Convolution	$128 \times 16 \times 32$	Convolution	$128 \times 16 \times 16$
Batch Normalization	$128 \times 16 \times 32$	Convolution	$128 \times 16 \times 16$
Convolution	$128 \times 32 \times 32$	Convolution	$256 \times 16 \times 8$
Batch Normalization	$128 \times 32 \times 32$	Batch Normalization	$256 \times 16 \times 8$
Convolution	$64 \times 64 \times 64$	Convolution	$256 \times 16 \times 4$
Convolution	$64 \times 64 \times 64$	Batch Normalization	$256 \times 16 \times 4$
Convolution	$32 \times 64 \times 64$	Convolution	$256 \times 16 \times 4$
Convolution	$32 \times 64 \times 64$	Batch Normalization	$256 \times 16 \times 4$
Convolution	$16 \times 64 \times 64$	Convolution	$256 \times 16 \times 2$
Convolution	$1 \times 64 \times 64$	Linear Layer	1×1

the discriminator which was trained five times than that of the generator. The performance of the model was assessed using the FID score, Geometric Score, and Recognition Accuracy, as shown in Table 3.

Performance of Model Trained on CLE and UCOM Database. The processed UCOM database consists of only 317 unique ligature formations as highlighted in Sect. 4.1, which makes it insufficient as an Urdu Handwriting Generator due to its limited data. To improve the dataset, 10,000 ligatures from the CLE database have been rendered and augmented to incorporate real handwriting variations. The model was trained on the CLE data and then transferred and fine-tuned on the UCOM database for an additional 50 epochs with a learning rate of $2e^{-6}$, drawing inspiration from the transfer learning approach used in GANs [17]. The performance was evaluated using the FID score, Geometric Score, and Recognition Accuracy, and the results are presented in Table 2.

Impact of Generated Data on Urdu OCR Performance. The objective of handwriting generation is to increase annotated data to improve handwriting recognition accuracy. To evaluate the improvement, an experiment was performed in which the OCR model was trained with both the generated data and the UCOM database. The performance of the proposed model was compared to

Sharif et al. [10] using the Character Error Rate (CER) as the evaluation metric, shown in Table 4.

5   Results and Discussion

The performance of our proposed method was evaluated using three quantitative metrics: Fréchet Inception Distance (FID) [22], Geometric Score (GS) [21], and Recognition Accuracy. FID was utilized to measure the similarity between the feature representations of the generated and real images. This was achieved by fitting two Gaussians on the feature representations obtained from an Inception Network and calculating the Fréchet distance between them. GS, on the other hand, compares the geometrical properties of the fundamental real and fake data manifolds and provides a means to quantify mode collapse.

In this study, a new evaluation metric, Recognition Accuracy, has been discussed as a more effective means of evaluating text image generation tasks. Unlike FID and Geometric Score, which evaluate the generated text images based on latent features, Recognition Accuracy evaluates the readability of the images through state-of-the-art generalized OCR systems. Although there may be limitations to this approach due to the limitations of OCR itself, it provides a standardized means of determining the quality of text images. Additionally, the FID score has its own limitations, as the Inception network it relies on is primarily trained on facial data, which may not accurately represent the target distribution for another target. For this reason, the use of Recognition Accuracy along with the FID score may provide a more comprehensive evaluation of the quality of content-controlled handwriting generation tasks.

In every experiment performed, the FID score was calculated for the entire dataset, comparing it with an equivalent number of generated samples, approximately 30,000 in total. The Geometric Score was determined through the analysis of 5,000 real and 5,000 generated samples using default parameter settings. Furthermore, the Recognition Accuracy was calculated for the complete dataset using a pre-trained recognizer network.

5.1   Results on CLE Database

The results of our study on the CLE database show an FID score of 69.01 and a Geometric Score of $7e^{-4}$, with a recognition accuracy of 77% as highlighted in Table 2. The recognition accuracy of 77% indicates that 77% of characters are readable in our generated images with the help of generalized OCR. Given that the CLE database is comprised of 10,000 unique ligatures and trained for 300 epochs, the results demonstrate the good performance of the proposed model.

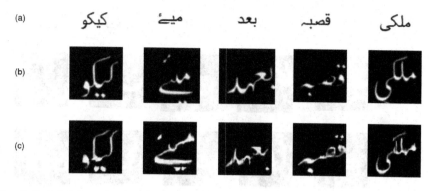

Fig. 5. Label, it's ground truth (printed text with Pango) from **CLE database** and generated sample through the proposed model, where **(a)** displays the input text, **(b)** presents ground truth and **(c)** showcases generated samples through the proposed model.

Table 2. Comparison of FID score, Geometric Score and Recognition Accuracy for proposed approach on **CLE database** and fine-tuned on **UCOM database**.

Training Data	FID Score	Geometric Score	Recognition Accuracy(%)
CLE Database	69.01	$7.87e^{-4}$	77
CLE Database + Fine-tuned on UCOM database	38.03	$8.81e^{-4}$	72.6

Furthermore, a qualitative assessment as shown in Fig. 5 confirms the validity of the results. The utilization of an augmented training dataset ensures that the generated images accurately depict natural handwriting, and the formation of ligatures in comparison with ground truth confirms the correctness of the model.

5.2 Results on UCOM Database

Results of the proposed approach on the UCOM database as explained in 4.4 demonstrate an FID score of 23.24 and a Geometric Score of $5.95e^{-}4$, with a recognition accuracy of 69.7% as illustrated in Table 3. The quality and accuracy of the images produced can also be confirmed by the visual representation in Fig. 6. As demonstrated, the majority of the generated samples contain recognizable characters, with the exception of one most right sample which is a five-letter ligature, and one out of five characters not being recognizable based on human evaluation. These results align with the quantitative recognition accuracy mentioned in Table 3, which is a mislabeling rate of 4-5% in the dataset and a limitation of the recognition model's accuracy.

Along with the quantitative comparison presented in Table 3, a qualitative analysis was also conducted, as shown in Fig. 7. The results demonstrate that the

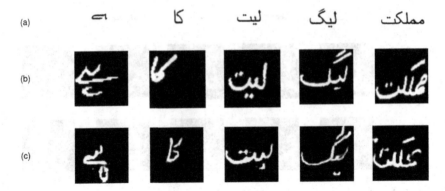

Fig. 6. Label, its ground truth from **UCOM database** and generated sample through proposed model, where **(a)** displays the input text, **(b)** presents ground truth and **(c)** showcases generated samples through proposed model.

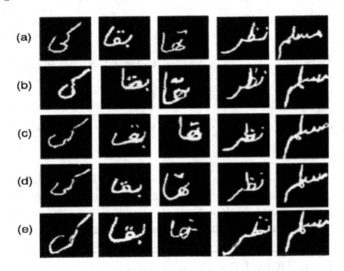

Fig. 7. Qualitative comparison of Proposed Model and different GAN variants from Sharif et al. [10] on UCOM database. **(a)** shows ground truth, **(b)** - **(d)** represents generated samples from Deep Convolutional GANs(DCGANs), Wasserstein GANs (WGANs), and Wasserstein GANs with gradient penalty (WGANs-GP), **(e)** showcase generated samples through the proposed model.

proposed model generates samples that are comparable in quality to those generated by other GAN variants from Sharif et al. [10]. It is worth noting that the proposed model generates content-controlled samples, while the samples from the previous works were the best-generated samples of the same class/label separated manually as they are not content-controlled generation. This superiority of the proposed model can be verified by examining Table 3, which shows no significant difference in FID and geometric score but a marked improvement

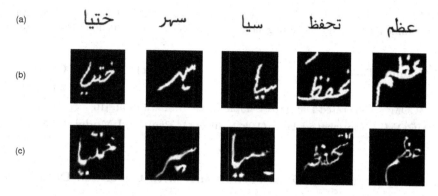

Fig. 8. Qualitative comparison of generated samples and ground truth when transfer learning is employed from **CLE database** to **UCOM database**, where (a) displays the input text, (b) presents ground truth and (c) showcases generated samples through the proposed model.

Table 3. Comparison of FID score, GS, and Recognition Accuracy for different GAN variants from previous works and Proposed Approach.

Model	FID Score	Geometric Score	Recognition Accuracy(%)
DCGANs	21.45	$7.82e^{-4}$	35.1
WGANs	17.97	$7.46e^{-4}$	37.6
WGANs-GP	**15.74**	$7.14e^{-4}$	39.2
Proposed Model	23.24	$5.95e^{-4}$	**69.7**

in recognition accuracy, reflecting the difference between controlled and uncontrolled generation. The results highlight the ability of the proposed model to generate annotated samples of equivalent quality to those generated by uncontrolled methods while still maintaining a close relationship to the input.

5.3 Result of Transfer Learning on UCOM Database

As detailed in Sect. 4.4, an attempt was made to apply transfer learning from large rendered and highly augmented data (CLE database) to real handwriting data, with the aim of improving ligature formation and increasing the generalization of the generator. The results are presented in Fig. 8, which shows that while the difference between generated samples and ground truth can still be distinguished, the generator is making progress toward replicating the smoothed strokes of real handwriting images. Table 2 reports the evaluation metrics, including the FID score of 38.03, the Geometric Score of $8.81e^{-4}$, and recognition accuracy of 72.6%, demonstrating the efficacy of this approach. These results are also in line with those obtained from training a model from scratch using only the UCOM database, as shown in Table 3. This highlights the potential of transfer learning in handwriting generation tasks when the training data is limited.

Table 4. Comparison of OCR performance in terms of CER(Character Error Rate), when trained with Synthetic data generated with WGANs-GP [10] and proposed model.

Training Data	Model	CER(%)
UCOM	-	7.12
UCOM + Generated Data(10k Samples)	Sharif et al. [10]	6.77
UCOM + Generated Data(10k Samples)	**Proposed Model**	**6.15**

5.4 Improvement in OCR Performance with Generated Data

Our proposed approach was comprehensively compared to the recent state-of-the-art in Urdu handwriting generation proposed by Sharif et al. [10]. The results, presented in Table 4, indicate that training with the generated samples from our proposed approach resulted in a reduction of the Character Error Rate (CER) from 7.12 to 6.15, compared to the CER of 6.77 from the previous approach. This improvement can be attributed to the better and content-controlled generation of ligatures through our proposed approach, as opposed to the uncontrolled generation in the previous approaches, which required manual labeling or separation of ligatures. The presence of annotated samples allows for more effective and faster training and improvement of our OCR.

6   Conclusion

In this study, a GAN-based model was proposed to generate handwriting samples with improved quality. The model uses a pre-trained recognition network and was trained on two datasets. The model was found to produce content-controlled samples with quality comparable to recent approaches.

The proposed model also demonstrated the potential for transfer learning in handwriting generation by utilizing rendered data. While the model showed slightly higher FID scores and limited variation in ligature formation, it indicates that incorporating a larger dataset of real handwriting data with the rendered data could lead to improved results. This study lays the foundation for further research in the direction that it can be extended to generate words and sentences dynamically. Supplementing GANs with language models instead of the simple embedding layer can enhance text image generation by leveraging the combined capabilities of language understanding and receptive power to improve ligature formation.

References

1. Naeem, M.F., Awan, A.A., Shafait, F., ul-Hasan, A.: Impact of ligature coverage on training practical urdu OCR systems. In: 2017 14th IAPR International Conference on Document Analysis and Recognition (ICDAR), vol. 1, pp. 131–136, IEEE (2017)

2. Wali, A., Hussain, S.: Context-sensitive Shape-substitution in Nastaliq writing system: analysis and formulation. In: Sobh, T. (ed.) Innovations and Advanced Techniques in Computer and Information Sciences and Engineering. Springer, Dordrecht (2007). https://doi.org/10.1007/978-1-4020-6268-1_10
3. Graves, A.: Generating sequences with recurrent neural networks. In: arXiv, preprint arXiv:1308.0850 (2013)
4. Aksan, E., Pece, F., Hilliges, O.: DeepWriting: making digital ink editable via deep generative modeling. In: Proceedings of the 2018 CHI Conference on Human Factors in Computing Systems, pp. 1–14 (2018)
5. Alonso, E., Moysset, B., Messina, R.: Adversarial generation of handwritten text images conditioned on sequences. In: 2019 International Conference on Document Analysis and Recognition (ICDAR), pp. 481–486, Sydney, Australia, IEEE (2019)
6. Ji, B., Chen, T.: Generative adversarial network for handwritten text. In: arXiv, preprint arXiv:1907.11845 (2019)
7. Haines, T.S., Mac Aodha, O., Brostow, G.J.: My text in your handwriting. ACM Trans. Graph. (TOG) 35(3), 1–18 (2016)
8. Fogel, S., Averbuch-Elor, H., Cohen, S., Mazor, S., Litman, R.: ScrabbleGAN: semi-supervised varying length handwritten text generation. In: Proceedings of the IEEE/CVF Conference on Computer Vision and Pattern Recognition, pp. 4324–4333 (2020)
9. Farooqui, F.F., Hassan, M., Younis, M.S., Siddhu, M.K.: Offline handwritten Urdu word spotting using random data generation. IEEE Access 8, 131119–131136 (2020)
10. Sharif, M., Ul-Hasan, A., Shafait, F.: Urdu handwritten ligature generation using generative adversarial networks (GANs). In: Frontiers in Handwriting Recognition: 18th International Conference, ICFHR 2022, Hyderabad, India, 4–7 December 2022, Proceedings, pp. 421–435, (2022)
11. El-Korashy, A., Shafait, F.: Search space reduction for holistic ligature recognition in Urdu Nastalique script. In: 12th IAPR International Conference on Document Analysis and Recognition (ICDAR), Washington, DC, USA, pp. 1125–1129 (2013)
12. Riaz, N., Arbab, H., Maqsood, A., Nasir, K.B., Ul-Hasan, A., Shafait, F.: Conv-transformer architecture for unconstrained off-lineUrdu handwriting recognition. Int. J. Document Anal. Recogn. (IJDAR) 25, 373–384 (2022)
13. Bin Ahmed, S., Naz, S., Swati, S., Razzak, I., Umar, A.I., Ali Khan, A.: UCOM Offline Dataset-an Urdu handwritten dataset generation. Int. Arab J. Inf. Technol. (IAJIT) 14, 239–245 (2017)
14. Taylor, O.: PANGO: an open-source unicode text layout engine. In: 25th Internationalization and Unicode Conference, Unicode Consortium, Washington DC, USA (2004)
15. Ioffe, S., Szegedy, C.: Batch normalization: accelerating deep network training by reducing internal covariate shift. In: International Conference on Machine Learning, pp. 448–456, Lille, France (2015)
16. Goodfellow, I., et al.: Generative Adversarial Networks. Commun. ACM 63(11), 139–144 (2020)
17. Fregier, Y., Gouray, J.-B.: Mind2Mind: transfer learning for GANs. In: Nielsen, F., Barbaresco, F. (eds.) GSI 2021. LNCS, vol. 12829, pp. 851–859. Springer, Cham (2021). https://doi.org/10.1007/978-3-030-80209-7_91
18. Zia, N.S., Naeem, M.F., Raza, S.M.K., Khan, M.M., Ul-Hasan, A., Shafait, F.: A convolutional recursive deep architecture for unconstrained Urdu handwriting recognition. In: Neural Computing and Applications, pp. 1635–1648 (2022)

19. Kingma, D.P., Welling, M.: Auto-encoding variational Bayes. In: arXiv, preprint arXiv:1312.6114 (2013)
20. Davis, B., Tensmeyer, C., Price, B., Wigington, C., Morse, B., Jain, R.: Text and style conditioned GAN for generation of offline handwriting lines. In: arXiv, preprint arXiv:2009.00678 (2020)
21. Khrulkov, V., Oseledets, I.: Geometry score: a method for comparing generative adversarial networks. In: International Conference on Machine Learning, pp. 2621–2629. Stockholm, Sweden (2018)
22. Heusel, M., Ramsauer, H., Unterthiner, T., Nessler, B., Hochreiter, S.: GANs trained by a two time-scale update rule converge to a local nash equilibrium. In: Advances in Neural Information Processing Systems, 30, California, USA (2017)
23. Khattak, I.U., Siddiqi, I., Khalid, S., Djeddi, C.: Recognition of Urdu ligatures - a holistic approach. In: 2015 13th International Conference on Document Analysis and Recognition (ICDAR), pp. 71–75, Washington, DC, USA. IEEE (2015)
24. Image and Text Corpora. https://www.cle.org.pk/clestore/index.htm

Weakly Supervised Information Extraction from Inscrutable Handwritten Document Images

Sujoy Paul[✉], Gagan Madan, Akankshya Mishra, Narayan Hegde,
Pradeep Kumar, and Gaurav Aggarwal

Google Research, Mountain View, USA
spaul003@ucr.edu

Abstract. State-of-the-art information extraction methods are limited by OCR errors. They work well for printed text in form-like documents, but unstructured, handwritten documents still remain a challenge. Adapting existing models to domain-specific training data is quite expensive, because of two factors, 1) limited availability of the domain-specific documents (such as handwritten prescriptions, lab notes, etc.), and 2) annotations become even more challenging as one needs domain-specific knowledge to decode inscrutable handwritten document images. In this work, we focus on the complex problem of extracting medicine names from handwritten prescriptions using only weakly labeled data. The data consists of images along with the list of medicine names in it, but not their location in the image. We solve the problem by first identifying the regions of interest, i.e., medicine lines from just weak labels and then injecting a domain-specific medicine language model learned using only synthetically generated data. Compared to off-the-shelf state-of-the-art methods, our approach performs $> 2.5\times$ better in medicine names extraction from prescriptions.

Keywords: handwriting · language model · prescription · weakly-supervised

1 Introduction

Optical character recognition (OCR) enables the translation of any image containing text into analyzable, editable and searchable format. Over the last decade, many large scale models [10,18,26] and sophisticated techniques [4,5,29] have been developed with neural network based architectures for OCR. These systems are not only limited to printed text but also work quite well on handwritten text, as they are trained on large amount of labeled as well as synthetic handwritten data. In the past, there have also been works around developing domain specific OCR models [6,21,41]. Most of these works develop these models for generic text lines [20,31], and require meticulously labeled data for learning. In this work, we primarily focus on how we can improve the quality of existing

© The Author(s), under exclusive license to Springer Nature Switzerland AG 2023
G. A. Fink et al. (Eds.): ICDAR 2023, LNCS 14190, pp. 445–463, 2023.
https://doi.org/10.1007/978-3-031-41685-9_28

OCR models on very hard to read, unstructured documents for specific entities of interest, with an application in handwritten medical prescriptions.

Medicines: Gluconorm, Forenza, Medrol, Formonide, Baro

Medicines: Litec, Deswell, Calamine, Albendazole, Flutec

Medicines: Monocef, Efcorlin, Foracort, Duolin, Corex, Doxy, Ivermectol

Fig. 1. Samples representative images from the prescription dataset used in this work. As we can see the handwriting is often inscrutable and does not follow any specific structure or format. The task we focus in this paper is to extract medicine names from such images.

In many countries, prescriptions are primarily delivered to patients in handwritten formats by doctors. A few billion prescriptions are generated every year world-wide [19]. Digitizing them would unlock numerous applications for many stakeholders and use cases in the healthcare ecosystem like e-pharmacies, insurance companies, creating electronic health records necessary for preventive healthcare, better diagnosis, analysis at a local and global level for policy making and so on. However, most of such documents, as shown in Fig. 1 are often hard to read for non-pharmacists [33]. Even pharmacists go through months/years of training to decipher such prescriptions. Existing state-of-the-art OCR models though trained on large amount of data, do not perform well on such inscrutable documents. Procuring large domain-specific datasets is not a cost-effective or scalable solution, as it involves annotation that too from domain experts which can become quite expensive. Although there have been some works [1,15,34] in extracting information from handwritten prescriptions, the algorithms are not generalizable, heavily hand-tuned and lack rigorous evaluations. With these problems in mind, we propose an approach that can significantly enhance the performance of existing state-of-the-art OCR systems by selectively infusing domain knowledge using only weak supervision.

Medical prescriptions consist of various information like data from lab reports, ordered tests, health vitals, observations along with medicine names. Our work focuses on the medicine section which is considered the most important from a consumer standpoint, but the techniques can be similarly applied to other sections or other types of documents beyond prescriptions, such as printed forms filled with handwriting. The medicine section of a prescription has a rough semantics consisting of medicine name, category, frequency of intakes and quantity (see Fig. 1). As these are non-form type of documents and quite unstructured, it is a challenge to extract medicine name entities from such documents.

Most OCR approaches [18, 26] take a two step approach - first localize the text regions by detecting bounding boxes around them, and then recognizing the text using line recognition models. The recognition model often consists of an optical recognizer and a language model (LM) to correct the optical model errors. The LM gives us the flexibility to infuse domain-specific knowledge. But, injecting such knowledge to all lines in the document may not be optimal, as different parts of the document can correspond to different entities, or even domains. For example, the pattern in which a medicine name is written is very different from the pattern in which normal text such as observations are written in the same prescription. Thus, in order to enhance the recognition of medicine names and extract them from the prescription, we first detect lines where medicine names are written. Then in the recognition model, we inject a LM which is specific to medicine names. For the rest of the image, we inject the vanilla LM.

Note that to learn the model which detects medicine lines, we do not use strong bounding polygon labels, but rather only weak labels, i.e., the medicine names present in the image. Such weak labels are much easier to obtain, as the annotators do not need to draw a bounding polygon and often labeling comes for free, for example, when a medicine bill is paired with a prescription. Apart from that, to learn the medicine LM, we do not use any annotated text lines, but rather generate synthetic text lines using a probabilistic programming approach. Our weakly supervised medicine line detector obtains 78% pixel mIoU with just weak labels, and helps to selectively infuse medicine LM, which in turn improves the overall performance from 19% to 48% jaccard index. The main contributions of this work are:

- Develop a weakly supervised segmentation method to detect specific text entities, such as medicine names in handwritten prescriptions.
- Learning a domain-specific medicine LM using synthetic medicine name lines generated by probabilistic programs and using it to enhance the performance of state-of-the-art OCR models.
- A model dependent unique way of enhancing the performance of matching with words from the vocabulary.

2   Related Works

Optical Character Recognition. OCR literature has seen tremendous improvements in the past decade. The successes [10, 18, 26] can be attributed to sophisticated models, synthetic data generation, various augmentation techniques, among others. An OCR system is made of multiple models, starting from text detection [29, 30, 43], script identification [12, 17], and finally line recognition [3, 10, 26, 27]. Even with all these advancements, recognition of handwritten lines still remains a challenging task as writing style can be a unique signature of the person, allowing room for huge variations. In our experiments, we found that off-the-shelf line recognition models, even though perform quite well for a lot of printed and handwritten datasets, they fail to perform equally well on handwritten images. In this work, we show how we can improve their accuracy by more than 2 times the baseline by first detecting specific entities of interest (rather than detecting all text) and then improving the line recognition model by injecting domain-specific LMs. We next discuss the existing literature around these topics.

Weakly-Supervised Detection. Detecting specific entities of interest in an image can be posed as detection or segmentation task. However, to learn these tasks, traditional methods would need strong labels, i.e., either pixel-wise [9,30] or bounding box labels [30,37,38]. In the recent past, there has been a lot of work in developing methods which can learn from only weak labels, such as weakly-supervised object detection [25,47], segmentation [22,44], action detection [32, 46], etc. These methods do not need access to strong labels such as bounding boxes, but can learn from just weak labels, i.e., image-level labels of the object categories present in the individual training images. Such a formulation reduces the manual labor needed to acquire strong labels, thus making it scalable to large datasets.

Motivated by these, we aim to learn a segmentation model to detect entities of interest in an image, such as medicine names from just weak labels, i.e., list of medicine names given an image. In this use case, the individual entities do not correspond to any underlying category unlike segmentation or detection of objects in natural scenes. Recently, it has been shown [23] that using weakly labeled data along with strong labels improves the performance of scene text recognition. In our task, we only have weakly labeled data without any strong labels (synthetic or real) and the text is primarily handwritten which is often inscrutable even if text detection is perfectly done. Moreover in our use case, we need to detect specific entities among other cluttered text, and not any generic text. There are also works on defining rules to derive weak labels from the data [36]. While that is quite challenging and not generalizable in our use case, we use the intuition to convert the weak labels to strong labels via labeling functions.

Domain-Specific Language Models. There has been a lot of work [24,35,45] which shows that injecting domain-specific knowledge in LMs helps to perform much better on those domains than models developed on generic text. Specifically for OCR, there have been some works [11,14] showing that having access to domain related text data helps to adapt existing LMs and thus improving final OCR performance. However, in our use case of decoding medicine names, it is non-trivial to acquire lines of medicine names written by doctors, as they are hardly available in normal text corpus. To solve that, we use domain knowledge to define a probabilistic program which can take in the medicine name and generates patterns of medicine lines as would be written by doctors in prescriptions. We show that using such a LM in the OCR decoder improves the performance significantly.

3 Methodology

3.1 Problem Statement

In this work, we focus on the problem statement of extracting textual entities from non-form type handwritten document images, which are often hard to read. We specifically focus on the challenging problem of extracting medicine names from handwritten prescriptions as shown in Fig. 1. Formally, given an image x, the output of the framework should be the medicine names $\{m_j\}_{j=1}^{n}$ that appear in the image, where $m_j \in V$, the vocabulary of medicines. n denotes the number of medicines in the prescription that varies from prescription to prescription. The training data that we use to solve this problem is only weakly labeled, i.e. for every image, we have a list of medicine names that appear in the image, and not their bounding box locations. Thus, our training data contains tuples of image and unordered set of medicine names as follows, $\mathcal{D} = \{(x_i, \mathcal{G}_i = \{m_j\}_{j=1}^{n_i})\}_{i=1}^{N}$, where n_i denotes the number of medicines in that image, N denotes the number of images in the training data and $\mathcal{G}_i$ is the ground truth list of medicines.

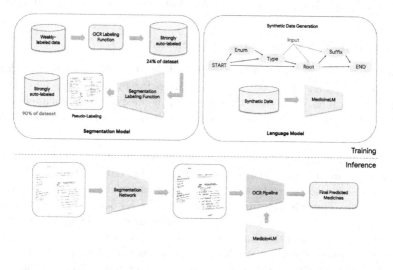

Fig. 2. Training and inference pipelines for medicine name extraction from prescriptions. The top-left block shows the weakly supervised medicine line segmentation pipeline. The top-right block shows the process of generating synthetic medicine lines using probabilistic programs and then using it to train a medicine LM. The bottom row shows the inference pipeline, that first localizes the medicine names using the segmentation network, and then injects the medicine LM while decoding the OCR outputs.

3.2 OCR Line Recognition Model

Most line recognition models have two parts - the encoder, often called the optical part of the model, which encodes the visual information, and the decoder, which is either trained end-to-end with the encoder, or CTC type decoder [13] where the encoder outputs are combined with LM scores to obtain the final text. We use the second option and train our network with CTC loss [13]. This allows us to decouple the optical and the LM, and replace it with domain specific LMs.

Encoder: The encoder or the optical part of the line recognizer consists of first 7 layers of inverted bottleneck conv layers [39] with 64 filters and stride of 1, followed by 12 layers of transformer encoder [42] with hidden size of 256 and 4 attention heads, and finally a fully connected symbol classification head. We use this backbone from [10], as it achieves state-of-the-art performance on various datasets. Our pre-trained model is also the same as [10]. It is interesting to note that our method is agnostic to the encoder used as it can be used to boost the performance of any OCR backbone.

Decoder: We use a CTC decoder [13] following [10], which combines scores from the encoder logits and a character n-gram LM. We set $n = 9$ unless otherwise mentioned. We will discuss how we train and use a medicine LM subsequently.

3.3   Weakly Supervised Line Segmentation

We next discuss our algorithm to detect medicine lines by just using weak labels while training, i.e., only the medicine names for every image, and not their bounding polygons. Note that while we use this method for medicine line detection, it can be also used for detecting other entities in other document types.

Labeling Functions. At the core of our algorithm is the idea of using labeling functions to automatically convert a weakly labeled dataset to strongly labeled. There have been some works [36] in literature where rules are defined as labeling functions. The labeling functions may not be as perfect as a human oracle and the strong labels they generate may have errors in them. There are often thresholds or rules used to reduce errors. Thus, while defining a labeling function we need to optimize coverage, which is the number of data points that can be labeled using such labeling functions and their error rate. Although there can be some noise in such labeling, this significantly reduces the annotation cost. We sequentially apply two labeling functions, as discussed next to convert a list of medicine names to bounding boxes. In our use case of assigning a bounding box to each medicine name, we can consider it as an assignment problem between the detected bounding boxes (p) by a generic text detector and the number of medicines in it n. Considering $p = 50$ and $n = 5$, the number of possible assignments turns out to be $^pC_n{}^pP_p \approx 2.5e8$. We solve this problem via two techniques - using the content of the boxes (via OCR Labeling Function), and using the visual features (via Segmentation Labeling Function).

OCR Labeling Function: As for every image, we have the list of medicines that appear on it, for each detected word in the image, we can naively find the closest medicine name (by edit distance) from the ground truth list, albeit applying a threshold. However, directly using the edit distance may not respect the model's predictions. For example, according to the OCR line recognition model, modifying an i to l may have lower cost than i to z, but it would be the same edit distance for both the cases. Thus, in order to utilize the model's predictions, we decode up to the top-k predictions, and stop when we find an exact match with a medicine name from the list of ground truth medicines, i.e., the weak labels. The bounding box associated with these matched words then can be used as the ground-truth bounding boxes of medicine names. We can define the labeling function as $\mathcal{F}(\boldsymbol{x}) = \{(t_j, l_j, h_j, w_j, r_j)\}_{j=1}^{q}$, where the bounding boxes of m medicines are in the rotated box format and t_j, l_j, h_j, w_j, r_j representing top, left, height, width, and rotation angle of each matched bounding box. Then, we can construct a training dataset as follows: $\mathcal{D}_{tr} = \{(\boldsymbol{x}_i, \mathcal{F}(\boldsymbol{x}_i))\}_{i=1}^{N}$.

The number of matching bounding boxes $q_i \leq n_i$, as in most cases the handwriting is so illegible that to decipher that even a higher number of top-k lines may not allow a match with the ground truth medicine names. This can happen for a sizable number of images, which in turn can introduce a significant noise in the data, leading to problems in learning the segmentation network. Thus, we only use those images to train our network where we find that at least 90% of the ground truth medicines have been matched. The reason behind setting such a high threshold is this set becomes the guiding signal for the rest of the algorithm. Thus our modified strongly-labeled training dataset can be represented as: $\mathcal{D}_{tr} = \{(\boldsymbol{x}_i, \mathcal{F}(\boldsymbol{x}_i)) \big| \frac{|\mathcal{F}(\boldsymbol{x}_i)|}{|\mathcal{G}_i|} \geq 0.9\}_{i=1}^{N}$. While increasing the number of top-k paths helps more images to pass this threshold, we find that it saturates after a point, specially for documents which are hard to read, such as prescriptions used in this work. While the 0.9 threshold allows us to reduce missing bounding boxes in the training set, it also reduces the number of images in the training set, as $|\mathcal{D}_{tr}| \leq |\mathcal{D}|$. We next discuss a second labeling function to alleviate this problem.

Segmentation Labeling Function. It may happen that even after decoding a high number of paths (k), we still are not able to match all the ground truth medicine names. This can happen when the handwriting is quite challenging for the model to predict accurately. In such a scenario, we leverage the visual appearance features via the segmentation model itself, rather than just labeling via OCR. Motivated by the success of self-training in domain adaptation [2,28] and semi-supervised [7,40], we use the segmentation model to pseudo-label the images in the rest of the dataset, i.e., $\mathcal{D}$ - $\mathcal{D}_{tr}$.

First, we train a segmentation network $\mathcal{M}$ using the relatively small training data $\mathcal{D}_{tr}$ obtained from the OCR Labeling Function outlined above. Then, we use it to predict the medicine lines on the images in $\mathcal{D}$ - $\mathcal{D}_{tr}$. We can consider the output of the model to be $\mathcal{M}(\boldsymbol{x}) = \{(t_j, l_j, h_j, w_j, r_j)\}_{j=1}^{l}$. Following our previous threshold, we add those images to the training dataset, where the union of the number of predicted medicine lines by the segmentation network and the OCR labeling function above, is at least 90% of total number of medicines in that image. We can represent the new training set as follows: $\mathcal{D}_{tr} = \{(\boldsymbol{x}_i, \mathcal{F}(\boldsymbol{x}_i) \cup \mathcal{M}(\boldsymbol{x}_i)) \big| \frac{|\mathcal{F}(\boldsymbol{x}_i) \cup \mathcal{M}(\boldsymbol{x}_i)|}{|\mathcal{G}_i|} \geq 0.9\}_{i=1}^N$.

Ideally, we can repeat this process, i.e. repeat pseudo-labeling the training images using a trained segmentation model and training a new model with the pseudo-labeled training set. The training set would grow over iterations. The two labeling functions can be generalized as: $\mathcal{D}_{tr}^T = \{(\boldsymbol{x}_i, \cup_{t=1}^T \mathcal{M}_t(\boldsymbol{x}_i)) \big| \frac{|\cup_{t=1}^T \mathcal{M}_t(\boldsymbol{x}_i)|}{|\mathcal{G}_i|} \geq 0.9\}_{i=1}^N$, where $\mathcal{M}_t = \mathcal{F}$ for $t = 1$, and the tth medicine line segmentation model for $t \geq 1$, and T represents the total number of iterations.

(a) Iter 1   (b) Iter 2   (c) Iter 3

Fig. 3. Evolution of labels from the labeling functions. Iter 1 represent the OCR Labeling Function and the subsequent ones represent the Segmentation Labeling Function for different iterations. The green highlighted regions denote the detected medicine names. (Color figure online)

Figure 3 shows how segmentation improves over iterations. Using only the OCR Labeling Function misses out some of the medicine names, as it is dependent on the ability of the underlying OCR model we use to decipher the medicine names. However, applying the Segmentation Labeling Function on top of it helps to predict the medicine patches which were missed, as it does not depend on OCR or the content, but rather on the visual features, such as strokes, indentation, etc. which we will discuss later in Sect. 4.

Segmentation Model. Given the bounding boxes obtained using the labeling functions, we can train a medicine line segmentation model. Our segmentation model is DeepLab [9] with a ResNet50 backbone [16]. Although we use this architecture, it can be replaced by any other state-of-the-art segmentation model. We convert the bounding boxes to label masks, and use them as supervision to train the segmentation network. The label mask has either 0 or 1 at each pixel location, denoting whether a pixel belongs to a medicine line. The segmentation model is trained with the above data using a semantic head with two output channels. The predicted medicine label masks obtained from this model may not always respect text boundaries, and hence we use a generic text detector in the OCR pipeline to detect text and refine the boundaries. Then, we crop out the detected bounding box from the original image x and send only those lines to the line recognizer. As these lines correspond to a special domain of medicine names, we can inject that knowledge to the OCR using a LM.

3.4   Domain-Specific Language Model

In OCR decoder, we can incorporate a LM to correct some of the OCR errors. Specifically, the decoded string Y^* can be obtained as follows:

$$Y^* = \arg\max_Y P(Y|X)P(Y)^\alpha \tag{1}$$

where $P(Y)$ is obtained from the LM denoting the probability of occurrence of a certain string Y in the dataset, α is the weight applied on the LM, and X is the input. In a generic OCR model, $P(Y)$ is trained on a large corpus of text such that it represents a diverse set of documents. Particular to our use case, once we have detected the medicine lines as discussed in the previous section, we need only medicine line specific knowledge while decoding the OCR output. However, medicine line patterns occurring in handwritten prescriptions often do not appear in normal text. It is also difficult and expensive to acquire and annotate such large corpora of handwritten prescriptions from which we can learn medicine line specific LMs. We inject domain knowledge to solve this problem.

In order to gather medicine line specific text data, we defined a probabilistic program from which we can sample data and learn a character based LM. Medicine lines written by doctors often have a few elements - a enumeration token (-, ., numbers, etc.), followed by the type of medicines (injection, tablet, etc.), the root name of the medicine, and then the suffix. These altogether comprise a single medicine name line. Note that some of these entities other than the root word may not appear in all prescriptions. With this domain knowledge, we can define a probabilistic program as shown in top-right portion of Fig. 2. The program starts from the START node and ends at the END node, and concatenates the output of each node with spaces in between. To sample a medicine name line, the program takes as input the medicine name and the type of the medicine, both of which appears in the vocabulary of medicines. We can create an exhaustive set of all possible medicine name lines, and then train a character based n-gram LM on that text corpus. Note that as we do not have the exact probabilities of the different transitions, we use equi-probable transitions between nodes, as well as for any choices in the nodes.

In OCR, as decoding is done at a character level, we need character LMs, unlike recent advanced large LMs which operate on word or sub-word tokens. There are also character LM using transformers, but those are generally useful for longer context. But, in our case, medicine names on average are only 7 characters long. Moreover, using such a large model takes a lot more inference time. Hence we stick to an n-gram model.

3.5   In-Vocabulary Prediction

In many entity extraction tasks, such as medicine name prediction studied in this paper, the entities often belong to either from a fixed vocabulary, or are defined by a regular expression. However, the OCR predictions will not be constrained to our medicine vocabulary. To constrain that, we can make a nearest neighbor edit distance search for each medicine line text and the medicine vocabulary. However, as we discussed before, it would not respect the model's confidence. Thus, we use the top-k path decoding as a robust method. Specifically, for each line, we decode the top-k predictions, and then find all the text which have an exact match with one of the medicine names from the vocabulary. Then, we take a majority voting of all these matched names, and that becomes the prediction for every line. It is possible that for some of the detected medicine lines, we do not find any match for any of the top-k prediction. These detected medicine lines would not have any output prediction. We find this method to be more effective compared to edit distance based matching with the top-1 prediction, or predicting only the first match from the top-k predictions, as shown in Sect. 4.

Table 1. (a) Statistics of the prescription dataset. (b) Coverage of different sections in prescriptions.

(a)	
# Images	9645
# Doctors	117
Avg. medicines / image	4.5
Avg. images / doctor	82.4

(b)	
Lab/Scan	70.4%
Medicine	100%
Observation	99.9%
Vital	40.5%

4   Experiments

We first introduce the dataset and implementation details, before sharing the experimental results and rigorous ablations to understand the efficacy of the framework.

Prescription Image Dataset: We use a dataset of handwritten prescriptions to validate the methodology outlined and evaluate the performance of the models. A few example images from the dataset are shown in Fig. 1. The dataset contains 9645 images written by 117 doctors. Table 1a outlines some of the details of the dataset, and Fig. 4a shows the distribution of prescription images per doctor. We use 80% of the dataset to train our models, and 20% for evaluation. There is no overlap between the doctors between the training and the test set at each iteration, ensuring that our results capture understanding across different handwriting styles. Each image in the dataset has a list of medicine names appearing in them, which we call weak labels, without any positional information. However, just for evaluation, we strongly annotate 500 images from the evaluation set to evaluate the segmentation performance. Prescriptions generally have multiple other sections as well (although unstructured in free-form), and Table 1b shows the percentage of images which have other sections such as lab/scans reported, observations and vitals. Also, note that any and all personally identifiable information was removed from the data prior to it being provided to the authors for this study.

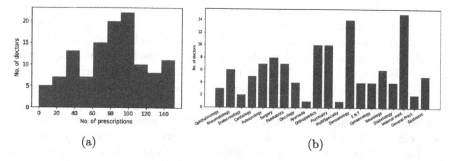

Fig. 4. (a) This plot shows the number of prescriptions per doctor in the dataset, (b) This plot shows the number of doctors per specialty.

Medicine Vocabulary: We also use a medicine name vocabulary consisting of more than 90,000 medicine names. We use this to generate synthetic medicine name lines and train the character based medicine LM. This vocabulary is also used to make the in-vocabulary predictions.

Evaluation Protocol: We evaluate all models on test set of the dataset mentioned above. To evaluate the performance of the segmentation model, we use mean intersection over union (mIoU) as used in the segmentation literature [8]. To evaluate the performance of the end-to-end medicine name prediction model, we use the mean jaccard index, over all the images. We also use two other metrics namely mean precision and mean recall, and the mean jaccard index can be considered as a combination of both these metrics. These are defined as follows

$$\text{Mean Jaccard Index (mJI)} = \frac{1}{M} \sum_{i=1}^{M} \frac{|P_i \cap G_i|}{|P_i \cup G_i|} \tag{2}$$

$$\text{Mean Precision (mP)} = \frac{1}{M} \sum_{i=1}^{M} \frac{|P_i \cap G_i|}{|P_i|}, \tag{3}$$

$$\text{Mean Recall (mR)} = \frac{1}{M} \sum_{i=1}^{M} \frac{|P_i \cap G_i|}{|G_i|} \tag{4}$$

where P_i, G_i are the predicted and ground truth list of medicines for the ith image. M is the number of evaluation images. The comparison between the prediction and ground-truths are not case-sensitive, as they are medicine names.

4.1   Results and Ablation Studies

Iterative Training Performance: As discussed in Sect. 3, our algorithm for converting weak labels (only medicine names) to strong labels (bounding box annotations for each medicine name) involves two labeling functions - OCR and Segmentation Labeling Function, where the latter can be applied iteratively. The number of images auto-labeled by the labeling functions increases with iterations, and hence the performance of both the medicine line segmentation model as well as the medicine name prediction model increases with subsequent iterations. We highlight this in Table 2. Iteration 1 shows the performance on only OCR Labeling Function, and Iteration ≥ 2 shows the performance on multiple iterations of Segmentation Labeling Function. For a significant number of prescriptions, it is difficult to decipher some of the medicine names, even when we use a high value of top-k (k=20,000 in our experiments) decoded outputs per line. For Iteration 1, the number of auto-labeled prescriptions is $< 25\%$ of the training set. This shows the difficulty level of the problem at hand. Note that the train sets are used to train only the medicine line segmentation model and not the lines recognizer of the OCR, thus it can be with any off-the-shelf OCR model.

The segmentation performance as well as the medicine name performance improve over iterations but saturates from Iteration 3. Note that mIoU computes the performance for every pixel, but normally a small change in the final bounding box do not have a lot of impact on the medicine name prediction, as long as they encapsulate the text within it. We also show the upper bound performance of medicine line recognition by using ground-truth medicine bounding boxes only while evaluating. As we can see, our algorithm with just using weak labels can reach within a few points of the strongest upper-bound with strong labels.

Table 2. Performance over iterations of the proposed framework. Iter 1 represents learning from only the OCR Labeling Function and iter ≥ 2 shows the performance after iteratively including the Segmentation Labeling Function. The medicine name performances are only for topk=1. GT bbox shows the performance when the groundtruth bounding boxes are provided for medicine names only during evaluation.

Iteration	1	2	3	GT bbox
Train data (%)	24.4	66.3	90.2	-
Segmentation (mIoU)	72.6	77.9	77.2	100%
Medicine Name (mJI)	44.8	45.9	45.9	49.8%

Cues for Medicine Name Segmentation: Unlike a generic text detector, specifically detecting medicine lines can be challenging, as handwritten prescriptions do not have any specific structure or location in the page. However, the segmentation model is still able to predict the location of the medicine lines with high performance as shown in Table 2. In order to understand the cues the segmentation model uses to segment the medicine names, we do the following experiment. Given a test image x, using a sliding window, we remove square patches from the image to remove potential cues, one at a time. Consider $x_{i,j}$ as the image when patch at location (i, j) is removed. We can run the segmentation model on this image, $\mathcal{M}(x_{i,j})$ and obtain the mIoU. For every location (i, j), on the image, we can obtain the model's performance drop when a patch around that is removed, and then display that as a heatmap. A drop in performance in certain regions of this image depicts the regions necessary for the segmentation model to segment the medicine names correctly. As we can see in Fig. 5, the model is clearly utilizing cues from visual features surrounding medicine lines such as starting of a line like Tab, Cap, hyphens, etc. These observations are aligned with what a pharmacist or even non-domain experts look to determine medicine lines, as in most cases the handwriting is illegible. These key demarcations serve as strong signals to recognize medicine lines, after which we can condition our knowledge to medicine names to enhance line recognition.

(a)   (b)   (c)

Fig. 5. Cues needed by the segmentation network. Deeper color denotes lower performance when a patch around that is removed. A few parts of the image other than the medicine names, such as hyphens, Tab, Cap, etc., also appear to be darker, which are some of the cues that the model looks at to determine whether it is a medicine line.

Contribution of Medicine LM and Segmentation Model: Here we show how selectively injecting medicine LMs can offer a significant improvement in performance. The vanilla LM is trained on a generic corpus of text from the Latin script. However, the medicine name LM is trained as discussed in Sect. 3.4. The performance improves with path length for both the models but for the medicine LM, the top-1 path itself performs much better than top-1000 path for the vanilla LM (Fig. 6). This also reduces the compute time in decoding the top-k paths from the logits, which is linear in the number of paths.

Moreover, segmenting and selectively injecting the LM plays a critical role on the performance, and MedLM + Segmented Lines perform the best. Applying the MedLM on the full image actually reduces the precision significantly, but improves the recall slightly as expected, but reducing the overall metric, i.e., jaccard index. This shows that selectively injecting the LM is important, otherwise it can mess up the rest of the prescription, and hallucinate medicine names from them.

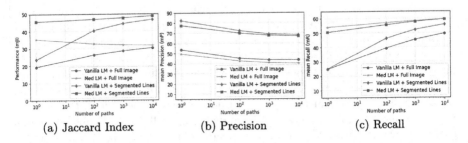

(a) Jaccard Index   (b) Precision   (c) Recall

Fig. 6. Jaccard index, precision and recall comparison using different language models and inputs (medicine line segmented and full page). The medicine LM on segmented medicine lines works the best, the top-1 of which is better than the top-1000 of the vanilla LM. Applying the medicine LM on the entire image decreases the precision of the predictions, as it hallucinates medicine names in the rest of the prescription.

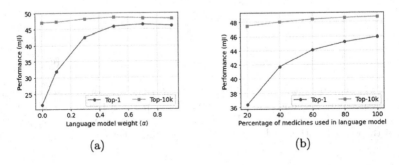

(a)   (b)

Fig. 7. (a) Ablation of performance with weight on the language model α. $\alpha = 0$ denote the performance of only the optical model. (b) Ablation of fraction of medicine names used to train the medicine language model. We present the performance when top-1 and top-10k paths are used to predict after vocabulary matching.

Performance with Varying Weight on LM: The weight α in Eq. 1 on the LM scores can have an impact on the final performance. A low weight may lead to no improvement beyond the optical model's prediction, and a high weight may not ground the output to the actual text on the image. Figure 7a shows an ablation of the medicine name prediction performance on the LM weight. Note that the changes in performance is much lower for top-10k paths than for top-1 path, as only the first path in the top-10k path is affected by the LM because for paths > 1, the predictions come from the top-k decoded paths which is based on only the logits without any LM scoring. Nonetheless, we see that the performance of both the models are very close after a certain value of α.

Varying the Vocabulary of the LM: The medicine names used in generating the synthetic lines can have an impact on the quality of the medicine name LM. Here we also show how the performance varies as we increase the number of medicine names used to train the medicine LM. Figure 7b presents the results for top-1 and top-10k with different size of medicine name dataset. The performance improves as we add more medicines, but starts saturating after a certain point.

Performance with Different N-Gram Models: The n-gram LM involves a parameter n, which is the number of history characters the model looks to obtain the score of the next character. We created multiple n-gram models on the synthetically generated medicine line text data, and show the results in Table 3. More context definitely helps in performance, but it saturates after $n = 7$. This is also intuitive as the length of the medicine names is around 7.9 on average.

Table 3. Ablation of different n-gram models trained on medicine line data.

	n=3	n=5	n=7	n=9
Top-1 (mJI) (%)	27.2	41.5	45.9	45.9
Top-10k (mJI) (%)	47.4	48.1	48.7	48.7

Table 4. Ablation of different algorithms to predict medicine names. We use $k = 1e4$.

	top-1	top-1-edit	top-k	top-k+majority
Jaccard Index	45.9	45.8	46.9	48.7
Precision	76.9	68.6	64.8	66.8
Recall	51.0	54.4	58.5	59.5

Predicting In-Vocabulary Words: In the final step of our algorithm to predict medicine names, we only predict those words where we find a direct match with one of the elements of the medicine vocabulary. As discussed before, finding a match for only the top-1 prediction may not be the best. Thus, we decode until top-k and find matches for all the text. As the top-k decoding is directly dependent on the output of the model, such a matching respects the model's predictions. We then take a majority voting of all the matches and that becomes the final predicted medicine for a line. Note that some lines may not have any prediction at all. In this section, we compare multiple strategies of predicting in-vocabulary words in Table 4. Top-1 represents an exact match with the first path, top-1 edit distance finds the nearest prediction from the vocabulary by edit distance, top-k denotes we decode the top-k outputs but stop when we find the first exact match, and finally top-k+majority is the algorithm we use, where we decode all the top-k lines and take a majority voting of all the exact matches.

Note that top-1-edit has the same jaccard index as top1, but the former has lower precision with higher recall than top-1, as expected, because it predicts beyond exact matches. We tried with multiple thresholds for edit distance, and found that 85% normalized distance performs the best. Increasing the threshold, i.e., allowing more matches significantly reduces the precision, at the gain of the recall, but hurting the overall performance. This is because of the intuition we discussed earlier that topk decodings respect the model's confidence, but edit distance treats every replacement with the same cost.

4.2   Error-Mode Analysis

The two types of errors possible are - medicine names predicted but not in the ground-truth (type I) and medicine names in the ground-truth but not predicted (type II). In our framework, there are two reasons behind the errors - segmentation network and OCR. If a medicine name is not segmented, then it leads to a type-II error. OCR errors contributes to the rest (type I and type II), a majority of which is contributed by misinterpreting very similar looking medicines such as emtel vs entel, eenosol vs eenasof, folvite vs folite, paro vs baro, zincovit vs zincort, aloliv vs alcoliv. Also we observe that the doctor can commit spelling mistakes, or vaguely write a medicine name, where only the first few characters are recognizable. To correct such errors, pharmacists generally use other contexts such as observation. Learning such contexts would need a lot more data, and injecting higher-level domain knowledge.

5   Conclusion

In this paper, we looked into the problem of extracting medicine names from inscrutable handwritten prescriptions. Our algorithm can selectively infuse domain knowledge to specific portions of a document to significantly improve the performance. We developed a framework that can learn to detect regions of interest from just weak labels, and also learn a medicine language model using synthetically generated text lines using probabilistic programs. The idea is generic enough to be applied to a variety of other types of documents, such as handwritten forms.

Acknowledgement. We thank Srujana Merugu, Ansh Khurana, Manish Gupta, Harsh Dhand and Shruti Garg for all the support and discussions during the course of this project. Without their effort, this project would not have been possible.

References

1. Achkar, R., Ghayad, K., Haidar, R., Saleh, S., Al Hajj, R.: Medical handwritten prescription recognition using CRNN. In: CITS. IEEE (2019)
2. Araslanov, N., Roth, S.: Self-supervised augmentation consistency for adapting semantic segmentation. In: CVPR (2021)
3. Bhunia, A.K., Sain, A., Chowdhury, P.N., Song, Y.Z.: Text is text, no matter what: unifying text recognition using knowledge distillation. In: ICCV (2021)
4. Bissacco, A., Cummins, M., Netzer, Y., Neven, H.: PhotoOCR: reading text in uncontrolled conditions. In: ICCV, pp. 785–792 (2013)
5. Breuel, T.M., Ul-Hasan, A., Al-Azawi, M.A., Shafait, F.: High-performance OCR for printed English and Fraktur using LSTM networks. In: ICDAR. IEEE (2013)
6. Bukhari, S.S., Kadi, A., Jouneh, M.A., Mir, F.M., Dengel, A.: anyOCR: an open-source OCR system for historical archives. In: ICDAR (2017)
7. Cascante-Bonilla, P., Tan, F., Qi, Y., Ordonez, V.: Curriculum labeling: revisiting pseudo-labeling for semi-supervised learning. In: AAAI (2021)
8. Chen, L.C., Papandreou, G., Kokkinos, I., Murphy, K., Yuille, A.L.: DeepLab: semantic image segmentation with deep convolutional nets, atrous convolution, and fully connected CRFS. IEEE Trans. Pattern Anal. Mach. Intell. **40**, 834–848 (2017)
9. Cheng, B., et al.: Panoptic-DeepLab: a simple, strong, and fast baseline for bottom-up panoptic segmentation. In: CVPR, pp. 12475–12485 (2020)
10. Diaz, D.H., Qin, S., Ingle, R., Fujii, Y., Bissacco, A.: Rethinking text line recognition models. arXiv preprint arXiv:2104.07787 (2021)
11. D'hondt, E., Grouin, C., Grau, B.: Generating a training corpus for OCR post-correction using encoder-decoder model. In: IJCNLP (2017)
12. Fujii, Y., Driesen, K., Baccash, J., Hurst, A., Popat, A.C.: Sequence-to-label script identification for multilingual OCR. In: ICDAR. IEEE (2017)
13. Graves, A., Fernández, S., Gomez, F., Schmidhuber, J.: Connectionist temporal classification: labelling unsegmented sequence data with recurrent neural networks. In: Proceedings of the 23rd International Conference on Machine Learning, pp. 369–376 (2006)
14. Gupta, H., Del Corro, L., Broscheit, S., Hoffart, J., Brenner, E.: Unsupervised multi-view post-OCR error correction with language models. In: EMNLP, pp. 8647–8652 (2021)
15. Gupta, M., Soeny, K.: Algorithms for rapid digitalization of prescriptions. Visual Inform. **5**, 54–69 (2021)
16. He, K., Zhang, X., Ren, S., Sun, J.: Deep residual learning for image recognition. In: CVPr, pp. 770–778 (2016)
17. Huang, J., et al.: A multiplexed network for end-to-end, multilingual OCR. In: CVPR (2021)

18. Ingle, R.R., Fujii, Y., Deselaers, T., Baccash, J., Popat, A.C.: A scalable handwritten text recognition system. In: ICDAR (2019)
19. Jayakumar, P.: Online doctor consultation market to grow (2021). https://www.businesstoday.in/lifestyle/health/story/online-doctor-consultation-market-to-grow-72-to-836-million-by-march-2024-study-304689-2021-08-19
20. Karatzas, D., et al.: ICDAR 2015 competition on robust reading. In: ICDAR. IEEE (2015)
21. Karthikeyan, S., de Herrera, A.G.S., Doctor, F., Mirza, A.: An OCR post-correction approach using deep learning for processing medical reports. IEEE Trans. Circuits Syst. Video Technol. **32**, 2574–2581 (2021)
22. Khoreva, A., Benenson, R., Hosang, J., Hein, M., Schiele, B.: Simple does it: weakly supervised instance and semantic segmentation. In: CVPR (2017)
23. Kittenplon, Y., Lavi, I., Fogel, S., Bar, Y., Manmatha, R., Perona, P.: Towards weakly-supervised text spotting using a multi-task transformer. In: CVPR (2022)
24. Lee, J., et al.: BioBERT: a pre-trained biomedical language representation model for biomedical text mining. Bioinformatics **36**, 1234–1240 (2019)
25. Li, D., Huang, J.B., Li, Y., Wang, S., Yang, M.H.: Weakly supervised object localization with progressive domain adaptation. In: CVPR (2016)
26. Li, M., et al.: TrOCR: transformer-based optical character recognition with pre-trained models. arXiv preprint arXiv:2109.10282 (2021)
27. Litman, R., Anschel, O., Tsiper, S., Litman, R., Mazor, S., Manmatha, R.: Scatter: selective context attentional scene text recognizer. In: CVPR (2020)
28. Liu, H., Wang, J., Long, M.: Cycle self-training for domain adaptation. Adv. Neural Inf. Process. Syst. **34**, 22968–22981 (2021)
29. Long, S., He, X., Yao, C.: Scene text detection and recognition: the deep learning era. Int. J. Comput. Vision **129**, 161–184 (2021)
30. Long, S., Qin, S., Panteleev, D., Bissacco, A., Fujii, Y., Raptis, M.: Towards end-to-end unified scene text detection and layout analysis. In: CVPR (2022)
31. Marti, U.V., Bunke, H.: The IAM-database: an English sentence database for offline handwriting recognition. Int. J. Doc. Anal. Recogn. **5**, 39–46 (2002)
32. Paul, Sujoy, Roy, Sourya, Roy-Chowdhury, Amit K..: W-TALC: weakly-supervised temporal activity localization and classification. In: Ferrari, Vittorio, Hebert, Martial, Sminchisescu, Cristian, Weiss, Yair (eds.) ECCV 2018. LNCS, vol. 11208, pp. 588–607. Springer, Cham (2018). https://doi.org/10.1007/978-3-030-01225-0_35
33. Pragnadyuti, M., Rabindranath, D., Suhrita, P., Kumar, S.A., Kumar, J.S.: Legibility assessment of handwritten OPD prescriptions of a tertiary care medical college and hospital in Eastern India. SJMPS (2017)
34. Rani, S., Rehman, A.U., Yousaf, B., Rauf, H.T., Nasr, E.A., Kadry, S.: Recognition of handwritten medical prescription using signature verification techniques. Comput Math Methods Med. (2022)
35. Rasmy, L., Xiang, Y., Xie, Z., Tao, C., Zhi, D.: Med-BERT: pretrained contextualized embeddings on large-scale structured electronic health records for disease prediction. Nature **4**, 86 (2021)
36. Ratner, A., Bach, S.H., Ehrenberg, H., Fries, J., Wu, S., Ré, C.: Snorkel: rapid training data creation with weak supervision. In: VLDB. NIH Public Access (2017)
37. Redmon, J., Divvala, S., Girshick, R., Farhadi, A.: You only look once: unified, real-time object detection. In: CVPR (2016)
38. Ren, S., He, K., Girshick, R., Sun, J.: Faster R-CNN: towards real-time object detection with region proposal networks. Adv. Neural Inf. Process. Syst. **28** (2015)
39. Sandler, M., Howard, A., Zhu, M., Zhmoginov, A., Chen, L.C.: MobileNetV2: inverted residuals and linear bottlenecks. In: CVPR (2018)

40. Sohn, K., et al.: FixMatch: simplifying semi-supervised learning with consistency and confidence. Adv. Neural Inf. Process. Syst. **33**, 596–608 (2020)
41. Thompson, P., McNaught, J., Ananiadou, S.: Customised OCR correction for historical medical text. In: 2015 digital heritage. IEEE (2015)
42. Vaswani, A., et al.: Attention is all you need. Adv. Neural Inf. Process. Syst. **30** (2017)
43. Wang, P., Li, H., Shen, C.: Towards end-to-end text spotting in natural scenes. IEEE Trans. Pattern Anal. Mach. Intell. **44**, 7266–7281 (2021)
44. Wei, Y., et al.: STC: a simple to complex framework for weakly-supervised semantic segmentation. IEEE Trans. Pattern Anal. Mach. Intell. **39**, 2314–2320 (2016)
45. Yang, X., et al.: GatorTron: a large clinical language model to unlock patient information from unstructured electronic health records. arXiv preprint arXiv:2203.03540 (2022)
46. Zhang, C., Cao, M., Yang, D., Chen, J., Zou, Y.: CoLa: weakly-supervised temporal action localization with snippet contrastive learning. In: CVPR (2021)
47. Zhang, D., Han, J., Cheng, G., Yang, M.H.: Weakly supervised object localization and detection: a survey. IEEE Trans. Pattern Anal. Mach. Intell. **44**, 5866–5885 (2021)

Author Index

Printed in the United States
by Baker & Taylor Publisher Services